Sustainable Approach to Protective Nanocoatings

Rajni Garg
Galgotias College of Engineering and Technology, India

Ansar Anjum
Galgotias College of Engineering and Technology, India

A volume in the Advances in Chemical and Materials Engineering (ACME) Book Series

IGI Global
PUBLISHER of TIMELY KNOWLEDGE

Published in the United States of America by
IGI Global
Engineering Science Reference (an imprint of IGI Global)
701 E. Chocolate Avenue
Hershey PA, USA 17033
Tel: 717-533-8845
Fax: 717-533-8661
E-mail: cust@igi-global.com
Web site: http://www.igi-global.com

Library of Congress Cataloging-in-Publication Data

CIP Data in progress

Title: Sustainable Approach to Protective Nanocoatings

ISBN: 9798369331361

This book is published in the IGI Global book series Advances in Chemical and Materials Engineering (ACME) (ISSN: 2327-5448; eISSN: 2327-5456)

British Cataloguing in Publication Data
A Cataloguing in Publication record for this book is available from the British Library.

For electronic access to this publication, please contact: eresources@igi-global.com.

Advances in Chemical and Materials Engineering (ACME) Book Series

ISSN:2327-5448
EISSN:2327-5456

Editor-in-Chief: J. Paulo Davim, University of Aveiro, Portugal

MISSION

The cross disciplinary approach of chemical and materials engineering is rapidly growing as it applies to the study of educational, scientific and industrial research activities by solving complex chemical problems using computational techniques and statistical methods.

The **Advances in Chemical and Materials Engineering (ACME) Book Series** provides research on the recent advances throughout computational and statistical methods of analysis and modeling. This series brings together collaboration between chemists, engineers, statisticians, and computer scientists and offers a wealth of knowledge and useful tools to academics, practitioners, and professionals through high quality publications.

COVERAGE

- Toxicity of Materials
- Numerical Techniques
- Computational methods
- Fracture Mechanics
- Composites
- Biomaterials
- Chemical engineering
- Multifuncional and Smart Materials
- Industrial Chemistry
- Materials to Renewable Energies

The Advances in Chemical and Materials Engineering (ACME) Book Series (ISSN 2327-5448) is published by IGI Global, 701 E. Chocolate Avenue, Hershey, PA 17033-1240, USA, www.igi-global.com. This series is composed of titles available for purchase individually; each title is edited to be contextually exclusive from any other title within the series. For pricing and ordering information please visit http://www.igi-global.com/book-series/advances-chemical-materials-engineering/73687. Postmaster: Send all address changes to above address.

Titles in this Series

For a list of additional titles in this series, please visit:
http://www.igi-global.com/book-series/advances-chemical-materials-engineering/73687

Machining Polymer Matrix Composites Tools, Techniques, and Sustainability
Francisco Mata Cabrera (Universidad de Castilla-La Mancha, Spain) and Issam Hanafi (Université Abdelmalek Essaâdi, Morocco)
Engineering Science Reference • © 2024 • 320pp • H/C (ISBN: 9781668499276) • US $300.00

Building a Low-Carbon Society Through Applied Environmental Materials Science
Takaomi Kobayashi (Nagaoka University of Technology, Japan)
Engineering Science Reference • © 2024 • 300pp • H/C (ISBN: 9798369300039) • US $255.00

FlexPDE and Finite Element Method Applications in Thermal Energy Storage and Cavities
Khashayar Hosseinzadeh (Sharif University of Technology, Iran) Mehdi Mahboobtosi (Babol Noshirvani University of Technology, Iran) and Davood Domairry Ganji (Babol Noshirvani University of Technology, Iran)
Engineering Science Reference • © 2024 • 300pp • H/C (ISBN: 9798369309322) • US $225.00

Advancements in Powder Metallurgy Processing, Applications, and Properties
Shashanka Rajendrachari (Bartin University, Turkey) and Rayappa Shrinivas Mahale (Jain College of Engineering and Research, India)
Engineering Science Reference • © 2024 • 300pp • H/C (ISBN: 9781668493854) • US $275.00

Innovative Multifunctional Nanomaterial for Photocatalysis, Sensing, and Imaging
Azad Kumar (Siddharth University, India)
Engineering Science Reference • © 2023 • 283pp • H/C (ISBN: 9781668487433) • US $270.00

For an entire list of titles in this series, please visit:
http://www.igi-global.com/book-series/advances-chemical-materials-engineering/73687

IGI Global
PUBLISHER of TIMELY KNOWLEDGE

701 East Chocolate Avenue, Hershey, PA 17033, USA
Tel: 717-533-8845 x100 • Fax: 717-533-8661
E-Mail: cust@igi-global.com • www.igi-global.com

Table of Contents

Preface xiv

Chapter 1
Introduction to Nanocoatings and Packaging 1
Ansar Anjum, Galgotias College of Engineering and Technology, India
Deepak Garg, Delhi Technological University, India
Rahul Singhal, Shivaji College, University of Delhi, India

Chapter 2
Applications of Nanocoatings: Smart and Sustainable Protection 28
Rishav Garg, Galgotias College of Engineering and Technology, India
Rajni Garg, Galgotias College of Engineering and Technology, India
Nnabuk Okon Eddy, University of Nigeria, Nigeria

Chapter 3
Advancements in Designing Smart and Intelligent Nanocoatings 57
Mukhtar Iderawumi Abdulraheem, Henan Agricultural University, China
Abiodun Yusuff Moshood, Henan Agricultural University, China
Yongqi Chen, Henan Agricultural University, China
Hongjun Chen, Henan Agricultural University, China
Haiyan Zhang, Henan Agricultural University, China
Jiandong Hu, Henan Agricultural University, China

Chapter 4
Nanocoatings for Energy Generation and Conservation of Solar Cells 88
Rafael Vargas-Bernal, University of Guanajuato, Mexico

Chapter 5
Perspectives of Biodegradable Nanocoatings in Food Packaging 113
Betül Bay Yılmaz, Yıldız Technical University, Turkey

Chapter 6
Nanocoatings for Cosmetotextiles 170
Bancha Yingngam, Ubon Ratchathani University, Thailand

Chapter 7
Properties and Characteristics of Nanocoatings for Medicinal Applications 203
John Tsado Mathew, Ibrahimm Badamasi Babangida Lapai, Nigeria
Abel Inobeme, Edo University, Uzairue, Nigeria
Abdulkadir Abdullahi, Federal University of Technology, Minna, Nigeria
Etsuyankpa Muhammad Bini, Federal University of Lafia, Nigeria
Elijah Yanda Shaba, Federal University of Technology, Minna, Nigeria
Monday Musah, Ibrahim BadamasiBabangida University, Lapai, Nigeria
Yakubu Azeh, Ibrahim Badamasi Babangida University, Lapai, Nigeria
Charles Oluwaseun Adetunji, Edo University, Uzairue, Nigeria
Aishetu Muhammad Ibrahim, Ibrahim Badamasi Babangida University, Lapai, Nigeria
Ezekiel Tanko, Federal University of Technology, Minna, Nigeria
Amos Mamman, Ibrahim Badamasi Babangida University, Lapai, Nigeria

Chapter 8
Healing and Health Perspectives of Nanocoatings for Medicinal Applications 228
Abhinay Thakur, Lovely Professional University, India
Ashish Kumar, Nalanda College of Engineering, Bihar Engineering University, India & Department of Science, Technology and Technical Education, Government of Bihar, India

Chapter 9
Nanocoatings in Medicine: Revolutionizing Healthcare Through Precision and Potential....................265
G. Boopathy, Vel Tech Rangarajan Dr. Sagunthala R&D Institute of Science and Technology, Chennai, India
K. Gurusami, St. Peter's Institute of Higher Education and Research, Chennai, India
K. R. Vijayakumar, Dr. MGR Educational and Research Institute, Chennai, India
Mohit Hemanth Kumar, Alliance University, India

Chapter 10
Sustainable Approach to Nanocomposites and Nanocoatings: Research and Industrial Applications of Hydrophobic Nanocoatings 290
Sana Dahmani, Independent Researcher, Germany

Compilation of References 347

About the Contributors 425

Index 431

Detailed Table of Contents

Preface xiv

Chapter 1
Introduction to Nanocoatings and Packaging 1
Ansar Anjum, Galgotias College of Engineering and Technology, India
Deepak Garg, Delhi Technological University, India
Rahul Singhal, Shivaji College, University of Delhi, India

The advent of nanotechnology has inaugurated new era of ingenuity, particularly evident in the domain of nanocoating with diverse applications across industries. The coatings herald a transformative shift in surface protection. In the automotive sector, they play crucial role in enhancing scratch resistance and bolstering durability, thereby extending lifespan of vehicles. The packaging industry reaps the benefits of nanotechnology through the integration of advanced barrier properties, augmenting food preservation and packaging integrity. In the realm of healthcare, nanocoating applied to medical devices act as deterrents to microbial adhesion, fostering a hygienic environment. On the other hand, the electronics industry embraces nanocoating to establish water-resistant and robust surfaces, ensuring the prolonged endurance of electronic components. Thus, nanocoating has applications ranging from self-cleaning surfaces to improved insulation in sustainable building practices.

Chapter 2
Applications of Nanocoatings: Smart and Sustainable Protection 28
Rishav Garg, Galgotias College of Engineering and Technology, India
Rajni Garg, Galgotias College of Engineering and Technology, India
Nnabuk Okon Eddy, University of Nigeria, Nigeria

The application of one material on another, often referred to as the substrate, is what is known as a coating. Their primary purpose is to shield the material from deterioration brought on by external factors. They serve as a conduit between the environment and the substrate. Furthermore, they serve as decorations as well. Nanocoatings are coatings in which the particles are at least one dimension apart and fall between 1 and 1000 nm in size. Because nanocoatings are tougher and harder on the substrate than other conventional coatings, they offer greater wear resistance. In addition, they offer antimicrobial, stain and wrinkle resistance, hydrophobic and hydrophilic qualities, UV protection, and antistatic qualities that affect the substrate material's bulk properties. Additionally, an attempt has been made to compile the different processing methods used in their fabrication.

Chapter 3
Advancements in Designing Smart and Intelligent Nanocoatings.......................57
Mukhtar Iderawumi Abdulraheem, Henan Agricultural University, China
Abiodun Yusuff Moshood, Henan Agricultural University, China
Yongqi Chen, Henan Agricultural University, China
Hongjun Chen, Henan Agricultural University, China
Haiyan Zhang, Henan Agricultural University, China
Jiandong Hu, Henan Agricultural University, China

Smart and intelligent nanocoatings have emerged as crucial components in various industries, offering enhanced functionalities and improved performance. This chapter explores the recent advancements in studying smart and intelligent nanocoatings, encompassing their design, synthesis, characterization, and applications. The chapter aims to provide a comprehensive understanding of the current state-of-the-art in this field and highlight the potential for future development and research. This chapter concluded that developing smart and intelligent nanocoatings has significantly enhanced various industries' performance, durability, and safety. However, future research focuses on reducing costs, scaling up production, and creating more advanced systems with existing technologies, such as sensors and actuators.

Chapter 4
Nanocoatings for Energy Generation and Conservation of Solar Cells..............88
Rafael Vargas-Bernal, University of Guanajuato, Mexico

Solar panels operating under outdoor environments during their useful life are exposed to various factors such as dust, humidity, snow, rain, thermal changes, abrasion, etc., which degrade their operation, decreasing the efficiency of converting light into voltage. Although various cleaning and/or conservation methods have been proposed and used for years to solve this problem, they offer results with low efficiency and high costs. The most effective functional method so far is the application of nanocoatings on the solar panel cover, which avoids these effects more comprehensively. This chapter presents the study of the advances that have been achieved so far through nanocomposites based on polymers and ceramic materials and proposes future trends that must be carried out to achieve their optimized design. The advances, although significant, allow us to predict the need to continue carrying out scientific and technological research to achieve results of high-added value using nanomaterials.

Chapter 5
Perspectives of Biodegradable Nanocoatings in Food Packaging 113
Betül Bay Yılmaz, Yıldız Technical University, Turkey

Nanotechnology has the potential to improve human health, economic prosperity, product innovation, and overall quality of life, including through food nanopackaging––a promising area focusing on biodegradable packaging solutions. This alternative to conventional packaging minimizes waste, extends food shelf life, and enhances overall quality. The production of biodegradable nanocoatings can contribute to industry sustainability by reducing water consumption, solid waste, electricity use, and emissions. Bio-based coatings, with compatibility and matrix properties, can incorporate antioxidant agents and antimicrobial compounds, enhancing product safety, functionality, and shelf life. Biodegradable polymers, including polysaccharides, proteins, lipids, and polyesters, offer innovative pathways for entirely bio-based nanocoatings. These cost-effective, biocompatible, and renewable materials can be sourced directly from marine organisms and plants or produced through fermentative processes by microorganisms, such as microbial polyesters or polysaccharides. However, challenges in handling biopolymers, such as their hydrophilicity, crystallization tendencies, brittleness or melting instabilities, necessitate blending them with other materials to enhance their coating performance. Integrating nanoparticles within biopolymers can address environmental concerns by reducing packaging materials and enhancing recyclability. This approach aligns with a more eco-sustainable approach to food packaging, resulting in reduced waste.

Chapter 6
Nanocoatings for Cosmetotextiles 170
Banĉha Yingngam, Ubon Ratchathani University, Thailand

In recent years, nanotechnology has found significant applications in the textile industry, especially in the development of cosmetotextiles. This chapter explores the role of nanocoatings in enhancing the functional and sustainable attributes of these textiles. Integrating insights from materials science, sustainability, and cosmetic technology, the chapter offers a comprehensive overview of current nanocoating techniques, materials, and their environmental impacts. Furthermore, it critically evaluated challenges and opportunities in implementing sustainable nanocoatings within cosmetotextiles and proposed strategies for developing eco-friendly nanocoatings and performing lifecycle analyses. Consequently, nanocoatings are essential for improving efficient, sustainable cosmetotextiles and offer potential for future research and commercial use. Key directions for future work include the creation of biodegradable nanomaterials, conducting toxicological studies to ensure user safety, and adopting circular economy principles within the cosmetic textile industry.

Chapter 7
Properties and Characteristics of Nanocoatings for Medicinal Applications 203
John Tsado Mathew, Ibrahimm Badamasi Babangida Lapai, Nigeria
Abel Inobeme, Edo University, Uzairue, Nigeria
Abdulkadir Abdullahi, Federal University of Technology, Minna, Nigeria
Etsuyankpa Muhammad Bini, Federal University of Lafia, Nigeria
Elijah Yanda Shaba, Federal University of Technology, Minna, Nigeria
Monday Musah, Ibrahim BadamasiBabangida University, Lapai, Nigeria
Yakubu Azeh, Ibrahim Badamasi Babangida University, Lapai, Nigeria
Charles Oluwaseun Adetunji, Edo University, Uzairue, Nigeria
Aishetu Muhammad Ibrahim, Ibrahim Badamasi Babangida University, Lapai, Nigeria
Ezekiel Tanko, Federal University of Technology, Minna, Nigeria
Amos Mamman, Ibrahim Badamasi Babangida University, Lapai, Nigeria

Nanotechnology has emerged as a revolutionary field with diverse applications, including nanocoatings for medicinal purposes. Nanocoatings, with their unique properties and capabilities, have opened new avenues for drug delivery, medical devices, and antimicrobial treatments. Nanocoatings can be engineered to encapsulate therapeutic agents, protecting them from degradation and facilitating targeted delivery to specific tissues or cells. The use of antimicrobial nanocoatings on medical surfaces has also shown promising results in preventing infections and combating antibiotic-resistant pathogens. While nanocoatings offer tremendous potential, concerns about potential toxicity, biocompatibility, and long-term effects require rigorous evaluation and testing. Furthermore, the study highlights ongoing research and future prospects in the field of medicinal nanocoatings. The development of multifunctional nanocoatings that combine drug delivery, sensing capabilities, and tissue regeneration holds promise for personalized medicine and advanced therapeutics.

Chapter 8
Healing and Health Perspectives of Nanocoatings for Medicinal Applications 228
Abhinay Thakur, Lovely Professional University, India
Ashish Kumar, Nalanda College of Engineering, Bihar Engineering University, India & Department of Science, Technology and Technical Education, Government of Bihar, India

Offering unprecedented opportunities to revolutionize healing and healthcare, this chapter delves into the cutting-edge advancements in nanocoatings tailored specifically for medicinal applications, presenting a comprehensive overview of their potential to enhance healing processes and improve health outcomes. A major focus of this chapter is on the multifaceted roles of nanocoatings in medicinal contexts. These include drug delivery systems, wound healing, antimicrobial coatings, and tissue engineering scaffolds, among others. Furthermore, it discusses the challenges and regulatory considerations associated with the integration of nanocoatings into clinical practice. By exploring the latest research findings, this chapter provides valuable insights into how nanocoatings are poised to transform the landscape of medicinal applications. It underscores their potential to revolutionize healing processes, mitigate health-related challenges, and pave the way for a healthier future.

Chapter 9
Nanocoatings in Medicine: Revolutionizing Healthcare Through Precision and Potential....................265
G. Boopathy, Vel Tech Rangarajan Dr. Sagunthala R&D Institute of Science and Technology, Chennai, India
K. Gurusami, St. Peter's Institute of Higher Education and Research, Chennai, India
K. R. Vijayakumar, Dr. MGR Educational and Research Institute, Chennai, India
Mohit Hemanth Kumar, Alliance University, India

The use of nanotechnology in healthcare and medicine is crucial, particularly for nanocoatings. The therapeutic potential of nanocoatings on medical implants and devices is examined in this chapter. It goes into the ingredients, production processes, and health benefits of nanocoatings. Device functionality and biocompatibility can be enhanced by nanocoatings. They are essential in addressing healthcare issues including cardiovascular stents and orthopedic implants. Additionally, nanocoatings are used in medication delivery systems, tissue engineering, and wound healing. This chapter offers important knowledge on the medical applications of nanocoatings to experts and researchers. Nanotechnology has the potential to alter the medical industry, which would be advantageous for patients.

Chapter 10
Sustainable Approach to Nanocomposites and Nanocoatings: Research and Industrial Applications of Hydrophobic Nanocoatings 290
Sana Dahmani, Independent Researcher, Germany

The global hydrophobic coatings market is projected to reach $3.0 billion by 2030, growing at a CAGR of 4.2%. These innovative coatings are in high demand across sectors such as automotive, marine, construction, healthcare, textile, environmental management, and further industries thanks to their properties including and not limited to water-repellency, corrosion-resistance, and self-cleaning. Challenges faced by hydrophobic nanocoatings in durability, adhesion, cost-effectiveness, and sustainability are being addressed through research on wetting control and scalability. Hydrophobic coating-as-a-service business model offers specialized services for applying hydrophobic coatings on various surfaces, with the highest efficiency. Choosing providers with advanced expertise and high-quality materials is essential for long-lasting product protection.

Compilation of References 347

About the Contributors 425

Index 431

Preface

Welcome to *Sustainable Approach to Protective Nanocoatings*, a comprehensive compilation edited by Rajni Garg and Ansar Anjum. The book is an exploration into the innovative realm of nanotechnology and its application in the development of environmentally conscious protective coatings. This book represents a collective effort to unravel the intricacies of sustainable nanocoatings, bridging the gap between research and practical, eco-friendly solutions.

In an era where environmental sustainability is at the forefront of technological advancements, the field of nanocoatings has emerged as a promising frontier. This book aims to provide readers with a deep dive into the sustainable aspects of protective nanocoatings, offering insights into the latest research, methodologies, and applications that align with the principles of environmental responsibility. The book reflects the compelling exploration of the multifaceted world of nanotechnology and its diverse applications in various domains and helps the reader to embark on a journey through pioneering research and practical insights, providing a comprehensive understanding of the dynamic field of nanocoatings. Readers will traverse the bioinspired synthesis of nanocoatings, witnessing their multidimensional applications in various industrial domains. Designed to address contemporary industrial needs, this book elucidates synthesis methodologies and real-world applications, offering insights into recent technological strides and their engineering implications.

Comprising rigorous research, experimental validations, and industrial applications, this book serves as an exhaustive exploration of nanotechnology and its potential. It navigates the fundamentals, current state, and future trajectories of nanotechnology, unveiling the diversity of nanocoatings, their properties, and prospective applications.

Its inclusivity extends to a general readership, presenting intricate concepts through illustrations and diagrams for easy comprehension. This book's significance spans academia, research, and industry, catering to students, researchers, academics, and industry professionals alike.

Spanning an array of topics, from photocatalytic applications to intelligent nanocoatings, from biomedical applications to environmental implications, this book serves as an invaluable resource for anyone intrigued by the vast landscape of nanotechnology and its practical manifestations.

We invite you on a journey through the realms of nanoscience and nanocoatings, where innovation meets sustainability, and cutting-edge research shapes a more resilient future.

ORGANIZATION OF THE BOOK

Chapter 1: "Introduction to Nanocoatings and Packaging"

Nanotechnology's transformative impact unfolds in this chapter, spotlighting the different types of nanocoatings and their diverse applications across industries. From automotive enhancements to healthcare and electronics, nanocoatings revolutionize surface protection, extending lifespans and preserving product integrity. The chapter compiles the recent innovations and breakthroughs with their market share and discusses the challenges and future prospects of the nanocoatings.

Chapter 2: "Applications of Nanocoatings: Smart and Sustainable Protection"

Exploring nanocoatings' role as shields and enhancements, this chapter delves into their impact on substrate materials. These coatings, tougher than conventional ones, offer superior wear resistance, antimicrobial properties, and UV protection. The chapter compiles various processing methods, offering insights into fabricating nanocoatings and their diverse functionalities.

Chapter 3: "Advancements in Designing Smart and Intelligent Nanocoatings"

Charting recent strides in smart and intelligent nanocoatings, this chapter navigates their design, synthesis, and applications. It highlights their contributions across industries, boosting performance, durability, and safety. It also emphasizes ongoing efforts to refine existing technologies and create more advanced systems, paving the way for future developments.

Chapter 4: "Nanocoatings for Energy Generation and Conservation of Solar Cells"

Focusing on solar panel preservation, this chapter underscores the degradation factors affecting solar panels. Nanocoatings' application emerges as a key strategy to combat these effects, elevating solar panels' resilience. The chapter presents current advancements in nanocomposite-based coatings, setting the stage for optimized designs through further scientific and technological research.

Chapter 5: "Perspectives of Biodegradable Nanocoating in Food Packaging"

Centering on biodegradable solutions in nano-packaging, this chapter spotlights biobased coatings' environmental benefits. These coatings, derived from biodegradable polymers, enhance product safety, functionality, and shelf life. Emphasizing the potential of polysaccharides and polyesters, it envisions innovative routes for entirely biobased nanocoatings.

Chapter 6: "Nanocoatings for Cosmetotextiles"

The convergence of nanotechnology and textile industry is the focus here, specifically in cosmetotextiles. This chapter explores how nanocoatings amplify sustainability and functionality in textiles. It evaluates current techniques, materials, and their environmental impacts while proposing strategies for eco-friendly nanocoatings. It identifies future avenues for biodegradable nanomaterials and lifecycle analyses.

Chapter 7: "Properties and Characteristics of Nanocoatings for Medicinal Applications"

Unveiling nanocoatings' potential in medicinal realms, this chapter examines their role in drug delivery, medical devices, and antimicrobial treatments. These coatings encapsulate therapeutic agents, ensuring targeted delivery while tackling antibiotic resistance. Despite promising outcomes, concerns about toxicity and biocompatibility warrant ongoing evaluation.

Chapter 8: "Healing and Health Perspectives of Nanocoatings for Medicinal Applications"

Offering an in-depth exploration, this chapter delves into nanocoatings' multifaceted roles in healing and healthcare. It details their impacts on drug delivery, wound healing, and tissue engineering while addressing challenges and regulatory considerations. It emphasizes how nanocoatings are poised to transform medicinal applications and healthcare outcomes.

Chapter 9: "Nanocoatings in Medicine: Revolutionizing Healthcare Through Precision and Potential"

This chapter unveils nanocoatings' pivotal role in medicinal contexts, particularly in enhancing medical implants and devices. It outlines the ingredients, production processes, and health benefits, emphasizing their transformative potential in addressing healthcare challenges.

Chapter 10: "Sustainable Approach to Nanocomposites and Nanocoatings: Research and Industrial Applications of Hydrophobic Nanocoatings"

Highlighting the booming hydrophobic coatings market, this chapter spotlights their demand across industries. It navigates through challenges in durability and scalability while emphasizing a Coating-as-a-Service business model. The chapter underscores the importance of reputable providers and high-quality materials for lasting protection.

Each chapter within this compilation serves as a window into the expansive realm of nanocoatings, illustrating their diverse applications, contributions to sustainability, and potential for future innovation and development.

IN SUMMARY

In concluding our exploration within these chapters, *Sustainable Approach to Protective Nanocoatings* stands out as a comprehensive guide illuminating the transformative prowess of nanotechnology and nanocoatings. This compilation collectively unravels the multidimensional potential of nanomaterials, showcasing their pivotal role in reshaping industries and driving sustainable solutions.

From the foundational insights in Chapter 1 on nanotechnology's transformative frontier to the intricate applications in biosensing and environmental remediation detailed in subsequent chapters, this compilation offers a panoramic view of nanotechnology's prowess. The chapters not only delve into the innovative materials and their applications but also navigate ethical considerations, acknowledging the imperative balance between progress and responsible deployment.

Chapter 2 and 3 take center stage, showcasing nanocoatings' role in intelligent and sustainable protection across various industries. The narrative progresses with recent advancements in designing smart and intelligent nanocoatings, delving into the latest breakthroughs that drive functionality and application innovation.

The exploration extends to Chapter 4 and 5 emphasizing crucial role of nanocoatings in enhancing solar cell efficiency and durability for sustainable energy solutions. The intersection of nanotechnology and textiles is explored in Chapter 6, highlighting innovative fabrics that merge fashion and functionality.

The journey also includes Chapter 7, 8 and 9, unveiling nanocoatings' transformative impact on medicine, from drug delivery systems to bioactive surfaces and finally concludes at Chapter 10, highlighting the booming hydrophobic coatings market, spotlighting their demand across industries.

This collaborative effort, featuring contributions from experts and innovators, provides a comprehensive resource on nanocoating technologies. As you go deeper into these chapters, the hope is that the diverse insights inspire a deeper appreciation for the transformative potential of nanotechnology in our world.

Through these chapters, the book aims to bridge the gap between scientific inquiry and practical applications, providing a comprehensive resource for students, researchers, and industry professionals. The intricate details, comprehensive research, and insightful analyses within these chapters aim to empower readers to delve deeper into the realm of nanotechnology and its impactful role in shaping a sustainable future.

As editors, our goal was to compile a compendium that not only informs but also inspires further exploration and innovation in the realm of nanocoatings and nanotechnology. We hope this book serves as a catalyst for interdisciplinary collaboration, propelling us toward a future where sustainability and technological advancements walk hand in hand, forging a path to a more resilient and responsible world.

Rajni Garg
Galgotias College of Engineering and Technology, India

Ansar Anjum
Galgotias College of Engineering and Technology, India

Chapter 1
Introduction to Nanocoatings and Packaging

Ansar Anjum
Galgotias College of Engineering and Technology, India

Deepak Garg
Delhi Technological University, India

Rahul Singhal
Shivaji College, University of Delhi, India

ABSTRACT

The advent of nanotechnology has inaugurated new era of ingenuity, particularly evident in the domain of nanocoating with diverse applications across industries. The coatings herald a transformative shift in surface protection. In the automotive sector, they play crucial role in enhancing scratch resistance and bolstering durability, thereby extending lifespan of vehicles. The packaging industry reaps the benefits of nanotechnology through the integration of advanced barrier properties, augmenting food preservation and packaging integrity. In the realm of healthcare, nanocoating applied to medical devices act as deterrents to microbial adhesion, fostering a hygienic environment. On the other hand, the electronics industry embraces nanocoating to establish water-resistant and robust surfaces, ensuring the prolonged endurance of electronic components. Thus, nanocoating has applications ranging from self-cleaning surfaces to improved insulation in sustainable building practices.

INTRODUCTION

Nanotechnology has ushered in a new era of innovation, particularly in the realm of nanocoating, where applications span across various industries. These coatings,

DOI: 10.4018/979-8-3693-3136-1.ch001

infused with nanomaterials, bring about a paradigm shift in surface protection (Anjum, 2017). In the automotive sector, nanocoating contribute to scratch resistance and heightened durability, ensuring a longer lifespan for vehicles. Nanotechnology has lately revolutionized diverse industries, redefining the landscape of innovation and functionality. In packaging, it introduces outstanding barrier properties, elevating food preservation and packaging integrity to unprecedented levels (Garg et al., 2023). Nanocoating is used in healthcare sector in medical devices, thwarting microbial and fostering pristine hygiene. The electronics sector benefits from water-resistant and durable surfaces, ensuring electronic components withstand the test of time. Aerospace industries leverage nanocoating for heightened corrosion resistance and superior thermal stability in challenging environments. In sustainable building practices, nanocoatings contribute to energy efficiency by enabling self-cleaning and improved insulation for windows. The renewable energy sector witnesses enhanced solar panel efficiency, majorly due to nanocoatings optimizing sunlight harnessing. Not only this, the textile industry embraces nanotechnology, as nanocoatings provide fabrics with stain resistance and UV protection, marking a significant leap in the multifaceted applications of the transformative technology.

Nanomaterials showcase exceptional properties, making them a highly viable choice for ballistic impact applications (Shakil et al., 2021). This has led to the development of advanced nanocomposites characterized by reduced density, impressive strength, stiffness, enhanced fracture resistance, and excellent absorption of impact energy (Jensin Joshua et al., 2020). Recently, nanoparticles such as carbon nanotubes (CNTs), nano-silica (Taş & Soykok, 2021), nano-clay (Wang et al., 2020), graphene nano-platelets (Khodadadi et al., 2021), milled nanofibers, Bucky paper, and nano-alumina have been employed to augment the ballistic impact behaviour of nanocomposites (Martínez-Morlanes et al., 2021).

The impact of nanotechnology resonates across sectors, promising a future where innovation knows no bounds. Over the past few years, global coating industries and scientific communities have introduced superhydrophobic coatings, showcasing remarkable water repellence applicable to marine, automotive, and medical sectors. Recently advancements and technical breakthroughs are explored in superhydrophobic coatings, emphasizing nanotechnology's role in innovative preparation strategies and unique property investigations. Lately, there has been a growing focus on crafting superhydrophobic coatings characterized by distinct structures and properties, with applications expanding into anti-corrosion, anti-icing, anti-fogging, self-cleaning, anti-fouling, and various other sectors. Notably, uptill 2010, superhydrophobic surfaces ranked 7th in material science's top 20 research fronts, as indicated by essential science indicators (Li et al., 2017). Recent strides in nanotechnology have spurred innovative developments in protective polymer nanocomposite (PNC) coatings, offering anti-corrosion, anti-fouling, and self-healing functionalities (Kashif

& Anjum, 2020). Additionally, it explores emerging market structures and novel applications (Idumah et al., 2020a). The polymer nanocoatings consist of densely packed grains, interphase boundaries, and dislocations, with inter-layer spacing approaching inter-atomic distances. This unique structure imparts diverse properties, allowing such coatings to surpass the mechanical and anti-corrosive capabilities of conventional coatings (Kaew-on et al., 2020). They are classified as metallic or ceramic based on their fundamental materials. Comprising two or more entities at the nanoscale, such coatings leverage the inherently fine dimensions of nanoparticles, enhancing coating efficiency by preventing voids and inhibiting the penetration of corrosive entities into the substrate's surface (Chen et al., 2020). In comparison to conventional coatings with micro-sized particles or thicknesses, these coatings are known to exhibit distinct corrosion behaviour. Leveraging their superior properties, PNCs find diverse applications in computers, cell phones, clothing, and eyeglasses. In construction, they enhance flame retardancy, wear resistance, scratch resistance, graffiti resistance, corrosion prevention, self-cleaning, and electrical conductivity on surfaces like air filters, windows, tiles, walls, and flooring. These coatings also improve adherence, anti-fogging, optical clarity, anti-fouling behaviour, and can serve as a substrate for photovoltaic applications (Huang et al., 2020). In the biomedical sector, metallic PNC coatings modify material surface characteristics when needed, playing a crucial role in medicine for protection, surface covering, and corrosion suppression. Additionally, they contribute to biocompatibility, drug delivery, and applications in fields such as military, automotive, energy, and the environment (Qian et al., 2020). Also, the nanomaterials are known to display unique optical, physical, and chemical features (Pathakoti et al., 2017). It is observed that the fundamental properties like flexibility, durability, flame resistance, barrier capabilities, and recyclability can be tailored through the incorporation of diverse nanomaterials (Ariyarathna et al., 2017). This adaptability smartly contributes to their increasing sophistication and widespread utilization in electronic, optical, magnetic devices, energy, and defense. Their influence has significantly extended to the industry of food packaging, with ongoing growth in this sector. Recent studies exploring their synthesis, characterization, applications, and assessments have propelled scientific progress, reshaping the landscape of the entire food industry (Cid & Simal-Gandara, 2020). Despite being a nascent technology, a steady and promising trajectory is anticipated for its continuous development in the foreseeable future.

Utilized across diverse sectors including hard coatings, biomedical applications, sensors, and aircraft manufacturing, nanocomposite coatings are witnessing substantial growth within the expansive realm of nanotechnology. The authors highlight their significance in aerospace, marine, automotive, sensor, dental implant, and electronics industries. This surge in application signifies the escalating role of nanocomposite coatings, indicating their pivotal position in advancing technological innovations across various fields (Xue et al., 2017). Nanocomposite coatings are coatings that contain well-dispersed nano-sized components generally added to a matrix phase, which is usually polymeric, metallic or ceramic.

TYPES OF NANOCOMPOSITE COATINGS

Metallic Matrix Nanocomposite Coating

Metallic nanocomposite coatings have captivated researchers due to their exceptional properties and diverse applications spanning industries like automotive, aerospace, energy production, and even biomedical engineering. The vast applications of such coatings are due to their sustainability, unique nanoarchitectures, superior mechanical characteristics, and outstanding wear resistance. However, among these remarkable features, corrosion resistance reigns supreme as the key determinant of metallic nanocomposite coatings lifespan.

Corrosion can significantly reduce the lifespan of any metal matrix nanocomposite coating, undermining its performance and rendering its other impressive attributes irrelevant. Therefore, understanding and enhancing the corrosion behaviour of these nanocomposite coatings is crucial to unlocking their full potential and ensuring their long-lasting functionality. This focus on corrosion resistance highlights the need for targeted research and development efforts aimed at creating metallic nanocomposite coatings that can withstand the harshest environments and deliver exceptional performance over extended periods. By prioritizing this aspect the researchers have been able to pave the way for the widespread adoption of such coatings in various sectors, enabling them to revolutionize industries and improve our everyday lives (Alizadeh Sani et al., 2022).

A study shows the deposition of Ni60-TiB2 composite coatings on steel substrates using the HVOF (high-velocity oxy-fuel) technique. The researchers then compared the mechanical and tribological properties of these nano-sized coatings to those of micro-sized Ni60-TiB2 coatings using a ball-on-disc configuration. The nano-sized Ni60-TiB2 coating exhibited a lower wear coefficient (indicating less volume loss) than the micro-sized coating. This improved performance was attributed to the higher fracture toughness, higher hardness, and higher strength of the nanocoating. Compared to microcoatings, the nanocoating displayed an 84% and 62% increase in fracture toughness and hardness, respectively. This improvement is due to their homogeneous microstructure and the strengthening effect of grain refinement (Wu et al., 2011; Yasin et al., 2020).

Silica has been the most prevalent material used for enhancing clear coatings in various applications, including acrylic eyeglass lenses, topcoats for automobiles, floor wear layers, and scratch-resistant polycarbonate sheets. Recent technological advancements have led to the availability of diverse inorganic nanoparticles for coating augmentation. Companies globally now offer colloidal silica in particle sizes ranging from 2 to 100 nm, in both aqueous and igneous forms. Highly concentrated nano-sized colloidal silica dispersed through UV-curable monomers has also become mainstream (Adeyemi & Fawole, 2023). This form of colloidal silica excels in preparing UV-cured coatings with superior chemical, abrasion, and scratch resistance. These benefits translate to improved film and gloss clarity, as well as enhanced durability. On the other hand, aluminium silicate nanoparticles have found their niche in automotive coatings. Patented by PPG Industries in 2002 under the brand name CeramiClear, this innovation revolutionized the industry. Mercedes-Benz even utilized the CeramiClear coating system in their vehicles, highlighting the improved scratch resistance achieved through the integration of silica nanoparticles (Awad Abouelata, 2019).

Technological advancements continue to expand the potential of nanoparticles in coating augmentation, offering exciting possibilities for improved performance and durability across various applications.

However, it is observed that there are some factors influence the functionality of nanocomposite coatings, including:

- Properties of the matrix and filler materials
- Spatial distribution of the fillers
- Surface morphology
- Deposition techniques

Polymer Matrix Nanocomposite Coating

Polymers excel in responding to stringent environmental regulations while ensuring consumer safety. Their ease of production, low cost, lightweight nature, and ductile behaviour make them the material of choice for various industrial applications. However, compared to metals and ceramics, they generally exhibit lower modulus and strength (Bagale et al., 2018). Many polymer composites embed nanocontainers loaded with corrosion inhibitors or repair agents that offer innovative solutions to corrosion. For example epoxy coatings with coconut-oil alkyd resin nanocontainers, urea-formaldehyde shells encapsulating *Azadirachta indica,* and surfactant-modified halloysite nanocapsules filled with benzotriazole (Malekkhouyan et al., 2020; Bagale et al., 2018).

Many techniques are employed to prepare polymer-based coatings on metals (Mashtalyar et al., 2019) through following techniques:

- Sol-gel process: The method allows for the formation of thin, uniform coatings at relatively low temperatures.
- Plasma-based coating method: This method utilize plasma to activate the surface of the metal, promoting strong adhesion of polymer coating.
- Electrodeposition: This technique is particularly suitable for conducting polymers including polyaniline and polypyrrole.
- Layer-by-layer deposition: This method helps in the precise control of coating thickness and composition by deposition of alternating layers of different materials.
- Spray coating: This method is widely used to apply polymer coatings on large surfaces.

A diverse range of polymers can be employed for metal corrosion protection and serve as matrices for composite materials. (figure 1) These include:

- Biopolymers: These naturally occurring polymers, such as chitosan and sodium alginate, offer eco-friendly alternatives to synthetic options.
- Synthetic polymers: They encompass a vast array of materials, including epoxies, polyaniline, and polypyrrole, each offering unique properties tailored to specific applications (Shahini et al., 2021).

The selection of polymers and coating techniques is crucial to ensure optimal performance and longevity of the metal coating. However, with careful consideration of important factors like environment, desired properties, and cost, polymer-based coatings can provide a powerful and versatile solution for protection of metals (Bansal et al., 2023).

Polymer nanocomposite coatings have emerged as powerful allies in the fight against corrosion due to their exceptional properties. These coatings are formed by incorporating nanofillers into a polymer matrix, resulting in a material with significantly enhanced performance compared to the individual components (Anjum et al., 2023). The key to the success of polymer nanocomposites lies in the remarkable effects of the nanofillers on the polymer matrix (Emmanuel et al., 2021). These effects include:

Enhanced mechanical and barrier properties:

- Increased rigidity and strength: The nanofillers act as reinforcement, improving the overall rigidity and strength of the coating.
- Improved heat resistance: The addition of nanofillers can significantly enhance the thermal stability of the polymer matrix.

Figure 1. Representative chitosan–filler combinations for anti-corrosion coatings

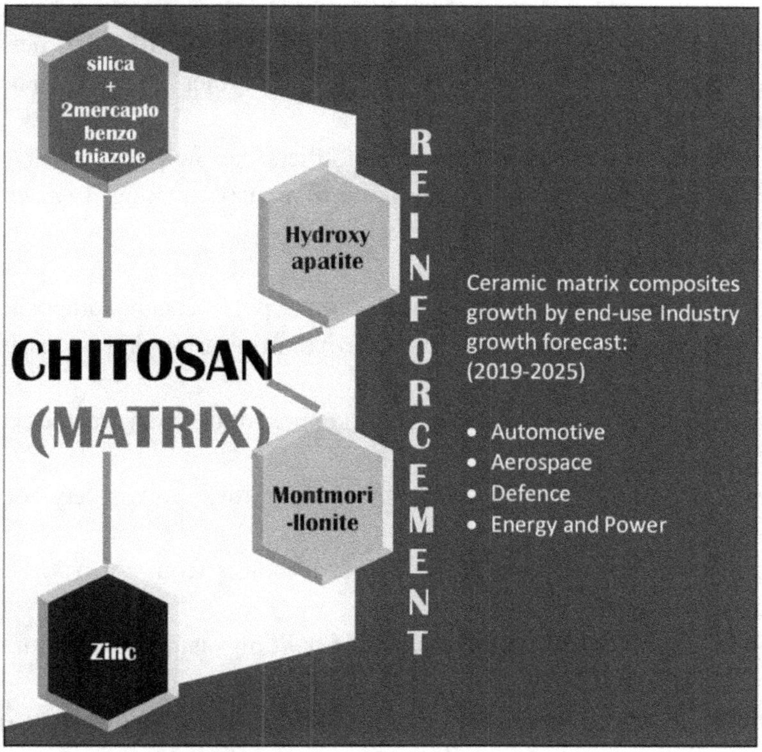

- Reduced flammability: Certain nanofillers can impart flame retardant properties to the coating, making it less susceptible to fire.
- Improved dimensional stability: The presence of nanofillers can help to reduce shrinkage and warping of the coating.
- Increased resistance to solvent attack: The nanofillers can create a tortuous path for corrosive agents, making it more difficult for them to penetrate the coating and reach the underlying metal.
- Improved water vapor barrier properties: This helps to prevent the formation of rust and corrosion beneath the coating.

Enhanced electrical and optical properties:

- Increased electrical conductivity: Certain nanofillers can be used to create conductive coatings with enhanced electrical properties.
- Improved transparency: Transparent nanocomposite coatings can be developed for applications where visibility is crucial.

Additional benefits:

- Reduced cost: By incorporating small amounts of high-performance nanofillers, the overall amount of expensive polymer material can be reduced without compromising performance.
- Improved processability: Certain nanofillers can help to improve the flow and processing properties of the polymer matrix, making it easier to apply the coating.

Due to the above mentioned remarkable benefits, polymer nanocomposite coatings have become a valuable tool for preventing corrosion in various industries, including:

- Automotive: Protecting car parts from exposure to harsh weather conditions and corrosive chemicals.
- Aerospace: Ensuring the long-term durability and safety of aircraft components.
- Marine: Protecting ships and offshore structures from corrosion caused by seawater.
- Construction: Extending the lifespan of buildings and infrastructure.

The development of polymer nanocomposite coatings continues to evolve, with researchers exploring new nanofillers and processing techniques to achieve even higher levels of performance and affordability. This ongoing innovation promises to further propel polymer nanocomposites into the future of corrosion protection.

Biopolymers have lately emerged as an eco-friendly alternative to corrosion protection. Acting as a barrier, they prevent corrosion with heteroatoms like oxygen and nitrogen. Their biodegradability, bioactivity, and non-toxicity make them ideal for medical implants. They can also be functionalized to deliver drugs and promote cell growth. Additionally, they can improve the corrosion and wear resistance of the substrate. Biopolymers offer a promising and sustainable solution for corrosion protection, particularly in sensitive applications.

Another significant advantage of biopolymers is their ability to improve the corrosion and wear resistance of the substrate. In some cases, they can even act as smart materials, mimicking biological surfaces in the human body (Christopher et al., 2016). Additionally, blending biopolymers with water-borne polymers can enhance the thermal, mechanical, and protective properties of the final material.

Among the various biopolymers, chitosan stands out as a particularly effective corrosion inhibitor for mild steel, copper, and zinc without causing environmental harm. This natural linear polysaccharide, derived from the shells of crustaceans, exhibits exceptional adsorption properties and film-forming abilities, making it an ideal candidate for eco-friendly corrosion protection, demonstrating its efficacy in mitigating corrosion without causing harm to the environment (Szőke et al., 2019) demonstrating its efficacy in mitigating corrosion without causing harm to the environment.

Biopolymers offer a notable advantage by enhancing the corrosion and wear resistance of substrates, showcasing versatility in various applications. Beyond their protective role, biopolymers can exhibit smart material characteristics, mimicking biological surfaces found in the human body. An additional benefit arises from blending biopolymers with water-borne polymers, leading to improvements in thermal, mechanical, and protective properties within the final material (Jafarzadeh et al., 2023).

Ceramic Matrix Nanocomposites Coatings

Ceramic nanocomposites are the emerging revolutionary class of materials, distinct from their macro-grained counterparts. The nanocrystals in which building blocks are not coarse grains, but tiny, nanometer-sized crystals often surrounded by a secondary nanocrystalline or amorphous matrix, create a unique structure with exceptional properties. Unlike their bulkier counterparts, ceramic nanocomposites have grains smaller than 100 nanometers. This diminutive scale has a profound impact because the smaller grains act like roadblocks, hindering the movement of dislocations which are the tiny defects within the material that weaken it that in turn leads to a remarkable increase in hardness and overall strength. The intricate network of grain boundaries in these nanocomposites plays a starring role. The boundaries, where the nanocrystals meet, are not mere passive interfaces but actively influence and even dominate the behaviour of material. This opens many opportunities to previously unimaginable properties, not found in their larger-grained counterparts.

Ceramic nanocomposites are not just stronger and harder. They can be tailored to exhibit enhanced toughness, wear resistance, and even electrical or thermal conductivity. This versatility makes them an ideal material for wide range of applications ranging from lightweight armour and cutting-edge engine components to next-generation batteries and medical implants. The potential of ceramic nanocomposites is still unfolding, but their unique properties are already sparking excitement across various industries. As research and development continue, we can expect these remarkable materials to reshape the landscape of materials science, paving the way for innovative technologies and advancements in diverse fields (Raveh et al., 2007).

When metals are subjected to high-temperature environments, their usual armour of corrosion resistance may melt away. However, for ceramic composite coatings a multifaceted shield is present to safeguard metals and alloys from corrosion. These coatings are no mere paints but are meticulously chosen allies, their thermal expansion perfectly in sync with the metal beneath. This minimizes the stress and cracking in metal, ensuring the shield to remain unbroken against the onslaught of heat. Their chemical inertness is a formidable barrier, repelling environmental corrosion and wear like a seasoned warrior.

Oxide-based ceramic composites can withstand high-temperature, offering unparalleled thermal resistance and phase stability. They can form an impregnable wall against the extreme temperatures, providing crucial insulation and structural integrity.

However, unlike their monolithic counterparts, these composites possess the adaptability of a legion, boasting enhanced toughness and wear resistance. This versatility arises from the ability to tailor the reinforcements within the matrix. The matrix strength of ceramic and its modulus are strategically deployed to create a shield with optimized fracture resistance. This ingenious design allows the coating to bend without breaking, a critical advantage of the composite in high-temperature environments. The adaptability is due to microstructure of the composite. By carefully manipulating the size, distribution, and morphology of the reinforcing phase within the oxide matrix, we can fine-tune the coating's mechanical properties, achieving a delicate balance between strength, toughness, and thermal resilience. This opens exciting avenues for researchers to engineer materials with properties for specific high-temperature applications, from aerospace engines to nuclear reactors.

Thus, oxide-based ceramic composite coatings are not just a passive barrier but they are strategically equipped to withstand the relentless assault of high-temperature environments. The understanding of their intricate design and adaptability is crucial for researchers pushing the boundaries of material science and unlocking new possibilities in this challenging realm.

The ceramic composite coatings are a multi-pronged approach that address the specific challenges of high-temperature corrosion with a combination of thermal harmony, chemical resilience, and adaptable toughness. This makes them the ideal protectors for a diverse range of applications. They stand as a testament to the power of innovation, safeguarding metals and unlocking possibilities in the hottest of environments (Kamalan Kirubaharan & Kuppusami, 2020). By incorporating these features, ceramic composite coatings provide superior protection for metals and alloys, ensuring their long-lasting performance under demanding conditions.

Examples of ceramic nanocomposites include oxides, carbides, and nitride matrices in combination with metallic or non-metallic elements (Balázsi et al., 2021). As already mentioned, commonly used ceramic matrices are aluminium oxide, titanium dioxide, silicon carbide, silicon nitride, Si3N4, etc., while mostly used reinforcement materials are clays, silica, metallic oxides, graphene, etc., under the form of nanoparticles, whiskers, fibres, or nanotubes (Kalisz et al., 2016).

Achieving precision in the shaping of high-performance ceramics is crucial to minimize the defects and ensuring consistency, a task challenging with current methods relying on pure nano powders. The solution to this emerges with micro-nanocomposites, a strategic blend of a submicron-sized matrix for stability and smaller nanosized additives for achieving desired functionality. This innovative combination facilitates the use of near-net-shape forming technologies such as injection moulding, allowing the production of high-performance ceramics at affordable costs while minimizing waste. The incorporation of micro-nanocomposites represents a significant leap in ceramic manufacturing, overcoming the limitations of traditional methods. This advancement not only addresses challenges associated with shaping but also opens avenues for cost-effective and efficient production of high-performance materials. As the ceramic industry grapples with these challenges, micro-nanocomposites stand out as an innovative solution, revolutionizing the landscape of ceramic manufacturing.

CHALLENGES AND FUTURE PROSPECTS

The role of nanocoating in food packaging is on the cusp of a revolution due to the innovative power of nanotechnology. The technology allows us to manipulate materials at the molecular level, unlocking a vast array of possibilities. The nature's building blocks are now being incorporated into polymer matrices that dramatically improves the functional properties, opening doors to enhanced barrier properties. The packaging can now breathe and protects simultaneously, preventing unwanted gases and vapours from infiltrating while allowing essential ones to escape. This translates to longer shelf life for food and reduced waste (Chausali et al., 2022). The smart and intelligent packaging consist of nanosensors embedded in packaging materials that have the potential to transform them into miniature sentinels. They can monitor various parameters like temperature, gas composition, and even food spoilage, providing real-time information about the condition of the food. However, as with any powerful tool, nanotechnology comes with its own set of considerations. Nanoparticles exhibit properties vastly different from their bulk counterparts. Their reduced size and increased surface area raise concerns about potential migration into food and subsequent exposure to consumers. These potential health risks need to be carefully evaluated and mitigated (He et al., 2019).

As we look into the fascinating world of nanotechnology in food packaging we also need to meticulously examine both its immense potential benefits and the need for responsible and cautious implementation. By striking a balance between innovation and safety, we can harness the power of nanotechnology to revolutionize food packaging, ensuring a future where food is not only delicious but also safe and sustainable (Figure 2).

Despite the exciting potential of nanotechnology to revolutionize the food industry, significant hurdles remain before its widespread adoption in novel food products and processes such as:

- Developing safe and effective delivery systems for nano-enabled foods presents a major challenge. These systems require sophisticated yet cost-effective processing to ensure efficient incorporation and controlled release of nanoparticles.
- The potential for nanoparticles to leach or migrate from packaging materials into food raises significant safety concerns. Comprehensive research is essential to understand and mitigate this risk, ensuring food safety and consumer health.
- At the nanoscale, materials exhibit unique properties that are often poorly understood. This necessitates further research to develop robust analytical methods to characterize and predict the behaviour of nanomaterials in food systems.

Figure 2. Representation of Challenges in nanocoating

- Understanding the nanoscale functions and potential toxicity of various nanoparticles is crucial for ensuring the safety of nanotechnology in food. This requires extensive research on potential risks, including environmental and health impacts.
- Developing a comprehensive and robust regulatory framework that addresses the unique challenges of nanotechnology in food is essential. This includes addressing issues like biosecurity, public perception, and environmental impact.
- Current methods for the synthesis of nanoparticles often generate harmful by-products that contribute to environmental pollution. Developing sustainable and eco-friendly synthesis methods is crucial for responsible implementation of nanotechnology in the food industry.
- Thorough in-vitro and in-vivo investigations are necessary to assess the safety and efficacy of nano-enabled antibacterial agents and other food applications and also needs to focus on nanoparticle interactions with living organisms and their potential environmental impact.
- Building public trust and addressing consumer concerns about nanotechnology in food requires open communication and transparent information sharing. Engaging with stakeholders and providing accurate scientific data is crucial to ensure responsible and sustainable development of nanotechnology in the food sector.

On addressing these challenges through coordinated research efforts and responsible development practices, nanocoatings have the potential to transform the food industry, leading to significant improvements in food safety, quality, and sustainability (Barage et al., 2022). While the use of nanomaterials in active, smart, and intelligent packaging offers exciting possibilities, achieving widespread social and commercial acceptance still remains a challenge. The concerns amongst industry and other consumers encompass:

- Health and Safety of Workers wherein the potential exposure to nanoparticles during manufacturing and handling raises concerns about their impact on worker health and safety.
- Environmental Implications of nanomaterials are still unclear that creates concern about the potential harm to ecosystems and wildlife.
- Studies suggest potential human health risks associated with nanoparticles, including oxidative damage, inflammation, and even cancer.
- The possibility of nanomaterials migrating from packaging into food is a major concern. Strict migration testing is crucial to ensure the safety of food products packaged with nanomaterials.

Focussing on these concerns is highly essential to gain public trust and promote the responsible development and adoption of nanotechnologies in food packaging (Sharifi et al., 2012). This however, can be deciphered by adopting the following:

- **Comprehensive Research:** The nanomaterials hold immense potential across various fields but at the same time, concerns about their impact on human health and the environment remain. To ensure their safe and responsible use, further research is crucial. This includes investigating potential risks like toxicity and inflammation, studying their environmental fate and impact on ecosystems, and exploring beneficial applications alongside risk mitigation strategies (Singh et al., 2023). By addressing these research needs, the true potential of nanotechnology can be unlocked while safeguarding human health and the environment.
- **Transparent Communication:** While nanotechnology offers immense potential, building public trust is crucial for its successful integration into society. This requires open and transparent communication with diverse stakeholders. The collaboration and engagement of researchers with industries in discussions about risk management and safety protocols fosters confidence and responsible use. Providing clear information empowers consumers to make informed choices and builds trust. Dialogues with regulatory bodies ensure guidelines are based on scientific evidence and facilitate robust frameworks for nanotechnology. Through open communication, a collaborative environment can be well created where concerns are addressed, risks are mitigated, and the potential of nanotechnology is harnessed for the benefit of all (Farris et al., 2023).
- **Strict Regulatory Frameworks:** Robust regulatory frameworks are necessary to ensure the safety of nanomaterials used in food packaging. This includes establishing clear guidelines for testing, risk assessment, and labelling.
- **Investment in Safe and Sustainable Technologies:** Funding research and development of safer and more sustainable nanomaterials and production processes is essential for minimizing environmental impact and ensuring public health.

By addressing these challenges and demonstrating a commitment to responsible development, the food industry can harness the potential of nanomaterials to create innovative packaging solutions that are both beneficial and safe. This will pave the way for a future where nanotechnology plays a positive role in the food industry, while ensuring the health and well-being of consumers and the environment (Ashfaq et al., 2022).

RECENT INNOVATIONS AND BREAKTHROUGHS

The world of materials science is constantly pushing boundaries, recent advancements with nanotechnology are particularly exciting (Singh et al., 2023). Some of the important and latest innovations in the field of nanocoating have been discussed below that possess the potential to revolutionize various fields:

1. **Fingerprint-Repelling coating for stainless steel:** The German researchers have offered a solution to stainless steel surfaces to remain pristine, untouched by fingerprints and grease with their revolutionary nanocoating. This breakthrough holds immense potential for applications ranging from household appliances and kitchenware to medical instruments and automotive components. It not only enhances aesthetics but also protects the metal from damaging grease, paving the way for a cleaner and more durable future for stainless steel (Kesmez et al., 2017). The coating works through two clever mechanisms:
    - Micro-roughness: The coating slightly roughens the stainless steel surface, increasing the overall surface area. When a finger touches the surface, it only interacts with the raised points, preventing grease from reaching the "valleys" and minimizing fingerprint formation.
    - Matching Refractive Index: The refractive index of coating is meticulously adjusted to match that of the natural oils found in human skin. This means light reflects off the coated surface in a similar way as it does from a finger-touched area, effectively making fingerprints invisible to the naked eye.
2. **Super-Black coatings that pushes the limits of absorption:** The quest for the blackest black has led to two remarkable achievements in recent years:
    - BMW's "Vantablack": This nanocoating, composed of carbon nanotubes, absorbs light almost completely. Objects coated with Vantablack appear to vanish against a deep black background, as the human eye perceives them as two-dimensional due to the lack of light reflection.
    - MIT's Carbon Nanotube Coating: Taking things even further, engineers have been able to develop a blacker-than-black coating that absorbs a staggering 99.995% of incoming light. This achievement, made possible by vertically aligned carbon nanotubes, marks a significant advancement in the field of light absorption.

The super-black coatings have diverse potential applications beyond aesthetics, including scientific instruments requiring minimal light interference, space telescopes for capturing faint celestial signals, and camouflage technology for military and aerospace purposes (https://ethz.ch/en/news-and-events/eth-news/news/2019/03/nanotechnology-and-sunlight-clear-the-way-for-better-visibility.html).

3. **Boron Nitride nanotube coatings for efficient heat management in turbines:** The jet engine relies on intricate components that operate under extreme temperatures. Protecting these components from heat damage is crucial for efficiency and longevity. While traditional ceramic coatings excel at this task, they lack control over how the turbine blades radiate heat, impacting overall performance. The researchers lately have developed a revolutionizing coating composed of these nanotubes (Table 1). This innovative approach involves:

    ◦ Polariton Excitation: As the turbine blade heats up, polaritons (oscillations of light and matter) within the ceramic coating become excited.
    ◦ Nanotube-Mediated Heat Radiation: The boron nitride nanotubes, acting as antennas, efficiently couple the excited polaritons to outgoing heat radiation. This allows for precise control over heat dissipation, directing it away from critical components and optimizing the turbine's performance (Xu et al., 2022).

The benefits of this technology are multifold:

- Increased Efficiency: Improved heat management leads to higher energy conversion efficiency, making jet engines and turbines more fuel-efficient.
- Extended Lifespan: Reduced thermal stress on engine components translates to longer lifespans and reduced maintenance needs.
- Reduced Environmental Impact: Increased efficiency translates to lower emissions and a smaller environmental footprint.

The boron nitride nanotube coating marks a significant leap forward in turbine technology. Its potential for enhanced efficiency, durability, and sustainability promises to revolutionize the aerospace and energy sectors (https://www.nonwovens-industry.com/issues/2019-03-1/view_features/smart-heating-a-flexible-conductive-coating-for-nonwovens).

These advancements in nanocoatings showcase the incredible potential of this technology to transform various fields in the coming future. As research continues, we can expect even more transformative innovations that push the boundaries of material science and shape the future of diverse industries.

Emerging Trends and Market Structure

The future trajectory of the nanocoating market looks highly promising. As the demand for advanced materials with superior performance continues to surge, nanocoatings are gaining substantial traction. The automotive industry, in particular,

Table 1. Summary of the key information for each type of coating

Type pf Coating	Functionality	Potential Applications	Benefits
Fingerprint-Repelling	Prevents fingerprints and grease	Household appliances, kitchenware, medical instruments, automotive components	Improved aesthetics, reduced maintenance, longer lifespan
Super-Black	Absorbs light almost completely	Scientific instruments, space telescopes, camouflage technology	Enhanced sensitivity, improved accuracy, reduced visibility
Boron Nitride Nanotube	Efficiently manages heat radiation	Jet engines, electricity-generating turbines	Increased efficiency, extended lifespan, reduced environmental impact

serves as a key driver, with nanocoatings playing a crucial role in enhancing both the durability and aesthetics of automotive parts. Furthermore, the increasing emphasis on energy-efficient buildings and infrastructure is propelling the demand for nanocoatings within the construction sector.

Moreover, the healthcare industry is emerging as a significant market for nanocoatings. These coatings find application in the healthcare sector for developing antimicrobial coatings on medical devices, effectively preventing infections. The rising awareness regarding health and hygiene is an additional factor contributing to the overall growth of the market. Antimicrobial nano-coatings represent advanced nanotechnology-based techniques designed to eliminate both gram-positive and gram-negative bacteria (Adeyemi & Fawole, 2023). These coatings adhere to bacterial cell membranes, disrupting bacterial metabolism and establishing a pathogen-free surface. Their efficacy extends beyond bacteria to include viruses, fungi, and specific microorganisms like methicillin-resistant Staphylococcus aureus (MRSA) and vancomycin-resistant Enterococci (VRE), contingent upon the specific antimicrobial agent employed. Projections indicate that the antimicrobial nanocoatings market is poised to encompass approximately 15% to 18% of the global nanocoatings industry, underlining the substantial impact and growing significance of these technologies in diverse applications (Farris et al., 2023).

The Future Market Insights report marks the global market for polymeric-based coatings, particularly PNCs, is projected to experience consistent growth with a CAGR (Compound Annual Growth Rate) within 2017-2022. By the end of 2022, revenues for polymeric coatings combating corrosion are expected to exceed US$ 19.5 billion (Idumah et al., 2020b). Additionally, the global conductive PNCs market is foreseen to achieve a CAGR of 9.2%, by the end of 2027 from USD 3.16 billion in 2016. The PNCs abrasion-resistant coatings (PNCs-ARCs) market, valued at USD 8.0 billion in 2019, is anticipated to grow to USD 11.1 billion by 2024, exhibiting a CAGR of 6.5% between 2019 - 2024. Latin America dominates the global abrasion-resistant coatings market, with an estimated USD 2.3 billion by 2024 and a CAGR of 6.3%

during the forecast period. PNCs-ARCs are expected to experience substantial growth in power generation, oil and gas, mining, transportation, marine, and construction industries in Latin America. Despite forecasting occurring before the COVID-19 pandemic, the values are expected to remain stable as the pandemic recedes. The increasing demand for semiconducting products, especially in smart textiles, solar, automotive, locomotive, electrical, and electronics industries, is poised to propel the growth of the global conductive polymer coating market. This surge is anticipated through the utilization of conductive polymer coatings in various applications such as sensors, fuel cells, cell phones, displays, touch panels, solid-state lighting, capacitors, batteries, and photovoltaic cells. (Future Market Insights) The field of food science stands on the precipice of a transformative journey, driven by the immense power of nanotechnology. This cutting-edge technology empowers us to manipulate matter at the atomic and molecular level, unlocking a vast array of possibilities for the food industry. From enhanced food quality and safety to extended shelf life and innovative food formulations, the benefits of nanotechnology are vast and far-reaching.

Beyond these immediate advantages, the future holds even greater promise. Imagine AI-powered intelligent packaging that monitors food quality in real-time, biodegradable nanocomposites for sustainable packaging solutions, and nanofiber-based delivery systems for targeted nutrition and flavour enhancement (Rani et al., 2023). However, realizing the full potential of nanotechnology requires overcoming key challenges. Developing safe and efficient delivery systems for nanoparticles, understanding their nanoscale behaviour, addressing potential toxicity concerns, and establishing a robust regulatory framework are crucial steps towards responsible and sustainable development. By embracing these challenges and harnessing the power of nanotechnology responsibly, we can unlock a future of enhanced food safety, reduced waste, improved nutrition, and a new era of innovative culinary experiences. The future of food lies at the intersection of science and technology, and by embracing the potential of nanotechnology, we can pave the way for a healthier, more sustainable, and delicious future for all.

ENVIRONMENTAL IMPACT AND SUSTAINABILITY IN FUTURE

Eco-Friendly Disposal Options

The modern food system, wrapped in convenience, often comes at a hidden cost, an environmental burden disguised as plastic packaging. The oil-based material has seen its global production soar, reaching a staggering 335 million tons in 2016 alone. In Europe, almost 23 million tons of plastic packaging is produced annually and the figure is anticipated to triple by 2050 (Jambeck et al., 2015). Out of this, a colossal 9 million tons, ends up in landfills that slowly leaches into the soil. Another 32% escapes the collection and sorting systems, finding its way into the oceans and soil (HLPE. Food Losses and Waste in the Context of Sustainable Food Systems, 2014).

Active packaging, in particular, warrants special attention. When volatile compounds are released within the package, the consumer exposure becomes a crucial element of the equation. We must assess not just the impact on food quality, safety, and shelf life, but also consider all potential sources of the volatiles, including their natural presence in the food itself. (Brody A. 2004) All stringent regulatory requirements must be met for active packaging to thrive in order to deliver on its intended efficacy, and ensure minimal impact on the sensory experience of the food (especially when volatiles are also used for flavouring). Ultimately, the success of active packaging hinges on its ability to demonstrate tangible benefits. We need to validate that its positive role in reducing food waste outweighs the potential cost of this new technology. This, in turn, contributes to a more sustainable food packaging ecosystem, respecting both the planet and human life. By prioritizing holistic design and rigorous benefit-cost analysis, we can steer in a new era of food packaging – one that is not just functional, but responsible, sustainable, and truly future-proof. During past decade, Europe's ambition in packaging technology soared high. Public and private coffers fuelled a surge in research, aiming to tap into a seemingly bottomless well of demand for bio- and smart packaging solutions. Forecasts paint a picture of explosive growth poised to witness the widespread adoption of these revolutionary technologies in. near future (Hall et al., 2009).

The plastic pollution degrades, transforming into micro- and nano-sized particles that infiltrate the animal and human life. Fishes ingest these microplastics, unknowingly feeding them up the food chain, potentially even to humans. The long-term consequences of this plastic paradox are yet to be fully understood, but the potential for harm is undeniable (Zhu et al., 2006a). The stark reality is that if such a trend of plastic usage and disposal is followed then by 2050, the weight of plastic in the oceans may surpass the weight of fish (Floras & Matsos, 2005). We must break free from the linear "take-make-dispose" model of plastic use and embrace a circular economy where packaging is not waste, but a valuable resource. Investing in biodegradable materials, robust recycling systems, and responsible consumer choices are crucial steps in this direction (Netherlands Organisation for Applied Scientific Research).

Our current food system carries with it a grim reality, that is, a staggering amount of food waste and plastic pollution. Around 100 million tons of food are wasted annually in the EU alone, translating to 495 million tons of CO_2 and a blue water footprint equivalent to half of Lake Geneva. By 2050, this waste could rise to over 200 million tons, occupying 210 million hectares of precious land. Also, a data shows that Europe produces 23 million tons of plastic packaging annually, a figure expected to triple by 2050. If left unchecked, the global plastic industry will reach 1.1 billion tons by 2050, consuming 20% of total oil production (Dahiya et al., 2018). A collaboration is required between researchers, manufacturers, policymakers, and consumers to create a food system that is both efficient and sustainable (Zhu et al., 2006b). Our oceans choke on 8 million tons of plastic annually, with packaging culprits accounting for a staggering 62%. While recycling seems like a beacon of hope, issues like safety and limited secondary markets for cascaded recycling hinder its effectiveness. Current bottle-to-bottle methods degrade PET material, necessitating virgin additions for food contact, further burdening the environment. By harnessing organic waste to create 100% biodegradable bioplastics, we can forge a resilient, productive food packaging economy aligned with circularity principles. This revolutionary approach promises a future where waste becomes a resource, not a burden, and our oceans will be able to breathe again (Bala & Bhagat, 2023). This bio-based bonanza (46 million tons by 2050) would be fully biodegradable, leaving our oceans and landfills pristine. Upcycling organic waste, from food scraps to packaging itself, would solve the plastic pollution crisis, seamlessly aligning with the EU's circular economy plan.

It is estimated that by 2050, after implementation of the packaging revolution, we can ensure a future food packaging that minimizes food waste by extending shelf life, optimizing storage conditions, and providing real-time monitoring. Thus, packaging can become a powerful weapon against food waste as it will be able to eliminate plastic reliance because biodegradable and edible materials, coupled with robust recycling systems, can break the stranglehold of plastic packaging and reduce its environmental footprint, it will be able to promote sustainability from resource utilization to greenhouse gas emissions (FAO, 2013).

By 2050, swapping just one in two food packages for organic waste-based alternatives could:

- **Slash Plastic Waste:** Ditching plastic for bio-benign materials could eliminate a staggering 46 million tons of plastic waste annually, easing the pressure on landfills and oceans.
- **Conserve Resources:** This transition would save an estimated 43 million tonnes of oil-based resources per year in Europe, and over 150 million tonnes globally.
- **Cut Greenhouse Gases:** Bio-packaging can potentially reduce greenhouse gas emissions by 120 million tons of CO_2-equivalent in Europe and 500 million tons globally (just considering direct emissions).

Thus, the bio-packaging revolution could deliver a 370 million ton reduction in CO_2-equivalent emissions at the EU level alone – a massive 10% cut, aligning with the 2050 goals of the Paris Agreement.

CONCLUSION

A decade of research has unveiled the outstanding potential of nanotechnology in the industry of food packaging, promising a future where food is healthier, tastier, and less wasteful. This will help in fresh fruits lasting weeks, not days, with intelligent packaging that adapts to their needs. Nanomaterials can enhance shelf life, monitor spoilage, and even release nutrients. But to achieve all we have to keep in mind that ethical concerns and potential health risks demand cautious development. The need of the hour suggests that scientists, companies, and consumers must work together, guided by robust regulations, to ensure safe and sustainable implementation. This fruitful partnership can revolutionize food packaging, making our food system not just delicious, but also safe and responsible. Thus the future of food packaging demands a shift from a reactive approach to a proactive approach. We must move beyond simply fulfilling immediate requirements and consider the full spectrum of factors that influence a material's success, starting right at the scaling-up stage. This involves incorporating food needs, consumer preferences, market demands, and legal regulations into the very fabric of the design process.

REFERENCES

Adeyemi, J. O., & Fawole, O. A. (2023). Metal-Based Nanoparticles in Food Packaging and Coating Technologies: A Review. *Biomolecules*, *13*(7), 1092. doi:10.3390/biom13071092 PMID:37509128

Alizadeh Sani, M., Maleki, M., Eghbaljoo-Gharehgheshlaghi, H., Khezerlou, A., Mohammadian, E., Liu, Q., & Jafari, S. M. (2022). Titanium dioxide nanoparticles as multifunctional surface-active materials for smart/active nanocomposite packaging films. *Advances in Colloid and Interface Science*, *300*, 102593. doi:10.1016/j.cis.2021.102593 PMID:34971916

Anjum, A. (2017). *Adsorption Technology for Removal of Toxic Pollutants*. doi:10.1007/978-3-319-61146-4_2

Anjum, A., Garg, R., Kashif, M., & Eddy, N. O. (2023). Nano-scale innovations in packaging: Properties, types, and applications of nanomaterials for the future. *Food Chemistry Advances, 3*, 100560. doi:10.1016/j.focha.2023.100560

Ariyarathna, I. R., Rajakaruna, R. M. P. I., & Karunaratne, D. N. (2017). The rise of inorganic nanomaterial implementation in food applications. In *Food Control* (Vol. 77, pp. 251–259). Elsevier Ltd. doi:10.1016/j.foodcont.2017.02.016

Ashfaq, A., Khursheed, N., Fatima, S., Anjum, Z., & Younis, K. (2022). Application of nanotechnology in food packaging: Pros and Cons. *Journal of Agriculture and Food Research, 7*, 100270. Advance online publication. doi:10.1016/j.jafr.2022.100270

Awad Abouelata, A. M. (2019). Nanocomposite coatings on aluminum surfaces for several applications. In *Advances in Smart Coatings and Thin Films for Future Industrial and Biomedical Engineering Applications* (pp. 163–205). Elsevier. doi:10.1016/B978-0-12-849870-5.00002-1

Bagale, U. D., Sonawane, S. H., Bhanvase, B. A., Kulkarni, R. D., & Gogate, P. R. (2018). Green synthesis of nanocapsules for self-healing anticorrosion coating using ultrasound-assisted approach. *Green Processing and Synthesis, 7*(2), 147–159. doi:10.1515/gps-2016-0160

Bala, K., & Bhagat, M. (2023). Nanomaterial-based smart coatings as antimicrobials. *Antiviral and Antimicrobial Smart Coatings: Fundamentals and Applications*, 413–429. doi:10.1016/B978-0-323-99291-6.00010-4

Balázsi, K., Furkó, M., & Balázsi, C. (2021). Ceramic Matrix Graphene and Carbon Nanotube Composites. Encyclopedia of Materials: Technical Ceramics and Glasses: Volume 1-3, 2, 243–259. doi:10.1016/B978-0-12-818542-1.00070-9

Bansal, H., Singh, S., Sharma, A., Singh, J., & Mehta, S. K. (2023). Polymer nanocomposite coatings and films: modern insights and emerging strategies to lengthen the lifespan of fruits and vegetables. *Food Materials Research, 3*(1). doi:10.48130/FMR-2023-0023

Barage, S., Lakkakula, J., Sharma, A., Roy, A., Alghamdi, S., Almehmadi, M., Hossain, M. J., Allahyani, M., & Abdulaziz, O. (2022). *Review Article Nanomaterial in Food Packaging: A Comprehensive Review*. doi:10.1155/2022/6053922

Chausali, N., Saxena, J., & Prasad, R. (2022). Recent trends in nanotechnology applications of bio-based packaging. *Journal of Agriculture and Food Research, 7*, 100257. doi:10.1016/j.jafr.2021.100257

Chen, Y., Ding, Y., & Zheng, J. (2020). A polymer nanocomposite coating with enhanced hydrophilicity, antibacterial and antibiofouling properties: Role of polymerizable emulsifier/anionic ligand. *Chemical Engineering Journal, 379*, 122268. doi:10.1016/j.cej.2019.122268

Christopher, G., Kulandainathan, M. A., & Harichandran, G. (2016). Biopolymers nanocomposite for material protection: Enhancement of corrosion protection using waterborne polyurethane nanocomposite coatings. *Progress in Organic Coatings, 99*, 91–102. doi:10.1016/j.porgcoat.2016.05.012

Cid, A., & Simal-Gandara, J. (2020). Synthesis, Characterization, and Potential Applications of Transition Metal Nanoparticles. In Journal of Inorganic and Organometallic Polymers and Materials (Vol. 30, Issue 4, pp. 1011–1032). Springer. doi:10.1007/s10904-019-01331-9

Dahiya, S., Kumar, A. N., Shanthi Sravan, J., Chatterjee, S., Sarkar, O., & Mohan, S. V. (2018). Food waste biorefinery: Sustainable strategy for circular bioeconomy. In *Bioresource Technology* (Vol. 248, pp. 2–12). Elsevier Ltd., doi:10.1016/j.biortech.2017.07.176

Emmanuel, O. A., Fayomi, O. S. I., Agboola, O., Ayoola, A. A., Oloke, O. C., & Amusan, L. M. (2021). Short review on nanocomposite coating advances in the industry. *IOP Conference Series. Materials Science and Engineering, 1107*(1), 012069. doi:10.1088/1757-899X/1107/1/012069

Farris, S., Silva De Sousa, M., Schlogl, A. E., Ramalho Estanislau, F., Gomes, V., Souza, L., Sélia, J., Coimbra, R., José, I., & Santos, B. (2023). Academic Editors: Swarna Jaiswal Nanotechnology in Packaging for Food Industry: Past, Present, and Future. doi:10.3390/coatings13081411

Floras, J. D., & Matsos, K. L. (2005). Introduction to modified atmosphere packaging. In *Innovations in Food Packaging*. Overiew. doi:10.1016/B978-012311632-1/50042-5

Future Market Insights. (n.d.). www.futuremarketinsights.com

Garg, R., Garg, R., Sharma, R., Anjum, A., & Eddy, N. O. (2023). Introduction to Nanotechnology and Nanomaterials. In *Nanotechnology for Sustainable Agriculture, Food and Environment*. CRC Press. doi:10.1201/9781003397861-1

Hall, K. D., Guo, J., Dore, M., & Chow, C. C. (2009). The progressive increase of food waste in America and its environmental impact. *PLoS One, 4*(11), e7940. Advance online publication. doi:10.1371/journal.pone.0007940 PMID:19946359

He, X., Deng, H., & Hwang, H. (2019). The current application of nanotechnology in food and agriculture. In Journal of Food and Drug Analysis (Vol. 27, Issue 1, pp. 1–21). Elsevier Taiwan LLC. doi:10.1016/j.jfda.2018.12.002

HLPE. (2014). *Food Losses and Waste in the Context of Sustainable Food Systems*. HLPE.

Huang, B. S., Lai, G. H., Yang, T. I., Tsai, M. H., & Chou, Y. C. (2020). A novel electroactive imide oligomer and its application in anticorrosion coating. *Polymers, 12*(1), 91. Advance online publication. doi:10.3390/polym12010091 PMID:31947895

Idumah, C. I., Obele, C. M., Emmanuel, E. O., & Hassan, A. (2020a). Recently Emerging Nanotechnological Advancements in Polymer Nanocomposite Coatings for Anti-corrosion, Anti-fouling and Self-healing. *Surfaces and Interfaces, 21*, 100734. doi:10.1016/j.surfin.2020.100734 PMID:34957345

Idumah, C. I., Obele, C. M., Emmanuel, E. O., & Hassan, A. (2020b). Recently Emerging Nanotechnological Advancements in Polymer Nanocomposite Coatings for Anti-corrosion, Anti-fouling and Self-healing. In *Surfaces and Interfaces* (Vol. 21). Elsevier B.V. doi:10.1016/j.surfin.2020.100734

Jafarzadeh, S., Nooshkam, M., Zargar, M., Garavand, F., Ghosh, S., Hadidi, M., & Forough, M. (2023). Green synthesis of nanomaterials for smart biopolymer packaging: challenges and outlooks. In *Journal of Nanostructure in Chemistry*. Springer Medizin. doi:10.1007/s40097-023-00527-3

Jambeck, J. R., Geyer, R., Wilcox, C., Siegler, T. R., Perryman, M., Andrady, A., Narayan, R., & Law, K. L. (2015). Plastic waste inputs from land into the ocean. *Science, 347*(6223), 768–771. doi:10.1126/science.1260352 PMID:25678662

Jensin Joshua, J., Venkatanarayanan, P. S., & Singh, D. (2020). A literature review on composite materials filled with and without nanoparticles subjected to high/low velocity impact loads. *Materials Today: Proceedings, 33*, 4635–4641. doi:10.1016/j.matpr.2020.08.294

Kaew-on, N., Katemake, P., & Prasongsuk, S. (2020). Primer formulations with antibacterial properties for murals. *Progress in Organic Coatings, 138*, 105395. doi:10.1016/j.porgcoat.2019.105395

Kalisz, M., Grobelny, M., Świniarskib, M., & Firek, P. (2016). Comparison of the structural and corrosion properties of the graphene/SiN(200) coating system deposited on titanium alloy surfaces covered with SiN transition layers. *Surface and Coatings Technology, 299*, 65–70. doi:10.1016/j.surfcoat.2016.04.064

Kamalan Kirubaharan, A. M., & Kuppusami, P. (2020). Corrosion behavior of ceramic nanocomposite coatings at nanoscale. In *Corrosion Protection at the Nanoscale* (pp. 295–314). Elsevier. doi:10.1016/B978-0-12-819359-4.00016-7

Kashif, M., & Anjum, A. (2020). Chemical and electrochemical corrosion studies of ricinus communis oil based poly(urethane-ricinoleamide) coatings. *Journal of Physics: Conference Series*, *1706*(1), 012034. Advance online publication. doi:10.1088/1742-6596/1706/1/012034

Kesmez, Ö., Tamsü Selli, N., Tunalı, A., Akarsu, E., Akarsu, M., & Arpaç, E. (2017). Fingerprint resistant coatings for stainless steel substrates. Progress in Organic Coatings, 112, 51–56. https://doi.org/10.1016/J.PORGCOAT.2017.07.003

Khodadadi, A., Liaghat, G., Taherzadeh-Fard, A., & Shahgholian-Ghahfarokhi, D. (2021). Impact characteristics of soft composites using shear thickening fluid and natural rubber–A review of current status. *Composite Structures*, *271*, 114092. doi:10.1016/j.compstruct.2021.114092

Li, Y., Zhang, Z., Wang, M., Men, X., & Xue, Q. (2017). Environmentally safe, substrate-independent and repairable nanoporous coatings: Large-scale preparation, high transparency and antifouling properties. *Journal of Materials Chemistry. A, Materials for Energy and Sustainability*, *5*(38), 20277–20288. doi:10.1039/C7TA05112C

Malekkhouyan, R., Nouri Khorasani, S., Esmaeely Neisiany, R., Torkaman, R., Koochaki, M. S., & Das, O. (2020). Preparation and Characterization of Electrosprayed Nanocapsules Containing Coconut-Oil-Based Alkyd Resin for the Fabrication of Self-Healing Epoxy Coatings. *Applied Sciences (Basel, Switzerland)*, *10*(9), 3171. doi:10.3390/app10093171

Martínez-Morlanes, M. J., Pascual, F. J., Guerin, G., & Puértolas, J. A. (2021). Influence of processing conditions on microstructural, mechanical and tribological properties of graphene nanoplatelet reinforced UHMWPE. *Journal of the Mechanical Behavior of Biomedical Materials*, *115*, 104248. doi:10.1016/j.jmbbm.2020.104248 PMID:33360486

Mashtalyar, D., Nadaraia, K., Sinebryukhov, S., & Gnedenkov, S. (2019). Polymer-Containing Layers Formed by PEO and Spray-Coating Method. *Materials Today: Proceedings*, *11*, 150–154. doi:10.1016/j.matpr.2018.12.123

Pathakoti, K., Manubolu, M., & Hwang, H. M. (2017). Nanostructures: Current uses and future applications in food science. *Yao Wu Shi Pin Fen Xi*, *25*(2), 245–253. doi:10.1016/j.jfda.2017.02.004 PMID:28911665

Qian, Y., Zhou, Y., Li, L., Liu, W., Yang, D., & Qiu, X. (2020). Facile preparation of active lignin capsules for developing self-healing and UV-blocking polyurea coatings. *Progress in Organic Coatings*, *138*, 105354. doi:10.1016/j.porgcoat.2019.105354

Rani, D., Singh, R., Kush, P., & Kumar, P. (2023). Nanomaterial-based smart coatings for antibacterial, antifungal, and antiviral activities. *Advances in Nanotechnology for Marine Antifouling*, 271–302. doi:10.1016/B978-0-323-91762-9.00008-3

Raveh, A., Zukerman, I., Shneck, R., Avni, R., & Fried, I. (2007). Thermal stability of nanostructured superhard coatings: A review. *Surface and Coatings Technology*, *201*(13), 6136–6142. doi:10.1016/j.surfcoat.2006.08.131

Shahini, M. H., Ramezanzadeh, B., & Mohammadloo, H. E. (2021). Recent advances in biopolymers/carbohydrate polymers as effective corrosion inhibitive macromolecules: A review study from experimental and theoretical views. *Journal of Molecular Liquids*, *325*, 115110. doi:10.1016/j.molliq.2020.115110

Shakil, U. A., Abu Hassan, S., Yahya, M. Y., Mujiyono, & Nurhadiyanto, D. (2021). A review of properties and fabrication techniques of fiber reinforced polymer nanocomposites subjected to simulated accidental ballistic impact. *Thin-walled Structures*, *158*, 107150. doi:10.1016/j.tws.2020.107150

Sharifi, S., Behzadi, S., Laurent, S., Forrest, M. L., Stroeve, P., & Mahmoudi, M. (2012). Toxicity of nanomaterials. *Chemical Society Reviews*, *41*(6), 2323–2343. doi:10.1039/C1CS15188F PMID:22170510

Singh, N., Gaur, S., Chawla, S., Singh, S., & Husen, A. (2023). Use of smart nanomaterials in food packaging. *Advances in Smart Nanomaterials and Their Applications*, 233–245. doi:10.1016/B978-0-323-99546-7.00008-2

Szőke, Á. F., Szabó, G. S., Hórvölgyi, Z., Albert, E., Gaina, L., & Muresan, L. M. (2019). Eco-friendly indigo carmine-loaded chitosan coatings for improved anti-corrosion protection of zinc substrates. *Carbohydrate Polymers*, *215*, 63–72. doi:10.1016/j.carbpol.2019.03.077 PMID:30981371

Taş, H., & Soykok, I. F. (2021). Puncture resistance of STF impregnated MWCNTs-grafted carbon fabrics. *Materials Today. Communications*, *28*, 102692. doi:10.1016/j.mtcomm.2021.102692

Wang, X., Zhang, J., Bao, L., Yang, W., Zhou, F., & Liu, W. (2020). Enhancement of the ballistic performance of aramid fabric with polyurethane and shear thickening fluid. *Materials & Design*, *196*, 109015. doi:10.1016/j.matdes.2020.109015

Wu, Y. S., Zeng, D. C., Liu, Z. W., Qiu, W. Q., Zhong, X. C., Yu, H. Y., & Li, S. Z. (2011). Microstructure and sliding wear behavior of nanostructured Ni60–TiB2 composite coating sprayed by HVOF technique. *Surface and Coatings Technology, 206*(6), 1102–1108. doi:10.1016/j.surfcoat.2011.07.096

Xu, T., Zhang, K., Cai, Q., Wang, N., Wu, L., He, Q., Wang, H., Zhang, Y., Xie, Y., Yao, Y., & Chen, Y. (2022). Advances in synthesis and applications of boron nitride nanotubes: A review. Chemical Engineering Journal, 431, 134118. https://doi.org/10.1016/J.CEJ.2021.134118

Xue, L., Wang, W., Guo, Y., Liu, G., & Wan, P. (2017). Flexible polyaniline/carbon nanotube nanocomposite film-based electronic gas sensors. *Sensors and Actuators. B, Chemical, 244*, 47–53. doi:10.1016/j.snb.2016.12.064

Yasin, G., Arif, M., Mehtab, T., Shakeel, M., Khan, M. A., & Khan, W. Q. (2020). Metallic nanocomposite coatings. In *Corrosion Protection at the Nanoscale* (pp. 245–274). Elsevier. doi:10.1016/B978-0-12-819359-4.00014-3

Zhu, L., Yao, K. L., & Liu, Z. L. (2006a). First-principles study of the polar (111) surface of Fe3 O4. *Physical Review B: Condensed Matter and Materials Physics, 74*(3), 035409. Advance online publication. doi:10.1103/PhysRevB.74.035409

ADDITIONAL READING

Enescu, D., Cerqueira, M. A., Fucinos, P., & Pastrana, L. M. (2019). Recent advances and challenges on applications of nanotechnology in food packaging. A literature review. *Food and Chemical Toxicology, 134*, 110814. doi:10.1016/j.fct.2019.110814 PMID:31520669

Porta, R. (2019). The plastics sunset and the bio-plastics sunrise. *Coatings, 9*(8), 526.

Sezgin, A. (2020) Food packaging: glass and plastic. *Researches on Science and Art in 21st Century Turkey*, 8.

Vasile, C., & Baican, M. (2021) Progresses in food packaging, food quality, and Safety—Controlled-Release antioxidant and/or antimicrobial packaging. *Molecules, 26*(5), 1263.

Chapter 2
Applications of Nanocoatings:
Smart and Sustainable Protection

Rishav Garg
https://orcid.org/0000-0001-6565-9234
Galgotias College of Engineering and Technology, India

Rajni Garg
https://orcid.org/0000-0002-9629-8141
Galgotias College of Engineering and Technology, India

Nnabuk Okon Eddy
https://orcid.org/0000-0001-7704-3082
University of Nigeria, Nigeria

ABSTRACT

The application of one material on another, often referred to as the substrate, is what is known as a coating. Their primary purpose is to shield the material from deterioration brought on by external factors. They serve as a conduit between the environment and the substrate. Furthermore, they serve as decorations as well. Nanocoatings are coatings in which the particles are at least one dimension apart and fall between 1 and 1000 nm in size. Because nanocoatings are tougher and harder on the substrate than other conventional coatings, they offer greater wear resistance. In addition, they offer antimicrobial, stain and wrinkle resistance, hydrophobic and hydrophilic qualities, UV protection, and antistatic qualities that affect the substrate material's bulk properties. Additionally, an attempt has been made to compile the different processing methods used in their fabrication.

DOI: 10.4018/979-8-3693-3136-1.ch002

INTRODUCTION

Nanotechnology serves as a pivotal catalyst for novel and cutting-edge coating applications, with nanocoatings demonstrating remarkable advancement in recent times. The properties exhibited by the nanomaterials are truly remarkable and extraordinary in nature (Kamal et al., 2023). Thin films, nanostructured coatings, and nanoengineered surfaces find extensive application across diverse industry sectors, exemplifying the transformative potential of nanotechnology in enhancing or perturbing established technological domains, while also fostering the emergence of novel sectors (Jahan, 2023). In case of coating applications, the paramount significance is placed upon the heightened characteristics, novel functionalities, and superior performance of the materials involved. The primary attribute conferred by nanostructured coatings, however, is the provision of safeguarding against various detrimental factors, encompassing but not limited to ice formation, pollutant intrusion, ultraviolet radiations, combustion, elevated temperatures, bacterial colonisation, marine organism adhesion, tactile interactions, and corrosion (Anjum et al., 2023). These factors incur substantial financial losses for the global industry due to maintenance expenses, loss of productivity, and operational downtime, amounting to a hefty amount annually. Furthermore, they also present a considerable public health risk (Caixeta, 2023).

Nanocoatings exhibit notable performance enhancements compared to conventional coatings, while also offering superior cost-effectiveness in the medium- to long-term. The coatings based on nanostructured materials have been found to exhibit enhanced properties in various areas, including anti-microbial activity, prolonged product lifespan, efficient thermal insulation, sustained glossiness, enhanced stability against ultraviolet radiation, effective resistance against dirt and water, efficient moisture absorption, increased hardness, superior corrosion resistance, reliable flame retardancy, improved energy efficiency, self-cleaning capabilities, and improved chemical and mechanical properties (Dhiman et al., 2021; Farag, 2020; Noeiaghaei et al., 2017; Parimalam et al., 2018). Nanocoatings display exceptional compatibility with diverse construction materials, including glass, concrete, sand limestone, and marble, rendering them highly adept at shielding said materials from detrimental environmental factors. Additionally, these nanomaterials effectively serve as corrosion inhibitors for the embedded and reinforced steel (Harilal et al., 2023). Commercially available paints, emulsions and surface coatings possess the capability to generate a low energy interface, resulting in a highly hydro- and oleophobic building surface. This advantageous property aids in extending maintenance intervals and minimising the need for cleaning. Dust-repellent protective paints and photocatalytic coatings are widely recognised as the foremost applications within the construction sector (Yang, 2021). The accumulation of particulate matter on the surfaces of building exteriors presents significant challenges in the realm of building maintenance. The process of cleaning building surfaces typically involves the utilisation of detergents in conjunction with mechanical agitation, such as scrubbing and wiping, as well as the application of high-pressure water jets (Fakhri & shahryari, 2021). These processes exhibit various limitations including the utilisation of chemical detergents, substantial energy consumption, and elevated labour expenses. This inevitably gives rise to elevated maintenance expenditures; hence, the pursuit of an efficient self-cleansing coating becomes imperative (Amor et al., 2022).

There has been a significant surge in the industrial interest surrounding self-cleaning coatings that utilise photocatalytic nanomaterials. With the aid of a minimal amount of UV light emitted by a fluorescence source or the radiant energy of sunlight, the nanomaterials exhibit two distinctive characteristics: (a) strong oxidative capabilities and (b) exceptional hydrophilic behavior (Gebreslassie & Gebretnsae, 2021). The utilisation of potent oxidising agents exhibits efficacy in eradicating bacterial colonies adhered to surfaces, as well as in the oxidation and elimination of malodorous compounds present in wastewater. Notably, commercially accessible options include tiles and glass coated with nano-titanium dioxide (nano-TiO_2) (Ubaid et al., 2019). The remarkable superhydrophilic characteristics of this coating facilitate the effortless removal of dirt and stains through the utilisation of water or precipitation, when it is meticulously applied onto external surfaces. The utilisation of nano-TiO_2 self-cleaning coating confers significant advantages to building maintenance, particularly in the context of skyscrapers, by mitigating the requirement for expensive surface cleaning procedures (Zhou et al., 2022). Photocatalytic coatings find application in enhancing the indoor air quality index by degrading the microorganisms, volatile organic compounds, and other hazardous chemicals encountered in diverse settings such as hotels, restaurants, commercial establishments, university laboratories, medical facilities, and residential spaces (Thakur et al., 2022). The inherent anti-adhesive characteristics of nanocoatings result in reduced staining and facilitate effortless cleaning on both external and internal surfaces. The clear impregnating nano-coatings that are water-based, devoid of volatile organic compounds (VOCs), and incorporate nanoscale UV absorbers in the form of nanoparticles have been recently commercialized. These products have been specifically engineered for application on masonry and concrete substrates, offering exceptional hydrophobic properties, mitigating the occurrence of efflorescence, and delivering a notable enhancement in abrasion resistance (Esteban-arranz et al., 2021).

Thus, nanocoating technologies augment the performance of current commercial products or introduce novel attributes to pre-existing technology and comprehend the substantial product performance and cost-saving benefits associated with nanocomposite-based coatings. This field is experiencing notable expansion and development leading to a strong surge in the demand for photocatalytic nanoparticles, which exhibit remarkable self-cleaning and antibacterial properties (Chougule et al., 2021; Jiang et al., 2018). These innovative aspects of nanotechnology, fostering an environment conducive to the proliferation and advancement of nanoparticle-based coatings is the need of the hour and has been explored in this chapter with special emphasis on titania based nanocoatings.

BACKGROUND

Nanomaterials are synthesised possessing a minimum dimension below 100 nm and can be zero, one, two or three-dimensional structures (Garg et al., 2023). Nanocoatings are materials with a two-dimensional structure that can be synthesised through various methods, including physical techniques such as plasma and physical vapour deposition (PVD), chemical processes like chemical vapour deposition (CVD) and sol-gel technologies (Figure 1) (Safavi et al., 2022; Strudwick et al., 2015). Nanocoatings can be classified into two primary categories: nanocrystalline coatings, which consist of multiple layers with nanometer-scale individual layer thickness, and nanocomposite coatings. In nanocoating systems, the outermost section comprises a transparent coating with a thickness ranging from 5 to 50 μm. This coating serves multiple purposes, such as safeguarding the underlying layers or substrate against chemical and ultraviolet deterioration (Parimalam et al., 2018). Additionally, it offers defence against mechanical harm that may lead to surface imperfections or abrasions. The design of nanostructured coating are heavily reliant on the selection of materials, the methods employed for deposition, the parameters involved in processing, and most importantly, the service conditions experienced by the coated components. The dimensions and distribution of nanoparticles within a coating formulation exert a significant impact on its functional attributes (Anjum et al., 2023).

The properties of a substance are essential in understanding its behaviour and characteristics. Nanostructured coatings possess a range of advantageous properties owing to their inherent characteristics at the nanoscale. These coatings are known to be multifunctional, often demonstrating one or more of the subsequent properties: resistance against scratches and abrasions, anti-static behaviour, oleophobicity, ease of cleaning, antireflective properties, antimicrobial activity, as well as sensor and catalytic capabilities (Farag, 2020). The utilisation of nanocoatings presents several notable advantages in the field of surface engineering. These include the enhancement of surface aesthetics, resulting in a superior visual appearance. Furthermore, nanocoatings exhibit commendable chemical resistance, thereby safeguarding the coated surfaces against potential chemical degradation (Parimalam et al., 2018) (Table 1).

Figure 1. Synthetic techniques for nanocoatings

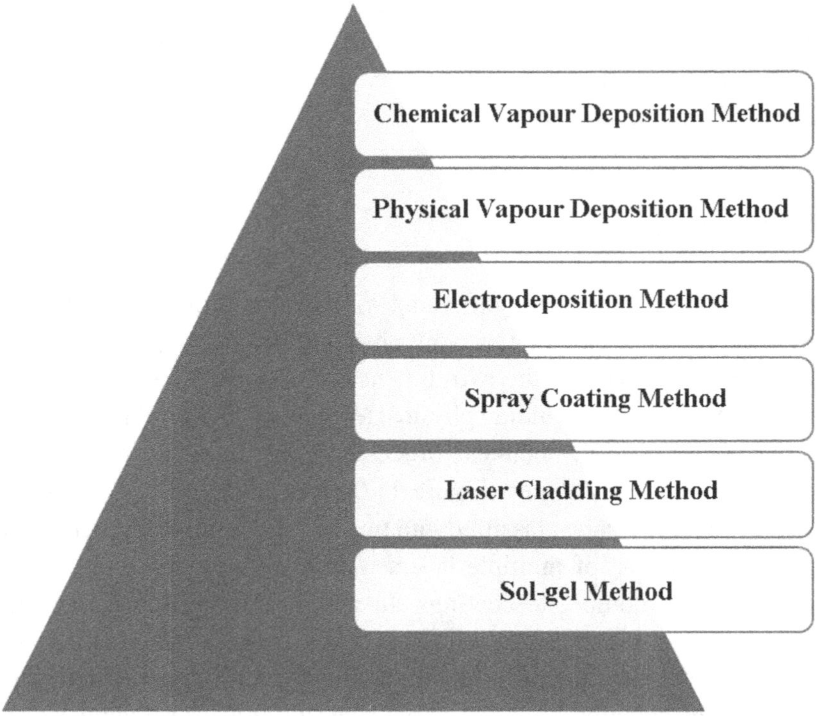

Table 1. Properties of nanocoatings with functional attributes

Property	Function
Enhanced barrier protection	Reduction in permeability to corrosive environments resulting in enhanced corrosion resistance properties.
Unique visual effects	Enhanced optical clarity properties Possesses inherent anti-reflective properties
Enhanced service life	Incorporation of additives to enhance the anti-skid, anti-fogging, and anti-fouling characteristics.
Resist scratches	Surface exhibiting facile cleanability
Photocatalytic properties	Antibacterial capabilities and protection against ultraviolet radiation
Augmented modulus and thermal stability characteristics	Enhancement of the mechanical properties, such as scratch resistance, through advanced formulation techniques.
Conduction of electricity	Optimisation of the thermal and electrical conductivity for improved performance.
Eco-friendly	Formulated without the inclusion of chromate and lead compounds
Self- repairing	Exhibits excellent adhesion characteristics across a wide range of materials.

NANOMATERIALS IN NANOCOATINGS

Nanoparticles facilitate the development of innovative chemical formulations and encompass a range of compounds, namely iron oxide, zinc oxide, titanium dioxide, silicon dioxide, and silver. Titanium dioxide exhibits remarkable photocatalytic characteristics. Due to its inherent properties, the utilisation of titanium dioxide at the nanoscale level is observed in wall paints to facilitate the eradication of organic pollutants present in the surrounding atmospheric environment (Garg et al., 2022). Nanostructured silicon dioxide primarily functions as a rheological additive, exhibiting notable efficacy in enhancing the flow properties and viscosity of various substances. Additionally, it finds application in the formulation of self-cleaning wall paints, where its unique properties contribute to the desired functionality of the coating (Garg et al., 2021). The paramount utilisation of iron oxide and nanoscale zinc oxide lies in their ability to confer ultraviolet (UV) shielding properties to transparent coatings. Nano-scale silver particles are incorporated into the formulation of wall paints used in hospital and food processing settings to effectively inhibit the growth and proliferation of bacteria and other microorganisms (Bhardwaj et al., 2022).

The nanoparticles exhibit remarkable characteristics owing to their substantial specific surface areas, which facilitate heightened reactivity, superior absorption capabilities, elevated solubility, reduced melting point, and enhanced electronic properties. Notably, particles with a size smaller than 10 nm demonstrate quantum effects, which hold significant importance in electronic and optoelectronic applications (Eddy et al., 2023). Due to the inherent characteristics of nanoparticles, it is possible to fabricate sophisticated products and coatings by incorporating nanocomposites within polymer matrices. Nanoparticles exhibit imperceptibility to the naked ocular senses when effectively incorporated within a polymeric matrix. They exhibit negligible impact on visible light transmission, rendering them highly suitable substances for sophisticated transparent coatings (Peng et al., 2013).

When the dispersion of nanoparticles within a polymer matrix occurs, it results in the creation of advanced transparent nanocomposites. These nanocomposites exhibit enhanced properties such as increased coating strength, heightened solar absorption and reflectance, improved thermal and scratch resistance, and enhanced performance of photochromic coatings (Beryl & Xavier, 2021). Polymer nanocomposite coatings exhibit remarkable practical implications across various domains, encompassing mechanical, optical, and electronic applications through various types of nanocoatings (Figure 2). There is no compromise in the transmission of light in optical applications. In a broader context, the enhancement of the mechanical characteristics of a polymer coating, such as its scratch resistance, necessitates the meticulous optimisation of the constituent elements of the polymer lacquer and the fortification of the coat's structure through the incorporation of minuscule fillers (Castro-Mayorga et al.,

2016). When employing nanopowders as fillers for the purpose of enhancing the strength of polymers, the increased interfacial surface area of these nanofillers leads to distinct behaviour compared to conventional fillers. Consequently, the nanocomposite exhibits noteworthy enhancements in mechanical and other desired properties even with a minimal amount of nanomaterial loading (El-Fattah et al., 2015). These properties are augmented while maintaining the integrity of visible light transmission. This is of utmost importance for optical applications. The subject of spectrally selective coatings composed of nanoparticles is currently a highly discussed area of interest in case of glass coatings (Xin, 2022). Nanoparticles exhibit superior spectral manipulation capabilities compared to alternative substances due to their exceptional transparency and clarity, thereby establishing their prominence in the market. In case of spectrally selective solar control films, the paramount factors under scrutiny encompass the attainment of elevated transparency within the visible light spectrum, while simultaneously ensuring the absorption and reflection of infrared light (Levchuk et al., 2022).

In order to enhance the competitive edge of commercial films, it is imperative to incorporate nanoparticles within the plastic films that are subsequently utilised for glass surface application. Nano films exhibit exceptional thermal absorption and reflectivity characteristics, all the while upholding commendable levels of optical transparency. The methodologies employed in the production of various substances and compounds are of paramount importance in the field of chemical engineering (Xin, 2022). These production methods encompass a wide range With the emergence of vacuum-based physical vapour deposition techniques, processes such as sputtering and pulse laser deposition have assumed prominence as the principal methodologies in nanomaterial research and development. Nevertheless, there is a burgeoning initiative to reintroduce the utilisation of chemical solution deposition for the fabrication of transparent ceramic oxides (Elshami & Ali, 2014).

TYPES OF NANOCOATINGS

Self-Cleaning Nanocoatings

Manufacturers exhibit significant interest in incorporating cutting-edge technology to ensure the sustained pristine appearance and longevity of their products throughout extended periods of utilisation. Ideally, these coatings will exhibit exceptional durability and facilitate significantly extended cleaning cycles, thereby minimising the need for frequent maintenance (Amor et al., 2022). This would entail occasional rinsing with water or gentle wiping with a damp cloth to effectively eliminate accumulated dirt and oils. Applications encompass a wide range of outdoor surfaces

Figure 2. Types of nanocoatings

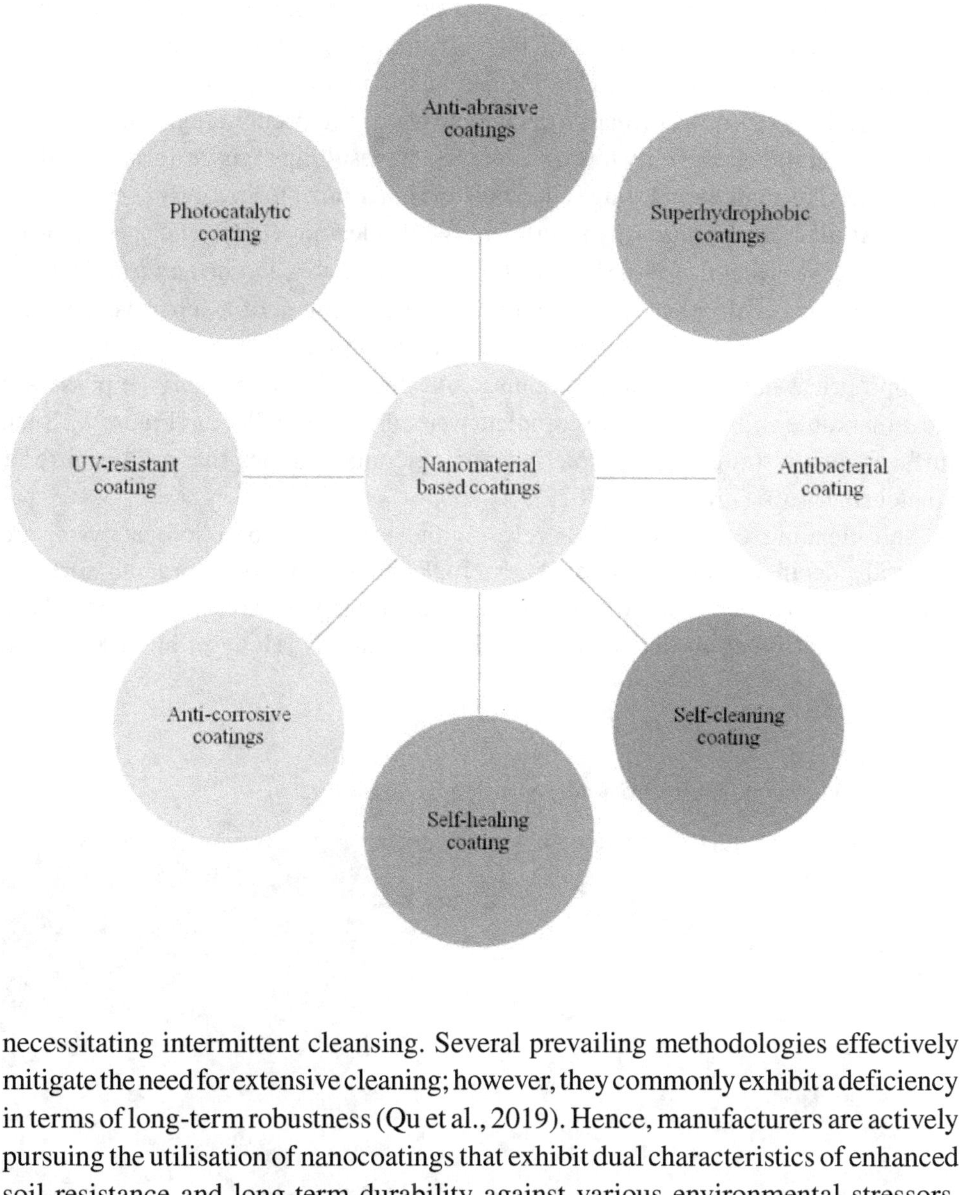

necessitating intermittent cleansing. Several prevailing methodologies effectively mitigate the need for extensive cleaning; however, they commonly exhibit a deficiency in terms of long-term robustness (Qu et al., 2019). Hence, manufacturers are actively pursuing the utilisation of nanocoatings that exhibit dual characteristics of enhanced soil resistance and long-term durability against various environmental stressors, including but not limited to abrasives like sand, dirt, and oil, fluctuations in pH levels, gusts of wind, freezing temperatures leading to ice formation, excessive moisture, corrosive effects of salt, prolonged exposure to sunlight, atmospheric pollution, the undesirable presence of bird droppings, as well as the adhesive nature of tree sap (Atta-ur-Rehman et al., 2019).

The attainment of clean or self-cleaning surfaces can be accomplished through two fundamentally distinct methodologies. A surface can be coated with a photocatalytic material, wherein the irradiation of the sun's ultraviolet rays initiates a catalytic reaction that facilitates the degradation of organic contaminants. Simultaneously, the surface undergoes a transformation into a state of superhydrophilicity, wherein water exhibits a heightened affinity towards the surface, resulting in its uniform spreading and reduced formation of drying remnants through droplet formation (Turalija et al., 2016). An alternative approach to fabricating a self-cleaning (bionic) surface involves employing the principles of the Lotus effect phenomenon, wherein the surface attains a state of super-hydrophobicity (Figure 3). The attainment of a superhydrophobic surface is facilitated by the amalgamation of hydrophobic chemical properties with an appropriate level of surface roughness. A superhydrophobic surface possesses the remarkable ability to achieve complete water droplet repellency (Figure 4). Such surfaces demonstrate water droplet advancing contact angles that are equal to or greater than 150º (Zhou et al., 2022).

Self-cleaning coatings, which rely on the utilisation of photocatalysts, are presently deemed most suitable for external applications on inorganic surfaces and coatings. However, there is also extensive ongoing research on photocatalytic materials that function under visible light wavelengths (Hong et al., 2006). The

Figure 3. Lotus effect
(CC BY-SA 3.0, httpscommons.wikimedia.orgwindex.phpcurid=364500)

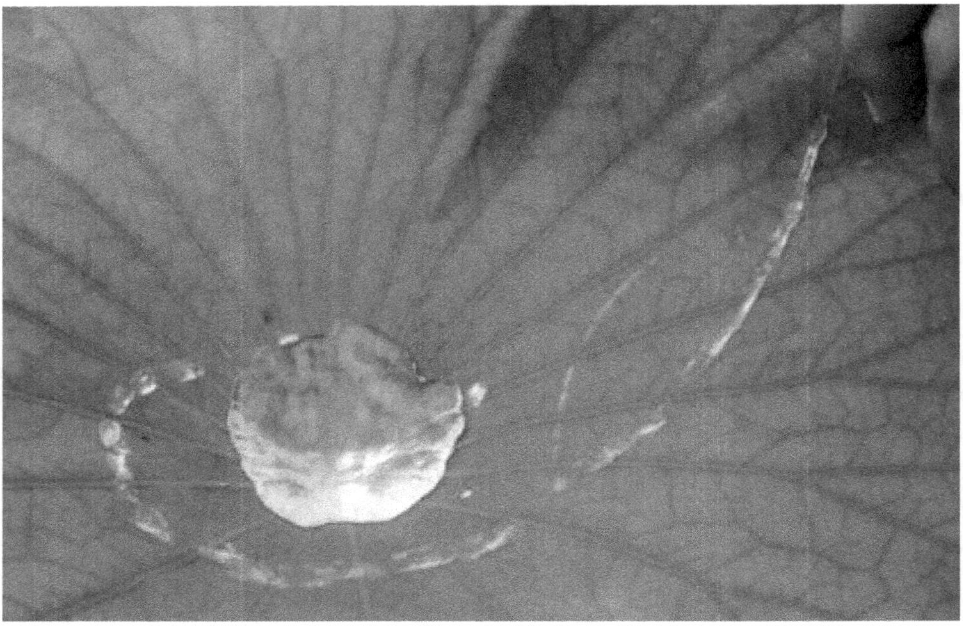

Figure 4. Hydrophobic and Superphydrophobic effect according to change in contact angle

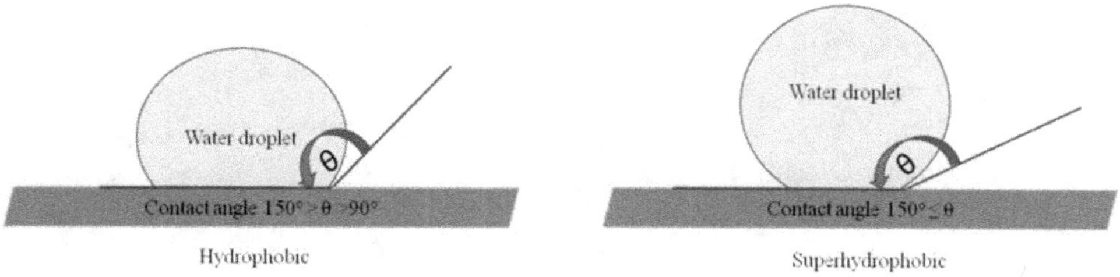

industries that experience notable effects from self-cleaning photocatalytic coatings encompass various domains such as Architectural glass and construction, including road infrastructure, Medical sector, Food production and packaging industry and energy sector (Bica et al., 2022; Garg et al., 2023; Oliani et al., 2022). Indoor air purification systems are designed to effectively remove contaminants and impurities from the air within enclosed spaces, ensuring a healthier and cleaner environment. Solar coatings refer to specialised coatings that are applied to various surfaces, such as solar panels or windows, to enhance their solar energy absorption and conversion capabilities (Tayel et al., 2022).

UV-Resistant Coatings

Nanoparticles have the potential to enhance the ultraviolet (UV) resistance of solvent-based, waterborne, and UV coatings. Transparent UV-absorbing or UV-blocking coatings presently find primary utility in two domains: firstly, as a UV-protective lacquer for wooden substrates, and secondly, as a UV-barrier coating applied onto the surfaces of polymer-based articles or apparatuses to impede their degradation kinetics (Abbas et al., 2021). With the escalating annual augmentation in the intensity of UV radiation, there is an imminent need to explore efficacious approaches for impeding UV rays in order to safeguard various polymers, such as plastics and timber, and other pertinent polymeric substances. UV radiation can induce detrimental effects in various materials, including textiles, plastics, paints, and timber products, manifesting as discoloration, chalking, and diminished mechanical characteristics (Figure 5). Hence, the advancement of efficacious UV-attenuating substances holds paramount significance in relation to our well-being, communal welfare, and ecological equilibrium (Channa et al., 2022).

Nanomaterials are a class of materials that possess unique properties and characteristics at the nanoscale Over the past decade, significant advancements have been made in the development of nanoparticles incorporating functional

Figure 5. UV-resistant action of nanocoatings

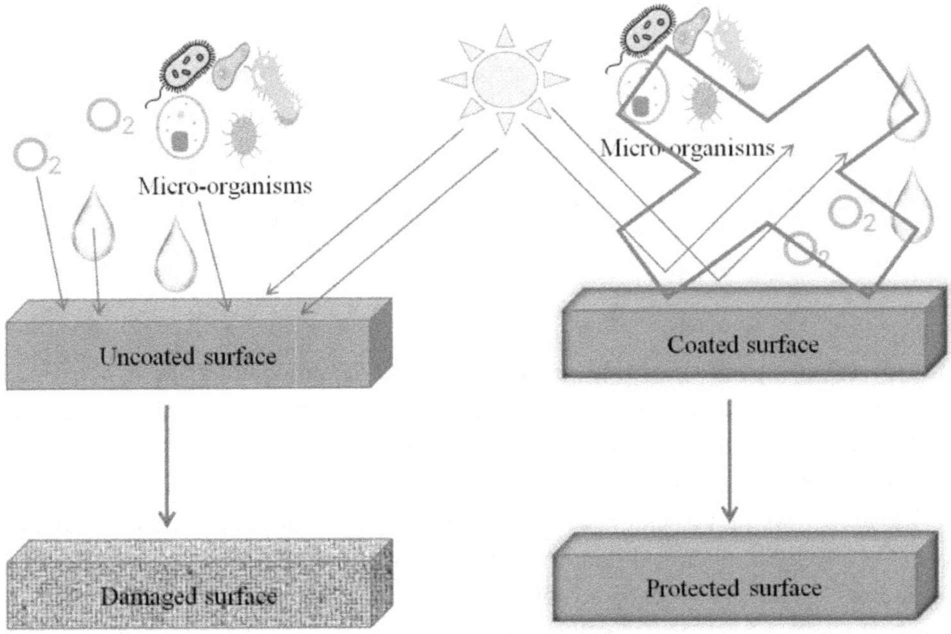

coatings to effectively attain UV-blocking capabilities. Zinc oxide (ZnO), titanium dioxide (TiO_2), and cerium oxide (CeO_2) nanoparticles find application in UV light protection formulations for lacquers (Yousefi et al., 2023). Nanoparticles exhibit a notable enhancement in UV light blocking efficacy in comparison to natural materials owing to their heightened surface area to volume ratio. The efficacy of a fabric's UV-blocking attribute is heightened through the incorporation of a dye, pigment, delustrant, or ultraviolet absorber finish (Mondal, 2022). This finish effectively absorbs ultraviolet radiation, thereby impeding its transmission through the fabric and preventing its interaction with the skin. Textiles play a pivotal role as crucial materials for safeguarding against UV radiation in numerous applications. The utilisation of nano-sized TiO_2 and ZnO is being employed for the purpose of providing UV protection in textile materials (Kathirvelu et al., 2009). ZnO nanoparticles have been successfully integrated onto the surfaces of cotton and wool fabrics. The incorporation of ZnO enhances the mechanical integrity of the textiles, leading to the development of a fabric with UV absorption properties. The UV absorption spectrum of ZnO exhibits a wider range compared to that of TiO_2 (Zamora-Ledezma et al., 2020). ZnO exhibits supplementary functionalities, encompassing antifungal and antistatic characteristics, which are not attainable with TiO_2. Nanoparticles have also been employed as additives in coatings aimed at mitigating the chromatic alteration of clear coated nano-subjected to outdoor exposure (Silva et al., 2023).

Anti-Corrosive Coatings

Corrosion poses a formidable challenge for proprietors of industrial equipment, installations, and plants. Due to its inherently sluggish nature, this process exhibits considerable versatility, manifesting in a multitude of configurations and exhibiting rapid kinetics upon activation by external stimuli, notably oxygen and water, both of which are indispensable constituents (Gopalakrishnan et al., 2023). It is conceivable, nonetheless, to mitigate corrosion, and nanocoatings hold significance as they possess the capability to redirect the attention from corrosion protection to corrosion prevention. The existence of corrosive substances and severe operational and environmental circumstances may lead to structural breakdown or containment loss, resulting in significant expenses related to repairs, compromised or polluted products, ecological harm, and potential hazards to personnel. The presence of corrosion poses a significant financial burden and exerts a substantial influence on the economic landscape of industrialised nations and their respective enterprises (Eddy et al., 2022).

Nanoparticles have been successfully integrated into protective coatings for various industrial applications, such as corrosion protection, thermal resistance, fire retardancy, water-based anti-corrosion properties, and fire retardant polymers. Nanocoatings have the potential to greatly enhance the cost-to-benefit ratio, offering economically viable solutions and enhanced performance capabilities (Bagherzadeh & Mousavinejad, 2012). Nanoparticles, including nano silica, clay, ZnO, Fe_2O_3, and TiO_2, are commonly employed in organic coatings to enhance the corrosion resistance properties. The nanomaterials exhibit an exceptionally elevated surface area. Upon functionalization of this particular surface, it exhibits the capability to effectively deliver substantial quantities of organic corrosion inhibitors (Anjum et al., 2023). Hence, custom-engineered nanoparticles serve as an ideal medium for the transportation of the requisite concentration of active corrosion inhibitors. Nanomaterials engineering further expands the potential for engineering coatings with intelligent capabilities, enabling the controlled release of corrosion inhibitors upon breach of the coating, application of stress, or activation of an electrical or mechanical control signal (Eddy et al., 2022). The applications of chemical engineering are vast and diverse. Nanocoatings possess the inherent capability to be meticulously tailored for utilisation on diverse steel substrates, thereby rendering them amenable for application across an extensive spectrum of markets encompassing various components including specialised components like seals, accessories, and bearings (Figure 5) (Verma, 2021).

Anti-Abrasive Nanocoatings

Through the process of nanoscale engineering, it becomes feasible to fabricate composite materials that exhibit exceptional hardness, comparable to the performance of diamond. In the realm of machining and wear-resistant applications, the incorporation of hard coatings is imperative to augment the wear resistance and toughness characteristics of cutting tools (Law et al., 2022). Traditional coatings fail to adequately fulfil the demands of contemporary machining and manufacturing prerequisites, in contrast to nanostructured coatings. The enhancement of surface hardness and wear resistance in materials can be effectively achieved by employing a nanoparticulate coating. This coating serves to augment the inherent hardness of the base material while concurrently enhancing the surface's ability to withstand wear. Nanoparticles exhibit enhanced wear resistance and toughness characteristics, thereby providing comprehensive corrosion inhibition capabilities while satisfying rigorous regulatory and safety prerequisites (Dai et al., 2022). This field of research is currently experiencing high levels of activity, and notable advancements have been successfully accomplished. The utilisation of nanoscale structuring employing nano-sized grains and nanolayers exhibits a profound impact on the mitigation of dislocations, consequently leading to a substantial augmentation in the wear-resistance characteristics (Xia et al., 2020). Nanocoatings have been observed to exhibit substantial enhancements in various characteristics, including toughness and thermal shock resistance, when applied to diverse conventional materials like metal alloys, ceramics, and composites (Yang et al., 2021). Nanocoatings exhibit a reduced wear rate in comparison to their commercial coarse-grained powder counterparts. The observed enhancement in wear resistance can be ascribed to the exceptional hardness and toughness exhibited by the nanomaterials, as well as the alteration in fracture behaviour and material removal mechanisms resulting from the ultrafine particle size (Figure 5) (Maeztu et al., 2017).

APPLICATIONS OF VARIOUS NANOMATERIALS-BASED COATINGS

ZnO nanoparticles function as inorganic UV filters through the process of light reflection. ZnO nanoparticles, characterised by their transparent nature, are being increasingly employed in various applications due to their enhanced protective capabilities resulting from their larger surface area (Sharma et al., 2020). This principle applies to the UV shielding of lacquer films and polymer surfaces, for instance. Nanocomposite coatings, which effectively integrate ZnO nanoparticles, exhibit enhanced UV radiation shielding capabilities when subjected to accelerated weathering and outdoor exposure, surpassing the protective performance of acrylic latex paint (Ahmed et al., 2017). The developed coatings find their application in safeguarding facades and wood against the deleterious effects of outdoor weathering induced by ultraviolet UV radiation (Parimalam et al., 2018). Nanocellulose is a fascinating material that exhibits unique properties due to its nanoscale dimensions. Nanocellulose is currently under development with the aim of enhancing the mechanical characteristics, specifically the abrasion and scratch resistance, of water-based polyurethane varnishes and paints (Siró & Plackett, 2010). The process of vigorously dispersing fragmented cellulose fibrils in aqueous medium yields a resilient gel, exhibiting stability even at relatively low solid concentrations, approximately around 2%. Cellulose fibril networks inherently manifest a notable aqueous reservoir capability (Silva et al., 2023). The inherent characteristics of these transparent fibrils render them highly suitable for incorporation into various paint formulations, thereby leading to significant improvements in dispersion behaviour and strength attributes (Honorato et al., 2015).

Alumina nanoparticles, specifically composed of aluminium oxide (Al_2O_3), are strategically incorporated into the coating formulations of various automotive components such as headlight parts, windscreens, and optical lenses. Additionally, they find application in the protective coatings for wood floors, safety glasses, electronic displays, and polycarbonate panels (Dhiman et al., 2021). The primary objective behind their inclusion is to enhance the scratch and wear resistance properties of these surfaces, thereby ensuring prolonged durability and improved performance. The applications encompass offshore oil and gas drilling systems, power transmission systems with enhanced efficiency, lighting systems of the next generation, buildings and transportation systems that are energy-efficient, and packaging solutions that are environmentally friendly (Kauppinen et al., 2021). In case of oil and glass drilling, the primary focus lies in the utilisation of coatings that exhibit exceptional resistance to corrosion, scratches, wear, and thermal barriers. These coatings offer exceptional long-term scratch resistance while minimally affecting crucial aspects such as optical clarity, gloss, colour, and physical properties. Nanocomposite alumina coatings exhibit remarkable potential for utilisation within the automotive sector as transparent topmost layers of paint coats (Ibrahim & El-Tayeb, 2021). Another innovative approach involves the application of wear-resistant nano-alumina coatings, specifically designed to withstand the challenging conditions encountered in harsh corrosive drilling environments. The anticipated outcome of applying the anti-wear nanocoating to the aluminium pipe is a notable increase in its durability, ranging from 30% to 40%, particularly when subjected to harsh and corrosive drilling conditions, as opposed to utilising uncoated aluminium pipes (Minh & Minh, 2020).

The experimental findings have showcased the efficacy of graphene coatings in impeding the occurrence of microbially-induced galvanic corrosion in metals over prolonged durations, specifically for a span of 2700 hours. This is primarily attributed to the ability of graphene coatings to obstruct the ingress of microbes onto the metal surface, thereby establishing a formidable protective barrier (Maeztu et al., 2017). Furthermore, these coatings effectively shield the surface from the detrimental impact of microbial byproducts, which have been observed to augment the dissolution of nickel. A recent investigation, however, has revealed that while graphene exhibits efficient short-term resistance against oxidation, it inadvertently facilitates heightened levels of wet corrosion over prolonged durations compared to an initially unshielded surface (Dhiman et al., 2021).

Silicon dioxide, also known as silica, is a compound that consists of one silicon atom bonded to two oxygen atoms. In nano- form, SiO_2 find extensive utilisation in various industrial sectors such as coatings, solar cell applications, high temperature insulators, gas sensors, plastics, polymers, and wires (Parimalam et al., 2018). They exhibit enhanced scratch and abrasion resistance, corrosion resistance, as well as superhydrophobic and easyclean characteristics. The application of hydrophilic silicon dioxide nanolayers onto diverse substrates, including steel, aluminium, plastics, and glass, imparts desirable surface functionalities, such as easy clean, anti-dirt, anti-fog, and anti-drip properties (Wimalasiri et al., 2017). The transparent nature of the coatings ensures that the visual appearance of the substrate remains unaffected. There is an absence of observable alterations in the chromaticity or lustre of pre-coated metallic substrates, such as steel and aluminium. The nanocoatings exhibit exceptional durability when subjected to outdoor environmental conditions. Upon the completion of the application and subsequent drying process, the inorganic silica nanoparticles undergo a transformation, leading to the creation of a uniform and interconnected three-dimensional structure (Sriramulu et al., 2016). This structure exhibits remarkable dirt-repelling properties, forming a robust and hydrophilic surface. Furthermore, this surface demonstrates exceptional capabilities in preventing moisture infiltration while simultaneously allowing the passage of water vapour. Nano surface-sealing products exhibit exceptional dirt-repellent properties and facilitate effortless cleaning (Ghodrati et al., 2023). These coatings effectively seal both untreated, absorbent surfaces as well as non-absorbent surfaces. The dispersions are synthesised utilising bead mills. Solvent and specially formulated surfactants are incorporated into the system to effectively suspend the nanoparticles and mitigate their propensity for re-agglomeration (Gong & He, 2020). Silica nanoparticles exhibit enhanced scratch and abrasion resistance properties when incorporated into coatings. These coatings exhibit enhanced micro hardness, storage modulus, and temperature resistance, while concurrently providing the desirable characteristic of transparency. The utilisation of surface-grafted silica nanoparticles serves to guarantee the optical transparency and reduced fluidity of the coatings (Liu et al., 2016).

Titania Based Nanocoatings

TiO_2 is a compound commonly used in various industrial applications necessitating exceptional opaqueness or whiteness, commendable resistance against corrosion, or commendable photocatalytic activity. Bulk titanium dioxide and nano titanium dioxide exhibit distinct material properties due to their differing particle sizes (Azeez et al., 2022). Titanium Dioxide serves as a prominent white pigment within the realms of the paint and cosmetic industries, predominantly in the form of micron-sized particles. Nano-TiO_2 exhibits a significantly amplified surface area per unit mass or volume of nanoparticles when compared to an equivalent mass or volume of conventional TiO_2 particles. This enhanced characteristic holds promising prospects for various properties, including catalytic activity and UV absorption at specific wavelengths (Liang et al., 2006). The utilisation of photocatalytic nano-TiO_2 has been successfully achieved and widely implemented owing to its inherent non-toxic nature, remarkable catalytic prowess, robust self-cleaning attributes, and economical viability. The photocatalytic process of TiO_2 entails the generation of electron-hole pairs through the absorption of photons. These cavities possess the capability to engage in chemical reactions with adjacent molecules, resulting in the generation of oxidising agents. The aqueous interaction of these particulates yields the hydroxyl radical (OH^-) that manifests as the superoxide molecule and the photocatalytic degradation takes place (Figure 6) (Nunzi & De Angelis, 2022).

Nano-TiO_2 finds application in self-cleaning surface coatings as well as coatings and paints that exhibit resistance towards ultraviolet radiation. Nano-TiO_2 exhibit variations in terms of particle size, surface area, purity (attributable to factors such as doping, coating, or quality control), surface properties, crystalline structure, chemical reactivity, and other pertinent attributes (Parimalam et al., 2018). Nano-

Figure 6. Photocatalytic action of TiO2 nanoparticles

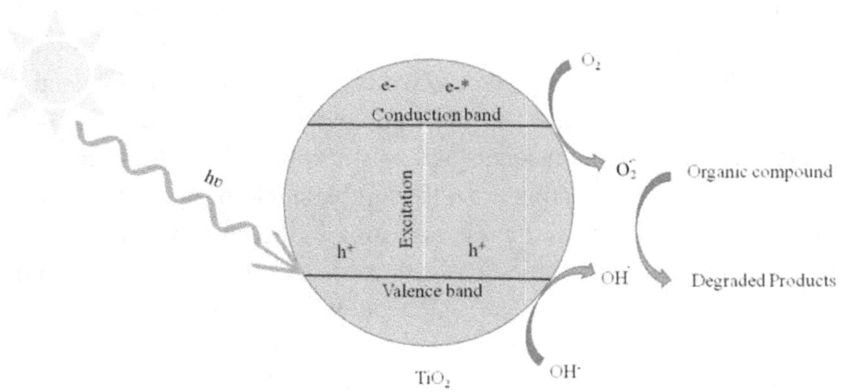

TiO_2 is obtainable in pristine anatase, pristine rutile, and amalgamations of anatase and rutile. In a broad sense, it can be observed that anatase nano-TiO_2 exhibits superior photocatalytic properties compared to the rutile counterpart. Conversely, nanoscale rutile demonstrates relatively lower photoreactivity when compared to both mixtures of anatase and rutile, as well as anatase in isolation (Awe et al., 2021). Nanostructured coatings exhibit exceptional compatibility with diverse construction materials, including but not limited to glass, concrete, sand limestone, and marble, thereby rendering them highly adept at safeguarding said surfaces against deleterious environmental factors, encompassing water staining, moss and algae proliferation, as well as soot and oil blemishes. Nano-TiO_2 finds application in construction materials such as paints and coatings on glass and concrete due to its remarkable properties including self-cleaning, air-cleaning, and antibacterial effects (Amor et al., 2022).

Upon exposure to solar radiation, Nano-TiO_2 exhibits catalytic properties, facilitating the photodecomposition of pollutant molecules that have been adsorbed onto its surface. This process leads to the conversion of said molecules into compounds that are devoid of toxicity. TiO_2 coatings possess the remarkable ability to exhibit inconspicuous characteristics, thereby refraining from inducing noticeable alterations in the colour or transparency of the underlying substrate (Hsissou et al., 2018). However, it is important to note that these coatings have the potential to undergo decomposition of organic matter upon its deposition. Hence, diverse surface substrates can be coated with TiO_2 to confer upon them the ability to undergo self-cleaning phenomena when exposed to both solar irradiation and indoor lighting conditions (Zhou et al., 2022). Surfaces incorporating paints, ceramics, glass, and cementitious materials that encompass active photocatalytic titania nanoparticles exhibit extensive utility in generating environmentally pristine regions in their immediate vicinity (Wang et al., 2022).

The Anti-Fog coating is formulated with nanoscale titanium dioxide and a hybrid polymer. The sol-gel methodology is employed for the fabrication of the coating, with an average coating thickness ranging from 30 to 100 nanometers, contingent upon the specific application at hand. These particles are characterised by their ultra-fine nature, surface treatment, and remarkable transparency and find their applications in various industries, including automotive coatings, wood lacquers, and plastics (Goffredo et al., 2017). These are also employed for the purpose of delivering UV shielding in transparent lacquers and plastic applications. In its capacity as an inorganic UV screen, this coating provides durable shielding against the undesirable darkening of wood substrates, thereby assisting in the preservation of the wood's inherent visual appearance. Specifically, it exerts a discernible impact on the flip/flop effect of the metallic pigment. Photocatalytic TiO_2 coatings have exhibited superior durability and efficacy compared to conventional antibacterial cleansers. The antibacterial efficacy of TiO_2 is not governed by its surface area, but rather by its capacity to generate active carriers, ultimately leading to the creation of potent chemical species like peroxides (Lai et al., 2022).

These coatings are engineered to optimise the utilisation of nanoparticles that act as photocatalysts and enhance the kinetics of chemical reactions when subjected to irradiation by light. This holds particularly true for the degradation of organic compounds, including lipids, triglycerides, or even microorganisms. The catalytic efficiency experiences a significant enhancement through the utilisation of nanoscale photocatalysts, with TiO_2 being the prevailing choice among photocatalysts (Jawad et al., 2016). Nanostructured TiO_2 exhibits exceptional characteristics such as a substantial specific surface area, considerable pore volume and size, remarkable activity, and a relatively low density. These properties enhance the surface area and mass transfer capabilities for adsorption of organic pollutants, thereby leading to improved photocatalytic performance. This is because photocatalytic reactions rely on surface-based chemical reactions. Applications utilise the inherent attributes of self-cleaning, antifogging, anti-microbial, or water cleaving properties (Levchuk et al., 2022). In the indoor environments, it is observed that a multitude of surfaces, including but not limited to ceramic tiles, windows glass, and paper, undergo a gradual process of being enveloped by organic substances, namely oils, dirt, and smoke residue, resulting in the phenomenon of fouling. Transparent TiO_2 coatings possess the remarkable ability to remain inconspicuous, exhibiting no readily observable alterations in the colour or transparency of the underlying material (Zhang et al., 2011). However, these coatings are capable of decomposing organic substances upon their deposition. 10-20 nm TiO_2 layer, exhibits remarkable self-cleaning properties, rendering it highly advantageous for building maintenance, particularly in the context of towering skyscrapers. This innovative coating effectively mitigates the requirement for expensive surface cleaning procedures (Yousefi et al., 2023). Diverse surface substrates have the potential to undergo a TiO_2 coating, thereby enabling them to exhibit self-cleaning properties when exposed to both solar irradiation and indoor lighting conditions. Henceforth, surfaces predicated upon coatings, ceramics, vitreous substances, and cementitious materials harbouring active photocatalytic titania nanoparticles exhibit extensive utility (Table 2) in fabricating environmentally pristine regions in their immediate vicinity (Levchuk et al., 2022).

FUTURE PROSPECTIVE

Nanocoatings exhibit remarkable surface modification capabilities, rendering them highly effective in conferring antibacterial, anti-fungal, and mold-resistant properties to treated surfaces. Additionally, this innovative coating demonstrates the ability to purify the surrounding air, further enhancing its multifunctional nature (Noeiaghaei

Table 2. Applications derived from harnessing the photocatalytic reaction of TiO_2

Aspect	Application
Utilisation of a photocatalyst coating as a hygienic surface treatment to effectively mitigate bacterial and viral contamination.	The application of this specialised coating facilitates the eradication of the essential organic constituents that serve as a growth medium for bacterial organisms. The robust oxidation mechanism effectively eradicates bacteria through the induction of cell membrane lesions and subsequent lysis of cellular contents.
A sustainable solution for maintaining the cleanliness of a building's exterior surface.	Photocatalytic coatings effectively eradicate contaminants, including oil residues from car exhausts, as well as unsightly mould and mildew stains.
Mitigate Indoor Air Pollution and Alleviate Sick Building Syndrome.	Photocatalyst coatings and products effectively regulate Sick Building Syndrome by eliminating Volatile Organic Compounds released from building materials and furniture, thereby fostering an environment conducive to enhanced comfort and productivity.
Air pollution reduction	Photocatalytic oxidation facilitates the mitigation of hydrocarbon waste generated by industrial facilities engaged in the combustion of fossil fuels or coal.
Waste Water Treatment	Photocatalysts possess the potential to supplant chlorine as a more efficacious and substantially safer alternative for oxidation techniques in the realm of water purification. Photo catalytic oxidation, a highly effective process in chemical engineering, possesses the remarkable capability to eliminate bacteria and organic toxins.

et al., 2017). The activation of self-cleaning mechanisms and other functionalities will be facilitated by the presence of light, thereby safeguarding the interior and exterior of a building against the detrimental effects of environmental contamination. The catalytic coating composed of nano-TiO_2 and Titanium Phosphate exhibits the ability to decompose organic sources of odour and bacteria. This coating is specifically designed for application on the walls and ceilings of building/home interiors, as well as on upholsteries, curtains/blinds and air conditioning filters, among other surfaces (Dhiman et al., 2021). During instances of intense precipitation, the visibility may be compromised due to the phenomenon of varying refraction caused by an uneven water film. Dioxins, as well as various chemical compounds, can be effectively mitigated within water systems without the introduction of additional chemical agents. Nano-TiO_2 coatings are extensively accessible in the commercial sector and have found significant utilisation primarily in the Japanese and European markets. These coatings have been employed for water and air purification, self-cleaning glass, concrete products, and various coating applications (Farag, 2020). Photocatalytic nanocoatings have demonstrated efficacy in mitigating Bird Flu, COVID-19 and SARS infections within regions of elevated susceptibility. Hospitals and buildings in Asia and, more recently, in Europe have undergone the application of nano-TiO_2 coatings. These coatings serve the purpose of safeguarding the property from pernicious infections and detrimental environmental pollution (Zamora-Ledezma et al., 2020).

CONCLUSION

Nanostructured coatings exhibit immense potential across a myriad of applications owing to their exceptional attributes that are not commonly observed in traditional coatings. They are presently and systematically employed in various materials viz. metals, ceramics, textiles, plastics, and glass. Nanoscale characterization has garnered escalating attention from manufacturers and regulators, as the former strive to attain the desired performance while maintaining cost-effectiveness, and the latter focus on evaluating the toxicological implications. The fascination with nanostructured materials for coatings arises from their extraordinary mechanical, electrical, magnetic, and optical characteristics, as well as the potential to fabricate materials with unparalleled physical-chemical properties. Advanced nanostructured coatings have the potential to yield a wide range of highly sophisticated surface properties, including but not limited to chemical, electronic, mechanical, optical, magnetic, catalytic, as well as tribological characteristics. These desirable attributes make such coatings particularly appealing for various modern industrial applications.

REFERENCES

Abbas, M., Buntinx, M., Deferme, W., Reddy, N., & Peeters, R. (2021). Oxygen gas and uv barrier properties of nano-ZnO-coated pet and phbhhx materials fabricated by ultrasonic spray-coating technique. *Nanomaterials (Basel, Switzerland)*, *11*(2), 1–14. doi:10.3390/nano11020449 PMID:33578826

Ahmed, J., Arfat, Y. A., Al-Attar, H., Auras, R., & Ejaz, M. (2017). Rheological, structural, ultraviolet protection and oxygen barrier properties of linear low- density polyethylene films reinforced with zinc oxide (ZnO) nanoparticles. *Food Packaging and Shelf Life*, *13*(March), 20–26. doi:10.1016/j.fpsl.2017.04.005

Amor, F., Baudys, M., Racova, Z., Scheinherrová, L., Ingrisova, L., & Hajek, P. (2022). Contribution of TiO2 and ZnO nanoparticles to the hydration of Portland cement and photocatalytic properties of High Performance Concrete. *Case Studies in Construction Materials*, *16*, e00965. Advance online publication. doi:10.1016/j.cscm.2022.e00965

Anjum, A., Garg, R., Kashif, M., & Eddy, N. O. (2023). Nano-scale innovations in packaging : Properties, types, and applications of nanomaterials for the future. *Food Chemistry Advances*, *3*(November), 100560. doi:10.1016/j.focha.2023.100560

Atta-ur-Rehman, Qudoos, A., Jakhrani, S. H., Kim, H. G., & Ryou, J.-S. (2019). Influence of Nano-silica on the Leaching Attack upon Photocatalytic Cement Mortars. *International Journal of Concrete Structures and Materials*, *13*(1), 35. Advance online publication. doi:10.1186/s40069-019-0348-x

Azeez, S. O., Saheed, I. O., Adekola, F. A., & Salau, S. S. (2022). Preparation of tio2-activated kaolinite composite for photocatalytic degradation of rhodamine b dye. *Bulletin of the Chemical Society of Ethiopia*, *36*(1), 13–24. doi:10.4314/bcse.v36i1.2

Bagherzadeh, M. R., & Mousavinejad, T. (2012). Preparation and investigation of anticorrosion properties of the water-based epoxy-clay nanocoating modified by Na +-MMT and Cloisite 30B. *Progress in Organic Coatings*, *74*(3), 589–595. doi:10.1016/j.porgcoat.2012.02.006

Bhardwaj, S., Lata, S., & Garg, R. (2022). Phyto-mediated green synthesis of silver nanoparticles using Acmella oleracea leaf extract: Antioxidant and catalytic activity. *Pharmacognosy Magazine*, *18*(77), 22. doi:10.4103/pm.pm_586_20

Bica, B. O., de Melo, J. V. S., & Gleize, P. J. P. (2022). Photocatalytic concrete blocks nanomodified with zinc oxide for paving: Mechanical performance and microstructural characteristics. *Journal of Testing and Evaluation*, *50*(1). Advance online publication. doi:10.1520/JTE20200326

Caixeta, M. P. (2023). Contaminants of Emerging Concern: A Review of Risk Assessment and Treatment Strategies. *U. Porto Journal of Engineering*, *9*(1), 191–228. doi:10.24840/2183-6493_009-001_001282

Castro-Mayorga, J. L., Fabra, M. J., & Lagaron, J. M. (2016). Stabilized nanosilver based antimicrobial poly(3-hydroxybutyrate-co-3-hydroxyvalerate) nanocomposites of interest in active food packaging. *Innovative Food Science & Emerging Technologies*, *33*, 524–533. doi:10.1016/j.ifset.2015.10.019

Channa, I. A., Ashfaq, J., Gilani, S. J., Shah, A. A., Chandio, A. D., & Jumah, M. N. (2022). UV Blocking and Oxygen Barrier Coatings Based on Polyvinyl Alcohol and Zinc Oxide Nanoparticles for Packaging Applications. *Coatings*, *12*(7), 897. Advance online publication. doi:10.3390/coatings12070897

Chougule, S. S., Gurme, S. T., Jadhav, J. P., Dongale, T. D., & Tiwari, A. P. (2021). Low density polyethylene films incorporated with Biosynthesised silver nanoparticles using Moringa oleifera plant extract for antimicrobial, food packaging, and photocatalytic degradation applications. *Journal of Plant Biochemistry and Biotechnology*, *30*(1), 208–214. doi:10.1007/s13562-020-00584-7

Dai, D., Zhou, D., Xie, H., Wang, J., & Zhang, C. (2022). The design, construction and application of graphene family composite nanocoating on dental metal surface. *Biomaterials Advances, 140*, 213087. doi:10.1016/j.bioadv.2022.213087 PMID:36029723

Dhiman, N. K., Sidhu, N., Agnihotri, S., Mukherjee, A., & Reddy, M. S. (2021). Role of nanomaterials in protecting building materials from degradation and deterioration. In Biodegradation and Biodeterioration at the Nanoscale (pp. 405–475). doi:10.1016/B978-0-12-823970-4.00024-5

Eddy, N., Odiongenyi, A. O., Garg, R., & Garg, R. (2022). Plant Wastes as Alternative Sources of Sustainable and Green Corrosion Inhibitors in Different Environments. SSRN *Electronic Journal,* 1–13. doi:10.2139/ssrn.4051680

Eddy, N. O., Ibok, U. J., Garg, R., Garg, R., Iqbal, A., Amin, M., Mustafa, F., Egilmez, M., & Galal, A. M. (2022). A Brief Review on Fruit and Vegetable Extracts as Corrosion Inhibitors in Acidic Environments. *Molecules (Basel, Switzerland), 27*(9), 1–18. doi:10.3390/molecules27092991 PMID:35566341

Eddy, N. O., Odiongenyi, A. O., Garg, R., Ukpe, R. A., Garg, R., El Nemr, A., Ngwu, C. M., & Okop, I. J. (2023). Quantum and experimental investigation of the application of Crassostrea gasar (mangrove oyster) shell–based CaO nanoparticles as adsorbent and photocatalyst for the removal of procaine penicillin from aqueous solution. *Environmental Science and Pollution Research International, 30*(23), 64036–64057. doi:10.1007/s11356-023-26868-8 PMID:37059957

El-Fattah, M. A., El Saeed, A. M., Dardir, M. M., & El-Sockary, M. A. (2015). Studying the effect of organo-modified nanoclay loading on the thermal stability, flame retardant, anti-corrosive and mechanical properties of polyurethane nanocomposite for surface coating. *Progress in Organic Coatings, 89*, 212–219. doi:10.1016/j.porgcoat.2015.09.010

Elshami, A., & Ali, A. H. (2014). Using new coating materials for corrosion prevention of reinforced concrete. *MATEC Web of Conferences, 11*. 10.1051/matecconf/20141101026

Emmanuel Awe, F., Faruruwa, M. D., Abba, H., Dahiru Faruruwa, M., & Abba, H. (2021). Green Synthesis, Characterization and Antibacterial Activity of Zinc Oxide and Titanium Dioxide Nanoparticles Using Terminalia Catappa and Cymbopogon Citratus Leaf Extract. *Communication in Physical Sciences, 7*(4), 563–572. https://journalcps.com/index.php/volumes

Esteban-arranz, A., de la Osa, A. R., García-lorefice, W. E., Sacristan, J., & Sánchez-silva, L. (2021). Long-term performance of nanomodified coated concrete structures under hostile marine climate conditions. *Nanomaterials (Basel, Switzerland)*, *11*(4), 869. Advance online publication. doi:10.3390/nano11040869 PMID:33805457

Fakhri, M., & shahryari, E. (2021). The effects of nano zinc oxide (ZnO) and nano reduced graphene oxide (RGO) on moisture susceptibility property of stone mastic asphalt (SMA). *Case Studies in Construction Materials*, *15*, e00655. Advance online publication. doi:10.1016/j.cscm.2021.e00655

Farag, A. A. (2020). Applications of nanomaterials in corrosion protection coatings and inhibitors. *Corrosion Reviews*, *38*(1), 67–86. doi:10.1515/corrrev-2019-0011

Garg, R., Garg, R., Sharma, R., Anjum, A., & Eddy, N. O. (2023). Introduction to Nanotechnology and Nanomaterials. *Nanotechnology for Sustainable Agriculture, Food and Environment*, 1–15. doi:10.1201/9781003397861-1

Garg, R., Rani, P., Garg, R., & Eddy, N. O. (2021). Study on potential applications and toxicity analysis of green synthesized nanoparticles. *Turkish Journal of Chemistry*, *45*(6), 1690–1706. doi:10.3906/kim-2106-59 PMID:38144602

Garg, R., Rani, P., Garg, R., Khan, M. A., Khan, N. A., Khan, A. H., & Américo-Pinheiro, J. H. P. (2022). Biomedical and catalytic applications of agri-based biosynthesized silver nanoparticles. *Environmental Pollution*, *310*(April), 119830. Advance online publication. doi:10.1016/j.envpol.2022.119830 PMID:35926739

Gebreslassie, Y. T., & Gebretnsae, H. G. (2021). Green and Cost-Effective Synthesis of Tin Oxide Nanoparticles: A Review on the Synthesis Methodologies, Mechanism of Formation, and Their Potential Applications. *Nanoscale Research Letters*, *16*(1), 97. Advance online publication. doi:10.1186/s11671-021-03555-6 PMID:34047873

Ghodrati, M., Mousavi-Kamazani, M., & Bahrami, Z. (2023). Synthesis of superhydrophobic coatings based on silica nanostructure modified with organosilane compounds by sol–gel method for glass surfaces. *Scientific Reports*, *13*(1), 548. doi:10.1038/s41598-023-27811-0 PMID:36631572

Goffredo, G. B., Terlizzi, V., & Munafò, P. (2017). Multifunctional TiO2-based hybrid coatings on limestone: Initial performances and durability over time. *Journal of Building Engineering*, *14*, 134–149. doi:10.1016/j.jobe.2017.10.006

Gong, X., & He, S. (2020). Highly Durable Superhydrophobic Polydimethylsiloxane/Silica Nanocomposite Surfaces with Good Self-Cleaning Ability. *ACS Omega*, *5*(8), 4100–4108. doi:10.1021/acsomega.9b03775 PMID:32149238

Gopalakrishnan, V., Balasubramanian, A., Subramanian, L., Muhammed, R. I., Garg, R., & Eddy, N. O. (2023). Experimental and Theoretical Analysis on Mild Steel Corrosion Inhibition by Two Novel Compounds (FD and ACP) in Acidic Media. *Portugaliae Electrochimica Acta, 41*(3), 223–246. doi:10.4152/pea.2023410304

Harilal, M., Anandkumar, B., George, R. P., Albert, S. K., & Philip, J. (2023). High-performance eco-friendly ternary blended green concrete in seawater environment. *Hybrid Advances, 3*(February), 100037. doi:10.1016/j.hybadv.2023.100037

Hong, R., Pan, T., Qian, J., & Li, H. (2006). Synthesis and surface modification of ZnO nanoparticles. *Chemical Engineering Journal, 119*(2–3), 71–81. doi:10.1016/j.cej.2006.03.003

Honorato, C., Kumar, V., Liu, J., Koivula, H., Xu, C., & Toivakka, M. (2015). Transparent nanocellulose-pigment composite films. *Journal of Materials Science, 50*(22), 7343–7352. doi:10.1007/s10853-015-9291-7

Hsissou, R., Benzidia, B., Hajjaji, N., & Elharfi, A. (2018). Elaboration and electrochemical studies of the coating behavior of a new pentafunctional epoxy polymer (pentaglycidyl ether pentabisphenol phosphorus) on E24 carbon steel in 3.5% NaCl. *Journal of Chemical Technology and Metallurgy, 53*(5), 898–905.

Ibrahim, S. H., & El-Tayeb, N. S. M. (2021). Effect of nano-silica/alumina hybrid coating on erosion resistance of glass fibre-reinforced polymer for the application of wind turbine blades. *Proceedings of the Institution of Mechanical Engineers. Part J, Journal of Engineering Tribology, 236*(10), 2013–2031. doi:10.1177/13506501211052724

Jahan, I. (2023). Advances and Applications of Nanomaterial-Assisted Phytoremediation. *Nanotechnology for Sustainable Agriculture, Food and Environment*, 159–170. doi:10.1201/9781003397861-12

Jawad, A. H., Mubarak, N. S. A., Ishak, M. A. M., Ismail, K., & Nawawi, W. I. (2016). Kinetics of photocatalytic decolourization of cationic dye using porous TiO2 film. *Journal of Taibah University for Science : JTUSCI, 10*(3), 352–362. doi:10.1016/j.jtusci.2015.03.007

Jiang, J., Pi, J., & Cai, J. (2018). The Advancing of Zinc Oxide Nanoparticles for Biomedical Applications. *Bioinorganic Chemistry and Applications, 18*, 1–18. Advance online publication. doi:10.1155/2018/1062562 PMID:30073019

Kamal, A., Saba, M., Ullah, K., Almutairi, S. M., AlMunqedhi, B. M., & Ragab abdelGawwad, M. (2023). Mycosynthesis, Characterization of Zinc Oxide Nanoparticles, and Its Assessment in Various Biological Activities. *Crystals, 13*(2), 171. Advance online publication. doi:10.3390/cryst13020171

Kathirvelu, S., D'Souza, L., & Dhurai, B. (2009). UV protection finishing of textiles using ZnO nanoparticles. *Indian Journal of Fibre and Textile Research, 34*(3), 267–273.

Kauppinen, C., Pasanen, T. P., Isakov, K., Serué, M., Heinonen, J., Vähänissi, V., Lipsanen, H., & Savin, H. (2021). Grass-like alumina coated window harnesses the full omnidirectional potential of black silicon photodiodes. *Applied Optics, 60*(33), 10415–10420. doi:10.1364/AO.442415 PMID:34807052

Lai, L., Wu, H., Mao, G., Li, Z., Zhang, L., & Liu, Q. (2022). Microstructure and Corrosion Resistance of Two-Dimensional TiO2/MoS2 Hydrophobic Coating on AZ31B Magnesium Alloy. *Coatings, 12*(10), 1488. Advance online publication. doi:10.3390/coatings12101488

Law, M. K., Zhao, Y., Zhang, W., Wang, R., Shi, M., Zhang, Y., Chen, S., & Yang, J. (2022). Highly transparent and super-wettable nanocoatings hybridized with isocyanate-silane modified surfactant for multifunctional applications. *Nano Materials Science, 4*(2), 151–168. doi:10.1016/j.nanoms.2021.09.001

Levchuk, I., Homola, T., Singhal, G., Rueda-Márquez, J. J., Vida, J., Souček, P., Svoboda, T., Villar-Navarro, E., Levchuk, O., Dzik, P., Lähde, A., & Moreno-Andrés, J. (2022). UVA and solar driven photocatalysis with rGO/TiO2/polysiloxane for inactivation of pathogens in recirculation aquaculture systems (RAS) streams. *Chemical Engineering Journal Advances, 10*(December). doi:10.1016/j.ceja.2022.100243

Liang, S. S., Makamba, H., Huang, S. Y., & Chen, S. H. (2006). Nano-titanium dioxide composites for the enrichment of phosphopeptides. *Journal of Chromatography. A, 1116*(1–2), 38–45. doi:10.1016/j.chroma.2006.03.012 PMID:16580007

Liu, J., Janjua, Z. A., Roe, M., Xu, F., Turnbull, B., Choi, K. S., & Hou, X. (2016). Super-hydrophobic/icephobic coatings based on silica nanoparticles modified by self-assembled monolayers. *Nanomaterials (Basel, Switzerland), 6*(12), 232. Advance online publication. doi:10.3390/nano6120232 PMID:28335360

Maeztu, J. D., Rivero, P. J., Berlanga, C., Bastidas, D. M., Palacio, J. F., & Rodriguez, R. (2017). Effect of graphene oxide and fluorinated polymeric chains incorporated in a multilayered sol-gel nanocoating for the design of corrosion resistant and hydrophobic surfaces. *Applied Surface Science, 419*, 138–149. doi:10.1016/j.apsusc.2017.05.043

Minh, D. Q., & Minh, H. N. (2020). Al2O3 Nano Coating and Hydrophobicity of Glass Surface. *Glass Physics and Chemistry*, *46*(6), 640–645. doi:10.1134/S1087659621010089

Mondal, S. (2022). Nanomaterials for UV protective textiles. *Journal of Industrial Textiles*, *51*(4), 5592S–5621S. doi:10.1177/1528083721988949

Noeiaghaei, T., Dhami, N., & Mukherjee, A. (2017). Nanoparticles surface treatment on cemented materials for inhibition of bacterial growth. *Construction & Building Materials*, *150*, 880–891. doi:10.1016/j.conbuildmat.2017.06.046

Nunzi, F., & De Angelis, F. (2022). Modeling titanium dioxide nanostructures for photocatalysis and photovoltaics. *Chemical Science (Cambridge)*, *13*(33), 9485–9497. doi:10.1039/D2SC02872G PMID:36091912

Oliani, W. L., Pusceddu, F. H., & Parra, D. F. (2022). Silver-titanium polymeric nanocomposite non ecotoxic with bactericide activity. *Polymer Bulletin*, (12), 10949–10968. Advance online publication. doi:10.1007/s00289-021-04036-7 PMID:35035006

Parimalam, M., Islam, M. R., & Yunus, R. M. (2018). Effects of nanosilica, zinc oxide, titatinum oxide on the performance of epoxy hybrid nanocoating in presence of rubber latex. *Polymer Testing*, *70*, 197–207. doi:10.1016/j.polymertesting.2018.07.008

Peng, C. W., Chang, K. C., Weng, C. J., Lai, M. C., Hsu, C. H., Hsu, S. C., Li, S. Y., Wei, Y., & Yeh, J. M. (2013). UV-curable nanocasting technique to prepare bio-mimetic super-hydrophobic non-fluorinated polymeric surfaces for advanced anticorrosive coatings. *Polymer Chemistry*, *4*(4), 926–932. doi:10.1039/C2PY20613G

Qu, L. L., Geng, Z. Q., Wang, W., Yang, K. C., Wang, W. P., Han, C. Q., Yang, G. H., Vajtai, R., Li, D. W., & Ajayan, P. M. (2019). Recyclable three-dimensional Ag nanorod arrays decorated with O-g-C3N4 for highly sensitive SERS sensing of organic pollutants. *Journal of Hazardous Materials*, *379*(January), 120823. doi:10.1016/j.jhazmat.2019.120823 PMID:31276918

Raja Beryl, J., & Xavier, J. R. (2021). Influence of silane functionalized nanoclay on the barrier, mechanical and hydrophobic properties by clay nanocomposite films in an aggressive chloride medium. *Colloids and Surfaces. A, Physicochemical and Engineering Aspects*, *630*, 127625. doi:10.1016/j.colsurfa.2021.127625

Safavi, M. S., Walsh, F. C., Visai, L., & Khalil-Allafi, J. (2022). Progress in Niobium Oxide-Containing Coatings for Biomedical Applications: A Critical Review. *ACS Omega*, *7*(11), 9088–9107. doi:10.1021/acsomega.2c00440 PMID:35356687

Sharma, R., Garg, R., & Kumari, A. (2020). A review on biogenic synthesis, applications and toxicity aspects of zinc oxide nanoparticles. *EXCLI Journal, 19*, 1325–1340. doi:10.17179/excli2020-2842 PMID:33192216

Silva, M. R. F., Alves, M. F. R. P., Cunha, J. P. G. Q., Costa, J. L., Silva, C. A., Fernandes, M. H. V., Vilarinho, P. M., & Ferreira, P. (2023). Nanostructured transparent solutions for UV-shielding: Recent developments and future challenges. *Materials Today Physics*, *35*(February), 101131. Advance online publication. doi:10.1016/j.mtphys.2023.101131

Siró, I., & Plackett, D. (2010). Microfibrillated cellulose and new nanocomposite materials: A review. *Cellulose (London, England)*, *17*(3), 459–494. doi:10.1007/s10570-010-9405-y

Sriramulu, D., Reed, E. L., Annamalai, M., Venkatesan, T. V., & Valiyaveettil, S. (2016). Synthesis and Characterization of Superhydrophobic, Self-cleaning NIR-reflective Silica Nanoparticles. *Scientific Reports*, *6*(1), 35993. doi:10.1038/srep35993 PMID:27824064

Strudwick, A. J., Weber, N. E., Schwab, M. G., Kettner, M., Weitz, R. T., Wünsch, J. R., Müllen, K., & Sachdev, H. (2015). Chemical vapor deposition of high quality graphene films from carbon dioxide atmospheres. *ACS Nano*, *9*(1), 31–42. doi:10.1021/nn504822m PMID:25398132

Tayel, S. A., Abu El-Maaty, A. E., Mostafa, E. M., & Elsaadawi, Y. F. (2022). Enhance the performance of photovoltaic solar panels by a self-cleaning and hydrophobic nanocoating. *Scientific Reports*, *12*(1), 1–13. doi:10.1038/s41598-022-25667-4 PMID:36481954

Thakur, N., Singh, S. B., & Anshuman. (2022). a Nanotechnology Approach on Self-Cleaning Material: a Review. *Indian Concrete Journal, 96*(4), 40–48.

Turalija, M., Bischof, S., Budimir, A., & Gaan, S. (2016). Antimicrobial PLA films from environment friendly additives. *Composites. Part B, Engineering*, *102*, 94–99. doi:10.1016/j.compositesb.2016.07.017

Ubaid, F., Radwan, A. B., Naeem, N., Shakoor, R. A., Ahmad, Z., Montemor, M. F., Kahraman, R., Abdullah, A. M., & Soliman, A. (2019). Multifunctional self-healing polymeric nanocomposite coatings for corrosion inhibition of steel. *Surface and Coatings Technology*, *372*, 121–133. doi:10.1016/j.surfcoat.2019.05.017

Verma, C. (2021). *Recent advancements in corrosion inhibitor systems through carbon allotropes : Past, present, and future*. doi:10.1002/nano.202100039

Wang, X., Pan, J., Li, Q., Dong, X., Shi, L., & Chang, S. (2022). Preparation, Corrosion Resistance, and Electrochemical Properties of MnO2/TiO2 Coating on Porous Titanium. *Coatings*, *12*(10), 1381. Advance online publication. doi:10.3390/coatings12101381

Wimalasiri, V. K., Weerathunga, H. U., Kottegoda, N., & Karunaratne, V. (2017). Silica Based Superhydrophobic Nanocoatings for Natural Rubber Surfaces. *Journal of Nanomaterials*, *2102467*, 1–14. Advance online publication. doi:10.1155/2017/2102467

Xia, F., Li, Q., Ma, C., & Guo, X. (2020). Preparation and characterization of Ni–AlN nanocoatings deposited by magnetic field assisted electrodeposition technique. *Ceramics International*, *46*(2), 2500–2509. doi:10.1016/j.ceramint.2019.09.244

Xin, M. (2022). Crystal Structure and Optical Properties of ZnO:Ce Nano Film. *Molecules (Basel, Switzerland)*, *27*(16), 1–9. doi:10.3390/molecules27165308 PMID:36014546

Yang, H. M. (2021). Role of organic and eco-friendly inhibitors on the corrosion mitigation of steel in acidic environments—A state-of-art review. *Molecules (Basel, Switzerland)*, *26*(11), 3473. Advance online publication. doi:10.3390/molecules26113473 PMID:34200445

Yang, J., Qian, H., Wang, J., Ju, P., Lou, Y., Li, G., & Zhang, D. (2021). Mechanically durable antibacterial nanocoatings based on zwitterionic copolymers containing dopamine segments. *Journal of Materials Science and Technology*, *89*, 233–241. doi:10.1016/j.jmst.2020.11.031

Yousefi, F., Mousavi, S. B., Heris, S. Z., & Naghash-Hamed, S. (2023). UV-shielding properties of a cost-effective hybrid PMMA-based thin film coatings using TiO2 and ZnO nanoparticles: A comprehensive evaluation. *Scientific Reports*, *13*(1), 1–19. doi:10.1038/s41598-023-34120-z PMID:37130903

Zamora-Ledezma, C., Clavijo, D. F. C., Medina, E., Sinche, F., Vispo, N. S., Dahoumane, S. A., & Alexis, F. (2020). Biomedical science to tackle the COVID-19 pandemic: Current status and future perspectives. *Molecules (Basel, Switzerland)*, *25*(20), 4620. Advance online publication. doi:10.3390/molecules25204620 PMID:33050601

Zhang, L., Bai, R., Li, B., Ge, C., Du, J., Liu, Y., Le Guyader, L., Zhao, Y., Wu, Y., He, S., Ma, Y., & Chen, C. (2011). Rutile TiO2 particles exert size and surface coating dependent retention and lesions on the murine brain. *Toxicology Letters*, *207*(1), 73–81. doi:10.1016/j.toxlet.2011.08.001 PMID:21855616

Zhou, C., Chen, Q., Chen, Q., Yin, H., Wang, S., & Hu, C. (2022). Preparation of TiO2 Superhydrophobic Composite Coating and Studies on Corrosion Resistance. *Frontiers in Chemistry*, *10*(July), 1–9. doi:10.3389/fchem.2022.943055 PMID:35873046

ADDITIONAL READING

Bhardwaj, S., Lata, S., & Garg, R. (2023). Application of nanotechnology for preventing postharvest losses of agriproducts. *The Journal of Horticultural Science & Biotechnology*, *98*(1), 31–44. doi:10.1080/14620316.2022.2091488

Garg, R., Garg, R., & Okon Eddy, N. (2022). *Handbook of Research on Green Synthesis and Applications of Nanomaterials*. IGI Global. doi:10.4018/978-1-7998-8936-6

Kumari, A., Garg, R., & Garg, R. (2023). *Nanotechnology for Sustainable Agriculture, Food and Environment*. CRC Press. doi:10.1201/9781003397861

Tan, Y., Tang, J., Zhao, N., Qi, F., & Ouyang, X. (2023). Effect of Fluorocarbon Polymers on Hydrophobicity, Wear Resistance and Corrosion Resistance of Epoxy Resins. *Coatings*, *13*(4), 685. Advance online publication. doi:10.3390/coatings13040685

Chapter 3
Advancements in Designing Smart and Intelligent Nanocoatings

Mukhtar Iderawumi Abdulraheem
https://orcid.org/0000-0002-6745-378X
Henan Agricultural University, China

Abiodun Yusuff Moshood
https://orcid.org/0000-0002-0917-8396
Henan Agricultural University, China

Yongqi Chen
Henan Agricultural University, China

Hongjun Chen
Henan Agricultural University, China

Haiyan Zhang
Henan Agricultural University, China

Jiandong Hu
Henan Agricultural University, China

ABSTRACT

Smart and intelligent nanocoatings have emerged as crucial components in various industries, offering enhanced functionalities and improved performance. This chapter explores the recent advancements in studying smart and intelligent nanocoatings, encompassing their design, synthesis, characterization, and applications. The chapter aims to provide a comprehensive understanding of the current state-of-the-art in this field and highlight the potential for future development and research. This chapter concluded that developing smart and intelligent nanocoatings has significantly enhanced various industries' performance, durability, and safety. However, future research focuses on reducing costs, scaling up production, and creating more advanced systems with existing technologies, such as sensors and actuators.

DOI: 10.4018/979-8-3693-3136-1.ch003

INTRODUCTION

Advanced nanocoatings that utilize nanotechnology and possess responsive or adaptive properties are known as competent and intelligent coatings (Abdel-Karim and Waheed, 2013; Girigoswami et al., 2022; Nile et al., 2020). These coatings are highly versatile and designed to sense and respond to changes in their surroundings, making them capable of performing various functions (Cao et al., 2022; Cerqueira et al., 2017; Malik et al., 2023). Intelligent nanocoatings have garnered significant attention in recent years, with potential applications across various industries (Cerqueira et al., 2017). To fully grasp the importance of studying smart nanocoatings, it is necessary to define key terms and concepts. Nanocoatings are thin films or layers, typically measured in nanometers (one billionth of a meter), and applied to surfaces to enhance their properties, such as durability, corrosion resistance, self-cleaning ability, or antimicrobial activity (Vasile, 2018). Nanotechnology involves manipulating materials at the nanoscale (typically between 1 and 100 nanometers) to achieve unique properties and functionalities. Advanced nanocoatings go beyond regular coatings by incorporating responsive elements into their composition. External factors such as temperature, light, humidity, pH, or mechanical stress can trigger these responsive elements. By responding to these factors, smart nanocoatings can change their properties in a controlled manner. The study of intelligent nanocoatings is vital for numerous reasons, including:

1. These coatings can revolutionize various industries by offering improved functionality and performance.
2. They provide solutions to challenges traditional coatings face, such as limited durability or a lack of adaptability.
3. Smart nanocoatings can contribute to sustainability efforts by reducing the need for frequent reapplications or maintenance. Intelligent nanocoatings have a wide range of applications across various industries. In the automotive industry, they can create self-healing surfaces that automatically repair minor scratches or damages, improving the aesthetics and extending the lifespan of vehicles (Thakur and Kumar, 2022).
4. These coatings can provide anti-fogging properties for windshields or windows, enhancing visibility and safety.

In food packaging, intelligent and innovative nanocoatings are quickly gaining popularity for their ability to enhance packaged food items' safety, quality, and longevity (Ashfaq et al., 2022; Singh et al., 2023). These coatings offer a range of functionalities, such as antimicrobial properties, gas barrier properties, and sensing capabilities. With advancements in developing these intelligent nanocoatings, the food packaging industry has revolutionized, providing improved protection against contamination, spoilage, and degradation (Girigoswami et al., 2022; Motelica et al., 2020). One of the most significant advancements in designing intelligent nanocoatings for food packaging is the integration of antimicrobial agents. These agents help to impede the growth of harmful bacteria, fungi, and other microorganisms on the surface of packaging materials, thereby minimizing the risk of foodborne illnesses (Jagtiani, 2022). Silver nanoparticles (AgNPs) are commonly used as antimicrobial agents due to their exceptional antibacterial properties (Singh et al., 2023). They can be integrated into the nanocoating formulation or applied as a separate layer on the packaging material. When silver ions are released from the nanoparticles, they disrupt the cellular processes of microorganisms, leading to their inactivation (Sharma et al., 2017).

Furthermore, intelligent nanocoatings can create flexible and stretchable electronics by providing a protective layer that can withstand mechanical stress. Smart nanocoatings also have significant potential in the healthcare sector (Ortega-Nieto et al., 2023). They can be used to develop antimicrobial coatings for medical devices or surfaces in healthcare facilities, thereby reducing the risk of infections (Blum et al., 2015; Sadabadi et al., 2022). Smart nanocoatings have potential applications in various industries, from energy-efficient buildings with self-regulating temperature control to anti-graffiti coatings (Fig. 1) that repel paint or ink and self-cleaning coatings for solar panels to maintain optimal efficiency (Idumah et al., 2020; Young et al., 2020). The advantages of these coatings are numerous, including enhancing the durability and lifespan of surfaces, reducing maintenance costs, improving energy efficiency, providing antimicrobial properties, enabling targeted drug delivery systems (Pham et al., 2020), and contributing to sustainability efforts.

Developing nanocoatings with gas barrier properties is a significant breakthrough in food preservation, where oxygen and moisture are two culprits that contribute to food spoilage and degradation (Girigoswami et al., 2022; Ros-Lis and Benitez Serra, 2023). Nanocoatings can be engineered to create a barrier against these gases, effectively preventing them from permeating packaged the food (Ashfaq et al., 2022). This results in preserving the freshness and quality of the food for an extended period. Nanocomposites, consisting of polymers and nanoparticles like clay or graphene oxide, have shown promising results in enhancing gas barrier properties (Singh et al., 2023; Xie et al., 2019).

Furthermore, intelligent nanocoatings have been developed to provide real-time monitoring and sensing capabilities in food packaging (Jagtiani, 2022). These coatings can detect changes in temperature, pH levels, or gas composition inside the package and provide feedback on the quality and safety of the food product (Motelica et al., 2020). For example, pH-sensitive nanocoatings change color when exposed to acidic or alkaline conditions, indicating potential spoilage or contamination (Ros-Lis and

Figure 1. Processes in smart and intelligent nanocoatings

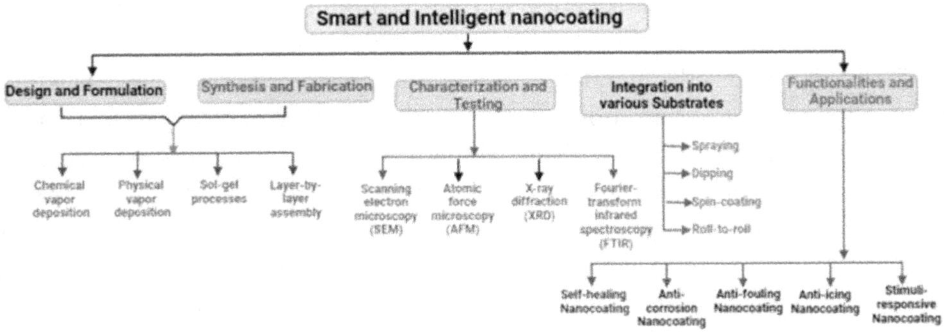

Benitez Serra, 2023; Sharma et al., 2017). Similarly, temperature-sensitive coatings can alert us if the food has been exposed to improper storage (Sharma et al., 2017). The smart and intelligent nanocoatings field holds immense promise for various industries. These advanced materials can actively sense and adapt to environmental changes by incorporating responsive elements into nanocoatings, benefiting the automotive and aerospace industries, electronics, healthcare sectors, and more (Han et al., 2023; Plazas-Tuttle et al., 2015). The benefits of these coatings include improved durability, enhanced functionality, reduced maintenance costs, increased safety, and sustainability contributions (Abdel-Karim and Waheed, 2013).

DESIGN AND SYNTHESIS OF NANOCOATINGS

Smart nanocoatings are a special kind of advanced material with unique features and abilities at the nanoscale. They are meant to enhance the performance and durability of different surfaces like metals, ceramics, polymers, and glass (Tang et al., 2023; Ulaeto et al., 2019). To create these coatings, precise control over the deposition process is needed to achieve the desired traits, such as better corrosion resistance, self-cleaning properties, anti-fouling behavior, or improved optical properties. Many techniques have been developed to synthesize these smart nanocoatings, including chemical vapor deposition (CVD), sol-gel methods, electrochemical deposition, and self-assembly processes (Figure 1). Table 1 summarizes the coating methods along with their process parameters.

Each of the synthesis technics are further discuss below:

1. **Chemical Vapor Deposition (CVD):** Chemical vapor deposition is commonly used to create accurate thin films and coatings, which involve introducing

Table 1. Comparison and summary of various nanocoatings technologies

Nanocoating Methods	Coating Materials	Process Parameters	Principles	Advantages	Disadvantages	References
CVD	Ceramic, Metallic	Size of nanoparticles, temperature, surface treatments, target materials, deposition time	Production of chemical reaction on solid substances Transferring of the volatile substance to the deposition area	Flexibility, Effectiveness, Purity of material, Processing cost and highly efficiency	Dangerous gas accrual, laborious and require experts	(Abbasi et al., 2020)
Sol-Gel	Nanocomposite, inorganic salt, Metallic	pH, Catalyst nature, initial material to produce sol, Temperature, Hydrolysis ratio	Heat treatment and hydrolysis reaction	Simple, Efficiency, reactions at low temperature, homologous coating with high purity.	Time consuming, high cost of raw materials	(Li et al., 2020; Mohammed and Hussein, 2019; Sun et al., 2019)
Electrochemical deposition	Metallic Matrix, Nanocomposite, Metallic	Current density, Bath temperature and composition, pH, Additive	Deposited on the substrate, materials reach melting state and atomized with airflow	Efficiency deposition, simple, flexibility, wide coating and substrate selection, varying coating thickness	Oxide inclusions due to high temperature, abrasion and hardness resistance of the coating	(Berger, 2015; Li et al., 2017)
Self-Assembly Process	Ceramic, Metallic	Laser power density, cooling density,	Deposition and melting, material are continuously fed	Simple, very fast, low coating dilution rate, effectiveness and efficient, unlimited powder selection	Poor repeatability, coating performance are affected due to porosity and cracks, poor cladding layers	(Li et al., 2017; More et al., 2017)

precursor gas and substrate material into a reaction chamber (Gu et al., 2020; Noah, 2020). The precursor gas undergoes chemical reactions on the substrate surface, producing a solid film. CVD can be performed under different conditions, such as low pressure (LPCVD) or atmospheric pressure (APCVD). Choosing the appropriate precursor gas and reaction conditions enables the synthesis of various smart nanocoatings with custom properties (He et al., 2020). For example, hydrophobic coatings can be created using CVD using precursor gases containing fluorine or silicon atoms (Ramasubramanian et al., 2019). These thin films exhibit water-repellent properties and can be helpful in self-cleaning surfaces or anti-fouling coatings. Although the application of CVD coatings to food processing equipment is limited due to the complex apparatus required and the size and throughput restrictions of the coatings, they offer exceptional results in certain areas (Bastarrachea et al., 2015a). These coatings are beneficial when coating small, irregularly shaped parts that endure significant wear and tear, such as sanitary valve components. The purity and strength of CVD coatings make them a valuable resource for specific applications (Noah, 2020).

2. **Sol-Gel Methods:** The sol-gel process synthesizes metal alkoxides or metal salts in a liquid solution, creating nanocoatings. This process entails producing a sol, a nanoparticle liquid (Vasile, 2018). The sol undergoes aging and drying to form a gel, which can be further processed to create a solid nanocoating (Mohammed and Hussein, 2019). The sol-gel approach provides a variety of advantages when producing smart nanocoatings, including the ability to include functional additives in the coating matrix, such as nanoparticles, organic molecules, or dyes (Farooq et al., 2022). This allows coatings to exhibit specific characteristics, such as increased mechanical strength, optical transparency, or customized surface chemistry. (Moncada et al., 2007; Prasad et al., 2018) has documented the production of hybrid layered aluminosilicate nanoparticles that incorporated octadecyl amine (ODA) as the organic component, along with SiO_2 nanoparticles that had a spherical shape and contained ODA or lacked it (Plazas-Tuttle et al., 2015). These nanoparticles were created using the sol-gel approach and were utilized to create nanocomposites with polypropylene. An instance of smart nanocoating generated through sol-gel methods is photocatalytic coatings that can decompose organic pollutants under UV light exposure (Abdulraheem et al., 2023; Bao et al., 2019). These coatings have numerous applications, including air purification systems and self-cleaning surfaces.
3. **Electrochemical Deposition**: This involves the reduction of metal ions from an electrolyte solution onto a conductive substrate, in which an electric current passing between the substrate and a counter electrode can be achieved (Bastarrachea et al., 2015a; Chandrasekar et al., 2010). By controlling factors like the current density, bath composition, and deposition time, thin films with specific properties can be produced (Barish and Goddard, 2013). When zinc nanocrystalline coatings are deposited on steel substrates using pulse electrodeposition, they are less porous and more corrosion-resistant than direct current electrodeposition (Chandrasekar et al., 2010; Chen et al., 2021; Li et al., 2016). There have been reports that the corrosion resistance of nanocrystalline zinc on copper substrates varies significantly depending on the current density used (Chen et al., 2021). Electrochemical deposition is frequently utilized to create corrosion-resistant coatings. For example, applying a layer of zinc or chromium to a metal substrate can enhance its corrosion resistance (Chandrasekar

et al., 2010). Additionally, it can be used to create composite coatings by adding nanoparticles or organic molecules to the electrolyte solution (Li et al., 2016).

4. **Self-Assembly Processes**: Smart nanocoatings can be synthesized through self-assembly processes, which involve the spontaneous organization of molecules or nanoparticles into ordered structures without external intervention (Cao et al., 2022). These processes rely on non-covalent interactions like van der Waals forces or hydrogen bonding to assemble building blocks into functional nanocoatings (Sarode et al., 2020). The polyelectrolyte film can be sensitive to different physical and chemical conditions, like changes in pH or mechanical pressure from the environment, and the sensitivity allows for the controlled release of inhibitors trapped within the multilayers (del Mercato et al., 2014; Naqvi et al., 2023; Plazas-Tuttle et al., 2015). Typically, polyelectrolytes are assembled on the surface of nanoparticles using a layer-by-layer (LBL) approach to create nanocontainers loaded with inhibitors (Andreeva et al., 2010; de Villiers et al., 2011). The LBL assembly is a self-assembly process that deposits positively and negatively charged molecules or nanoparticles onto a substrate, forming a multilayer coating with tailored properties (Bastarrachea et al., 2015b). This technique incorporates functional components like enzymes, dyes, or nanoparticles into the coating structure (Mustafa and Andreescu, 2020). In addition to self-assembly processes, other techniques like chemical vapor deposition, sol-gel methods, and electrochemical deposition can also synthesize smart nanocoatings (Bastarrachea et al., 2015a). Each technique offers unique advantages and allows for coating synthesis with tailored properties and functionalities (Cao et al., 2022). These smart nanocoatings have many applications in various industries, including automotive, aerospace, electronics, energy, and healthcare (Singh et al., 2018).

CHARACTERIZATION TECHNIQUES

The different characterization techniques employed to analyze the properties of smart nanocoatings include scanning electron microscopy (SEM), atomic force microscopy (AFM), X-ray diffraction (XRD), Fourier-transform infrared spectroscopy (FTIR), and ellipsometry.

1. **Scanning Electron Microscopy (SEM):** Analyzing the morphology and structure of smart nanocoatings can be done with the help of SEM because it uses a focused beam of electrons to detect the images of the surface of materials with high resolution (Li et al., 2021; Singh et al., 2018). Through SEM, researchers can visualize nanoscale features of nanocoatings with precision

and observe nanoparticle size, shape, and distribution (Farooq et al., 2022). SEM also enables the examination of the porosity and roughness of the coating surface (Wadullah et al., 2022; Xia et al., 2020). Furthermore, the effects of wear and tear on the coating over time can be studied using SEM, providing valuable insights into the durability and stability of the material.

2. **Atomic Force Microscopy (AFM):** The AFM technique plays a crucial role in analyzing the properties of intelligent nanocoatings because it involves using a pointed tool to examine the surface of the coating, creating a 3D image of the material's terrain and mechanical traits (Dufrêne et al., 2017; Voigtländer, 2019). AFM helps researchers understand the coating's functional qualities by measuring its Young's modulus, adhesion strength, and friction coefficient (Dufrêne et al., 2017; Reggente et al., 2015). Furthermore, AFM can be used to study changes in the coating's surface properties after different or multiple exposures (Reggente et al., 2015; Voigtländer, 2019).
3. **X-ray Diffraction (XRD):** The XRD technique is highly versatile and helpful in analyzing smart nanocoatings' crystal structure and composition. XRD provides information about the arrangement of atoms within the material by measuring the intensity of X-ray reflections from the coating (Martinez et al., 2014; Noah, 2020; Xia et al., 2020). Through XRD, researchers can determine if any specific phases or defects are present in the coating, the degree of crystallinity, and the orientation of crystal grains (Bugnicourt et al., 2016; Xia et al., 2020). This information is crucial in understanding the coating's mechanical, thermal, and electrical properties (Li et al., 2021).
4. **Fourier-Transform Infrared Spectroscopy (FTIR):** The FTIR technique is a non-destructive way to analyze smart nanocoatings' molecular structure and composition and detect the infrared radiation absorption by the coating to gather information regarding the types of chemical bonds in the material (Kokate et al., 2013; Li et al., 2021). With FTIR, researchers can detect the presence of specific functional groups or molecules in the coating and determine the degree of cross-linking between them (Koslowski et al., 2018). This knowledge is essential for comprehending the chemical properties of the coating and its interactions with other materials (Wadullah et al., 2022).
5. **Ellipsometry:** It measures the polarization state of light reflected from the coating to provide data on the material's thickness, refractive index, and optical constants (Li et al., 2021; Petrik et al., 2020). Researchers use this method to determine the coating's optical properties across various wavelengths and angles, which helps understand its performance in various applications (Petrik et al., 2020; Wadullah et al., 2019). Moreover, ellipsometry can also observe how the coating's optical properties change after exposure to different environmental

conditions like temperature, humidity, or chemicals (Bugnicourt et al., 2016; Wadullah et al., 2019).

FUNCTIONALITIES AND APPLICATIONS OF SMART AND INTELLIGENT NANOCOATINGS

Nanocoatings have recently received much attention because of their distinct properties and potential applications. Among them, smart and intelligent nanocoatings have emerged as an encouraging technology with advanced features like self-healing, anti-fouling, anti-corrosion, anti-icing, and stimulus-responsive behavior (de Villiers et al., 2011; Khan et al., 2020; Xie et al., 2019). Nanocoatings consist of one or more molecular layers and can be applied to various materials such as metals, glass, ceramics, and polymers (Fig. 2) (Bastarrachea et al., 2015b; Cagri et al., 2004; Smirnova et al., 2012; Vasile, 2018). Certain nanocoatings are polymers that are either polymerized before or during application. “Smart coatings” offer a range of functions, including thermal insulation, controlled release of active ingredients, and self-healing capabilities (Vasile, 2018). They also provide additional features such as antimicrobial properties, wrinkle and stain resistance, hydrophobic and hydrophilic properties, UV protection, and antistatic properties, which contribute to the overall characteristics of the substrate material (Singh et al., 2018). These coatings can adjust to changes in their surroundings, which enhances their performance and lifespan (Abdel-Karim and Waheed, 2013).

1. **Self-Healing Nanocoatings:** These coatings incorporate healing agents like polymer chains or nanoparticles, which react when the coating is compromised and create a new layer that restores it to its original state (Sarode et al., 2020; Wei et al., 2015). Smart and intelligent nanocoatings have the crucial feature of self-healing, which means that the coating can repair itself automatically in case of damage or scratches (Shah et al., 2022; Vasile, 2018). This is made possible by specialized molecules that bond and fill in any cracks or gaps in the coating (Gu et al., 2020). Such nanocoatings have numerous potential uses, including protective coatings for bridges, buildings, vehicles, medical devices, and implants (Coyle et al., 2007). Self-healing nanocoatings contain microcapsules with healing agents that react when the coating is damaged (Vishwakarma, 2020), allowing it to repair itself without external assistance (Fig. 2). The coatings are ideal for situations where access is limited or frequent maintenance is impractical, such as protecting car exteriors from scratches or safeguarding aircraft surfaces from corrosion (Thakur and Kumar, 2022; Vishwakarma, 2020). Self-healing nanocoatings safeguard electronic devices

from harm and extend their longevity (Farooq et al., 2022; Idumah et al., 2020; Trentin et al., 2022). Additionally, they can enhance the efficiency and durability of solar panels by repairing surface damage. Nanocoatings with self-healing properties are revolutionary for enhancing durability across multiple industries (Kausar et al., 2023; Trentin et al., 2022). These coatings effectively safeguard building exteriors and infrastructure from damage, thereby minimizing the need for frequent repairs and maintenance (Kausar et al., 2023; Sadabadi et al., 2022). They have proven especially useful in the automotive, aerospace, electronics, energy, construction, and infrastructure industries (Kausar et al., 2023; Khan et al., 2020; Tang et al., 2023). Their unique ability to self-repair makes them an ideal option in situations where coated materials are hard to access or frequent maintenance could be more practical (Bastarrachea et al., 2015b; Idumah et al., 2020; Ortega-Nieto et al., 2023).

2. **Anti-Fouling Nanocoatings**: Nanocoatings for anti-fouling purposes are created to stop the accumulation of small organisms like bacteria and algae on surfaces (Nguyen-Tri et al., 2019). They can be used in several industries, such as aerospace, automotive, and healthcare, to protect against the growth of dangerous microorganisms (Barish and Goddard, 2013; de Villiers et al., 2011). Thin layers of anti-fouling nanocoatings can effectively prevent the accumulation of fouling substances on surfaces, providing benefits (Bhatt et al., 2015). These coatings can increase durability, decrease the need for maintenance, and minimize friction between surfaces (Zhou et al., 2022). Furthermore, they can improve fuel efficiency and corrosion resistance in marine applications while enhancing the resistance of devices and systems to biofouling (Abdeen et al., 2019). Anti-fouling nanocoatings find application in a broad spectrum of industries, including marine, medical devices, food processing, and water treatment. In marine industries, these coatings help enhance fuel efficiency, reduce maintenance costs, and extend the hull's lifespan (Zhou et al., 2022). In medical devices, nanocoatings prevent bacteria and other microorganisms from building up on devices like catheters and implants (Blum et al., 2015; Nguyen-Tri et al., 2019). The food processing industry uses anti-fouling nanocoatings to prevent the accumulation of bacteria and microorganisms on equipment and surfaces, enhancing food safety and minimizing contamination risks (Abdeen et al., 2019). Similarly, water treatment systems benefit from these coatings by preventing the buildup of fouling substances and improving the efficiency of the treatment process (Bhatt et al., 2015).
3. **Anti-Corrosion Nanocoatings**: Many industries face the challenge of corrosion-damaging surfaces, but anti-corrosion nanocoatings have been created to combat this issue. These special coatings safeguard metals such as steel and aluminum from corrosion, even in harsh surroundings. For instance,

two methods exist for creating intelligent coatings: directly adding inhibitors to the coating or encapsulating inhibitors in micro/nanocontainers distributed throughout the coating (Chen et al., 2023). The direct doping approach can cause problems like inhibitor leakage and interactions that harm the barrier properties of the coating (Wei et al., 2015). Adding inhibitors directly can also lead to the formation of electrolyte pathways and reduced barrier properties. Encapsulation of active materials in micro/nanocontainers that are compatible with the coating matrix and sensitive to local changes caused by corrosion can overcome these problems (Nazeer and Madkour, 2018). Additionally, incorporating nanoparticles into polymers can improve the durability of the coatings by filling cavities and reducing cracking (Khan et al., 2020). Also, nanocrystals embedded in a polymer matrix of corrosion-resistant materials like zinc oxide will prevent damage to surfaces by reacting with corrosive substances (Farag, 2020).

4. **Anti-Icing Nanocoatings**: Nanocoatings that prevent surfaces ice formation are called anti-icing nanocoatings, commonly utilized in cold climates to improve safety and performance in numerous industries, such as aerospace and automotive (Lin et al., 2018; Saini and Bhatt, 2020). This process is achieved by utilizing specific molecules that can lower the freezing point of water and prevent ice from forming on the coating's surface, even at freezing temperatures (Coyle et al., 2007). Anti-icing nanocoatings have numerous potential uses, such as on aircraft wings and control surfaces, wind turbines, other renewable energy systems, roofs, and gutters (Fig. 2) (Singh et al., 2023). It comprises a nanostructured surface that creates a slippery surface, making it challenging for ice to form and adhere to the surface (Lin et al., 2018; Xie et al., 2019). Furthermore, these coatings can prevent and release ice, lower maintenance needs, improve safety, increase energy efficiency, and may have environmental advantages (Xie et al., 2019). Transportation, infrastructure, renewable energy, marine, electronics, and healthcare sectors use anti-icing nanocoatings in practical applications (He and Guo, 2021; Lin et al., 2018; Xie et al., 2019)
5. **Stimuli-Responsive Nanocoatings**: Stimuli-responsive nanocoatings have gained popularity due to their adaptability and capacity to respond to environmental factors (Han et al., 2023; Plazas-Tuttle et al., 2015). They employ active ingredient release in a controlled manner according to specific stimuli, like pH or temperature, making them highly beneficial in many industries. They can be tailored to dispense drugs (Pham et al., 2020), antimicrobial agents, corrosion inhibitors, or other bioactive compounds in a targeted and localized fashion, ensuring precise delivery of therapeutic agents or safeguarding against corrosion (Plazas-Tuttle et al., 2015). Nanocoatings that respond to external

Figure 2. Self-healing and anti-icing properties of nanocoatings

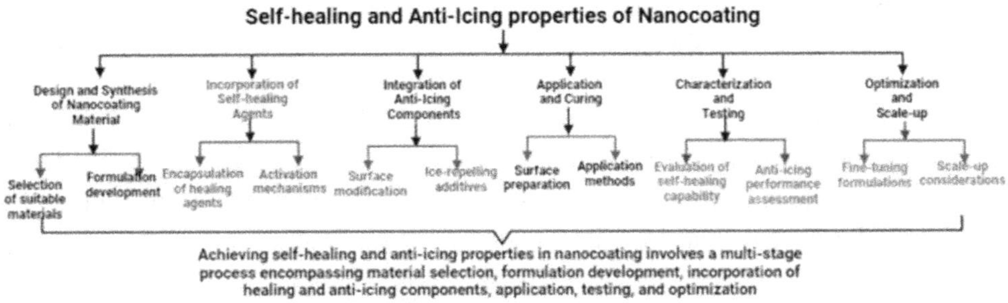

stimuli are a game-changer due to their surface properties that can be altered from hydrophilic to hydrophobic, making them ideal for anti-fog coatings and self-cleaning surfaces (Cordeiro, 2011). Even nanocoatings that can change their optical properties, which has led to the creation of smart windows that optimize light transmission for energy-efficient buildings and vehicles (Plazas-Tuttle et al., 2015; Saha et al., 2021). Stimuli-responsive nanocoatings offer a wide range of applications across various industries. They can be utilized for drug delivery in biomedicine (Cabane et al., 2012; Pham et al., 2020), antibacterial coatings for medical devices (Blum et al., 2015), self-healing coatings for electronics, and anti-fogging and self-cleaning coatings in the automotive sector (Saha et al., 2021). Nanocoatings that are reactive to external factors have many uses in energy storage and conversion (Cordeiro, 2011). For instance, they can function as safeguarding coatings for batteries and fuel cells (Saha et al., 2021). Moreover, these coatings can enhance solar panel efficiency and may become more hydrophilic, which boosts its capacity to repel water and prevent the proliferation of microorganisms as the temperature increases.

Real-World Applications

Nanocoatings have many practical applications across various industries, providing distinctive features and advantages that make them suitable for specific uses. These coatings have the potential to enhance the effectiveness, longevity, and safety of different products and devices, and their usage is projected to expand. Smart and intelligent nanocoatings have numerous real-world applications in various industries, including:

i. **Aerospace**: self-healing, anti-icing, and stimuli-responsive nanocoatings can enhance aircraft and spacecraft performance and safety (Xie et al., 2019). Self-healing nanocoatings repair damaged surfaces, while anti-icing nanocoatings prevent ice formation on wings and other surfaces (He and Guo, 2021).

ii. **Automotive**: Anti-corrosion, anti-fouling, and stimuli-responsive nanocoatings influence the effectiveness of vehicles (Chen et al., 2023). For instance, anti-corrosion nanocoatings can safeguard car bodies from rust, while anti-fouling nanocoatings can prevent the build-up of microorganisms on surfaces (Zhou et al., 2022).

iii. **Electronics**: Nanocoatings that respond to stimuli can enhance the efficiency and dependability of electronic devices (Sathe et al., 2017). Temperature-sensitive nanocoatings can help regulate the heat of electronic devices, while humidity-sensitive nanocoatings can prevent the growth of microorganisms on surfaces.

iv. **Clothing**: Silica, silver, and titanium dioxide nanoparticles are frequently used to manufacture fabrics to deter moisture, eliminate bacteria, and inhibit creases. These beneficial properties make them popular for various clothing items, including raincoats, T-shirts, pants, and socks (Farooq et al., 2022).

v. **Healthcare**: Medical devices can benefit from nanocoatings that are anti-fouling, anti-icing, and stimulus-responsive (Blum et al., 2015). Anti-fouling nanocoatings prevent microorganism accumulation, while anti-icing nanocoatings prevent surface ice formation (He and Guo, 2021; Lin et al., 2018).

vi. **Food Industry**: Nanocoatings improves the taste and appearance of food, aiding in processing, and preventing harmful microorganisms from growing, which extends the shelf life (Bao et al., 2019). Additionally, it contributes to the development of safer packaging solutions, reduces the use of pesticides, and provides essential nutrients to crops. Food science has become increasingly reliant on nanotechnology and its benefits (Cui et al., 2023; Nile et al., 2020).

vii. **Energy**: Self-healing, anti-corrosion, and stimuli-responsive nanocoatings can enhance the performance and lifespan of energy-related infrastructure, including wind turbines and solar panels (Sathe et al., 2017). For instance, self-healing nanocoatings can fix damaged surfaces, while anti-corrosion nanocoatings can shield metal structures from corrosion (Saini and Bhatt, 2020).

viii. **Furniture**: Household furniture can be safeguarded from pests and fungi using silver, copper, and zinc nanoparticles and titanium dioxide particles to effective repel of dust and contaminants (Abdel-Karim and Waheed, 2013). Coatings that contain nanomaterials have the added benefit of extending furniture lifespan and reducing maintenance expenses. Furthermore, using carbon nanofibers can decrease upholstered furniture's flammability by as much as 35%

Smart and intelligent nanocoatings can transform numerous industries with advanced features. These coatings have the potential to enhance performance, safety, and durability while reducing maintenance expenses and environmental effects. As this technology progresses, we anticipate widespread implementation of these coatings in various applications.

CHALLENGES AND SOLUTIONS OF IMPLEMENTING NANOCOATINGS

Nanocoatings have gained significant attention recently due to their unique properties and potential applications in various industries. However, implementing these coatings on a large scale poses several challenges that must be addressed. This section will discuss the challenges and solutions of implementing nanocoatings in different industries.

1. **Scalability and Cost-Effectiveness**: Scaling up the production of nanocoatings poses a significant challenge in meeting the demands of industries. The cost of production and the requirement for large-scale manufacturing facilities are significant barriers to their widespread adoption. To overcome these challenges, developing cost-effective and scalable manufacturing processes is necessary.
2. **Toxicity and Environmental Impact**: Using nanomaterials in coatings has raised concerns about potential toxicity and environmental impact. One challenge is to ensure that nanocoatings are thoroughly tested and evaluated for their effects on human health and the environment before they can be widely adopt. It is important to address potential risks and ensure that nanocoatings are safe (Plazas-Tuttle et al., 2015).
3. **Interoperability and Standardization**: Ensuring the interoperability and compatibility of nanocoatings is a significant challenge due to varying specifications and requirements among manufacturers and industries. To address this, industry standards and specifications for nanocoatings need to be developed.
4. **Regulatory Framework**: The rules and regulations concerning nanocoatings are still developing, and there is a necessity for more guidance on their usage in various industries. Governments and regulatory agencies must collaborate to establish a comprehensive regulatory framework for safe, effective, and environmentally friendly.
5. **Public Awareness and Acceptance**: Increasing public knowledge and acceptance of nanocoatings is necessary. Some individuals must know these coatings' advantages and possible uses, resulting in limited usage. To overcome

this issue, we require more investment in educational and awareness initiatives that encourage the use of nanocoatings across various industries.

Despite facing some obstacles, the outlook for utilizing nanocoatings in multiple sectors appears promising. These coatings present many benefits, such as increased toughness, scratch resistance, and self-cleaning properties, rendering them an attractive option for various applications. Moving forward, we will delve into the potential implementation of nanocoatings across various industries.

EVALUATING THE PERFORMANCE OF SMART NANOCOATINGS

Smart nanocoatings are advanced materials that offer a range of properties and functions, such as self-healing, anti-microbial, and temperature-sensing capabilities. Evaluating the performance of these coatings is crucial to ensure their effectiveness and reliability in various applications. This section discusses the methodologies used to evaluate the performance of smart nanocoatings, including mechanical testing, adhesion testing, durability testing, and environmental stability assessment (Plazas-Tuttle et al., 2015). Additionally, standardized testing protocols are highlighted to ensure accurate and reliable results.

1. **Mechanical Testing**: Mechanical testing is essential to evaluate the physical properties of smart nanocoatings, such as their hardness, elasticity, and fracture toughness (Saha et al., 2021). These properties are critical in determining the coating's potential to sustain adverse effects and tear and its resistance to scratches and cracks. Standard mechanical testing methods include:
    i. Indentation testing: This method uses a sharp probe to indent the coating's surface and measure its resistance to deformation.
    ii. Scratch testing: This method involves scratching the coating's surface with a sharp object and measuring the depth and angle of the scratch.
    iii. Flexure testing: This method involves bending the coated surface to measure flexural strength and modulus.
    iv. Tensile testing: This method stretches the coated surface to measure its tensile strength and elongation.
2. **Adhesion Testing**: Adhesion testing is crucial to evaluating the bonding strength between the smart nanocoating and the substrate material. This testing is essential for applications where the coating must adhere to a specific surface, such as in medical devices or aerospace applications (Blum et al., 2015). Standard adhesion testing methods include:

   i. Peel testing: This method measures the force required to peel the coating off the substrate.
   ii. Shear testing: This method measures the force required to shear the coating of the substrate.
   iii. Adhesion testing using a scratch test involves scratching the coating's surface and measuring the force required to remove the scratch.
3. **Durability Testing**: Durability testing is essential to evaluating the long-term performance of smart nanocoatings under various environmental conditions. This testing is essential for applications where the coating must remain effective over an extended period, such as in construction or automotive applications. Standard durability testing methods include:
   i. Accelerated aging testing: This method involves exposing the coating to various environmental conditions, such as temperature, humidity, and UV radiation, to simulate the effects of aging (Bao et al., 2019).
   ii. Cyclic testing: This method involves subjecting the coating to repeated cycles of stress and strain to simulate real-world conditions.
   iii. Environmental stability assessment: This method involves evaluating the coating's stability under different environmental conditions, such as temperature, humidity, and chemical exposure (Plazas-Tuttle et al., 2015).
4. **Environmental Stability Assessment**: Environmental stability assessment is critical to evaluating the performance of smart nanocoatings under various environmental conditions. This testing is essential for applications where the coating must remain effective over an extended period, such as in construction or automotive applications. Standard environmental stability assessment methods include:
   i. Temperature testing: This method involves exposing the coating to different temperatures to evaluate its thermal stability.
   ii. Humidity testing: This method involves exposing the coating to different humidity levels to evaluate its resistance to moisture.
   iii. Chemical exposure testing: This method involves exposing the coating to different chemicals to evaluate its resistance to chemical corrosion

Importance of Standardized Testing Protocols

When evaluating smart nanocoatings, having standardized testing protocols in place is essential. These protocols help ensure that results are accurate and reliable while minimizing variability and ensuring the safety of testers and the environment. Testing methods like mechanical testing, adhesion testing, durability testing, and environmental stability assessment are all essential for evaluating the performance of these coatings. Manufacturers and researchers can use these methods and protocols to ensure that smart nanocoatings are effective and reliable, ultimately improving safety and performance in various applications.

COMPUTATIONAL MODELING AND SIMULATION TECHNIQUES IN STUDYING SMART NANOCOATINGS

Smart nanocoatings exhibit unique properties and functions, such as self-healing, adaptive behaviour, and responsiveness to environmental stimuli. Computational modelling and simulation techniques have played a crucial role in understanding the behaviour and performance of these nanocoatings at the molecular or atomic level. This part will explore the challenges and opportunities associated with computational approaches and highlight the benefits of using these tools to study smart nanocoatings.

Challenges and Opportunities of Computational Modeling

Smart nanocoatings pose a significant challenge in computational modeling due to their inherent complexity. These coatings consist of multiple layers of materials with varying properties and behaviors, with several factors affecting their functioning. Additionally, their molecular and atomic structures are susceptible to their surroundings, making it difficult to model their behavior precisely. However, despite these challenges, computational modeling and simulation techniques provide numerous opportunities for studying and understanding smart nanocoatings. These techniques allow researchers to:

- Design and optimize nanocoatings with specific properties and functions that can be used in simulations to test different material combinations and structures, identifying the most promising candidates for further experimentation.
- Study the behavior of nanocoatings under various stages, which can help researchers understand how the coatings respond to their environment and how they can be optimized for specific applications.
- Predict the performance of nanocoatings in different scenarios, such as exposure to chemicals or extreme temperatures, which can help manufacturers and engineers design and test coatings better suited for specific applications (Thakur and Kumar, 2022).
- Identify defects and weaknesses in nanocoatings, allowing researchers to develop strategies to improve their performance and longevity.

Benefits of Computational Modeling

Using computational modeling and simulation techniques to study smart nanocoatings has numerous benefits. Some of the key benefits include:

- Cost-effective way to study the behavior of nanocoatings, as it eliminates the need for expensive and time-consuming experiments.
- Fast and accurate methods allow researchers to study nanocoatings' behavior quickly under a wide range of conditions.
- Scalability is measured by scaling up or down to study the behavior of nanocoatings at different length scales, from the molecular to the macroscale.
- Insights into the material properties and behavior of nanocoatings at the molecular and atomic levels allow researchers to understand how the materials respond to their environment.

In conclusion, computational modeling and simulation techniques have proven crucial in understanding the behavior and capabilities of intelligent nanocoatings on a molecular or atomic level. These methods offer many possibilities, including designing and enhancing these coatings, comprehending their actions, predicting their performance, and identifying faults or vulnerabilities. The advantages of employing these techniques to investigate smart nanocoatings are numerous, such as cost-effectiveness, rapid and accurate results, scalability, and the acquisition of knowledge regarding material properties.

SMART AND INTELLIGENT NANOCOATING MARKET OPPORTUNITIES

The demand for advanced materials and technologies in various applications has rapidly expanded the market for smart and intelligent nanocoatings (Cui et al., 2023; Nile et al., 2020). According to reports by (Idumah et al., 2020; Thakur and Kumar, 2022), the global market for smart nanocoatings is projected to increase by 18.5% from 2020 to 2025, reaching $10.3 billion. This growth is driven by an increasing need for improved performance, durability, and sustainability in various applications that require advanced materials and technologies (Barhoum et al., 2019). The use of nanotechnology in industries like electronics, energy, and healthcare is also fueling demand for nanocoatings (Cui et al., 2023). Moreover, governments and companies invest heavily in research and development to develop new and improved nanocoatings with advanced properties. Additionally, the demand for surfaces that can self-heal or self-clean drives the smart and intelligent nanocoating market (Vijayan and Puglia, 2019).

The smart and intelligent nanocoatings market comprises three categories: type, application, and region (Akhila and Badwaik, 2022; Nguyen-Tri et al., 2019). The types of nanocoatings available are smart, intelligent, and hybrid, and they offer a range of benefits, such as sensing and communication capabilities, self-healing properties, and the ability to adapt to changing conditions (Barhoum et al., 2019). The market is divided around the globe, with many players striving to capture market share, and the competition is fierce. However, it is to be noted that some of the essential players include PPG Industries, Inc., AkzoNobel N.V., DuPont, Sherwin-Williams Company, Nippon Paint Holdings Co., Ltd., Jotun, RPM International Inc., Axalta Coating Systems, BASF SE, and Henkel AG & Co. KGaA. As technology advances, there is a growing emphasis on investing in research and development to make nanocoatings more accessible (Akhila and Badwaik, 2022). This has created lucrative market possibilities since nanocoatings have many applications, including construction, vehicles, defense equipment, airplanes, and boats. Thanks to extensive research and development, nanocoatings that are self-cleaning, water-resistant, and reduce friction have been developed (Ulaeto et al., 2019).

Conclusively, the smart and intelligent nanocoating market is experiencing rapid expansion, attributed to the increasing demand for advanced materials and technologies. This sector is categorized by type, application, and region and is highly competitive, with numerous significant players competing for a market share. As interest surges for surfaces that possess self-healing and self-cleaning properties, coupled with escalated investment in research and development, this industry is predicted to sustain growth in the future.

CRITICAL ANALYSIS OF EXISTING LITERATURE ON SMART AND INTELLIGENT NANOCOATINGS

In recent years, there has been a growing scientific interest in smart and intelligent nanocoatings because of their potential applications in various fields like energy, healthcare, and environmental protection. However, a thorough analysis of the existing literature has revealed gaps in knowledge, findings that need to be more consistent, and areas that require further research.

Gaps in Knowledge

- Lack of Standardization: There needs to be more standardization in the definition, synthesis, and characterization of smart and intelligent nanocoatings, which hinders the comparability and reproducibility of results across different studies (Ulaeto et al., 2019).

- Limited Understanding of Nanocoating-Substrate Interactions: The interactions between nanocoatings and substrate materials need to be better understood, which can affect the performance and stability of the nanocoatings.
- Insufficient Data on Toxicity and Environmental Impact: There is a need for more data on the toxicity and environmental impact of smart and intelligent nanocoatings, particularly in the long term.

Inconsistencies in Findings

- Variability in Nanocoating Properties: The properties of smart and intelligent nanocoatings can vary depending on the synthesis method, composition, and surface modification, leading to inconsistent results in different studies.
- Discrepancies in Characterization Techniques: Different characterization techniques can produce conflicting results, highlighting the need for standardized methods and protocols.
- Inconsistent Definition of Smart and Intelligent Nanocoatings: The terms “smart” and “intelligent” are often used interchangeably, but they have different meanings and implications for the properties and applications of nanocoatings.

Controversies and Debates

- Nanotoxicity: There is ongoing debate about the potential toxicity of nanomaterials, including nanocoatings, which can affect their use in various applications.
- Environmental Impact: The environmental impact of nanocoatings, especially their potential to leach or dispose needs further attention.
- Patent Issues: The patenting of nanocoating technologies can create controversy and hinder the development of new and improved coatings.

Different Perspectives

- Materials Science Perspective: Researchers from a materials science perspective may focus on the synthesis, characterization, and properties of nanocoatings while neglecting their potential applications and environmental impact.

- Application-Oriented Perspective: Application-oriented researchers may focus on the specific uses of nanocoatings, such as in energy or healthcare, without considering their broader implications and limitations.
- Environmental Perspective: Environmental researchers may focus on the potential environmental impact of nanocoatings, such as their disposal and leaching, without considering their potential benefits and applications.

In conclusion, while smart and intelligent nanocoatings have shown great promise in various fields, a critical analysis of the existing literature reveals several gaps in knowledge, inconsistencies in findings, and areas that require further research. Addressing these issues will be crucial for developing safe, effective, and sustainable nanocoatings for various applications

CONCLUSION AND FUTURE TRENDS

In conclusion, this book chapter provides a detailed overview of current research in the smart and intelligent nanocoatings where we emphasize the importance of this research and suggest new directions for further investigation. Developing smart and intelligent nanocoatings has significantly enhanced various industries' performance, durability, and safety. For instance, self-healing coatings could prolong the lifespan of structures and lower maintenance costs. Self-cleaning coatings could enhance solar panel efficiency and reduce water usage. Anti-icing coatings could make aircraft and cars safer during icing conditions. Sensing coatings could enable real-time monitoring of environmental conditions and structural health. The prospects for smart and intelligent nanocoatings also involve addressing scalability, cost-effectiveness, and regulatory compliance challenges. As research progresses, it is anticipated that advancements in manufacturing processes and formulation techniques will lead to commercializing these coatings on a larger scale, making them more accessible to various industries. Therefore, the authors recommend that future research focus on reducing costs and scaling up production as well as integrating smart and intelligent nanocoatings with existing technologies, such as sensors and actuators, to create more advanced systems. This could be achieved through the use of affordable materials or the development of new synthesis methods.

Funding: This study was funded by the Major Science and Technology projects of Henan Province (221100320200) and supported by Henan Center for Outstanding Overseas Scientists (No. GZS2021007).

REFERENCES

Abbasi, H., Antunes, M., & Velasco, J. I. (2020). Effects of Graphene Nanoplatelets and Cellular Structure on the Thermal Conductivity of Polysulfone Nanocomposite Foams. Polymers.

Abdeen, D. H., El Hachach, M., Koc, M., & Atieh, M. A. (2019). A Review on the Corrosion Behaviour of Nanocoatings on Metallic Substrates. Materials. doi:10.3390/ma12020210

Abdel-Karim, R., & Waheed, A. F. (2013). Nanocoatings. In Modern Surface Engineering Treatments. IntechOpen.

Abdulraheem, M. I., Zhang, W., Li, S., Moshayedi, A. J., Farooque, A. A., & Hu, J. (2023). Advancement of Remote Sensing for Soil Measurements and Applications: A Comprehensive Review. Sustainability. doi:10.3390/su152115444

Akhila, V., & Badwaik, L. S. (2022). Recent advancement in improvement of properties of polysaccharides and proteins based packaging film with added nanoparticles: A review. *International Journal of Biological Macromolecules*, *203*, 515–525. doi:10.1016/j.ijbiomac.2022.01.181 PMID:35122798

Andreeva, D. V., Skorb, E. V., & Shchukin, D. G. (2010). Layer-by-Layer Polyelectrolyte/Inhibitor Nanostructures for Metal Corrosion Protection. *ACS Applied Materials & Interfaces*, *2*(7), 1954–1962. doi:10.1021/am1002712

Ashfaq, A., Khursheed, N., Fatima, S., Anjum, Z., & Younis, K. (2022). Application of nanotechnology in food packaging: Pros and Cons. *Journal of Agriculture and Food Research*, *7*, 100270. doi:10.1016/j.jafr.2022.100270

Bao, W., Deng, Z., Zhang, S., Ji, Z., & Zhang, H. (2019). Next-Generation Composite Coating System: Nanocoating. *Frontiers in Materials*, 6.

Barhoum, A., Pal, K., Rahier, H., Uludag, H., Kim, I. S., & Bechelany, M. (2019). Nanofibers as new-generation materials: From spinning and nano-spinning fabrication techniques to emerging applications. *Applied Materials Today*, *17*, 1–35. doi:10.1016/j.apmt.2019.06.015

Barish, J. A., & Goddard, J. M. (2013). Anti-fouling surface modified stainless steel for food processing. *Food and Bioproducts Processing*, *91*(4), 352–361. doi:10.1016/j.fbp.2013.01.003

Bastarrachea, L. J., Denis-Rohr, A., & Goddard, J. M. (2015a). Antimicrobial food equipment coatings: Applications and challenges. *Annual Review of Food Science and Technology*, *6*(1), 97–118. doi:10.1146/annurev-food-022814-015453 PMID:25422880

Bastarrachea, L. J., Wong, D. E., Roman, M. J., Lin, Z., & Goddard, J. M. (2015b). Active Packaging Coatings. Coatings, 771-791. doi:10.3390/coatings5040771

Berger, L.-M. (2015). Application of hardmetals as thermal spray coatings. *International Journal of Refractory & Hard Metals*, *49*, 350–364. doi:10.1016/j.ijrmhm.2014.09.029

Bhatt, S., Pulpytel, J., & Arefi-Khonsari, F. (2015). *Low and atmospheric plasma polymerisation of nanocoatings for bio-applications*. Academic Press.

Blum, A. P., Kammeyer, J. K., Rush, A. M., Callmann, C. E., Hahn, M. E., & Gianneschi, N. C. (2015). Stimuli-Responsive Nanomaterials for Biomedical Applications. *Journal of the American Chemical Society*, *137*(6), 2140–2154. doi:10.1021/ja510147n PMID:25474531

Bugnicourt, E., Kehoe, T., Latorre, M., Serrano, C., Philippe, S., & Schmid, M. (2016). Recent Prospects in the Inline Monitoring of Nanocomposites and Nanocoatings by Optical Technologies. Nanomaterials. doi:10.3390/nano6080150

Cabane, E., Zhang, X., Langowska, K., Palivan, C. G., & Meier, W. (2012). Stimuli-Responsive Polymers and Their Applications in Nanomedicine. *Biointerphases*, *7*(1), 9. doi:10.1007/s13758-011-0009-3 PMID:22589052

Cagri, A., Ustunol, Z., & Ryser, E. T. (2004). Antimicrobial edible films and coatings. *Journal of Food Protection*, *67*(4), 833–848. doi:10.4315/0362-028X-67.4.833 PMID:15083740

Cao, L., Huang, Y., Parakhonskiy, B., & Skirtach, A. G. (2022). Nanoarchitectonics beyond perfect order–not quite perfect but quite useful. *Nanoscale*, *14*(43), 15964–16002. doi:10.1039/D2NR02537J PMID:36278502

Cerqueira, M. Â., Pinheiro, A. C., Ramos, O. L., Silva, H., Bourbon, A. I., & Vicente, A. A. (2017). Chapter Two - Advances in Food Nanotechnology. In R. Busquets (Ed.), *Emerging Nanotechnologies in Food Science* (pp. 11–38). Elsevier. doi:10.1016/B978-0-323-42980-1.00002-9

Chandrasekar, M. S., Shanmugasigamani, & Malathy, P. (2010). Synergetic effects of pulse constraints and additives in electrodeposition of nanocrystalline zinc: Corrosion, structural and textural characterization. *Materials Chemistry and Physics, 124*(1), 516–528. doi:10.1016/j.matchemphys.2010.07.004

Chen, X., Chen, J., Zhu, J., Cai, T., & Guo, Z. (2021). Investigation on the Electrochemical Deposition of Nanocrystalline Zinc with Cationic Polyacrylamide (CPAM)-ZnSO(4) Electrolyte. *Micromachines*, 12. PMID:34577763

Chen, Z., Scharnagl, N., Zheludkevich, M. L., Ying, H., & Yang, W. (2023). Micro/nanocontainer-based intelligent coatings: Synthesis, performance and applications – A review. *Chemical Engineering Journal, 451*, 138582. doi:10.1016/j.cej.2022.138582

Cordeiro, A. L. (2011). Stimuli-Responsive Polymer Nanocoatings. Nanotechnologies for the Life Sciences.

Coyle, S., Wu, Y., Lau, K.-T., De Rossi, D., Wallace, G., & Diamond, D. (2007). Smart Nanotextiles: A Review of Materials and Applications. *MRS Bulletin, 32*(5), 434–442. doi:10.1557/mrs2007.67

Cui, H., You, Y., Cheng, G.-W., Lan, Z., Zou, K.-L., Mai, Q.-Y., Han, Y.-H., Chen, H., Zhao, Y.-Y., & Yu, G.-T. (2023). Advanced materials and technologies for oral diseases. *Science and Technology of Advanced Materials, 24*(1), 2156257. doi:10.1080/14686996.2022.2156257 PMID:36632346

de Villiers, M. M., Otto, D. P., Strydom, S. J., & Lvov, Y. M. (2011). Introduction to nanocoatings produced by layer-by-layer (LbL) self-assembly. *Advanced Drug Delivery Reviews, 63*(9), 701–715. doi:10.1016/j.addr.2011.05.011 PMID:21699936

del Mercato, L. L., Ferraro, M. M., Baldassarre, F., Mancarella, S., Greco, V., Rinaldi, R., & Leporatti, S. (2014). Biological applications of LbL multilayer capsules: From drug delivery to sensing. *Advances in Colloid and Interface Science, 207*, 139–154. doi:10.1016/j.cis.2014.02.014 PMID:24625331

Dufrêne, Y. F., Ando, T., Garcia, R., Alsteens, D., Martinez-Martin, D., Engel, A., Gerber, C., & Müller, D. J. (2017). Imaging modes of atomic force microscopy for application in molecular and cell biology. *Nature Nanotechnology, 12*(4), 295–307. doi:10.1038/nnano.2017.45 PMID:28383040

Farag, A.A. (2020). *Applications of nanomaterials in corrosion protection coatings and inhibitors*. Academic Press.

Farooq, S. A., Raina, A., Mohan, S., Arvind Singh, R., Jayalakshmi, S., & Irfan Ul Haq, M. (2022). Nanostructured Coatings: Review on Processing Techniques, Corrosion Behaviour and Tribological Performance. *Nanomaterials (Basel, Switzerland)*, *12*(8), 12. doi:10.3390/nano12081323 PMID:35458032

Girigoswami, A., Thirumalai, A., Karthick, H., Pallavi, P., Gowtham, P., & Girigoswami, A. (2022). Nanotechnology driven improvement of smart food packaging. *Materials Research Innovations*.

Gu, Y., Xia, K., Wu, D., Mou, J., & Zheng, S. (2020). Technical Characteristics and Wear-Resistant Mechanism of Nano Coatings: A Review. Coatings. doi:10.3390/coatings10030233

Han, J., Ma, Q., An, Y., Wu, F., Zhao, Y., Wu, G., & Wang, J. (2023). The current status of stimuli-responsive nanotechnologies on orthopedic titanium implant surfaces. *Journal of Nanobiotechnology*, *21*(1), 277. doi:10.1186/s12951-023-02017-8 PMID:37596638

He, H., & Guo, Z. (2021). Superhydrophobic materials used for anti-icing Theory, application, and development. *iScience*, *24*(11), 103357. doi:10.1016/j.isci.2021.103357 PMID:34825136

He, Q., Paiva, J. M., Kohlscheen, J., Beake, B. D., & Veldhuis, S. C. (2020). An integrative approach to coating/carbide substrate design of CVD and PVD coated cutting tools during the machining of austenitic stainless steel. *Ceramics International*, *46*(4), 5149–5158. doi:10.1016/j.ceramint.2019.10.259

Idumah, C. I., Obele, C. M., Emmanuel, E. O., & Hassan, A. (2020). Recently Emerging Nanotechnological Advancements in Polymer Nanocomposite Coatings for Anti-corrosion, Anti-fouling and Self-healing. *Surfaces and Interfaces*, *21*, 100734. doi:10.1016/j.surfin.2020.100734 PMID:34957345

Jagtiani, E. (2022). Advancements in nanotechnology for food science and industry. *Food Frontiers*, *3*(1), 56–82. doi:10.1002/fft2.104

Kausar, A., Ahmad, I., Maaza, M., & Bocchetta, P. (2023). *Self-Healing Nanocomposites&mdash.* Advancements and Aerospace Applications. In Journal of Composites Science.

Khan, S. A., Sezer, N., & Koç, M. (2020). Design, synthesis, and characterization of hybrid micro-nano surface coatings for enhanced heat transfer applications. *International Journal of Energy Research*, *44*(15), 12525–12534. doi:10.1002/er.5685

Kokate, M., Garadkar, K., & Gole, A. (2013). One pot synthesis of magnetite–silica nanocomposites: Applications as tags, entrapment matrix and in water purification. *Journal of Materials Chemistry. A, Materials for Energy and Sustainability*, *1*(6), 2022–2029. doi:10.1039/C2TA00951J

Koslowski, L. A. D., Nogueira, A. L., Licodiedoff, S., Comper, A. T., & Folgueras, M. V. (2018). Silver nanoparticles impregnated with polyamide-66 to disinfect drinking water. *Revista Ambiente & Água*, *13*(6), e1947. doi:10.4136/ambi-agua.1947

Li, C., Xia, F., Ma, C., & Li, Q. (2021). Research on the Corrosion Behavior of Ni-SiC Nanocoating Prepared Using a Jet Electrodeposition Technique. *Journal of Materials Engineering and Performance*, *30*(8), 6336–6344. doi:10.1007/s11665-021-05891-1

Li, M., Su, B., Zhou, B., Wang, H., & Meng, J. (2020). One-pot synthesis and self-assembly of anti-wear octadecyltrichlorosilane/silica nanoparticles composite films on silicon. *Applied Surface Science*, *508*, 145187. doi:10.1016/j.apsusc.2019.145187

Li, Q., Lu, H., Cui, J., An, M., & Li, D. (2016). Electrodeposition of nanocrystalline zinc on steel for enhanced resistance to corrosive wear. *Surface and Coatings Technology*, *304*, 567–573. doi:10.1016/j.surfcoat.2016.07.056

Li, T., Liu, Y., Liu, B., Guo, W., & Xu, L. (2017). Microstructure and Wear Behavior of FeCoCrNiMo0.2 High Entropy Coatings Prepared by Air Plasma Spray and the High Velocity Oxy-Fuel Spray Processes. Coatings.

Lin, Y., Chen, H., Wang, G., & Liu, A. (2018). Recent Progress in Preparation and Anti-Icing Applications of Superhydrophobic Coatings. Coatings. doi:10.3390/coatings8060208

Malik, S., Muhammad, K., & Waheed, Y. (2023). Nanotechnology: A Revolution in Modern Industry. *Molecules (Basel, Switzerland)*, 28. PMID:36677717

Martinez, G., Shutthanandan, V., Thevuthasan, S., Chessa, J. F., & Ramana, C. V. (2014). Effect of thickness on the structure, composition and properties of titanium nitride nano-coatings. *Ceramics International*, *40*(4), 5757–5764. doi:10.1016/j.ceramint.2013.11.014

Mohammed, M. T., & Hussein, S. M. (2019). Synthesis and characterization of nanocoatings derived by sol–gel onto a new surgical titanium surface. *Materials Research Express*, *6*(7), 076426. doi:10.1088/2053-1591/ab1722

Moncada, E., Quijada, R., & Retuert, J. (2007). Nanoparticles prepared by the sol–gel method and their use in the formation of nanocomposites with polypropylene. *Nanotechnology*, *18*(33), 335606. doi:10.1088/0957-4484/18/33/335606

More, S. R., Bhatt, D. V., & Menghani, J. V. (2017). Resent Research Status on Laser Cladding as Erosion Resistance Technique - An Overview. *Materials Today: Proceedings*, *4*(9), 9902–9908. doi:10.1016/j.matpr.2017.06.291

Motelica, L., Ficai, D., Oprea, O. C., Ficai, A., & Andronescu, E. (2020). Smart Food Packaging Designed by Nanotechnological and Drug Delivery Approaches. Coatings. doi:10.3390/coatings10090806

Mustafa, F., & Andreescu, S. (2020). Nanotechnology-based approaches for food sensing and packaging applications. *RSC Advances*, *10*(33), 19309–19336. doi:10.1039/D0RA01084G PMID:35515480

Naqvi, S. M. Z. A., Zhang, Y., Tahir, M. N., Ullah, Z., Ahmed, S., Wu, J., Raghavan, V., Abdulraheem, M. I., Ping, J., Hu, X., & Hu, J. (2023). Advanced strategies of the in-vivo plant hormone detection. *Trends in Analytical Chemistry*, *166*, 117186. doi:10.1016/j.trac.2023.117186

Nazeer, A. A., & Madkour, M. (2018). Potential use of smart coatings for corrosion protection of metals and alloys: A review. *Journal of Molecular Liquids*, *253*, 11–22. doi:10.1016/j.molliq.2018.01.027

Nguyen-Tri, P., Tran, H. N., Plamondon, C. O., Tuduri, L., Vo, D.-V. N., Nanda, S., Mishra, A., Chao, H.-P., & Bajpai, A. K. (2019). Recent progress in the preparation, properties and applications of superhydrophobic nano-based coatings and surfaces: A review. *Progress in Organic Coatings*, *132*, 235–256. doi:10.1016/j.porgcoat.2019.03.042

Nile, S. H., Baskar, V., Selvaraj, D., Nile, A., Xiao, J., & Kai, G. (2020). Nanotechnologies in Food Science: Applications, Recent Trends, and Future Perspectives. *Nano-Micro Letters*, *12*(1), 45. doi:10.1007/s40820-020-0383-9 PMID:34138283

Noah, N. M. (2020). Design and Synthesis of Nanostructured Materials for Sensor Applications. *Journal of Nanomaterials*, *2020*, 8855321. doi:10.1155/2020/8855321

Ortega-Nieto, C., Losada-Garcia, N., Pessela, B. C., Domingo-Calap, P., & Palomo, J. M. (2023). Design and Synthesis of Copper Nanobiomaterials with Antimicrobial Properties. *ACS Bio & Med Chem Au*, *3*(4), 349–358. doi:10.1021/acsbiomedchemau.2c00089 PMID:37599792

Petrik, P., Racz, A. S., & Menyhard, M. (2020). Complementary physicochemical analysis by ellipsometry and Auger spectroscopy of nano-sized protective coating layers. *Applied Surface Science*, *534*, 147593. doi:10.1016/j.apsusc.2020.147593

Pham, S. H., Choi, Y., & Choi, J. (2020). Stimuli-Responsive Nanomaterials for Application in Antitumor Therapy and Drug Delivery. *Pharmaceutics*, *12*(7), 12. doi:10.3390/pharmaceutics12070630 PMID:32635539

Plazas-Tuttle, J., Rowles, L. S., Chen, H., Bisesi, J. H., Sabo-Attwood, T., & Saleh, N. B. (2015). *Dynamism of Stimuli-Responsive Nanohybrids*. Environmental Implications. In Nanomaterials.

Prasad, S., Kumar, V., Kirubanandam, S., & Barhoum, A. (2018). Engineered nanomaterials: nanofabrication and surface functionalization. In A. Barhoum & A. S. H. Makhlouf (Eds.), *Emerging Applications of Nanoparticles and Architecture Nanostructures* (pp. 305–340). Elsevier. doi:10.1016/B978-0-323-51254-1.00011-7

Ramasubramanian, K., Arunachalam, N., & Ramachandra Rao, M. S. (2019). Wear performance of nano-engineered boron doped graded layer CVD diamond coated cutting tool for machining of Al-SiC MMC. *Wear*, *426-427*, 1536–1547. doi:10.1016/j.wear.2018.12.004

Reggente, M., Rossi, M., Angeloni, L., Tamburri, E., Lucci, M., Davoli, I., Terranova, M. L., & Passeri, D. (2015). Atomic Force Microscopy Techniques for Nanomechanical Characterization: A Polymeric Case Study. *Journal of the Minerals Metals & Materials Society*, *67*(4), 849–857. doi:10.1007/s11837-015-1340-9

Ros-Lis, J. V., & Benitez Serra, M. (2023). Recent advancements in smart and intelligent packaging technologies: an overview. In B. P. Singh, S. Agnihotri, G. Singh, & V. K. Gupta (Eds.), *Postharvest Management of Fresh Produce* (pp. 241–263). Academic Press. doi:10.1016/B978-0-323-91132-0.00010-1

Sadabadi, H., Allahkaram, S. R., Kordijazi, A., & Rohatgi, P. K. (2022). Self-healing Coatings Loaded by Nano/microcapsules: A Review. *Protection of Metals and Physical Chemistry of Surfaces*, *58*(2), 287–307. doi:10.1134/S2070205122020162

Saha, D. R., Kumar, K. R., Ansari, F., Ghosh, S. K., Datta, A., & Chakravorty, D. (2021). Stimuli-responsive coating by simple physical blending route. *Ceramics International*, *47*(18), 26357–26365. doi:10.1016/j.ceramint.2021.06.046

Saini, J., & Bhatt, R. (2020). Global warming-causes, impacts and mitigation strategies in agriculture. *Current Journal of Applied Science and Technology*, *39*, 93–107. doi:10.9734/cjast/2020/v39i730580

Sarode, A., Annapragada, A., Guo, J., & Mitragotri, S. (2020). Layered self-assemblies for controlled drug delivery: A translational overview. *Biomaterials*, *242*, 119929. doi:10.1016/j.biomaterials.2020.119929 PMID:32163750

Sathe, P., Laxman, K., Myint, M. T. Z., Dobretsov, S., Richter, J., & Dutta, J. (2017). Bioinspired nanocoatings for biofouling prevention by photocatalytic redox reactions. *Scientific Reports*, *7*(1), 3624. doi:10.1038/s41598-017-03636-6 PMID:28620218

Shah, M. A., Pirzada, B. M., Price, G., Shibiru, A. L., & Qurashi, A. (2022). Applications of nanotechnology in smart textile industry: A critical review. *Journal of Advanced Research*, *38*, 55–75. doi:10.1016/j.jare.2022.01.008 PMID:35572402

Sharma, C., Dhiman, R., Rokana, N., & Panwar, H. (2017). Nanotechnology: An Untapped Resource for Food Packaging. *Frontiers in Microbiology*, *8*, 8. doi:10.3389/fmicb.2017.01735 PMID:28955314

Singh, A., Mittal, S., Mudgal, D., & Gupta, P. (2018). Design, Development and Application of Nanocoatings. In Z. H. Khan (Ed.), *Nanomaterials and Their Applications* (pp. 191–207). Springer Singapore. doi:10.1007/978-981-10-6214-8_7

Singh, R., Dutt, S., Sharma, P., Sundramoorthy, A. K., Dubey, A., Singh, A., & Arya, S. (2023). Future of Nanotechnology in Food Industry: Challenges in Processing, Packaging, and Food Safety. *Global Challenges (Hoboken, NJ)*, *7*(4), 2200209. doi:10.1002/gch2.202200209 PMID:37020624

Smirnova, V. V., Krasnoiarova, O. V., Pridvorova, S. M., Zherdev, A. V., Gmoshinskiĭ, I. V., Kazydub, G. V., Popov, K. I., & Khotimchenko, S. A. (2012). Characterization of silver nanoparticles migration from package materials destined for contact with foods. *Voprosy Pitaniia*, *81*, 34–39. PMID:22774476

Sun, Z., Yang, L., Zhang, D., & Song, W. (2019). High performance, flexible and renewable nano-biocomposite artificial muscle based on mesoporous cellulose/ ionic liquid electrolyte membrane. *Sensors and Actuators. B, Chemical*, *283*, 579–589. doi:10.1016/j.snb.2018.12.073

Tang, K., Xue, J., Zhu, Y., & Wu, C. (2023). Design and synthesis of bioinspired nanomaterials for biomedical application. *WIREs Nanomedicine and Nanobiotechnology*.

Thakur, A., & Kumar, A. (2022). Self-healing nanocoatings for automotive application. In H. Song, T. A. Nguyen, G. Yasin, N. B. Singh, & R. K. Gupta (Eds.), *Nanotechnology in the Automotive Industry* (pp. 403–427). Elsevier. doi:10.1016/B978-0-323-90524-4.00019-0

Trentin, A., Uvida, M. C., de Araújo Almeida, A., de Souza, T. A. C., & Hammer, P. (2022). Self-healing nanocoatings. In H. Song, T. A. Nguyen, G. Yasin, N. B. Singh, & R. K. Gupta (Eds.), *Nanotechnology in the Automotive Industry* (pp. 371–401). Elsevier. doi:10.1016/B978-0-323-90524-4.00018-9

Ulaeto, S. B., Pancrecious, J. K., Rajan, T. P. D., & Pai, B. C. (2019). Smart Coatings. In S. Mohapatra, T. A. Nguyen, & P. Nguyen-Tri (Eds.), *Noble Metal-Metal Oxide Hybrid Nanoparticles* (pp. 341–372). Woodhead Publishing. doi:10.1016/B978-0-12-814134-2.00017-6

Vasile, C. (2018). *Polymeric Nanocomposites and Nanocoatings for Food Packaging: A Review*. Materials. doi:10.3390/ma11101834

Vijayan, P. P., & Puglia, D. (2019). Biomimetic multifunctional materials: A review. *Emergent Materials*, *2*(4), 391–415. doi:10.1007/s42247-019-00051-7

Vishwakarma, V. (2020). Synthesis, design, and characterization of controlled nanoparticles for the construction industry. In C. M. Hussain (Ed.), *Handbook of Nanomaterials for Manufacturing Applications* (pp. 323–341). Elsevier. doi:10.1016/B978-0-12-821381-0.00015-6

Voigtländer, B. (2019). Technical Aspects of Atomic Force Microscopy. In B. Voigtländer (Ed.), *Atomic Force Microscopy* (pp. 35–67). Springer International Publishing. doi:10.1007/978-3-030-13654-3_3

Wadullah, H. M., Ajeel, S. A., & Abbass, M. K. (2019). Synthesis and Characterization of Nanocoatings Thin films by Atomic Layer Deposition for Medical Applications. *IOP Conference Series. Materials Science and Engineering*, *518*(3), 032057. doi:10.1088/1757-899X/518/3/032057

Wadullah, H. M., Talib Mohammed, M., & Khalid Abdulrazzaq, T. (2022). Structure and characteristics of Nb2O5 nanocoating thin film for biomedical applications. *Materials Today: Proceedings*, *62*, 3076–3080. doi:10.1016/j.matpr.2022.03.229

Wei, H., Wang, Y., Guo, J., Shen, N. Z., Jiang, D., Zhang, X., Yan, X., Zhu, J., Wang, Q., Shao, L., Lin, H., Wei, S., & Guo, Z. (2015). Advanced micro/nanocapsules for self-healing smart anticorrosion coatings. *Journal of Materials Chemistry. A, Materials for Energy and Sustainability*, *3*(2), 469–480. doi:10.1039/C4TA04791E

Xia, F., Li, Q., Ma, C., & Guo, X. (2020). Preparation and characterization of Ni–AlN nanocoatings deposited by magnetic field assisted electrodeposition technique. *Ceramics International*, *46*(2), 2500–2509. doi:10.1016/j.ceramint.2019.09.244

Xie, Q., Hao, T., Zhang, J., Wang, C., Zhang, R., & Qi, H. (2019). Anti-Icing Performance of a Coating Based on Nano/Microsilica Particle-Filled Amino-Terminated PDMS-Modified Epoxy. Coatings.

Young, E., Mirosa, M., & Bremer, P. (2020). A Systematic Review of Consumer Perceptions of Smart Packaging Technologies for Food. *Frontiers in Sustainable Food Systems*, *4*, 63. doi:10.3389/fsufs.2020.00063

Zhou, Z., Pourhashem, S., Wang, Z., Duan, J., Zhang, R., & Hou, B. (2022). Distinctive roles of graphene oxide, ZnO quantum dots, and their nanohybrids in anti-corrosion and anti-fouling performance of waterborne epoxy coatings. *Chemical Engineering Journal*, *439*, 135765. doi:10.1016/j.cej.2022.135765

Chapter 4
Nanocoatings for Energy Generation and Conservation of Solar Cells

Rafael Vargas-Bernal
https://orcid.org/0000-0003-4865-4575
University of Guanajuato, Mexico

ABSTRACT

Solar panels operating under outdoor environments during their useful life are exposed to various factors such as dust, humidity, snow, rain, thermal changes, abrasion, etc., which degrade their operation, decreasing the efficiency of converting light into voltage. Although various cleaning and/or conservation methods have been proposed and used for years to solve this problem, they offer results with low efficiency and high costs. The most effective functional method so far is the application of nanocoatings on the solar panel cover, which avoids these effects more comprehensively. This chapter presents the study of the advances that have been achieved so far through nanocomposites based on polymers and ceramic materials and proposes future trends that must be carried out to achieve their optimized design. The advances, although significant, allow us to predict the need to continue carrying out scientific and technological research to achieve results of high-added value using nanomaterials.

INTRODUCTION

Solar energy, being the most abundant and clean renewable green energy, has the fastest-growing use at the industrial, agricultural, commercial, and residential

DOI: 10.4018/979-8-3693-3136-1.ch004

levels (Shaju, 2018; Hasan, 2022). The sun delivers 1367 W/m^2 to the atmosphere and the global absorption is in the order of 1.8×10^{11} MW worldwide. The interest in studying, designing, and improving global power sources lies in the fact that their production in 2017 was 368 GW while it is projected that in 2030 production will be in the range of 3000 to 10,000 GW (Kumar, 2018). The emerging need for electrical energy generation around the world has allowed researchers to achieve, with their advances in research and installed capacity, that solar panels are the second fastest-growing technology and the third largest as a renewable technology (Ehsan, 2022). Solar cells convert light energy from the sun into electrical power using semiconductor materials. The main disadvantage of the energy that solar cells produce is that it is intermittent as it varies depending on climatic and environmental conditions. Furthermore, at the level of the photovoltaic panel, its performance in power delivery is diminished by other factors, among which are the technology and materials used in its manufacture, the mismatch of the components that make up the solar cell array, surface shape, orientation, angle of inclination, temperature, inverter efficiency, panel fouling, encapsulation materials, electrical charges, etc. as illustrated in Figure 1. Photovoltaic modules based on solar cells may be subjected to hazardous environmental conditions such as soiling during outdoor operation (Sun, 2018; Ehsan, 2021; Cherupurakal, 2021; Hasan, 2022). The particles may be wet or dry, may contain low or high organic content, etc. so the adhesion of them to the solar panel cover is completely different depending on the environment found in the place where the panel is installed. In the desert, the wind carries both sand and dust particles towards the solar panel, which reduces energy conversion efficiency. For example, environmental dust that accumulates on the surface of the module prevents sunlight from penetrating the solar cells, which reduces the electrical power produced by the panel. To increase the power conversion efficiency (PCE) of a solar cell, it is required that its Shockley-Queisser limit be optimized to reduce reflection losses at the transparent glass-air interface (Cherupurakal, 2021; Vasilopoulou, 2022). Approximately one-third of the incident light received by a solar cell is reflected, so its efficiency needs to be increased for solar cells to be attractive as renewable energy sources. Within the technological possibilities that have been proposed to solve this technical problem, it has been proposed to place coatings on the transparent side of the solar cell (Ehsan, 2021). Among the functions that nanocoatings must improve the performance of solar cells are redirecting incident light to trap it, improving light absorption in a wide range of wavelengths, and optimizing light handling. Among the proposals that have been used to improve the efficiency of solar cells are photonic crystals, diffraction gratings, surface textures based on honeycomb structures or inverted pyramids, and metamaterials. Among the changes being introduced are a reduction in the thickness of the solar cell from 400 to 200 microns and an expansion of the solar cell area from 100 to 240 cm^2

(Kumar, 2018; Cherupurakal, 2021). Advances in solar cell technology have allowed the Shockley-Queisser limit to be overcome, making it possible to achieve a power conversion efficiency of 41% for third-generation solar cells and even a reported 65% efficiency for solar cells based on silicon nanocrystals, cadmium telluride (CdTe) or copper indium gallium selenide (CIGS). Despite the advances achieved so far, solutions based on nanomaterials are being researched to offer simpler, lighter, and more cost-effective nanocoatings. The purpose of this chapter is to review the state of the art of the main contributions based on nanomaterials to implement nanocoatings to optimize the performance of solar cells.

Nanocoatings must be multifunctional to offer multiple properties to the solar cell such as anti-reflection, hydrophobicity, anti-moisture, self-cleaning, etc. Solar cells are susceptible to the accumulation of dust and contaminants on the glass surface where the incident light reaches the solar cell (Zhang, 2019; Wang, 2021). To prevent this glitch, a coating must be added to the glass so that the solar cell can self-clean to prevent this buildup (Maharjan, 2020). Some solar cells may be exposed to corrosive environments, subjected to acid rain, alkaline or saline solutions, or even extreme temperature cycles. Therefore, the coatings must protect the solar cells against these dangerous effects. Exposure of solar cells to outdoor environments in very hot or high-temperature environments reduces both the efficiency and power output of the photovoltaic modules (Ehsan, 2022). The synergistic effect of both radiation absorption and elevated external temperature due to incident sunlight on the solar panels produces high operating temperatures of the photovoltaic cells. The objective of this chapter is to highlight the role that nanocoatings based on polymer matrices and nanomaterials have had, have, and will have in optimizing the multifunctional performance of solar cells, to prevent the energy conversion efficiency of the cells from being reduced during their lifetime.

The chapter has been organized as follows: In the Section titled "*Technical problems in solar cells without nanocoatings*", the main performance problems encountered during the practical operation of solar cells without nanocoatings are discussed to point out the areas of opportunity in the solar cell design. Subsequently, the properties that are improved through the application of nanocoatings for solar cells are presented in the section titled "*Properties that should be improved using nanocoatings*" to learn about the technical alternatives that could be adapted to improve the performance of the cells. The section titled "*Solutions and recommendations using nanocoatings to improve solar cells*" discusses some of the results achieved through practical implementations of nanocoatings on solar cells. The challenges that this area of research has in the present century are summarized in the section titled "*Future research directions*" to highlight the needs that must be met to achieve a guaranteed performance of nanocoatings as a high-value-added technology. Finally, the conclusions of the chapter regarding nanocoatings for energy generation and conservation of solar cells are given in the section titled "*Conclusions*".

TECHNICAL PROBLEMS IN SOLAR CELLS WITHOUT NANOCOATINGS

Solar Cell Performance Parameters

During long-term operation, solar cell panels are prone to dust and contaminants accumulating on their surface (Wang, 2021). The work of cleaning solar cell panels without functional coatings turns out to be expensive, laborious, and time-consuming and it also must be scheduled. This buildup reduces both the efficiency and reliability of solar panels for generating electricity. In practice, although the absorption of light in a solar cell is not restricted to one wavelength or range of wavelengths, the truth is that a significant portion of this radiation also produces heating of the solar cell, which must be restricted to maintain a voltage conversion efficiency suitable for use in electricity generation (Ehsan, 2022). It has been estimated that under standard operating conditions for monocrystalline silicon panels, there is a degradation of 0.4% per year in their efficiency and that for each degree Celsius lower in the operating temperature, 2 years are added to the life of the solar panel. The glass used in solar panel covers experiences a shift in its transmittance range towards shorter frequencies with increasing temperature, therefore the coating must prevent this shift in the light transmission range from occurring. Furthermore, the increase in the temperature of the solar cell reduces the bandgap of the semiconductor material used as the active material for the solar cell. This fact reduces parameters such as efficiency (η), fill factor (*FF*), peak power (P_{max}), and open circuit voltage (V_{oc}); but increases the short circuit current (I_{sc}).

In 1961, the theoretical upper limit of the conversion efficiency in a *p-n* solar cell was determined, through what was called the detailed balance limit, which was estimated for the ideal case of solar radiation (Shockley, 1961). The maximum efficiency was 30% for a bandgap of 1.1 eV. Since then, this value has been called the Shockley-Queisser limit. Recently, the tabulated values of this limit were tabulated along with other performance parameters of single junction solar cells (Rühle, 2016). Figure 2 illustrates in part (a) the Shockley-Queisser limit of solar cells and in part (b) the possible trapping light spectral regions for solar cells. As can be seen, the capture efficiency of the solar cell varies according to the bandgap of the material used in its design. Likewise, the wavelength of the light incident on the solar cell is associated with the bandgap of the material used in the implementation. The design of a solar cell that can capture all three ranges of light in the same device continues to be a challenge for all researchers around the world. The solar cell performance parameters that are directly related to its operation are summarized in Table 1.

PERFORMANCE LIMITS OF A SOLAR CELL

During the light capture process, the glass/air interface presents reflection losses which produce the Shockley-Queisser limit that prevents reaching the maximum power conversion efficiency (Vasilopoulou, 2022). It is estimated that one-third of the sunlight incident on a solar cell is reflected externally away from the cell. To reduce the amount of reflected light, anti-reflective coatings, and textured surfaces have been designed These arrays are placed on the receiving side of the interface. Three main functions can be highlighted for these arrangements. The first consists of redirecting or trapping the light beams to achieve total internal reflection. The second is to increase the light absorption of semiconductor materials at wavelengths with wide bandgap energies. The third is to optimize the retention of light within the solar cell. Material options to achieve the above improvements include the use of structures such as inverted pyramids, surface textures based on honeycomb structures, scattering surfaces with random texturing, metamaterials, nanowires, plasmonic nanostructures, diffraction gratings, and photonic crystals. Some of these structures make use of volumetric inorganic materials that require vacuum deposition methods or complex nanopatterning arrangements, which complicate achieving a successful and useful repetitive process for the large-

Table 1. The main performance parameters of a solar cell

Parameter	Symbol	Unit
Short-circuit current density	J_{sc}	mA/cm^2
Maximum current density	J_{max}	mA/cm^2
Current density at the maximum power point	J_{mpp}	mA/cm^2
Short-circuit current	I_{sc}	mA
Open-circuit voltage	V_{oc}	V
The voltage at the maximum power point	V_{mpp}	V
Peak power	P_{max}	W
The bandgap of the semiconductor material	E_g	eV
The wavelength of the incident light	λ_g	nm
Filling factor	FF	%
The conversion efficiency of light to electrical power	η	%

scale electronics industry. Currently, solar panel covers use glass thicknesses of 4.0, 3.2, and 2.0 mm (Syafiq, 2022). These thicknesses reduce the transmittance by 8 to 10%, which reduces the electrical power generated by the solar panels. Transmittance allows us to determine the ratio of the amount of light that passes through a material concerning the amount of incident light that was applied to it. Its values range from 0 to 1, with high transmittance being when the value tends to be 1 or 100% (Ehsan, 2022).

Ambient air contains particles in the range of 1 to 500 microns (Shaju, 2018; Aljdaeh, 2021; Hasan, 2022). Typically, particles suspended in the air or deposited on a photovoltaic panel come from pollen, environmental pollution, sand, dust from the construction industry, agricultural waste, leaves, animal excrement, pedestrian, and vehicular movements, or even lichen growth in forested areas. Dust particles can be made up of iron (Fe), silicon (Si), oxygen (O_2), calcium ions (Ca^{+2}), magnesium ions (Mg^{+2}), potassium ions (K^+), and sodium ions (Na^+). The regular period for cleaning the solar panel covers is 13 to 15 days. These buildup problems are made more complicated by wind, rain, snow, dew, humidity, and dirt sources: dust storms and volcanic eruptions, or the texture of the photovoltaic panel cover. Particles settle more on panels placed horizontally than on those that are inclined. The gravitational effect causes the largest particles to roll from the top to the bottom, accumulating at the bottom of an inclined panel. The direction of the wind has a direct influence on the angle of inclination of the panel for the accumulation or not of dust. To solve this problem, it is necessary to carry out cleaning by passive or active means and where man must regularly intervene. Active methods are classified as mechanical and electrostatic (Hasan, 2022; Syafiq, 2022). In the case of electrostatic methods, the use of power sources will allow triboelectric charging and the production of dielectric forces to remove dust. To remove the dust particles on the surface, it is necessary to apply strong electric fields to overcome the gravitational force and cohesive force between the particles and between the particles and the solar panel cover.

Environmental conditions of low humidity and high temperature facilitate the transport of dust thanks to the wind (Shaju, 2018; Hasan, 2022). At low temperatures water vapor may condense, forming water droplets on the surface of the panel cover. At high relative humidity levels, dust adhesion increases, forming a sticky surface on the cover and causing the surface to dull. When the wind speed is high and under conditions of low atmospheric pressure, less dust accumulation occurs. Under humid and dusty conditions, the power conversion efficiency (PCE) is reduced in the range of 60 to 70%. Irradiance is the ratio of the incident power per unit area expressed in W/m^2, which varies with time (Hasan, 2022). Its variation is due to the state of the weather, position of the sun in the sky, time of day, geographic location, and seasonal changes. Throughout the day, the inclination of the panel to sunlight must be modified to achieve the maximum irradiance since it is related to the latitude angle of the specific location, ranging between -15 degrees in summer and +15 degrees in winter. The output voltage of the solar cell increases according to the increase in irradiance. Humidity modifies the irradiance which has a slight effect on the open-circuit voltage and a huge effect on the short-circuit current (Shaju, 2018; Hasan, 2022). Depending on the geographical location of the photovoltaic panel, different types of dust can be found, so using a single type of coating to produce self-cleaning is not possible. The dust that settles on the solar panel cover has very different physical, electrostatic, chemical, and biological properties. These clouds of dust do not have the same weight, shape, or size. Electrically charged dusts tend to accumulate more effectively than those that are not electrically charged. Flat solar panel covers reduce the possibility of dust buildup, while uneven or textured covers tend to increase dust buildup. Concerning technology, the solar cells in photovoltaic panels can be monocrystalline, polycrystalline, and amorphous, which have different transmittance spectra that can be affected due to the presence of dust.

To reduce dust accumulation, photovoltaic panels require periodic cleaning, which can cause some inconveniences. (Ehsan, 2021). Water can be used for the purpose but in some places on the planet, it is a scarce resource. Rain is a possible solution, but not very effective in removing long-accumulated dust or bird droppings. This partial cleaning produces partial shading which reduces the conversion efficiency of the solar cell. The effect that the dust particles have on the light irradiance on the solar panel cover is the reflection and dispersion of sunlight, which prevents the transmission of light to the solar cell and thus a decrease in the electrical power generated (Alamri, 2020; Maharjan, 2020). When the glass cover is moistened, the -OH groups present on the surface attract dust and dirt particles, which favors their accumulation. The use of salt water to clean the cover of a solar panel is not feasible because it would leave salt particles on the surface which would cause shading and hot spots on the solar cell. Even the use of mechanical or robotic cleaning systems is impractical due to damage to the glass surface, their high cost, and their inability to be used during heavy rain events (Syafiq, 2022). Manual cleaning processes involve the use of detergents and water, which makes them not environmentally friendly and very expensive. In practice, however, these methods become inadequate due to the use of additional sources with high energy consumption as well as poor performance in complex environmental climatic conditions, which are very common in many installations around the world. In the case of mechanical methods, windshield wipers, spray nozzles, or cleaning robots can be used. Unfortunately, its use involves the use of enormous electrical power sources and the formation of micro-scratches on the surface of the solar panel cover. Dust cleaning methods could also include unmanned aerial vehicle (UAV)-based, semi-automated, and automated methods, all of which involve high operating costs that are not economical or completely viable for mass use (Khalid, 2023).

Fouling can cause hard or soft shading on the surface of the panel cover which reduces the power output of the solar cell (Hasan, 2022). Mud or a mass of dirt produces hard shading, and smog favors soft shading. Fouling also varies according to geographic location because the contaminants are diverse and have different effects on the transparency of the panel cover. Fouling on the panel surface can reach up to 50% annual loss in some areas, which reduces the power output of the solar cell (Ehsan, 2022). Additionally, shading causes some solar cells to restrict the flow of current to other cells, causing overheating or even melting of the cells, which is called hot spot generation. These hot spots raise the temperature of the solar panel, reducing its performance and increasing the rate of degradation of the solar cell. To regulate the temperature of photovoltaic panels, various alternatives have been explored such as thermoelectric cooling, phase change materials, the use of heat pipes, and air cooling. Unfortunately, these techniques are expensive to implement in practical applications since they involve piping, assemblies, installations, etc. which are not profitable in the investment for light capture by solar cells as a renewable energy means of generating electrical energy.

The solar panel cover is made of two glass films, one thin and one thick. The first prevents light reflection, while the second protects against environmental conditions. The glass reflects 8 to 10% of the incident light, and of the transmitted light, about 35% of this is reflected by the active materials of the solar cell. Therefore, research on nanocoatings to avoid this reduction of light due to reflection is of interest to increase the conversion efficiency of solar cells (Cherupurakal, 2021). The glass offers protection of the active materials of the solar cell against ultraviolet light irradiation, abrasion, temperature changes, hazardous chemical environments, abrasion, and scratching.

The design of transparent coatings for solar cells also implies water resistance, transmittance, and high mechanical and environmental resistance (Nguyen-Tri, 2019). Silicon solar cells are arranged in arrays placed behind glass in a solar panel. Solar panels with a concentrator system are more susceptible to the adhesion of dust deposited on their covers due to the electrically polarized reflective surfaces. The surfaces covered with dust act as electrical resistors that produce heating in the solar cells, which can permanently damage the panel and, in addition to reducing their efficiency, force them to be replaced sooner or later by new ones without having reached a relevant useful life.

ENHANCING MATERIAL PROPERTIES THROUGH NANOCOATINGS

Role of Nanomaterials and Nanocoatings

With the introduction of nanomaterials, it is possible to place functional nanocoatings on the surface of photovoltaic panels to solve the problems mentioned in the previous section (Ehsan, 2022; Alarifi, 2023). The main nanocoatings that must be incorporated to improve the performance of solar panels are of two types, namely, anti-reflective and self-cleaning. The first type of coating should be used to improve the transmittance and thereby increase the efficiency of the solar cell by avoiding reflection on the surface of the panel. The second type of coating should be used to avoid shading, dirt, and hot spots on the surface of the panel and thereby increase the efficiency of the solar cell. To solve the problem of the increase in temperature in the photovoltaic panel, it is necessary to apply a superhydrophobic nanocoating to counteract the thermal effect, reduce the degradation of the solar cell, and guarantee a longer lifetime of the solar cell (Lu, 2022). The transmittance spectra for ceramic oxides such as silicon dioxide (SiO_2), titanium dioxide (TiO_2), and aluminum oxide (Al_2O_3), used as anti-reflective nanocoatings are illustrated in Figure 3. As can be seen in this figure, these Materials have better transmittance for the ultraviolet range, while for the visible range, their transmittance is more variable for the different wavelengths of the spectrum.

The inclusion of nanocoating on the protective glass of the solar cell panel could offer an ideal solution to avoid the cleaning problem mentioned above (Wang, 2021). Among the properties that are desired to be improved using coatings are anti-reflective properties, self-cleaning, hydrophilic antifogging, hydrophobicity, etc. A self-cleaning coating can be implemented using two possible approaches (Maharjan, 2020; Zhao, 2021). The first consists of creating hierarchical structures based on micro or nanomaterials on hydrophobic substrates. The second involves modifying the surface of a substrate to achieve low surface-free energies through the chemical modification of micro- or nanomaterials. Coating synthesis involves the use of methods such as layer-by-layer assembly, chemical vapor deposition, or micro-patterning which do not easily allow for application as large surface layers.

The use of nanocoatings can offer one or more functional properties to conventional protective materials used in solar cell panels (Wang, 2021). In addition, with the coatings in the protective layers, the amount of sunlight captured by the solar cells can be increased, the accumulation of dust is reduced, and the mechanical robustness is increased, leading to operation in outdoor environments for a long time. With a functional material only one of the properties can be achieved, however, through a multifunctional material two or more properties are obtained thanks to the use of materials with synergistic performance between its components. Despite the progress achieved so far, the current challenge remains to identify materials that can work synergistically by making each of the required properties more efficient. For example, although commercial application implies that high robustness needs to be achieved for hydrophilic or hydrophobic behavior of the surface, it is also important to remember that this leads to vulnerable rough textures that limit the practical application of the protected material. Therefore, the ideal properties for nanocoating for solar cell panels must seek a balance and optimization of robustness, humidity, and optical performance. The ideal nanocoating should be able to operate under the irradiance spectrum of the AM 1.5G standard solar spectrum and in the ultraviolet range under continuous illumination.

TYPES OF NANOCOATINGS AND MOST COMMON MATERIALS

A superhydrophobic coating is a material that has the property of preventing water from being transported through it (Si, 2015; Das, 2018; Li, 2019; Liang, 2020). Water shielding can be obtained naturally or artificially through physical or chemical processes. During its design, the resistance of the bond between the coating and the substrate where it will be deposited, the mechanical stability, thickness, and roughness must be considered, which directly influence its practical operation. Superhydrophobic nanocoatings can be classified as inorganic, organic, and hybrid (Huang, 2021). Due to the wetting behavior, the main types of nanocoatings are defined as hydrophilic when the contact angle between water and the surface has a value between 10 and 90 degrees, while a coating is considered hydrophobic when the contact angle between them is between 90 and 150 degrees. To illustrate the effect of the hydrophobic nanocoating, Figure 4 (a) illustrates an unprotected solar panel cover, which favors the adhesion of water, while the application of a nanocoating promotes the movement or displacement of water droplets, preventing their accumulation on the solar panel cover, as sketched in Figure 4(b). The passive method to produce self-cleaning involves the use of either hydrophobic or hydrophilic coatings which do not require power sources during the process, making them economical thanks to the use of natural raindrops to remove particles or organic contaminants from the solar panel cover (Cherupurakal, 2021). For the implementation of hydrophobic coatings, the use of processable, low-cost polymers, with the possibility of being molecularly designed, and flexible, is necessary. Furthermore, polymers offer versatile mechanical and thermal properties that can be exploited at the surface level to be deposited with high durability. Polymeric materials that possess hydrophobic properties are polystyrene (PS), polydimethylsiloxane (PDMS), polymethyl hydroxy siloxane (PMHS), polyurethane (PU), fluorinated polymers, and silane-based polymers. In practical application, nanocoatings extend the operational life of solar cells, depending on the incorporated functional properties. The final properties of the composite material used as a nanocoating depend on both the material used as a matrix and the reinforcing materials incorporated. Extreme environmental conditions produce abrasion damage, cementation, changes in surface chemistry, reduction in coating thickness, and blistering.

The inclusion of the hydrophobic coating on the solar cell reduces light transmission losses by reducing dust accumulation both in terms of particle composition as well as deposition densities (Ehsan, 2021). The presence of dust on the solar panel cover prevents the irradiation of sunlight, reducing transmittance which reduces the electrical power generated by 30%. The use of the hydrophobic coating reduced dust accumulation on the panel cover, improving the electrical power generated by 11% after prolonged exposure to the environment. Various groups of researchers have proposed coatings based on micro- or nanostructures using either hydrophobic or hydrophilic substrates to be deposited on the glass covers of photovoltaic panels. It has been demonstrated that the application of the coating offers better results when it is applied during the solar cell manufacturing process than when it is placed on a solar cell previously manufactured in a separate independent process. Self-cleaning coatings have a variety of applications in paints, cement, textiles, and window glass due to the savings in cleaning costs and the wide variety of potential applications such as the building industry, automotive industry, and energy industry, due to their use as functional materials. With water present on the surface of the coating of either the hydrophilic or hydrophobic type, both options are capable of self-cleaning (Syafiq, 2022). Like the lotus leaf, which, thanks to its hydrophobicity, does not allow water to pass through it, hydrophobic coatings have a contact angle with water greater than 90 degrees. On the surface of these coatings, water is strongly repelled because the droplets take a spherical shape, which makes them roll away, taking the dirt with them as well. To achieve contact angles greater than 150 degrees or superhydrophobicity, it is necessary to control the topographic and chemical properties of the surface. Therefore, producing rough surfaces at the micro- and nanoscale will also reduce the adhesion of dirt on the coating. On the other hand, self-cleaning can be achieved through hydrophilic type coatings using photocatalysis to exploit UV light in the chemical breakdown of impurities found on a surface (Afzal, 2021). During the photocatalysis process, electron-hole pairs are generated, which oxidize the chemical species adsorbed on the surface, causing them to break before they are accumulated. Once the rupture is produced, the water removes the dirt forming thin films with small contact angles, which can be easily removed by sliding. This is achieved when the Van der Waals forces of the particles on the surfaces are overcome. To prevent water from being attracted to a surface, the presence of polar functional groups such as hydroxyls, carbonyls, or carboxyls must be eliminated or reduced. The use of photocatalysts such as titanium dioxide (TiO_2) has spread thanks to the fact that the photoresponse to visible light of this material can be extended using dopants such as silicon dioxide (SiO_2), zirconium dioxide (ZrO_2), niobium oxide (Nb_2O_3), tungsten (W), iron (Fe), nitrogen (N_2) and polyaniline (Sarkin, 2020). With increases in the temperature of a solar cell due to the absorption of solar radiation, an increase in the reverse saturation current occurs as well as a reduction in the open voltage and the bandgap (Ahmadi, 2019). Temperature increases cause power conversion efficiency to decrease because only 20% of solar energy is converted electrically and 80% is converted thermally. Also, the increased temperature reduces the bandgap energy, and increases the absorption of higher wavelength photons, the lifetime of the minority carriers, and the electric current generated by the light, thereby reducing the open circuit voltage and the fill factor (Hasan, 2022). For maximum conversion power the fill factor, open circuit voltage, and short circuit current must be increased to the limit. The addition of thermal insulating materials and photocatalysts such as titanium dioxide (TiO_2), tungsten oxide (WO_3), and vanadium IV oxide (VO_2) to the solar panel cover glass decreases the increase in temperature when exposed to sunlight and decomposing dust particles by the photocatalytic reaction (Syafiq, 2022). Of these types of coatings, it has been pointed out that rolling movement is more effective than sliding movement in producing self-cleaning of a surface. A self-cleaning coating must also be transparent or have high transmittance to light incident on the panel to allow maximum absorption of sunlight. The addition of rare earth particles in a polymer matrix used as a coating for the solar panel cover can improve efficiency by 2 to 4% in addition to improving the spectral response to light incident on the solar cell. These chemical elements can modify the mismatch of bandgap photon energies in semiconductor materials either through downconversion or upconversion processes or using conducting polymers and engineered light-absorbing materials (Cherupurakal, 2021).

Therefore, using a covering material that is transparent for long periods and not too thick increases the efficiency of the solar panel (Syafiq, 2022). The solar panel cover should include by default an anti-reflective cover with a refractive index between 1.2 and 1.3 to offer higher optical transmission properties. To implement the anti-reflective coating, it is necessary to use acid-catalyzed ceramic materials. However, there are some difficulties, for example, the use of titanium dioxide which has a refractive index between 2.2 and 2.6 would tend to reflect sunlight even more without prior treatment of the material. To reduce the refractive index of ceramics it is necessary to add homogeneous porosity by choosing the pore size less than 100 Å which is smaller than the wavelength of the incident light. For materials such as alumina with a refractive index of 1.7, it can be reduced to a material with an index of 1.31 when the porosity is 64% and with a thickness of 0.1 microns, but unfortunately, this is susceptible to absorbing water and thus increasing its refractive index. Synthesizing porous and highly durable oxide films is a challenge because they regularly present weak mechanical properties due to the Van der Waals interaction forces presented when they are deposited on a glass substrate. Ceramic materials composed of titanium dioxide (TiO_2) and silicon dioxide (SiO_2) can offer high optical transmission and low refractive indices of 94% and 1.3, respectively, for the wavelength range of 300 to 1200 nm.

Among the polymers most used as self-cleaning coatings because they are hydrophobic are polytetrafluoroethylene (PTFE), polymethylmethacrylate (PMMA), and polydimethylsiloxane (PDMS) (Syafiq, 2022). This last polymer is used mainly because it has a transmittance greater than 85% in the ultraviolet-visible region, it has excellent adhesion on glass panel covers because it contains methyl groups as linkers, but unfortunately its abrasion resistance must be improved. The dip coating method can be applied to add silica microspheres to PDMS to achieve abrasion resistance, transparency, uniformity, and homogeneity. If the same method and reinforcer are used but with PMMA, the coating loses hydrophobicity as well as transparency and is damaged by abrasion. A coating based on the heteropolymer containing PDMS and polyurethane has a transmittance of 99%, and greater abrasion resistance, but its adhesion to glass and synthesis process are complicated to work with glass on a large scale, which prevents its application on a commercial level.

Anti-reflection materials used in micro- or nanocoatings for solar cells include silicon oxynitride (SiON), silicon nitride (SiN), silicon dioxide (SiO_2), titanium dioxide (TiO_2), zinc sulfide (ZnS), aluminum isopropoxide ($C_9H_{21}AlO_3$), polyaniline, single-wall carbon nanotubes (SWCNT), and multi-wall carbon nanotubes (MWCNT) (Kumar, 2018; Pan, 2019). The main purpose of coatings based on these materials is to increase power conversion efficiency (PCE).

Recently, the use of light-harvesting biomimetic structures based on plants or animals has been proposed (Vasilopoulou, 2022). In the case of animals, it has been proposed to mimic the anti-reflective surfaces of the corneas of moths' eyes or the transparent wings of cicadas. The natural structures of moth eyes have protrusions with a height of 250 nm and spaces of 200 to 250 nm, which have been synthesized by lithographic techniques for use in solar cells. These structures offer self-cleaning properties and exhibit adhesion to water droplets with a contact angle of more than 150 degrees, which is not the maximum reached naturally since lotus leaves have a contact angle of more than 170 degrees. In the case of plants, it is possible to mimic either the leaves or the flowers to be used in the design of solar cells. These structures allow light to be trapped and behave hydrophobically.

NANOCOATINGS FOR ENHANCED SOLAR CELL PERFORMANCE

Multifunctionality

Nanocoatings based on nanocomposites with multifunctional properties should be used to increase the conversion efficiency of solar panels not only of the photoactive layers of the solar cells but also of the protective layers of the solar panels (Müller, 2017). Polymeric matrices are regularly reinforced by nanoparticles as nanocoatings for solar cells. To improve the conversion efficiency and durability of solar cells, nanocoatings must fulfill the following functions: electrical insulation, flexibility, anti-fouling, transparency, thermal stability, as well as a barrier against water, oxygen, and ultraviolet light, as shown in Figure 5.

Among the emerging materials that are offering functional alternatives for various scientific and technological applications are nanomaterials (Alarifi, 2023; Hussein, 2023). Its performance advantages come from its unique biological, physical, and chemical properties, which researchers around the world have been able to exploit favorably. The interesting applications that nanomaterials can offer are the practical implementation of sensors of various physical and chemical variables, the design of solar cells, and the implementation of intelligent coatings such as electrochromic, photochromic, and thermochromic coatings. These materials can be classified into four groups: nanomaterials based on composites, nanomaterials based on carbon, nanomaterials based on organic materials, and nanomaterials based on inorganic materials. The inorganic nanomaterials most used as functional coatings in solar cells are those based on ceramic oxides, including silicon dioxide (SiO_2) or silica, titanium dioxide (TiO_2) or titania, and aluminum oxide (Al_2O_3) or alumina. Nanomaterials can be used to reduce costs and increase the efficiency of electrical energy generation, transmission, distribution, and storage systems. For energy applications, the photovoltaic, catalytic, triboelectric, thermoelectric, and piezoelectric properties of nanomaterials must be exploited. Inorganic nanomaterials offer large surface area, excellent chemical stability, as well as unprecedented thermal and electrical conductivity. The main advantages of nanomaterials for renewable energy applications are their high efficiency for storage and heating, high electrical storage capacity, and significant reduction of contaminants.

The covers of solar panels should be covered using multifunctional thin films to improve their transmittance, anti-fog, anti-reflection, and self-cleaning properties (Mozumder, 2019). To achieve this objective, it is necessary to use materials that can serve as photoactive, hydrophilic, and/or hydrophobic coatings. In the design of coatings, it is necessary to control light transparency, electrical conductivity, wettability, and surface morphology. Among the most recent functionalities that have been introduced in nanocoatings are antimicrobial and self-healing properties. All these properties must be tuned to the properties of the materials used in the active layers and the electrodes possibly on both sides of the solar panel. To develop anti-reflective, anti-fog, and self-cleaning coatings, hydrophobic materials are necessary. It is also possible to obtain anti-fog nanocoatings using materials with photocatalytic and hydrophilic properties even on rainy days or under darkness. Silica-based nanocoatings can develop anti-reflective and anti-fog properties. Composite nanocoatings based on silica and titania have exhibited antimicrobial, anti-fog, anti-cleaning, hydrophobic, and photocatalytic properties. Tin dioxide (SnO_2) nanocoatings exhibit durable, homogeneous, transparent, antimicrobial, and self-cleaning properties. The antimicrobial properties are greatly improved with the use of silver (Ag) nanoparticles as a reinforcing material. Bacteria can accumulate on the surfaces of the glass used as the cover of solar panels which reduces conversion efficiency and transmittance. Other alternative options for preparing antimicrobial coatings include films containing copper nanoparticles.

Due to the absorption of contaminants or humidity, the covers of the solar panels can deteriorate when exposed to the sun (Mozumder, 2019). Likewise, sand particles can be abrasive on the covers when exposed to outdoor environments. Both effects decrease wettability and transmittance. The regenerative capacity of these materials could reverse the original hydrophobicity and optical transmittance properties of the cover material. The use of materials such as epoxy resins, silica nanoparticles, and perfluorooctyl triethoxysilane offers corrosion resistance and hydrophobicity or self-healing materials. Another material proposal to develop a self-healing coating is the combination of polyisobutylene, borosilicate, molybdenum disilicide, and silica.

SELF-CLEANING, TRANSPARENT NANOCOATINGS

A self-cleaning hydrophobic nanocoating based on trichloro(1H,1H,2H,2H-perfluorooctyl)silane (TCPFOS) was deposited on a glass substrate previously polished with an abrasive based on micro- and nanoparticles of diamond to generate tortuous trenches with covalent bonds (Maharjan, 2020). This coating showed resistance to a set of environmental factors including abrasion, temperature changes, and exposure to acidic, saline, and alkaline media. A minimum surface tension between the self-cleaning surface and water droplets with a volume of 0.1 mL resulted in a sliding angle of 14 degrees which allowed the dust particles to be removed by sliding or rolling.

A sol-gel composite formed between silica (SiO_2) nanoparticles with acid catalysis and base catalysis was proposed to deposit a 0.9 m by 1.6 m nanocoating capable of offering antifogging properties, mechanical robustness, and broadband transmittance for glass panel solar cells (Wang, 2021). This coating requires a versatile manufacturing spray or dip deposition process without requiring post-treatment. Additionally, the manufacturing process is cost-effective and allows the deposition of thin films over large areas. The coating presented good adhesion, porosity, and surface roughness, high transmittance, high resistance to water flows, and hydrophilic and antifogging properties.

One of the strategies to achieve the multifunctionality of a coating for a solar cell consists of imitating the natural structures that capture sunlight such as flower petals through a photonic nanostructure printed at room temperature and low cost (Vasilopoulou, 2022). The coating was light polymerized and operated as an adaxial epidermis. The nanocoating was used on a perovskite solar cell to protect it from moisture by repellency, and it has improved light trapping because it can capture ultraviolet light in addition to visible light (Luo, 2020). The authors replicated epidermal cells from floral petals of species such as *Antirrhinum majus*, *Geranium endressii*, *Aubrieta deltoidea*, *Anchusa officinalis*, and *Ranunculus repens*. The researchers used polydimethylsiloxane (PDMS) and ultraviolet light-curable acrylate urethane adhesives to synthesize petal epidermal cells. For the structures to be used in solar cells, it is necessary to produce composite materials using as reinforcement materials layers of mesoporous titanium dioxide (TiO_2) to transport electrons, a layer of Spiro-MeOTAD to transport holes, and layers of graphene to passivate the interface between the charge transport layers. These structures have high transmittance, do not have photocatalysis properties, and have self-cleaning and water-repellency properties.

From an industrial point of view, nanocoatings using titanium dioxide (TiO_2) can be applied as dust repellents on solar panel covers to increase their photovoltaic conversion efficiency as well as reduce the amount of corrective maintenance required (Hussein, 2023). Titanium offers a low-cost, highly transparent, environmentally friendly, and non-toxic alternative to be used as a nanocoating in dye-sensitized solar cells. Recently, anti-reflective superhydrophobic nanocoatings formed by double-layer films are being introduced and based on silica (Wang, 2022). The first layer interacts with the top of the substrate using covalent bonds. The second layer interacts with the bottom part of an embedded structure to increase the strength of the physical connection between layers of the solar cell. This type of layer arrangement increases transmittance, and hardness, and provides excellent self-cleaning properties.

In perovskite solar cells, critical semitransparent inorganic buffer layers are added to protect the cell from damage when the transparent electrodes are placed in contact with the panel cover (Jaqt, 2020). These layers are based on *p*-type copper oxide (CuO_x) which is grown on the perovskite as the active material of the solar cell or on the organic hole transport layers, without causing damage to the materials used as a substrate during the sputtering. The layers have superior hole mobility, ionization potential for hole extraction, quality, as well as transmittance, to increase the conversion efficiency of the solar cell. In addition to copper oxide, other *n*-type ceramic materials acting as transparent conducting oxide have been used to protect the perovskite active material and the cell electrodes for light capture, including indium-doped zinc oxide (IZO), aluminum-doped zinc oxide (AZO), zinc oxide (ZnO), titanium dioxide (TiO_2), zinc tin oxide (ZTO), and tin dioxide (SnO_2). A schematic of the layers commonly included in a perovskite-based solar cell is shown in Figure 6.

The use of nanocomposites based on functional nanoparticles is enabling high-performance energy applications such as solar cells (Saleh, 2020). Nanostructured materials allow the blocking of ultraviolet light, gases, and moisture from the interior of the solar cell when they are integrated into the outer layer of its packaging. Among the nanomaterials that can be used for this reinforcement of the solar cell encapsulation are graphene, carbon nanotubes, and nanoclays.

ANTI-REFLECTIVE NANOCOATINGS

SiO_2 nanocoatings have hydrophobic, anti-reflection, and anti-static properties on the covers of solar panels and can improve panel efficiency by 5 to 15% (Alamri, 2020). Dust prevents sunlight from reaching the solar panels, causing photons to operate free of electrons, and therefore, the power output of the panel is reduced because the conversion efficiency is lower. The antistatic effect occurs thanks to the fact that the water droplets are removed from the cover, preventing the attraction of charge between the water and the glass surface, and this facilitates the access of photons to the active materials of the solar cell (Walz, 2023). Nanocoatings based on silicon oxide and titanium oxide prevent wetting behavior and soiling deposition thanks to the action of photocatalysts.

Graphene and its derivatives can be used to implement devices for energy applications thanks to their properties such as flexibility, mechanical strength, optical transparency, high thermal and electrical conductivities, as well a large specific surface area (Díez-Pascual, 2018). As reinforcing nanomaterials, they can be incorporated into polymer matrices to produce materials with functional and structural properties operating synergistically. These nanocomposites can be used to design interfacial layers, active layers, or transparent conductive electrodes for polymer-based solar cells. Furthermore, these solar cells must have a low ecological impact and low processing costs. The performance in the intrinsic state of graphene is low, however, it can be improved through doping or functionalization to increase charge carrier densities, tune the band gaps of solar cells, as well as improve electrical stability during the useful life of the solar cell.

A nanocomposite based on polydimethylsiloxane (PDMS) and silicon dioxide nanoparticles (SiO_2) has been proposed as a nanocoating to provide hydrophobic and self-cleaning properties to photovoltaic panels (Tayel, 2022). The wide variation in the density of the dust accumulated on the cover of a solar panel favors the formation of hot spots, which deteriorates the performance of the cells, reducing their conversion efficiency. The coating was tested for forty days in an outdoor environment and presented an efficiency improvement of 30.7% over an uncovered panel.

Photonic nanostructures on front covers can enable more efficient trapping of light in a solar cell through advanced light management that optimizes conventional finishing texture (Santos, 2023). Arrangements in the form of titanium dioxide honeycomb panels with various geometries can be placed on a silicon solar cell for this purpose (Thongsuwan, 2022). This arrangement can offer anti-reflection properties to increase the efficiency of the solar cell. These nanostructures have photocatalytic, superhydrophilic, and anti-reflective properties.

FUTURE RESEARCH DIRECTIONS

One of the great challenges that scientists and technologists have related to self-cleaning coatings is the lack of standardized methods and specific parameters that must be established to evaluate the performance of this type of coatings (Syafiq, 2022). This type of coating faces very diverse environmental conditions which deteriorate its performance in an uncontrolled manner. So far performance has been evaluated under controlled and simulated laboratory tests with amounts of dust as well as common contaminants, however, real environments are challenging because they do not have regularity and controllability. Test standards should be proposed and tested to evaluate the performance of self-cleaning coatings. Among the parameters that have been detected should be included in the performance standards of the coatings are the static wettability contact angle (WCA) and the optical transmittance. The synthesis methods as well as the coating deposition methods should be simple, economical, applicable to any thickness and dimension of solar panel covering substrate, and preferably be cured at room temperature.

One of the areas of opportunity in the development of nanocoatings is the need to develop mathematical models that can predict the behavior of the impact of dust, bacteria, abrasion, and humidity on the performance of a solar panel. Among the parameters that can be affected by dust on the cover of a solar panel are energy conversion efficiency, dirt loss rate, derating factor, projection ratio, and total radiation. Mastering mathematical models could optimize the performance of nanocoatings by considering the conditions of environments to include the best methods and materials associated with their installation and high value-added application. The models must be able to make correlations between the various factors to establish the best strategy to achieve maximum performance from the nanocoatings to be used on the cover of a solar panel.

CONCLUSION

The need to have reliable renewable energy generation devices continues to motivate the optimization of their performance through emerging technologies that reduce costs and that can be integrated from the initial manufacturing process to extend the useful lifetime. One of these emerging technologies is the application of nanocoatings to protect solar panel casings from hostile environments by providing functional properties that work synergistically to increase or at least maintain the original conversion efficiency of the solar cells. Among the properties that nanocoatings have been able to optimize are transmittance, anti-reflective, self-cleaning, antibacterial, anti-abrasive, hydrophobic, and hydrophilic. The use of nanocomposites based on polymers as well as reinforcing materials based on metals and ceramics can offer multifunctional properties to protect solar cells against the adhesion of dust, humidity, snow, animal- or agricultural waste, etc., the penetration of ultraviolet light or abrasion wear that degrades the operation of solar panels and reduces conversion power during operation in outdoor environments. Also, the self-healing property can be achieved if the appropriate polymers are used, which expands the versatility of functional properties available through nanocoatings. Despite the progress achieved so far, researchers and entrepreneurs will continue to promote the development and research of nanocoatings for solar panel covers, so both new materials and new synthesis processes will be proposed to make use of novel functional materials that allow for optimizing the efficiency of solar energy conversion.

REFERENCES

Afzal, A., Habib, A., Ulhasan, I., Shahib, M., & Rehman, A. (2021). Antireflective Self-Cleaning TiO_2 Coatings for Solar Energy Harvesting Applications. *Frontiers in Materials*, *8*, 687059. doi:10.3389/fmats.2021.687059

Ahmadi, M. H., Ghazvini, M., Nazari, M. A., Ahmadi, M. A., Pourfayaz, F., Lorenzini, G., & Ming, T. (2019). Renewable Energy Harvesting with the Application of Nanotechnology: A Review. *International Journal of Energy Research*, *43*(4), 1387–1410. doi:10.1002/er.4282

Alamri, H. R., Rezk, H., Abd-Elbary, H., Ziedan, H. A., & Elnozahy, A. (2020). Experimental Investigation to Improve the Energy Efficiency of Solar PV Panels using Hydrophobic SiO_2 Nanomaterial. *Coatings*, *10*(5), 503. doi:10.3390/coatings10050503

Alarifi, I. M. (2023). Advanced Selection Materials in Solar Cell Efficiency and Their Properties – A Comprehensive Review. *Materials Today: Proceedings*, *81*(Part 2), 403–414. doi:10.1016/j.matpr.2021.03.427

Aljdaeh, E., Kamwa, I., Hammad, W., Abuashour, M. I., Sweidan, T., Khalid, H. M., & Muyeen, S. M. (2021). Performance Enhancement of Self-Cleaning Hydrophobic Nanocoated Photovoltaic Panels in a Dusty Environment. *Energies*, *14*(20), 6800. doi:10.3390/en14206800

Cherupurakal, N., Mozumder, M. S., Mourad, A.-H. I., & Lalwani, S. (2021). Recent Advances in Superhydrophobic Polymers for Antireflective Self-Cleaning Solar Panels. *Renewable & Sustainable Energy Reviews*, *151*, 111538. doi:10.1016/j.rser.2021.111538

Das, S., Kumar, S., Samal, S. K., Mohanty, S., & Nayak, S. K. (2018). A Review on Superhydrophobic Polymer Nanocoatings: Recent Development, Application. *Industrial & Engineering Chemistry Research*, *57*(8), 2727–2745. doi:10.1021/acs.iecr.7b04887

Díez-Pascual, A. M., Sánchez, J. A. L., Capilla, R. P., & Díaz, P. G. (2018). Recent Developments in Graphene/Polymer Nanocomposites for Application in Polymer Solar Cells. *Polymers*, *10*(2), 217. doi:10.3390/polym10020217

Ehsan, R. M., Simon, S. P., Kinattingal, S., Kumar, K. A., & Srihasrsha, I. (2022). Effects of Nanocoatings on the Temperature-dependent Cell Parameters and Power Generation of Photovoltaic Panels. *Applied Nanoscience*, *12*, 3945–3962. doi:10.1007/s13204-022-02633-0

Ehsan, R. M., Simon, S. P., Sundareswaran, K., Kumar, K. A., & Sriharsha, T. (2021). Effect of Soiling on Photovoltaic Modules and Its Mitigation using Hydrophobic Nanocoatings. *IEEE Journal of Photovoltaics*, *11*(3), 742–749. doi:10.1109/JPHOTOV.2021.3062023

Hasan, K., Yousuf, S. B., Tushar, M. S. H. K., Das, B. K., Das, P., & Islam, M. S. (2022). Effects of Different Environmental and Operational Factors on the PV Performance: A Comprehensive Review. *Energy Science & Engineering*, *10*(2), 656–675. doi:10.1002/ese3.1043

Huang, Z.-S., Shen, C., Fan, L., Ye, X., Shi, H., Zhang, Y., Lai, Y., & Quan, Y.-Y. (2021). Experimental Investigation of the Anti-Soiling Performances of Different Wettability of Transparent Coatings: Superhydrophilic, hydrophilic, Hydrophobic and Superhydrophobic Coatings. *Solar Energy Materials and Solar Cells*, *225*, 111053. doi:10.1016/j.solmat.2021.111053

Hussein, H. S. (2023). The State of the Art of Nanomaterials and Its Applications in Energy Saving. *Bulletin of the National Research Center*, *47*, 7. doi:10.1186/s42269-023-00984-4

Jaqt, R. A., Huq, T. N., Hill, S. A., Thway, M., Liu, T., Napari, M., Roose, B., Gałkowski, K., Li, W., Lin, S. F., Stranks, S. D., MacManus-Driscoll, J. L., & Joye, R. L. Z. (2020). Rapid Vapor-Phase Deposition of High-Mobility p-Type Buffer Layers on Perovskite Photovoltaics for Efficient Semi-Transparent Devices. *ACS Energy Letters*, *5*(8), 2456–2465. doi:10.1021/acsenergylett.0c00763

Khalid, H. M., Rafique, Z., Muyeen, S. M., Raqeeb, A., Said, Z., Saidur, R., & Sopian, K. (2023). Dust Accumulation and Aggregation on PV Panels: An Integrated Survey on Impacts, Mathematical Models, Cleaning Mechanisms, and Possible Sustainable Solution. *Solar Energy*, *251*, 261–285. doi:10.1016/j.solener.2023.01.010

Kumar, K. R. T., Ramakrishna, M., & Sukumar, G. D. (2018). A Review on PV Cells and Nanocomposite-Coated PV Systems. *International Journal of Energy Research*, *42*(7), 2305–2319. doi:10.1002/er.4002

Li, W., Tan, X., Zhu, J., Xiang, P., Xiao, T., Tian, L., Yang, A., Wang, M., & Chen, X. (2019). Broadband Antireflective and Superhydrophobic Coatings for Solar Cells. *Materials Today. Energy*, *12*, 348–355. doi:10.1016/j.mtener.2019.03.006

Liang, Z., Zhou, Z., Zhao, L., Dong, B., & Wang, S. (2020). Fabrication of Transparent, Durable and Self-Cleaning Superhydrophobic Coatings for Solar Cells. *New Journal of Chemistry*, *44*(34), 14481–14489. doi:10.1039/D0NJ01402H

Lu, H., & Zheng, C. (2022). Comparison of Dust Deposition Reduction Performance by Super-Hydrophobic and Super-Hydrophilic Coatings for Solar PV Cells. *Coatings*, *12*(4), 502. doi:10.3390/coatings12040502

Luo, J., Yang, H. B., Zhuang, M., Liu, S., Wang, L., & Liu, B. (2020). Making Fully Printed Perovskite Solar Cells Stable Outdoor with Inorganic Superhydrophobic Coating. *Journal of Energy Chemistry*, *50*, 332–338. doi:10.1016/j.jechem.2020.03.082

Maharjan, S., Liao, K.-S., Wang, A. J., Barton, K., Haldar, A., Alley, N. J., Byrne, H. J., & Curran, S. A. (2020). Self-Cleaning Hydrophobic Nanocoating on Glass: A Scalable Manufacturing Process. *Materials Chemistry and Physics*, *239*, 122000. doi:10.1016/j.matchemphys.2019.122000

Mozumder, M. S., Mourad, A.-H. I., Pervez, H., & Surkatti, R. (2019). Recent Developments in Multifunctional Coatings for Solar Panel Applications: A Review. *Solar Energy Materials and Solar Cells*, *189*, 75–102. doi:10.1016/j.solmat.2018.09.015

Müller, K., Bugnicourt, E., Latorre, M., Jorda, M., Sanz, Y. E., Lagaron, J. M., Miesbauer, O., Bianchin, A., Hankin, S., Bölz, U., Pérez, G., Jesdinszki, M., Lindner, M., Scheuerer, Z., Castelló, S., & Schmid, M. (2017). Review on the Processing and Properties of Polymer Nanocomposites and Nanocoatings and Their Applications in the Packaging, Automotive and Solar Energy Fields. *Nanomaterials (Basel, Switzerland)*, *7*(4), 74. doi:10.3390/nano7040074

Nguyen-Tri, P., Tran, H. N., Plamondon, C. O., Tuduri, L., Vo, D.-V. N., Nanda, S., Mishra, A., Chao, H.-P., & Bajpai, A. K. (2019). Recent Progress in the Preparation, Properties and Applications of Superhydrophobic Nano-based Coatings and Surfaces: A Review. *Progress in Organic Coatings*, *132*, 235–256. doi:10.1016/j.porgcoat.2019.03.042

Pan, A., Lu, H., & Zhang, L.-Z. (2019). Experimental Investigation of Dust Deposition Reduction on Solar Cell Covering Glass by Different Self-Cleaning Coatings. *Energy*, *181*, 645–653. doi:10.1016/j.energy.2019.05.223

Rühle, S. (2016). Tabulated Values of the Shockley-Queisser Limit for Single Junction Solar Cells. *Solar Energy*, *130*, 139–147. doi:10.1016/j.solener.2016.02.015

Saleh, T. A., Shetti, N. P., Shanbhag, M. M., Reddy, K. R., & Aminabhavi, T. M. (2020). Recent Trends in Functionalized Nanoparticles Loaded Polymeric Composites: An Energy Application. *Materials Science for Energy Technologies*, *3*, 515–525. doi:10.1016/j.mset.2020.05.005

Santos, I. M., Alexandre, M., Mihailetchi, V. D., Silva, J. A., Mateus, T., Mouquinho, A., Boane, J., Vicente, A. T., Nunes, D., Menda, U. D., Águas, H., Fortunato, E., Martins, R., & Mendes, M. J. (2023). Optically-Boosted Planar IBC Solar Cells with Electrically-Harmless Photonic Nanocoatings. *Advanced Optical Materials*, *11*(15), 2300276. doi:10.1002/adom.202300276

Sarkin, A. S., Ekren, N., & Sağlam, S. (2020). A Review of Anti-Reflection and Self-Cleaning Coatings on Photovoltaic Panels. *Solar Energy*, *199*, 63–73. doi:10.1016/j.solener.2020.01.084

Shaju, A., & Chacko, R. (2018). Soiling of Photovoltaic Modules – Review. *IOP Conference Series. Materials Science and Engineering*, *396*, 012050. doi:10.1088/1757-899X/396/1/012050

Shockley, W., & Queisser, H. J. (1961). Detailed Balance Limit of Efficiency of p-n Junction Solar Cells. *Journal of Applied Physics*, *32*(3), 510–519. doi:10.1063/1.1736034

Si, Y., & Guo, Z. (2015). Superhydrophobic Nanocoatings: From Materials to Fabrications and to Applications. *Nanoscale*, *7*(14), 5922–5946. doi:10.1039/C4NR07554D

Sun, K., Lu, L., Jiang, Y., Wang, Y., Zhou, K., & He, Z. (2018). Integrated Effects of PM2.5 Deposition, Module Surface Conditions and Nanocoatings on Solar PV Surface Glass Transmittance. *Renewable & Sustainable Energy Reviews*, *82*(Part 3), 4107–4120. doi:10.1016/j.rser.2017.10.062

Syafiq, A., Balakrishnan, V., Ali, M. S., Dhoble, S. J., Rahim, N. A., Omar, A., & Bakar, A. H. A. (2022). Application of Transparent Self-Cleaning Coating for Photovoltaic Panel: A Review. *Current Opinion in Chemical Engineering*, *36*, 100801. doi:10.1016/j.coche.2022.100801

Tayel, S. A., El-Maaty, A. E. A., Mostafa, E. M., & Elsaadawi, Y. F. (2022). Enhance the Performance of Photovoltaic Solar Panels by a Self-Cleaning and Hydrophobic Nanocoating. *Scientific Reports*, *12*, 21236. doi:10.1038/s41598-022-25667-4

Thongsuwan, W., Sroila, W., Kumpika, T., Kantarak, E., & Singjai, P. (2022). Antireflective, Photocatalytic, and Superhydrophilic Coating prepared by Facile Sparking Process for Photovoltaic Panels. *Scientific Reports*, *12*, 1675. doi:10.1038/s41598-022-05733-7

Vasilopoulou, M., da Silva, W. J., Soultati, A., Kim, H. P., Kim, B. S., Reo, Y., Gavim, A. E. X., Conforto, J., Schneider, F. S., Felippi, M., & Palilis, L. C. (2022). Photonic Nanostructures Mimicking Floral Epidermis for Perovskite Solar Cells. *Cell Reports. Physical Science*, *3*(9), 101019. doi:10.1016/j.xcrp.2022.101019

Walz, K. A., Hoege, T. D., Duensing, J. W., Zeltner, W. A., & Anderson, M. A. (2023). Field Tests of a Self-Sintering, Anti-Soiling, Self-Cleaning, Nanoporous Metal Oxide, Transparent Thin Film Coating for Solar Photovoltaic Modules. *Solar Energy Materials and Solar Cells*, *262*, 112560. doi:10.1016/j.solmat.2023.112560

Wang, D., Li, Y., Wen, Y., Li, X., & Du, X. (2021). Simple and Low Cost Fabrication of Large Area Nanocoatings with Mechanical Robustness, Enhanced Broadband Transmittance and Antifogging. *Colloids and Surfaces. A, Physicochemical and Engineering Aspects*, *629*, 127522. doi:10.1016/j.colsurfa.2021.127522

Wang, P., Yan, X., Zeng, J., Luo, C., & Wang, C. (2022). Anti-reflective Superhydrophobic Coatings with Excellent Durable and Self-Cleaning Properties for Solar Cells. *Applied Surface Science*, *602*, 154408. doi:10.1016/j.apsusc.2022.154408

Zhang, L.-Z., Pan, A.-J., Cai, R.-R., & Lu, H. (2019). Indoor Experiments of Dust Deposition Reduction on Solar Cell Covering Glass by Transparent Super-Hydrophobic Coating with Different Tilt Angles. *Solar Energy*, *188*, 1146–1155. doi:10.1016/j.solener.2019.07.026

Zhao, W., & Lu, H. (2021). Self-Cleaning Performance of Super-Hydrophilic Coatings for Dust Deposition Reduction on Solar Photovoltaic Cells. *Coatings*, *11*(9), 1059. doi:10.3390/coatings11091059

ADDITIONAL READING

Abdin, Z., Alim, M.A., Saidur, R., Islam, M.R., Rashmi, W., Mekhilef, S. & Wadi, A. (2013). Solar Energy Harvesting with the Application of Nanotechnology. Renewable and Sustainable Energy Reviews, 26, 837-852. DOI: 10.1016/j.rser.2013.06.023

Gebretinsae, H. G., Tsegay, M. G., Welegergs, G. G., Maaza, M., & Nuru, Z. Y. (2022). Effect of Rotational Speed on the Structural, Morphological, and Optical Properties of Biosynthesized Nickel Oxide Thin Films for Selective Solar Absorber Nanocoatings. *Energies*, *15*(23), 8960. doi:10.3390/en15238960

Mallikarjuna, K., Reddy, Y. S., Reddy, K. H., & Kumar, P. V. S. (2021). A Nanofluids and Nanocoatings Used for Solar Energy Harvesting and Heat Transfer Applications: A Retrospective Review Analysis. *Materials Today: Proceedings*, *37*(Part 2), 823–834. doi:10.1016/j.matpr.2020.05.833

Olorunfemi, T.R., Nwulu, N.I. &Ismail, S.O. (2022). Composites as Candidate Materials for Photovoltaic Cells. *Advances in Materials and Processing Technologies.* 8, sup. 4, 2378-2397. doi:10.1080/2374068X.2022.2046328

Serrano, E., Rus, G. & García-Martínez, J. (2009). Nanotechnology for Sustainable Energy. Renewable and Sustainable Energy Reviews, 13, 2373-2384. DOI: 10.1016/j.rser.2009.06.003

Ueno, S., & Fujihara, S. (2012). Effective Sol-Gel Nanocoatings on ZnO Electrodes for Suppressing Recombination in Dye-Sensitized Solar Cells. *International Journal of Photoenergy*, 2012, *268173*. doi:10.1155/2012/268173

Welegergs, G. G., Akoba, R., Sacky, J., & Nuru, Z. Y. (2021). Structural and Optical Properties of Copper Oxide (CuO) Nanocoatings as Selective Solar Absorber. *Materials Today: Proceedings*, *36*(Part 2), 509–513. doi:10.1016/j.matpr.2020.05.298

KEY TERMS AND DEFINITIONS

Antiicing Nanocoating: A thin layer that reduces or prevents the adhesion of ice on the surface of a material.

Antifogging Nanocoating: A thin layer offering hydrophilic properties that prevent the formation of dispersed water droplets by evenly spreading water over the surface of a material.

Functional Properties: Physical characteristics of some materials to exploit them in high-added-value applications of a mechanical, magnetic, thermal, electrical, optical, and/or acoustic nature.

Hydrophobic Nanocoating: A thin layer that repels water using the property of low surface energy.

Nanocoating: A thin film based on nanometric materials used to protect and/or improve the performance of a material using functional properties.

Self-Cleaning Nanocoating: A thin layer offering anti-static, photocatalytic, and hydrophilic or hydrophobic properties used to keep the surface of a material clean.

Substrate: Material on which a coating is deposited to improve the efficiency of a solar cell.

Transparent Nanocoating: A thin layer that allows light to pass through or has high transmittance when placed on the surface of a material.

Chapter 5
Perspectives of Biodegradable Nanocoatings in Food Packaging

Betül Bay Yılmaz
Yıldız Technical University, Turkey

ABSTRACT

Nanotechnology has the potential to improve human health, economic prosperity, product innovation, and overall quality of life, including through food nanopackaging—a promising area focusing on biodegradable packaging solutions. This alternative to conventional packaging minimizes waste, extends food shelf life, and enhances overall quality. The production of biodegradable nanocoatings can contribute to industry sustainability by reducing water consumption, solid waste, electricity use, and emissions. Bio-based coatings, with compatibility and matrix properties, can incorporate antioxidant agents and antimicrobial compounds, enhancing product safety, functionality, and shelf life. Biodegradable polymers, including polysaccharides, proteins, lipids, and polyesters, offer innovative pathways for entirely bio-based nanocoatings. These cost-effective, biocompatible, and renewable materials can be sourced directly from marine organisms and plants or produced through fermentative processes by microorganisms, such as microbial polyesters or polysaccharides. However, challenges in handling biopolymers, such as their hydrophilicity, crystallization tendencies, brittleness or melting instabilities, necessitate blending them with other materials to enhance their coating performance. Integrating nanoparticles within biopolymers can address environmental concerns by reducing packaging materials and enhancing recyclability. This approach aligns with a more eco-sustainable approach to food packaging, resulting in reduced waste.

DOI: 10.4018/979-8-3693-3136-1.ch005

INTRODUCTION

The implementation of nanoscience and nanotechnology in the fields of food, agriculture, and food safety is a recent innovation compared with its well-established utilization in sectors such as paints and coatings, cosmetics, personal care products, drug delivery and pharmaceuticals, diagnostics and medical therapeutics, molecular computing, energy production, and structural materials (Vasile, 2018). The food industry has been prompted to reassess its approach and prioritize secure, sustainable, and user-friendly packaging due to evolving consumer demands, driven not only by stringent EU-wide regulations, but also by shifting demographics, such as the rise of single households, aging populations, and an increasing demand for convenient "food on the run" (European Parliament & Council, 2004). As per Regulation (EC) No. 1935/2004, an effective packaging solution is required to fulfill various functions, encompassing the safeguarding of food from contaminants, such as oxygen, moisture, dirt, light, dust, pathogenic microorganisms, and other potentially deleterious substances. Additionally, packaging must demonstrate safety under its designated usage conditions, including inertness, cost-effectiveness in production, lightweight properties, ease of disposal or reusability, resilience to rigorous circumstances encountered throughout the processing or filling stages, impermeability to diverse sustainable storage and transport conditions, and durability against physical stress (European Parliament & Council, 2004; Vasile, 2018).

Modern manufacturing, commercialization trends, transformations in retailing and distribution methods, and the global nature of the food industry require longer product shelf-lives in order to accommodate the complexities of long-distance transportation (Janjarasskul & Suppakul, 2016). Therefore, it is imperative for packaging materials to exhibit the desirable optical, thermal, and mechanical properties to fulfill their protective functions. Additionally, the antimicrobial characteristics of food packaging materials and their capacity to serve as barriers against aromas and gases are of paramount importance in the realm of nanopackaging (Rai et al., 2019).

The use of bio-based materials in food packaging is constrained by the inadequate barrier and mechanical properties of natural polymers. To address these limitations, natural polymers have commonly been combined with synthetic counterparts, or have been chemically modified, aiming to enhance their applicability in packaging (Kuswandi, 2017). Nanomaterials exhibit exceptional functional attributes, encompassing a broad specific surface area, improved barrier properties, and antibacterial activity. These features offer advantages in overcoming the limitations associated with traditional approaches to extending the shelf life of food products (Manikandan & Min, 2023). Growing consumer awareness of environmental concerns has encouraged the industry to explore "green" alternatives with renewed interest (Vilarinho et al., 2018). In this context, nanotechnology has the potential to revolutionize food preservation and can play a vital role in ensuring food security for the future (Wypij et al., 2023). With growing global consciousness regarding the environmental implications of carbon emissions, the requirement for sustainable nanocomposite materials is on the rise. Nanocomposites are rapidly emerging as compelling substitutes for conventional materials due to their distinctive attributes, which render them well-suited for deployment in ecologically conscious technological endeavors (Saharudin et al., 2023).

According to recent studies, the food packaging industry is a rapidly expanding market, with projected growth exceeding billions of dollars. Anticipated overall expansion is expected to achieve a 6% increase by the year 2024 (Mohammad & Ahmad, 2024). Current challenges in the food and packaging sectors include unsustainable production, limited recyclability, and inadequate mechanical and barrier properties. Despite the effectiveness of barrier materials, i.e. metals and glass in preventing undesirable mass transport in food packaging, the enduring preference for plastics, driven by their lightweight, formable, cost-effective, and versatile nature, has led to the packaging industry utilizing over 40% plastics, half of which is dedicated to food packaging (Mihindukulasuriya & Lim, 2014). Plastic packaging materials have low heat resistance, contain toxic plasticizers, and contribute to environmental pollution due to their non-biodegradable nature. Furthermore, annually, hundreds of thousands of tons of plastic waste are disposed of in the environment, leading to its accumulation in the oceans. While one approach to mitigate this accumulation is through incineration, this is both expensive and hazardous, emitting carbon dioxide (CO_2) and detrimental chemicals into the atmosphere (Vilarinho et al., 2018). Plastic pollution incorporates two key aspects, first during the production phase, primarily through carbon emissions, and second during disposal, leading to the introduction of contaminants and physical hazards that affect the environment and ecosystems. A promising remedy to this issue involves the substitution of petroleum-derived polymers with bio-based alternatives, which offer a considerable reduction in carbon emissions and energy demands throughout the production process (Meereboer et al., 2020). Polymers have played a prominent role in the realm of food packaging for several decades. Conventional polymers originate from non-renewable fossil sources, the pervasive utilization of plastics in food packaging being attributed to their exceptional attributes, advancements in technology, and their comparatively economical production (Nilsen-Nygaard et al., 2021). Biodegradable films present a viable alternative, offering to mitigate or lessen reliance on non-biodegradable packaging, primarily driven by escalating apprehensions regarding the ecological ramifications associated with conventional non-biodegradable materials. The shift toward biodegradable alternatives underscores a growing recognition in both academic and industrial spheres of the imperative need for sustainable practices in packaging, aligning with broader environmental conservation objectives (Assis et al., 2017).

Nanotechnology offers significant improvements by providing packaging materials with enhanced mechanical and barrier properties (Majid et al., 2018). Current scientific research is focusing on the advancement of nanocoatings and films using bionanocomposites, or incorporating nanoparticles (NPs) as a matrix to enhance the shelf life and preservation quality of fruits and vegetables during storage (Primožič et al., 2021). Food commodities, being biological systems, undergo rapid deterioration following harvesting or slaughtering, necessitating specialized handling practices to mitigate the spoilage risks associated with contamination. The preservation of optimal food quality is contingent on the judicious application of appropriate processing and packaging methodologies. Thus, the imperative lies in the conscientious management of food and its products, employing packaging materials characterized by non-toxicity, safety, and environmental friendliness. This underscores the significance of employing meticulous care in the selection and application of packaging materials in the context of food preservation and quality maintenance (Adeyeye & Ashaolu, 2021). Nanomaterials inherently exhibit distinctive physical, optical, and chemical properties, thereby presenting a paradigm shift in materials science. The introduction of nanomaterials has provided a means of tailoring fundamental material properties, including, but not limited to, flexibility, durability, flame resistance, barrier characteristics, and recyclability (Bumbudsanpharoke et al., 2015). The utilization of nanocoating technology in the food industry has significant potential, particulary in adressing food safety concerns and prolonging the shelf life of perishable fruits and vegetables. Additionally, it offers opportunities to enhance quality and mitigate financial losses. Numerous research studies have underscored the efficacy of edible nanocoatings—employing diverse NPs in extending the shelf life and preserving the nutritional integrity of fruits and vegetable throughout their storage (Mohammad & Ahmad, 2024). The recent incorporation of food-grade materials endowed with enhanced properties, coupled with innovative deposition techniques, has facilitated the application of edible coatings in exceptionally thin layers on food surfaces. This process establishes an imperceptible physical barrier that effectively guards against oxygen exposure, moisture absorption/desorption in food, and external microbial contamination, all while preserving its organoleptic properties (Pirozi et al., 2020). An edible coating is characterized as a thin layer of edible material. Edible films are thin, solid layers or sheets employed as wrappings for food products. By contrast, edible coatings are administered in a liquid state onto the food to be coated, typically through immersion in a solution containing structural matrix-forming substances, such as proteins, lipids, carbohydrates, or combinations of these (Ju et al., 2019).

In the realm of nanotechnology, polymer nanocomposites are defined as solid materials composed of two or more phase-separated components, where one or more dispersed phases exist at the nanoscale, embedded within a polymeric matrix.

Polymer nanocomposites and NPs are also applicable as nanocoatings, signifying the deposition of a nanoscale layer onto specific substrates to achieve the desired surface characteristics. Materials are considered nano-scale when their size falls within a range of approximately 1 to 100 nm in at least one of the three external dimensions (Müller et al., 2017). Recently, bionanomaterials, consisting of polysaccharides and proteins infused with small quantities of nanostructured substances, have been attracting considerable interest as potential novel, eco-friendly, and active packaging solutions for the food industry (Sothornvit, 2019). Biopolymers, which are cost-effective, biodegradable and biocompatible, are a viable alternative to synthetic plastics (de Azeredo, 2009). With growing consumer demands for long-lasting goods, exploring the creation of biocompatible materials with functional traits, such as edible, active, and biodegradable films, presents a suitable strategy for elevating product quality and improving nutritional value (Alizadeh-Sani et al., 2018). The development of biopolymers and sustainable packaging solutions is underway around the world, with researchers devoting considerable effort to generating raw materials from natural sources, including proteins, fats, carbohydrates, and microbial derivatives, such as polylactic acid and polyhydroxybutyrate, for coating or film applications (Achachlouei & Zahedi, 2018). The quest for more-natural and eco-friendly packaging systems involves various measures, such as replacing synthetic materials with biodegradable or edible alternatives, reducing overall material usage, and increasing the use of recyclable and reusable materials. Ideally, to sustain the protective features of food packaging, a holistic approach should be embraced. This involves integrating research findings on the carbon footprint and life-cycle studies of diverse materials to implement effective sustainable strategies (Han et al., 2012). Incorporating NPs into the matrices of synthetic and biopolymer films is proving to be a highly effective approach to this enhancement. These engineered nanocomposites show improvements in barrier, mechanical, and thermal properties when compared to pure biopolymers and traditional composites. These advancements are yielding cutting-edge, high-performance materials, applicable across diverse industrial sectors, and notably in the food packaging arena (Achachlouei & Zahedi, 2018).

Bio-based materials are classified based on their origin: (1) polymers directly derived from natural sources, such as polynucleotides, polysaccharides, and proteins; (2) polymers synthesized from biomonomers, chemically produced from bio-based monomers or chemically produced using bio-based monomers, including biopolyesters, poly(butylene succinate) (PBS), polylactic acid (PLA), polyglycolic acid, and polyvinyl alcohol (PVA); (3) polymers derived from genetically modified organisms or microorganisms, such as xanthan, including polyhydroxyalkonoates (PHAs), polyhydroxybutyrate (PHB), poly(3-hydroxybutyrate-co-3hydroxyvalerate) (PHBV), and bacterial cellulose (Chausali et al., 2022; Kuswandi & Moradi, 2019). Biodegradable polymers are identified as materials capable of undergoing degradation to methane, CO_2, water, biomass and/or inorganic compounds, predominantly through the enzymatic activities of microorganisms (Khaneghah et al., 2018). Biopolymers have a range of applications, notably in coatings and films. Coatings involve direct application onto the surface of food items, while films are structures crafted separately and subsequently applied (Popovic et al., 2018). The utilization of edible coatings has gained prominence, particularly through the integration of bioactive compounds. These compounds, comprising natural antimicrobials, antioxidants, minerals, and vitamins, are strategically incorporated to not only elevate food safety standards and uphold quality attributes, but also to confer substantive health benefits to consumers. This paradigm underscores a nuanced approach toward enhancing both the intrinsic and extrinsic qualities of edible coatings for fortified food products (Pirozzi et al., 2021). A diagram illustrating the key characteristics of bio-based food nanopackaging is provided in Figure 1. These characteristics include biocompatibility, antimicrobial activity, biodegradability, structural stability, mechanical strength, and high surface area. Materials derived from biopolymers need to meet fundamental criteria, including health safety, mechanical and chemical resilience, and long-lasting performance. Consequently, producers of biopolymer-based packaging prioritize attributes that extend beyond just biodegradability, focusing on ensuring these materials offer functionality that can rival traditional polymers (Popovic et al., 2018).

NATURAL BIODEGRADABLE NANOCOATINGS

Researchers and industries have redirected their focus toward the exploration of biodegradable materials for food packaging purposes to alleviate environmental issues. Lebreton et al. (2018) reported that the Great Pacific Garbage Patch holds more than 79,000 tons of plastic waste, and marine microplastics have sharply increased from 0.4 kg km^{-2} in the 1970s to 1.23 kg km^{-2} in 2015 (Zhao et al., 2021). Simultaneously, the production of bio-based non-biodegradable and biodegradable plastics, forecasted from 2020 to 2023, is projected to grow at a rate of 13% annually. Leading plastic packaging manufacturers aim to achieve a goal of incorporating 100% recycled, biodegradable or reusable plastics in their products by 2025, as part of their sustainability initiatives (Meereboer et al., 2020). The global production capacity of bioplastics is anticipated to experience significant growth, forecasted to rise by over 15%, from around 2.11 million metric tonnes in 2019 to approximately 2.43 million metric tonnes by 2024 (Primožič et al., 2021).

Bio-based materials can serve as a protective barrier, safeguarding food from environmental factors that could lead to quality deterioration, including microbial contamination, alterations in gas composition, and fluctuations in relative humidity. Notably, these materials have the advantage of being degradable through the action

Figure 1. Key characteristics of bio-based food nano-packaging
Source: Modified from Perera et al. (2023)

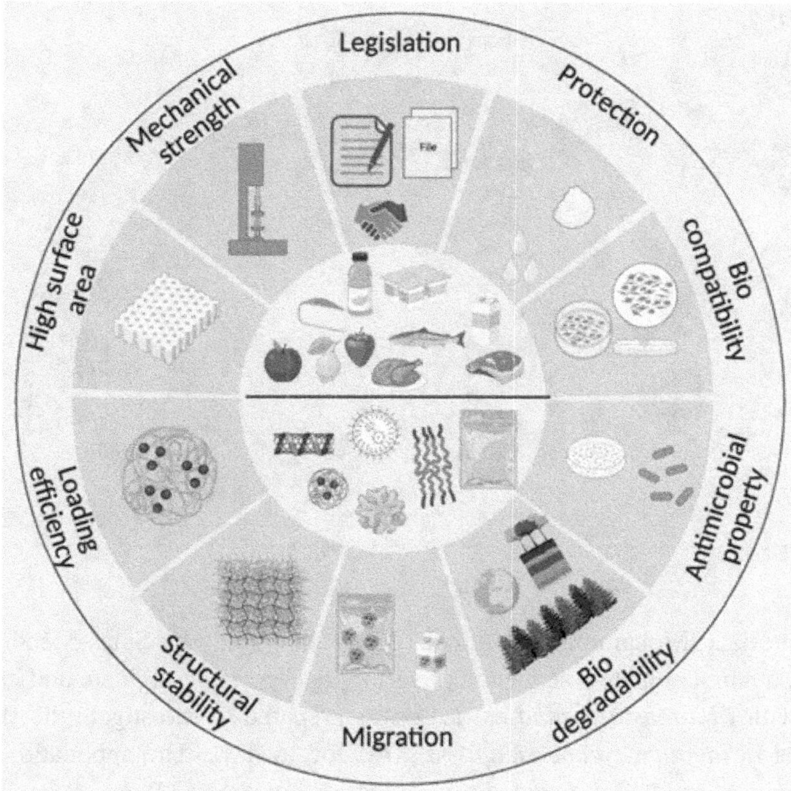

of living microorganisms, rendering bio-based packaging more environmentally sustainable in comparison to conventional plastic packaging (Primožič et al., 2021). Bio-based polymers, derived from bio-based resources, can be directly obtained from marine organisms or plants, such as proteins and polysaccharides, or manufactured by microorganisms through fermentative processes, exemplified by microbial polysaccharides or microbial polyesters (PHAs). Notably, PHB acid and PLA represent two biopolymers targeted for deployment in the food industry as biodegradable packaging components (Nešić et al., 2020). Various nano-based materials and products have been designed for food storage, coating, and packaging. Figure 2 shows a selection of commercially available items dedicated to food packaging and safety.

The utilization of edible coatings containing NPs, particularly those crafted on a nano-scale ranging from 100 to 1,000 nm, has promoted a notable enhancement in prolonging the shelf life of fruits compared to conventional coatings lacking

Figure 2. Commercially accessible biodegradable food nano-packaging in the market

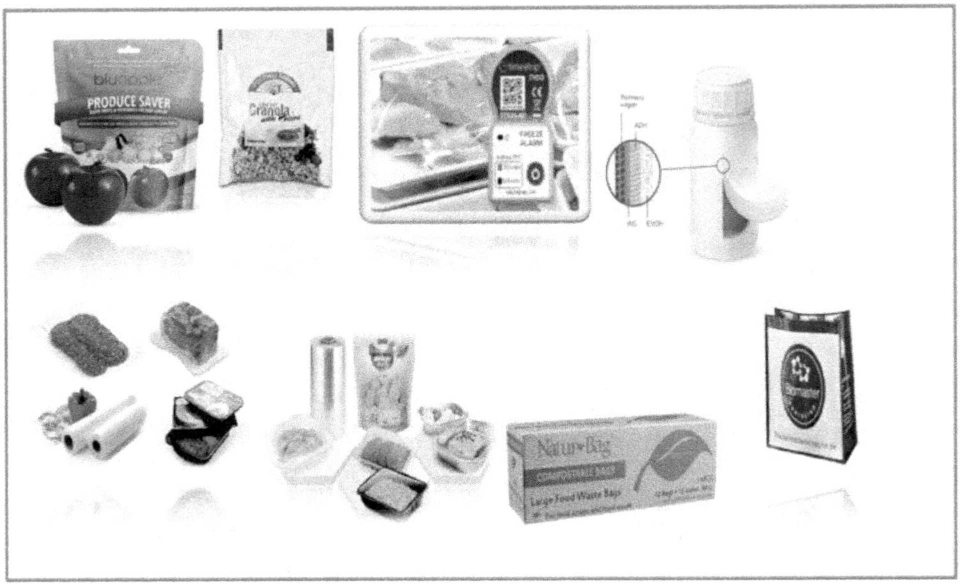

nanomaterials (Mohammad & Ahmad, 2024). In a study by Salas-Méndez et al. (2019), a nanolaminate—sequential layering incorporating alginate and chitosan, along with *Flourensia cernua* extract—was prepared to investigate the shelf life of tomato (*Solanum lycopersicum*) at 20°C for 15 days. The application of this nanolaminate resulted in reduced weight loss and oxygen/CO_2 gas exchange rates. The coating exhibited a decline in ethylene production and effectively hindered the proliferation of aerobic mesophilic microorganisms, as well as suppressing yeast and mold growth. Dwivany et al. (2020) formulated a coating for bananas, applying treatments with carrageenan concentrations of 0.5%, 1.0%, and 1.5%, followed by storage at temperatures of 26°C and 20°C. The findings indicated that the optimal condition for prolonging shelf life and preserving fruit quality involved treating the bananas with 1.5% carrageenan and storing them at the lower temperature of 20°C. In the preservation of strawberries stored at 35–40% relative humidity and 20°C, the application of chitosan-based coatings, including formulations with chitosan as a monolayer, as a three-layered coating with separate layers of beeswax, beeswax, and chitosan, and a crosslinked three-layered coating with sodium tripolyphosphate (TPP), chitosan, and a composite coating, proved beneficial. Velickova et al. (2013) reported that these coatings contributed to a reduction in fungal infection, weight loss, and respiration rate, and enhanced maintenance of firmness, color, pH, titratable acidity, and sugars and soluble solids.

Biodegradability refers to the capacity of materials to decompose and re-enter the natural environment shortly after disposal. These materials, sourced from both natural and fossil origins, undergo biodegradation facilitated by microorganisms in their inherent surroundings. The European standard EN 13432 outlines the definition of biodegradable bioplastic as the process wherein, under the influence of microorganisms in the presence of oxygen, an organic chemical compound undergoes decomposition, resulting in CO_2, water, mineral salts, and the generation of new biomass (Baghi et al., 2022). When producing biodegradable films for food packaging, it is crucial to select biopolymers that fulfill specific criteria, such as renewability, cost-effectiveness, and preferably being sourced from waste materials. Among the materials investigated for biodegradable films and coatings, polysaccharides, such as starch, cellulose, carrageenan, chitosan, pectin, alginate, kefiran, and pullulan, have been thoroughly examined (Cazón et al., 2017; Primožič et al., 2021; Wang et al., 2019). Nanocomposite films derived from biopolymers are referred to as bionanocomposites (Rai et al., 2019). The use of bio-based nanofilms and hydrogels as immediate coatings offers the potential to extend shelf life and safeguard against pathogens and pests. The use of biopolymers in food packaging has considerable potential. Their inherent properties, which include mechanical strength and being barriers to gases, water, and microbes, can be significantly enhanced by the integration of nanofillers, and particularly NPs (Wypij et al., 2023). However, the handling of most biopolymers can be challenging due to their inherent characteristics, such as their comparatively high molecular weight, fragility, viscosity, crystallization activity, hydrophobicity, and melting nature (Chaudhary et al., 2020). Natural polymers, such as chitosan, face challenges in mechanical strength compared to their synthetic counterparts, and their inherently hydrophilic nature, while contributing to high moisture barrier properties, has prompted the exploration of various strategies to address these limitations in biodegradable packaging films. Among these strategies is the incorporation of plasticizers, such as glycerol, designed to augment the flexibility of the final product (Srinivasa et al., 2007).

The adoption of biomaterials, such as edible and biodegradable films derived from renewable sources, is crucial to mitigate the environmental impact of non-degradable plastics and enhance the sustainability of food packaging. Typically, these substances demonstrate ecological friendliness and possess the ability to undergo complete degradation into CO_2, water, and biomass. Packaging nanomaterials derived from renewable sources—known as bio-based materials—are manufactured, leading to potential energy conservation through the incineration of these sustainable materials (Kuswandi & Moradi, 2019). Bio-based materials, which can be sourced from renewable biomass or non-renewable resources, exhibit characteristics of both biodegradability and non-biodegradability (Primožič et al., 2021). Below, bio-based materials characterized by biodegradability are specifically addressed. Figure 3 outlines the essential categories for classifying biodegradable nanocoating materials.

Figure 3. Classification of biodegradable nanocoating materials

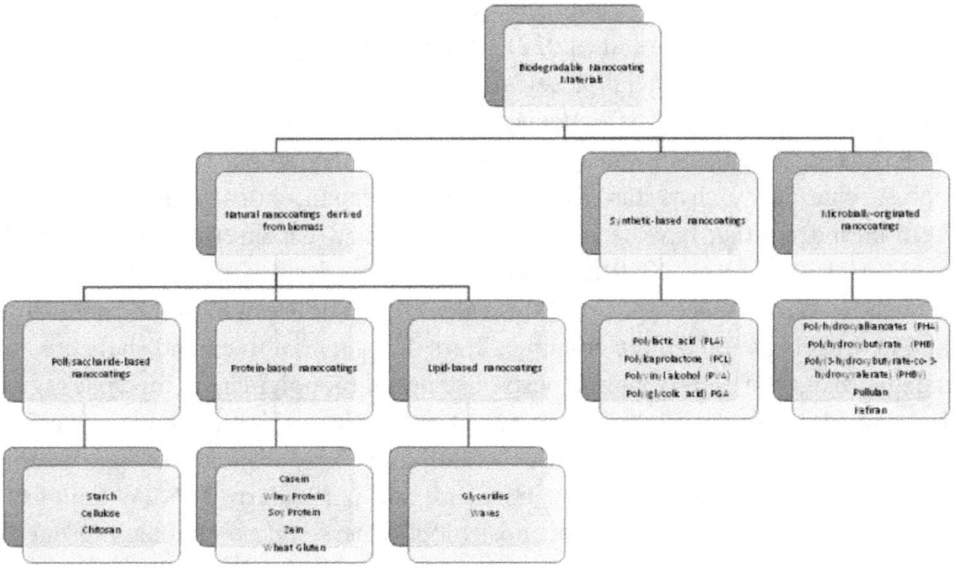

Figure 3. Classification of biodegradable nanocoating materials

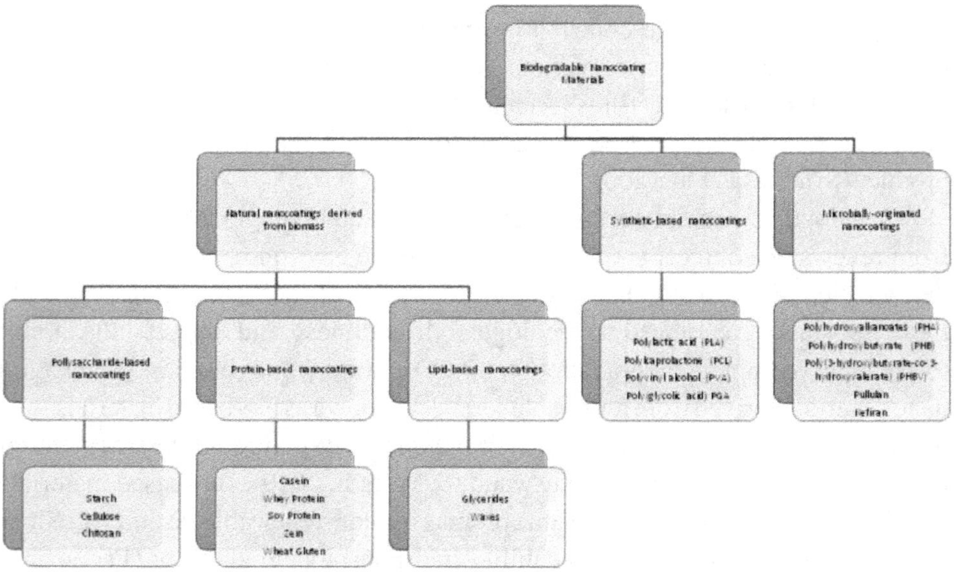

Polysaccharide-Based Nanocoatings

The need for biodegradable packaging films has been fueled by concerns regarding environmental pollution, solid packaging waste and the depletion of petroleum reserves. As a result, there has been a shift toward the advancement of biodegradable materials that can undergo degradation in the environment through the action of microorganisms. These substitutes are derived from renewable biopolymers, including proteins, polysaccharides, lipids, and their composites. Biopolymer-based packaging materials offer a promising alternative to the disposal of non-biodegradable plastic materials derived from petrochemicals. Researchers have focused on the utilization of naturally occurring biopolymers, such as proteins and lipids (Rai et al., 2019).

Starch-based Nanocoatings: Starch—a natural polysaccharide—plays a crucial role in food production, both as a high-quality component and as an additive. Structurally, starch consists of two main macromolecules—amylose, a linear polysaccharide formed by poly(α-1,4-glucopyranosyl), and amylopectin, a poly(α-1,4-glucopyranosyl) with multiple α-1,6-glucopyranosyl branches (Mesgari et al., 2022). Polysaccharide-based films offer numerous advantages, such as biodegradability, non-toxicity, relatively effective gas barrier properties, and natural abundance. However, they have limited resistance to water permeation because of their inherent hydrophilic nature. This, in turn, affects their mechanical, thermal, and gas permeability properties (Noshirvani et al., 2017).

Starch, in contrast to synthetic materials, has two main disadvantages. Firstly, its high hydrophilicity, due to its three hydroxyl groups per D-glycosylic unit, makes it sensitive to moisture changes, affecting its physical and mechanical properties (Versino et al., 2016). Secondly, starch is not inherently thermoplastic. It can be transformed into a continuous polymeric phase by the addition of sufficient water or non-aqueous plasticizer—often polyols, such as glycerol (Angellier et al., 2006). In the production of thermoplastic starch, plasticizers are used to reduce intramolecular hydrogen bonding, thereby stabilizing the properties of the product (Kuswandi & Moradi, 2019). Another feasible option is to enhance thermoplastic starch by incorporating mineral or cellulosic fillers. The use of cellulosic fillers as reinforcing agents in thermoplastic starch helps to maintain its biodegradability (Angellier et al., 2006; Shi et al., 2013). However, plasticizers can also negatively impact the resistance properties of the film, leading to heightened permeability to gases, as well as a decrease in tensile strength (Versino et al., 2016).

Starch NPs—also known as starch nanocrystals (SNCs)—possess distinctive functional attributes that render them suitable for various food applications, particularly in the realm of biodegradable food packaging. These NPs are derived from starch through processes such as acid or enzymatic hydrolysis, as well as alternative mechanical techniques (Kim et al., 2015). Platelet-like SNCs were utilized in crafting a composite membrane in conjunction with biodegradable poly(ε-caprolactone) (PCL), demonstrating possible applications in packaging. The inclusion of SNCs significantly enhances gas barrier properties, tear strength, and creep resistance in the PCL membrane, particularly at relatively low loading levels. When acetylated SNCs (1 wt%) are incorporated, notable improvements are observed, including a 70% reduction in oxygen transmission rate and a 68% increase in tear strength, attributed to enhanced phase adhesion (Xu et al., 2018). The introduction of NPs into the polymer matrix resulted in a reduction in water uptake. Therefore, one approach to mitigate the hydrophilic nature of materials is to create composites by blending them with biopolymers (Noshirvani et al., 2017). Shi et al. (2013) documented the successful production of starch NPs using the water-in-oil mini-emulsion crosslinking technique with the aid of a high-pressure homogenizer. In this context, starch-based films represent a highly valuable category of bio-based packaging materials, showcasing excellent performance, adaptability to various products, processing capabilities, and cost-effectiveness (Mesgari et al., 2022). Eco-friendly bionanocomposites made from starch and titanium dioxide (TiO_2) were created with varying concentrations of nano-TiO_2 (1%, 3%, and 5% by weight). The findings revealed that a higher TiO_2 content led to an enhanced hydrophobic nature of the starch/TiO_2 films. This increase in TiO_2 content also correlated with a reduction in the permeability of water vapor through the bionanocomposites. In addition, the starch/TiO_2 nanocomposites demonstrated a strong ability to shield products from UV light, suggesting their potential application as protective packaging materials against UV radiation (Goudarzi et al., 2017).

The utilization of biodegradable packaging materials also enables the creation of antimicrobial packaging solutions. Antimicrobial food packaging has gained attention for its ability to inhibit or slow down the microbial decay of food products. This approach involves integrating antimicrobial substances into the packaging system to mitigate the risk of pathogen contamination. One effective method in antimicrobial packaging development is the application of antimicrobial agents as coatings on polymer surfaces, providing a large surface area for exposure, while minimizing the release of biocides into the packaged food (Al-Naamani et al., 2018). Figure 4 illustrates various sources of naturally derived biodegradable packaging materials, as well as the processes involved in their production. Corn starch (CS) was incorporated into a PCL/pomegranate-peel extract matrix to produce antimicrobial packaging materials. Bacteriostatic experiments showed that the PCL/CS/pomegranate-peel extract composite film exhibited remarkable inhibitory antimicrobial activity against *Staphylococcus aureus*. In addition, the incorporation of starch enhanced the antimicrobial efficacy of the pomegranate-peel extract, while also acting as a conduit for the release of polyphenols by mitigating interactions between the PCL and the pomegranate-peel extract. Given the biodegradable and food-safe nature of all materials employed, the resulting product has promise as a food-grade antimicrobial packaging material (Khalid et al., 2018). Babaee et al. (2022) focused on plasticized starch (PS)-based films, which incorporated chitosan NPs (CNPs) at concentrations of 1%, 2%, 3%, and 4%, serving as reinforcing and antibacterial agents. The escalation in CNP content yielded enhanced mechanical and barrier properties. Notably, PS-CNP composite films demonstrated inhibition zones against *Escherichia coli* and *S. aureus*, leading to a 100% reduction in colony-forming units (CFUs) against *S. aureus*. The study underscored the heightened antibacterial efficacy of PS-CNP films compared to pure PS, signifying their potential application in antimicrobial food packaging.

Nanoparticles have the capability of strengthening packaging materials, which can lead to a reduction in water-vapor permeability (WVP) and an improvement in its functional properties. Zein NPs encapsulating 10% (w/w) of clove essential oil (CEO) were electrosprayed onto, and subsequently incorporated into, a starch matrix at varying levels (5%, 10%, and 15% w/w). Remarkably, up to 10% incorporation of electrosprayed zein NPs loaded with CEO significantly reduced WVP. Similarly, the 10% inclusion of electrosprayed zein NPs loaded with CEO notably enhanced the tensile strength and Young's modulus and diminished elongation at break. The study indicated sustained CEO release, primarily in a 10% ethanol medium, showcasing its potential use as a biopreservative and edible coating for perishable food products (Alinaqi et al., 2021). In a study by Jayakumar et al. (2019), composite films based on starch and PVA were created. The films were enriched with zinc oxide (ZnO) NPs and phytochemicals, using the solvent casting method. The resulting nanocomposite films exhibited improved attributes in terms of water resistance, mechanical strength, and antimicrobial efficacy. Notably, the nanocomposite's ability to indicate pH suggested its promise in the realm of food packaging (Jayakumar et al., 2019). Novel bioactive food packaging materials were created by incorporating zein, CS, and orange-peel oil. The integration of CS, orange-peel oil, and zein nanocapsules (known as OZN) resulted in improved mechanical strength and moisture barrier

Figure 4. Diagram illustrating various sources of naturally derived biodegradable packaging materials, as well as the processes involved in their production

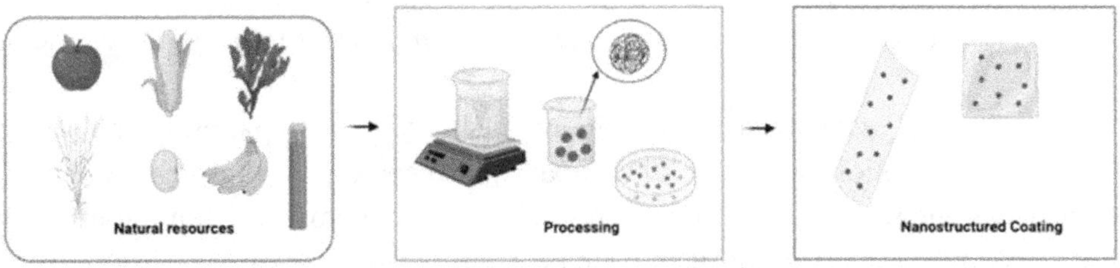

properties, and prolonged duration of orange-peel oil release. Optimal properties were observed at an OZN/CS ratio of 5:5, where the material exhibited the highest elongation at break (30.91%) and tensile strength (12.19 MPa). In addition, the composite film had favorable oxidation resistance and 1,1-diphenyl-2-picrylhydrazyl (DPPH) radical scavenging activity (30.16%) (Wang et al., 2019).

Cellulose-based Nanocoatings: Cellulose is a natural polymer comprising repeating units of homopolysaccharides connected by β-1,4 linkages of D-glucose (Azizi-Lalabadi et al., 2020). Cellulose—a widely available and renewable raw material—stands out as an ideal candidate for polymer nanocoating processing due to its reinforcing capabilities, mechanical properties, low density, abundance, and biodegradability (Vilarinho et al., 2018). Cellulosic nanomaterials possess expansive specific surface areas and the capability of establishing hydrogen bonds. This bonding capacity enables the material to construct a robust and compact network, rendering it highly resistant to the permeation of different molecules (Ferrer et al., 2017). Significant research has been conducted on various potential applications of cellulose nanocrystals (CNCs) in the food industry. These applications include roles as food stabilizers, functional ingredients for improved nutrition, delivery systems for bioactive compounds, and, in particular, components of environmentally friendly biodegradable packaging composites (Xiao et al., 2020). Cellulose NPs are commonly employed to reinforce polymer composites in the context of food packaging applications. Cellulosic nanocoatings possess several advantageous characteristics, including widespread availability, environmental friendliness, ease of recycling through incineration, and low energy consumption during production. These qualities make cellulose nanocoatings an attractive category of nanomaterials for the production of cost-effective, lightweight, and high-strength nanocoatings (de Azeredo, 2009). Nanocellulose can be synergistically combined with other materials through various ex-situ and in-situ techniques to improve its mechanical, physical, moisture, and gas properties, and to incorporate antimicrobial, antioxidant and UV-blocking properties (Perdani & Gunawan, 2021). In a recent study, the mechanical strength of the composite was significantly improved by incorporating dextran-coated silver (Ag) NPs into a cellulose nanofiber (CNF) film (Barage et al., 2022). In addition, the nanocomposite film exhibited high oxygen permeability and demonstrated time-dependent antimicrobial activity against *E. coli*, making it a viable and environmentally friendly option for various food packaging applications. However, its effectiveness against *S. aureus* was limited due to the slow release of Ag ions.

Cellulose is recognized for its biodegradability, biocompatibility, non-toxic nature, and exceptional water-holding capacity, and bacterial cellulose has garnered significant interest in the additive manufacturing, biomedical, environmental, and food industries. Cellulose is sourced from various origins, such as plants, microbes, animals, and select algae, often through enzymatic processes. Bacterial cellulose can be produced by aerobic bacteria, such as *Aerobacter*, *Agrobacterium*, *Rhizobium*, *Azotobacter*, *Gluconacetobacter*, *Pseudomonas*, *Alcaligenes*, *Sarcina*, and *Rhodobacter* (Jang et al., 2023). Bacterial-cellulose-based packaging films were fabricated using cashew apple juice as a precursor, and further enriched with lignin (0–15 wt%) and CNCs (0–8 wt%) sourced from cashew tree pruning fiber. The incorporation of these components resulted in improved tensile properties and reduced WVP in the films (Sá et al., 2020). Nanocomposite films made from soy-protein isolate strengthened with CNCs, ZnO NPs, CNC/ZnO NP mixtures, and CNC/ZnO NP nanohybrids exhibited enhanced water resistance, barrier, mechanical, and thermal properties. Films containing ZnO effectively impeded the proliferation of foodborne pathogens, such as *S. aureus* and *E. coli*, leading to a reduction in total viable counts. This study underscores the potential of these nanocomposite films for diverse applications, emphasizing their improved mechanical, barrier, antimicrobial, and thermal characteristics (Xiao et al., 2020).

The impact of adding different concentrations (0.1%, 0.2%, and 0.3% w/v) of pectin to the growth medium of *Gluconacetobacter xylinus* on the properties of the resulting biocomposite was investigated. The bacterial cellulose–pectin biocomposites were evaluated for structural, mechanical, crystalline, morphological, color, and opacity characteristics. Generally, the inclusion of pectin led to reduced crystallinity, tensile strength, tear strength, and lightness in the biocomposites. However, the tensile strength (23.05–25.72 MPa) and tear strength (13.30–15.72 g μm^{-1}) of the cellulose–pectin biocomposites were comparable to conventional non-biodegradable plastics, such as high-density polyethylene, hinting at their potential use as green food packaging with a semi-translucent appearance (Agustin et al., 2021).

Cellulose nanocrystals also find potential applications in the active food packaging field. Coconut shell CNFs were successfully synthesized from agro-waste coconut shells and incorporated into a biodegradable packaging film through a multi-step process. These CNFs were combined with PVA and enriched with lemon oil and linseed oil. The inclusion of essential oils notably enhanced the film's antioxidant properties, promoting $31.52 \pm 0.08\%$ free-radical scavenging activity. The resulting composite film also exhibited strong antimicrobial activity against foodborne pathogens, presenting possible applications in the active food packaging domain, and displayed excellent biodegradability, with $87.34 \pm 0.91\%$ degradation achieved within 45 days (Arun et al., 2022).

Carboxymethyl cellulose (CMC), which is derived from cellulose, possesses favorable attributes for forming films and results in transparent films with strong barriers against oxygen and lipids (Noshirvani et al., 2017). Several biodegradable films have been created utilizing CMC, either independently or in conjunction with diverse carbohydrates, proteins, lipids, and additional functional additives. While these CMC-based films exhibit strong barriers and relatively favorable mechanical properties, they show significant susceptibility to water-vapor transmission. The inclusion of hydrophobic substances has been found to effectively reduce the WVP of nanocomposite films (Achachlouei & Zahedi, 2018). Some examples of starch- and cellulose-based biodegradable nanocoating materials are provided in Table 1.

Chitosan-based Nanocoatings: Chitosan—a heteropolysaccharide—is recognized for its biocompatibility, biodegradability, and ability for metal complexation (Sharma et al., 2017). Chitosan's structure comprises ß-(1,4)-linked N-glucosamine and N-acetylglucosamine units. It exhibits solubility in dilute acids, including acetic acid and formic acid (Phisalaphong & Jatupaiboon, 2008). Chitosan, resulting from the deacetylation of chitin, ranks as the second most plentiful polysaccharide in nature, following cellulose. It has excellent potential as a biopolymer for food packaging due to its advantageous film-forming and antimicrobial properties, and its ability to selectively permeate gases (Mitelut et al., 2015). To improve both the functional and physical characteristics of chitosan films, the idea of incorporating other natural biopolymers has been proposed. The concept of chitosan-blended films has been explored in numerous studies in the literature. Examples include films combining chitosan with whey protein, methylcellulose, gelatin, and pea starch (Ashori & Bahrami, 2014).

The integration of chitosan with metallic NPs presents a pragmatic method for creating antimicrobial packaging systems. This approach offers benefits such as thermal stability, superior barrier properties, and the provision of antimicrobial and antioxidant features (Mesgari et al., 2022). Due to its ability to effectively disperse TiO_2 NPs, the CS/TiO_2 composite photocatalyst has become a highly studied topic, with a focus on elucidating the components' synergistic interactions. The widespread exploration of the immobilization of TiO_2 within CS films is due to their easy miscibility, facilitated by the inherent hydrophilicity of TiO_2, making it a readily achievable process (Ainali et al., 2021).

Chitosan finds extensive applications across diverse fields, including the realm of edible films and coatings. Chitosan possesses inherent natural attributes,

Table 1. Starch-based and cellulose-based food coating applications and suitability

Bio-Based Material Type	Application	Suitability	References
Starch Based	Food wrap, Food container, Grocery plastic bag, Sachet for food product	Suitable for diverse packaging solutions (Low water vapor permeability, integrity of packaging, enhanced thermal stability)	Ayu et al., 2018; Rommak et al., 2021
Cellulose Based	Coating for fresh fruit and vegetable	Enhance shelf life, reduce physical damage	Basumatary et al., 2022

showcasing remarkable strength and elongation properties. Chitosan films exhibit selective permeability to gases such as CO_2 and O_2, along with outstanding mechanical characteristics (Elsabee & Abdou, 2013; Pereda et al., 2012). Despite these positive qualities, the application of chitosan films is encumbered by a number of limitations, the foremost concern being their notable brittleness and rigidity. Additionally, the hydrophilic nature of chitosan films results in a limited moisture barrier, constraining its potential applications (Azaza et al., 2022). Typically, the incorporation of plasticizers, both hydrophilic to augment and hydrophobic to diminish WVP, is a common practice in the development of films derived from proteins and polysaccharides. The addition of plasticizers serves to enhance the mechanical characteristics of films, leading to a reduction in density, hardness, tension, and viscosity in the materials. The inclusion of plasticizers also contributes to improved polymer-chain flexibility and resistance to fracture (Primožič et al., 2021).

Several biodegradable films have been created utilizing CMC, either independently or in conjunction with diverse carbohydrates, proteins, lipids, and additional functional additives. An edible film based on emulsion was formulated by incorporating CNCs into a CNP matrix. The oil-in-water emulsion was formed by dispersing beeswax into an aqueous colloidal CNP suspension using a high-speed homogenizer. By incorporating 2.5% w/w CNCs, the tensile strength of the composite film peaked at 74.9 MPa—a significant improvement of 32.5% compared to pure chitosan films. Notably, the optimum tensile strength of 62.5 MPa was achieved by adding 25% w/w beeswax. A comprehensive analysis indicated that the interaction between the chitosan molecules and the CNCs enhanced the physical and thermal characteristics of the film (Wardhono et al., 2019). Phisalaphong and Jatupaiboon (2008) investigated the incorporation of chitosan into an *Acetobacter xylinum* cellulose cultivation. Their objective was to create a novel nanostructured film combining chitosan and cellulose. The study involved characterizing the microstructure and mechanical properties of the resulting films to offer insights for enhancing bacterial cellulose–chitosan film modifications. The reference sample for the bacterial cellulose film had a 0% chitosan content. Comparative analysis with the bacterial cellulose film revealed an increase in tensile strength for bacterial cellulose–chitosan films corresponding to a higher chitosan content.

Among the most crucial nanomaterials utilized in food packaging are montmorillonite, kaolinite, ZnO (ZnO NPs), coated silicate, Ag NPs (Ag NPs), and TiO_2 (TiO_2 NPs). These nanomaterial-coated films create effective barriers against O_2, CO_2, and other favorable compounds (Chaudhary et al., 2020). Mesgari et al. (2022) demonstrated the prospective antibacterial efficacy of chitosan/TiO_2 composites in the context of food preservation. Sani et al. (2019) developed an antibacterial chitosan biopolymer nanofilm incorporating ZnO particles (0–3% w/v) and Melissa essential oil (0–0.5% w/v) for biopolymer enhancement. Increased concentrations of ZnO and Melissa essential oil enhanced film tensile strength and reduced water solubility and WVP, increased opacity, and improved antibacterial efficacy, suggesting the potential application of the resulting composite film as an innovative antimicrobial and biodegradable packaging solution (Sani et al., 2019).

Fish gelatin and chitosan were utilized as the base materials for film formation, with various concentrations of TiO_2-Ag added to generate composite films. The results revealed a significant increase in water solubility due to the addition of the TiO_2-Ag. At a TiO_2-Ag concentration of 0.5%, the film exhibited its best antibacterial efficacy and its lowest light transmittance, at 54.6%. However, this concentration led to the lowest tensile strength—a decrease from 17.39 to 9.014 MPa. Initially, there was a decrease, followed by an increase in WVP, reaching a minimum value of 2.63 $\times 10^{-12}$ g·cm cm^{-2}. This composite film holds promise for applications in the food packaging industry (Lin et al., 2020). The WVP is the measure of a biodegradable film's capacity to counteract water loss from environmental exposure and its ability to withstand atmospheric moisture. In packaging materials, a lower WVP is the preferred characteristic. Azaza et al. (2022) developed and characterized packaging films using chitosan and *Sardinella* protein isolate to assess the impact of the chistosan films on the quality and shelf life of shrimps. The chitosan film, with its hydrophilic nature, exhibited a high WVP. In composite films, there was a noticeable reduction in the WVP value, declining from 4.18 to 3.11 g·m^{-1}·s^{-1}·Pa x 10^{-10} for the chitosan-containing films. The physical and antimicrobial features of banana flour and chitosan biodegradable and self-sealing films, used for preserving fresh-cut vegetables, were investigated. The results revealed that the chitosan/starch film-forming suspensions and chitosan solutions both exhibited pseudoplastic behavior. The WVP of the starch film ranged from 13.2 to 21.2 $\times 10^{-11}$ g·m^{-1}·s^{-1}·Pa, while for the chitosan film, it was 4.5 $\times 10^{-11}$ g·m^{-1}·s^{-1}·Pa and for the chitosan/starch film, 3.8–4.5 $\times 10^{-11}$ g·m^{-1}·s^{-1}·Pa. This improvement in barrier properties stemmed from the interactions between the hydroxyl groups in the starch and the amino groups in the chitosan. In a different study, an innovative edible film was developed by combining the antimicrobial attributes of chitosan with the self-sealing property of Kluai Namwa banana starch (Abdul Khalil et al., 2018).

Carrageenan-Based Nanocoatings: Carrageenan is a sulfated polysaccharide that dissolves in water and derived from red seaweed (Rhodophyta). It has found extensive applications in the food and pharmaceutical industries as a gelling, stabilizing, and emulsifying agent. Carrageenan is made of a copolymer of α-(1-3)-D-galactose and β-(1-4)-3,6-anhydro-d or L-galactose units (Kanmani & Rhim, 2014). As with other biopolymers, carrageenan's distinctive characteristics, such as its bioactivity, biocompatibility, and biodegradability, have led to its widespread utilization in various industries, including food, dairy, cosmetics, and pharmaceuticals. It serves as a gelling, emulsifying, and stabilizing agent, as well as in the creation of edible films and coatings (Nouri et al., 2020). α-carrageenan has demonstrated superior film-forming attributes, delivering impressive water-barrier and mechanical properties. However, there is a need for enhancement in the mechanical characteristics and water-vapor barrier attributes of carrageenan films to adapt them for use in the food packaging sector (Shojaee-Aliabadi et al., 2013). The inclusion of additional polysaccharides, such as starch, notably enhances the film's physical, thermal, and mechanical attributes (Perera et al., 2023).

Liu et al. (2020) developed an intelligent food packaging material by incorporating pomegranate-flesh or -peel extracts into κ-carrageenan. This addition resulted in increased tensile strength, reaching a maximum of 30.94 MPa, and reduced WVP, to as low as 3.47×10^{-11} g m^{-1} s^{-1} Pa^{-1}, compared to pure κ-carrageenan film. Notably, films containing pomegranate flesh extract exhibited pH-sensitive characteristics, attributed to the abundance of anthocyanins. Shojaee-Aliabadi et al. (2013) assessed the biodegradable composite films of κ-carrageenan infused with *Satureja hortensis* regarding their physical, mechanical, barrier, and antioxidant characteristics. The incorporation of *S. hortensis* notably enhanced the water-vapor barrier properties of the films, while making them more flexible, less resistant to breakage, and more opaque. Films with *S. hortensis* displayed promising antioxidant properties, especially when the proportion of *S. hortensis* added was 3%, and effectively inhibited the growth of the five microorganisms tested. These findings suggest the potential use of *S. hortensis* as a natural antibacterial agent in packaging various food products, particularly those prone to oxidation and microbial spoilage.

Low-cost biodegradable films were created by blending commercial κ-carrageenan and cassava starch at various weight ratios. The films exhibited high transparency, as evidenced by their apparent opacity and color analyses. They varied in water solubility, swelling degree, and WVP, with the 50 κ-carrageenan film displaying the lowest WVP. The oil permeability ranged between 0.0033 and 0.0043 mm m^2 day^{-1}. Films with higher κ-carrageenan content had increased stiffness, while those with more starch exhibited improved thermal stability (de Lima Barizão et al., 2020). Recent research has extended to the utilization of carrageenan biopolymers, with the introduction of glycerol into the carrageenan solution prior to film formation resulting in a remarkable 56% increase in tensile strength, albeit with a 61% reduction in elongation at break. Notably, the studied films exhibited effectiveness against *E. coli* and *Pseudomonas aeruginosa*.

Protein-Based Nanocoatings

Biopolymers are gaining heightened attention for their application in food packaging, particularly protein-based biopolymers, which exhibit excellent film-forming capabilities. These proteins, owing to their organized and densely packed hydrogen-bonded network structures, function as effective barriers against O_2 in film applications (Hu et al., 2022). Among the significant categories of bio-based degradable polymers, proteins represent a pivotal group sourced primarily from both plants and animals. These proteins are composed of various polar and non-polar α-amino acids. Plant-derived proteins are obtained from diverse sources, including wheat gluten, corn zein, soy, peanuts, rice-bran, cottonseed, barley, and sunflowers. Conversely, proteins from animal origins include gelatin, collagen, casein, whey, and fish myofibrillar protein (Baghi et al., 2022).

Wheat Gluten-Based Nanocoatings: Wheat gluten is attracting considerable interest in the field of biopolymers because of its functional attributes, its abundant availability, and its cost effectiveness as a by-product of starch production (e.g., for biofuel ethanol). In addition, wheat gluten exhibits favorable material properties and good biodegradability. The combination of polysaccharides and proteins holds promise for creating structural complexes that offer enhanced physical and chemical attributes, making them valuable for the advancement of film and coating applications (Sartori et al., 2018). However, a drawback lies in its increased susceptibility to moisture, resulting in a significant reduction in gas-barrier efficacy under elevated relative humidity conditions. Rovera et al. (2020) focused on the advantages of applying wheat-gluten coatings onto paperboard substrates for food packaging. The investigation involved wheat-gluten/silica (SiO_2) hybrid coatings, and assessments of their morphological, structural, thermal, surface, and water-vapor barrier characteristics. The results indicated that higher SiO_2 contents in the hybrid coatings correlated with an increased glass transition temperature and enhanced wettability compared to pure wheat-gluten coatings. In addition, the deposition of these coatings led to a significant reduction in water-vapor transmission rate by approximately four times.

Soy Protein-Based Nanocoatings: Soy proteins find diverse applications across various industries, including the food sector, where they are utilized in roles such as adhesives, composites, plastics, and more. These proteins exhibit biodegradability and remarkable capabilities for forming films. In the literature, the inclusion of antimicrobial agents into soy-protein films has been demonstrated to be effective, underscoring the potential of soy protein as a material for producing packaging films (Baghi et al., 2022).

Yu et al. (2019) investigated innovative soy-protein-based packaging materials that included CNFs, *Cedrus deodara* pine needle extract, and lactic acid. The research revealed that introducing CNFs into the composite led to heightened tensile strength due to the beneficial effects of CNFs as fillers. Notably, films infused with pine needle extract exhibited robust antioxidant capabilities and notably improved antimicrobial properties against foodborne pathogens, such as *E. coli* O157:H7, *S. aureus*, *Salmonella enterica* serovar *typhimurium*, and *Listeria monocytogenes*.

The outcomes underscored the nanocomposites' exceptional attributes, including antioxidant, antimicrobial, and light-blocking functionalities, positioning them as promising candidates for active and biodegradable food packaging materials.

Zein-Based Nanocoatings: Zein, derived from corn kernels, is both economically viable and widely available (Wani et al., 2023; Zhou & Wang, 2021). Primarily derived from natural plant sources containing polysaccharides and proteins, zein aligns well with the criteria for food packaging materials. Being a hydrophobic macromolecular protein, zein boasts advantages over other proteins because of its ease of extraction and cost-effectiveness (Jiang et al., 2022). The effectiveness of zein coatings relies on their specific attributes, including their ability to control the permeability of gases, water, lipids, and odors. However, these attributes are intricately linked to the structure of the polymers used and their chemical composition, the nature of the products, and the conditions under which they are stored (Wypij et al., 2023). It exhibits favorable biocompatibility. However, it suffers from notable brittleness and has a somewhat limited ability for electrospinning. As a result, it is often deemed essential to enhance the characteristics of zein by blending it with synthetic polymers known for their superior mechanical strength (Jiang et al., 2022). This is achieved through alterations to the supramolecular structure of zein, reducing interactions between the zein protein chains or employing other methods to augment the mechanical properties of zein films (Xu et al., 2012). The primary plasticizers frequently employed with zein are glycerol, polyethylene glycol (PEG), and fatty acids. These plasticizers, including PEG-400, tend to enhance the films' gas permeability and WVP. Notably, PEG-400 has proven effective in significantly augmenting the elongation at break, whereas oleic acid accomplishes a reduction in WVP by permeating the protein network (Zhou & Wang, 2021). A novel bioactive food packaging material, OZN (CS/orange-peel oil/zein nanocapsules), was developed using zein, CS, and orange-peel oil. The study found that the incorporation of OZN led to improvements in both the mechanical strength and moisture barrier properties of the composite films, extending the release time of the orange-peel oil. Specifically, the best outcomes concerning elongation at break and tensile strength (30.91% and 12.19 MPa, respectively) were achieved at a 5:5 ratio of OZN to CS. The composite films exhibited good oxidation resistance, with the 5:5 ratio of CS to OZN displaying the highest DPPH free-radical scavenging activity (30.16%) (Wang et al., 2019).

Lipid-Based Nanocoatings

Lipids encompass a broad category of molecules classified as either hydrophobic or amphiphilic, which include waxes, acylglycerols, and fatty acids. Coating biomaterials derived from lipids offers desirable characteristics for enhancing the sheen of confectionery items and providing an effective moisture barrier. Conversely, waxes are conventionally utilized to coat vegetables and fruits, primarily to prolong their freshness by slowing down respiration and minimizing moisture loss (Kuswandi & Moradi, 2019).

Beeswax is a naturally occuring, complex substance derived from the honeycomb of bees. It consists of a diverse mixture of extended-chain alcohols, fatty acids, and their corresponding esters (Zhang et al., 2015). Due to its ingestible nature and significant hydrophobic characteristics, beeswax has received considerable attention as a superb moisture barrier in the food sector in recent years (Zhang et al., 2016).

SYNTHETIC-BASED BIODEGRADABLE NANOCOATINGS

Synthetic biodegradable biopolymers involve PLA, polyglycolic acid (PGA), polylactic-co-glycolic acid (PLGA), PBS, PCL, polyethylene adipate, poly(p-dioxanone), and their corresponding copolymers (Wypij et al., 2023; Zhou et al., 2021). Numerous bioplastics, including PLA, PCL, and PVA, can be sourced from bio-based materials. While PLA can be synthesized from fossil fuels, it is mainly manufactured by the fermentation of renewable resources, such as starch and sugarcane, earning it the designation of a bioplastic (Ibrahim et al., 2021).

Polylactic Acid-Based Nanocoatings: Polylactic acid belongs to the polyester family that is typically produced by the ring-opening polymerization of lactides or the condensation polymerization of lactic acid monomers (Kumari et al., 2022). Polylactic acids, which are cost-effective polymers sourced from renewable materials, such as corn, sugar beet, wood residues, and various biomass resources, have garnered significant attention due to their myriad of promising physical and mechanical properties (Ahmed et al., 2017).

In contrast to other biopolymers, PLA has attracted greater interest in the packaging sector because it is compatible with existing processing technologies used for conventional plastics and does not require any modifications. In addition, PLA has a reduced O_2 transmission rate compared to polyethylene terephthalate (PET) and PS, and a comparable tensile strength and process stability to PS (Kumari et al., 2022). In general, the incorporation of nanocellulose into PLA biocomposites leads to enhancements in both tensile strength and elasticity (Nazrin et al., 2020). Currently, PLA-based packaging materials are being explored as a viable alternative to address the environmental concerns stemming from petroleum-based plastic packaging, given their biodegradable nature (Ahmed et al., 2017).

In a study carried out by Zhang et al. (2019), CNCs were uniformly dispersed in PLA using the Pickering emulsion technique. Alterations in the rheological properties of the resulting composites indicated that the CNCs facilitated the transition of the composites from a fluid-like to a solid-like behavior at elevated temperatures.

The incorporation of 5 wt% CNCs notably enhanced the crystallinity of PLA, and elevated the initiation temperature of crystallization by approximately 10°C. The storage modulus of the composites exhibited a consistent increase across the entire temperature range during testing, contributing to a substantial improvement in the flexural modulus.

In another study (Hegyesi et al., 2019), it was determined that CNCs accelerated the degradation of PLA. The enzymatic degradation of PLA and its nanocomposites, enhanced with CNCs, was catalyzed using lipase from *Candida rugosa* and proteinase K from *Tritirachium album*. The composites were formulated through the Pickering emulsion process, incorporating nanocellulose at 5%, 10%, and 15% weight ratios. The presence of the CNCs as reinforcement in the PLA accelerated the degradation process, resulting in a swift disintegration of the samples (within three days). Employing a model, a quantitative assessment of the degradation and denaturation kinetics was conducted, revealing a doubling of the rate for both processes conducted in the presence of CNCs (Hegyesi et al., 2019). In their study, Talebi et al. (2018) formulated antimicrobial films by incorporating different concentrations of *Mentha piperita* essential oil, *Bunium percicum* essential oil, and nanocellulose into PLA films. These films were stored at 4°C for 12 days, and their antimicrobial properties were assessed against Enterobacteriaceae, *S. aureus,* and *Pseudomonas*. The addition of *B. percicum* essential oil was found to improve the water-vapor barrier of the prepared films (Talebi et al., 2018).

Swaroop and Shukla (2018) aimed to develop biofilms for food packaging by reinforcing PLA biopolymer with magnesium oxide (MgO) NPs through the solvent casting method. Among the prepared biocomposite films, those with 2 wt% MgO reinforcement displayed the most significant improvements in tensile strength and O_2-barrier properties, with enhancements of up to 29% and 25%, respectively, compared to pure PLA films. However, the water-vapor barrier properties decreased by around 25%, attributed to interfacial interactions and the presence of voids surrounding the MgO NPs. The PLA/MgO films also demonstrated notable antibacterial effectiveness, causing substantial damage to, and mortality of, approximately 46% of an *E. coli* bacterial culture after a 12-hour treatment period.

Poly(Caprolactone)-Based Nanocoatings: Poly(caprolactone) is a renowned synthetic biodegradable polymer characterized by a glass transition temperature of approximately –60ºC and a melting point of also 60ºC (Verma et al., 2023). Poly(caprolactone) is a semi-crystalline polymer characterized by a considerable level of crystallinity, reaching approximately 50%. This polymer is known for its impressive elongation at break and low modulus. These physical characteristics, combined with its widespread commercial availability, render it an exceedingly appealing choice for various packaging applications (Kuswandi, 2017). In the literature, authors have reported combining PCL with filler nanocomposites, notably those involving PCL and organically modified layered silicates. These nanocomposites exhibit improved physical properties. In one study (Di et al., 2003), nanocomposites were created by blending biodegradable PCL with organically modified layered silicates (organoclay) through melt mixing. The study revealed that the successful exfoliation of organoclay depended on factors such as the type of organic modifier, organoclay content, and processing temperature. The incorporation of organoclay into PCL resulted in an increase in the crystallization temperature, enhancing the mechanical properties and thermal stability of the nanocomposites. Additionally, these nanocomposites displayed higher complex viscosity, significant shear-thinning behavior at lower frequencies, and reduced frequency dependence in viscoelastic properties compared to pure PCL, attributed to interactions between the PCL molecules and organoclay layers, along with the effective dispersion of exfoliated organoclay platelets within the PCL matrix (Di et al., 2003). Another study was aimed at developing superhydrophobic antimicrobial films for food packaging using a combination of cellulose acetate, PCL, and copper (Cu) NPs produced through fungal synthesis by *Aspergillus terreus* AH20. The Cu NPs were found to be stable at a particle size below 50 nm and with a zeta potential exceeding –30 mV. Incorporating Cu NPs into the film resulted in a rough surface due to NP deposition, improving mechanical properties and air permeability. The films loaded with CuNPs exhibited exceptional antimicrobial activity against a range of microorganisms, including *P. aeruginosa*, *E. coli*, *S. aureus*, *Aspergillus niger*, *Bacillus subtilis*, *Penicillium expansum*, *Candida albicans*, *A. terreus*, and *Fusarium oxysporum*, but no cytotoxicity to normal cells (El-Naggar et al., 2022).

Polyvinyl Alcohol-Based Nanocoatings: Polyvinyl alcohol is a recognized biodegradable polymer, valued for its ease of processing, as well as its ability to offer low O_2 permeability and maintain stability when exposed to UV radiation (Chandio et al., 2021). These qualities make it suitable for diverse applications, particularly in the field of flexible paper coatings. Despite its water solubility, PVA films exhibit a notable resilience to moisture. The water-vapor transmission rate in PVA films is equilavent to that of commonly used packaging polymers, such as low-density polyethylene and polyvinyl butyral. However, PVA films boast several other advantages, including ease of processing, biodegradability, and a strong barrier against O_2 diffusion (Chandio et al., 2021). Renowned for its exceptional gas-barrier properties, PVA has one of the lowest levels of gas permeability among all polymers. This outstanding O_2-barrier characteristic, coupled with its flexibility, transparency, toughness, non-toxicity, water solubility, and chemical resistance, makes it a preferred polymer for high-barrier food packaging. However, it is worth noting that the O_2-barrier effectiveness of PVA can be influenced by high humidity levels. By way of a solution, PVA layers are often strategically placed behind layers of polypropylene or polyethylene in typical multi-layered flexible packaging designs to enhance overall barrier performance (Schiessl et al., 2022).

Wen et al. (2016) investigated the utilization of a PVA/CEO/β-cyclodextrin nanofibrous film in preserving strawberries stored at 4°C. In comparison to the control group, the nanofilm and fresh-keeping film reduced weight loss and maintained firmness, extending the strawberries' shelf life. By the 18th day, the control group had lost 33.6% of its weight, while the nanofilm-packed strawberries had only lost 15.2%. Firmness loss was notably delayed in the nanofilm group. This suggests the nanofilm has potential in enhancing strawberry preservation during refrigerated storage.

Recent studies have been dedicated to developing methods for creating antibacterial nanocoatings for polymer-based food packaging. Nano-Ag, known for its potent antimicrobial properties, has been widely utilized in antimicrobial films. In one particular study, a nano-Ag-based PVA coating was prepared and applied to polypropylene food packaging material. The coating was tested against *E. coli* and revealed significant antibacterial activity. Polyvinyl-alcohol-based nano-Ag coatings are under exploration as a novel solution for antimicrobial food packaging materials, with the potential to improve food quality, safety, and shelf life by effectively inhibiting microbial proliferation (Egodage et al., 2017).

MICROBIALLY ORIGINATED NANOCOATINGS

Polyhydroxyalkanoate-Based Nanocoatings: Polyhydroxyalkanoates—bio-derived polymers naturally synthesized by various microorganisms, such as *Cupriavidus necator* and *Bacillus* spp.—emerge under specific growth conditions characterized by an abundance of carbon sources and limited nitrogen sources. In addition, PHAs exhibit superior biodegradability and biocompatibility, making them a promising alternative to conventional plastics (Boey et al., 2021; Lhamo & Mahantly, 2023). They are recognized for their properties that include effective UV resistance, high tensile strength, printability, and resistance to grease and oil. These properties make PHAs well suited for use in food packaging (Kumari et al., 2022).

Polyaliphatic biodegradable materials—PHAs—are synthesized by various bacterial strains during a specific biological process. Notably, the production of PHA is triggered in these microbes when they find themselves in an environment with an excess of carbon sources and a significant absence of other large molecules. This particular state serves as a stimulus for the bacteria to generate substantial quantities of PHA, with certain soil microorganisms exhibiting an impressive capacity to synthesize up to 1,000 grams of PHA (Samrot et al., 2020). Polyhydroxyalkanoates represent a class of biodegradable thermoplastics that have the potential to substitute traditional polymers in various applications. These innovative materials are generated through the microbial fermentation of carbon-based feedstocks, making them unique as the only polymers produced by industrial fermentation processes. Thanks to their renewability, biodegradability, and ability to act as a barrier to water vapor, PHAs offer significant advantages when used in food packaging. In addition to their biodegradable nature, PHAs exhibit desirable attributes in terms of reducing greenhouse gas emissions (Yadav et al., 2018). They also possess some impressive characteristics, including robust tensile strength, printability, the capacity to serve as a barrier against flavors and odors, heat-sealing capability, resistance to grease and oil, temperature resilience, and simplicity in the dyeing process. These attributes enhance their suitability for various applications in the food industry (Yadav et al., 2018). These novel nanocoating materials have the potential to prolong the shelf life of food products and potentially improve health attributes. Bioactive films were formulated by combining citrus pectin with spent coffee grounds extract. In response to the limitations associated with pectin-based materials, in particular their hydrophilic nature and limited water-barrier capabilities, an innovative multi-layered film was developed. This involved a two-step process in which functionalized pectin was immersed in PHA solutions. The introduction of the PHA coating affected the film's rigidity and opacity, while increasing its water resistance and improving its WVP (Mirpoor et al., 2023).

Polyhydroxybutyrate-Based Nanocoatings: Polyhydroxybutyrate—known for its eco-friendly attributes—has been the subject of extensive research as the most prominent PHA used in packaging materials. This polymer falls within the category of polyesters, making it a compelling choice for biodegradable and bio-based plastic applications (Kuswandi & Moradi, 2019). In a general context, PHAs can be produced via three different pathways, using a variety of carbon sources. Pathway I is mainly utilized by PHB-producing organisms, such as *C. necator* and *Bacillus* sp., whereas Pathways II and III are found in mcl-PHA-producing *Pseudomonas* sp. This process results in the production of different monomers based on the specific bacterial strains and the types of carbon substrates they utilize. As a result, the ratios of the copolymers can be tailored to suit the intended applications of the final product (Boey et al., 2021).

One of most-studied and simplest members of the group is PHBV packaging solutions. Through the solvent casting technique, a pioneering bioplastic blend of PHBV and tannin films was meticulously crafted by Ferri et al. (2023). The obtained findings illuminate the adaptability of these films, rendering them viable for diverse scenarios, spanning from refrigeration conditions to even elevated temperatures suitable for food heating, reaching as high as 200°C. The PHBV/tannin composite demonstrated an intriguing capability for colorimetrically detecting ammonia vapors. This revelation propels the material into the realm of potential applications as an intelligent indicator for monitoring food spoilage.

Abbas et al. (2021) investigated the potential of ultrasonic spray-coating as a deposition method for applying ultrathin ZnO NP coatings onto PET and poly(3-hydroxybutyrate-co-3-hydroxyhexanoate) (PHBHHx) films. In the case of PHBHHx substrates, no reduction in O_2 transmission rate was observed for any number of coating passes when compared to the pure PHBHHx substrate. However, UV testing on PET and PHBHHx substrates coated with ZnO NPs at 3×, 5×, and 10× concentrations yielded a significant decrease in percentage transmission for the 10×-coated PET and PHBHHx substrates compared to their 3× and 5× ZnO NP-coated counterparts. This study highlighted that, although PET substrates with a 50× ZnO NP coating exhibited only a minor variation in O_2 transmission rate compared to the reference substrate, the ultrasonic spray-coating technique resulted in a substantial UV barrier effect for the PET and PHBHHx substrates coated with ZnO NPs at 3×, 5×, and 10×. Lee et al. (2002) employed a melt intercalation technique to create nanocomposite films using the aliphatic polyester PBS and organoclay. Their findings indicated a reduction in the biodegradability rate of the nanocomposites compared to the pure PBS polymer. Interestingly, they observed that, as more organoclay was introduced, the biodegradability of the nanocomposite decreased further (Rhim et al., 2013). Hablot et al. (2008) conducted a study that similarly observed a decrease in the biodegradation tendency of PHB nanocomposites. They investigated how the presence of fermentation residues and quaternary ammonium salts affected the thermal and thermomechanical degradation of bacterial PHB. The findings derived from the analyses indicate that ammonium cations significantly accelerated the degradation process, leading to a considerable reduction in the molecular weight of the PHB (Hablot et al., 2008).

Pullulan-Based Nanocoating: Pullulan is a water-soluble polysaccharide that is naturally generated by the *Aureobasidium pullulans* fungus. It is renowned for its exceptional capacity to create films, such as oral film strips. Furthermore, it has found utility in the food sector, particularly for tasks involving thickening and encapsulation (Tonyali et al., 2020). The presence of α-(1-6)-linkages in pullulan disrupts what would otherwise be a linear amylose chain. This unique combination of α-(1-4)- and α-(1-6)-linkages in a single compound classifies the pullulan structure as an intermediary form between the amylose and dextran structures. Consequently, pullulan boasts considerably higher water solubility in comparison to amylose, which features exclusively α-(1-4)-linkages (Farris et al., 2014). Polysaccharides generated by microorganisms, such as pullulan, gellan, and xanthan, have gained widespread use across different industries, including food and pharmaceuticals, owing to their distinctive functional qualities (Sapper et al., 2019).

Pullulan possesses significant mechanical resilience and a range of functional attributes, including adhesive qualities, an ability to be formed into films and fibers, and susceptibility to enzymatic degradation. Notably, it remains impervious to the digestive enzymes found in the human gastrointestinal system. Moreover, this exopolysaccharide exhibits additional noteworthy characteristics. Its aqueous solutions maintain stability across a wide spectrum of pH levels, and it exhibits lower viscosity in contrast to various other polysaccharides. This substance does not have the capacity to form gels, and when used in film and coating derivatives, it demonstrates commendable O_2-barrier properties (Farris et al., 2014). Coatings derived from pullulan exhibit transparency, edibility ("generally recognized as safe"), neutrality in taste, inertness to food components, non-toxicity, adhesive characteristics, significant mechanical strength, and limited permeability to O_2 and CO_2 gases (Wypij et al., 2023). However, it has been proposed that pullulan, serving as a challenging carbon source for bacteria and fungi, can inherently hinder microbial growth on the surface of food, ultimately leading to an extension of its shelf life (Kraśniewska et al., 2017).

The utilization of pullulan films for packaging remains constrained, primarily, by its elevated cost of $25 per kilogram and its lack of inherent active functionalities, such as antimicrobial or antioxidant properties (Kraśniewska et al., 2019). To address these constraints and produce a flexible packaging material with enhanced mechanical strength, improved barrier properties, and added bioactive capabilities, pullulan can be combined with various natural functional polymers (Table 2). This blend can incorporate diverse active agents, such as bacteriocins, essential oils, and plant extracts, and reinforcement additives, such as NPs and nanofibers. An eco-friendly and biodegradable nanocomposite, comprising pullulan and graphene, was created and utilized as a coating (Pan et al., 2014). The results indicated that increasing the graphene content in the pullulan coating led to improved opacity, surface hydrophobicity, tensile strength, O_2 and water-vapor transmission rates, and significant antibacterial effectiveness against *E. coli* and *S. aureus* (Shanmugam et al., 2022).

A film created by blending pullulan and alginate, fortified with propylene glycol alginate and glycerol, exhibited significant reductions in WVP (by 70%) and O_2 permeability (by 40%) compared to the control film. This material displayed enhanced mechanical characteristics, with the film's tensile strength increasing from 69.34 to 103.68 MPa, and the elongation at break rising from 5.71% to 46.35% (Pan et al., 2014). The introduction of *Zataria multiflora* and *Atremisia biennis* essential oils into the pullulan matrix endowed the resulting films with antimicrobial properties. The study aimed to investigate the impact of adding essential oils on the physical characteristics of the pullulan films. Notably, their inclusion led to a reduction in WVP. Films lacking this additive exhibited a WVP of 5.01×10^{-11} g·m^{-1}·s^{-1}·Pa, while films containing 35% of essential oil (the highest concentration tested) displayed decreased WVP values of 2.63×10^{-11} g·m^{-1}·s^{-1}·Pa for *Z. multiflora* oil and 2.86×10^{-11} g·m^{-1}·s^{-1}·Pa for *A. biennis* oil (Rad et al., 2016).

Table 2. Bioplastics applications in food packaging (adopted from Zhao et al., 2020)

BioBased Material	Flexible Packaging	Manufacturers
PLA	Snack bag options, features for paper bag, films for fresh fruit and vegetable	NatureWorks, BIOFRONT, PURAC, HiSun, PepsiCo, Pyramid
PLA/PHB blends	Coffee capsules, coffee pouches	Ingeo, BIOFRONT, HiSun, Pyramid, PURAC
PLA/Cellulose blends	Films	FKur
PLA/PHB/PBAT blends	Shopping bags, waste bags	Ingeo, BIOFRONT, HiSun, Pyramid, PURAC
Cellulose	Coating for fresh fruit and vegetable, wrapping films, confectionary	Innovia
Cellulose	Metalized cellulose film, snack packing	Boulder Canyon Thomton, QualityStreet
Starch blends with PLA/PHB	Blown films	BiologiQ
Bio-PET	Bottles, bricks	Braskem, FKur, Avery Dennison, Asda (retailer)
PLA	Bottles, bricks, bowls, cups, trays, jars	Mosburger, Danone
Starch based	Trays	Cadbury Schwepps

APPLICATIONS OF BIO-BASED NANOCOATINGS IN FOOD PACKAGING

Nanomaterials find application in packaging across three main categories: (1) enhanced packaging, addressing factors such as improved barrier properties against gases, including O_2, CO_2, and UV rays, along with the addition of strength, stiffness, dimensional stability, and heat resistance; (2) active packaging, incorporating properties such as antimicrobial features, antioxidants, and UV absorbance, intentionally released to impact the packaged food's taste, freshness, and shelf life; and (3) smart/intelligent packaging, encompassing indicators for O_2 levels and freshness. These three primary applications of nanomaterials in food packaging represent the ongoing research trends in utilizing nanotechnology to advance packaging (Kuswandi, 2017). Traditional food packaging systems primarily serve as passive barriers, protecting food by creating a separation between the food and its external environment. Improved packaging utilizing a blend of nanomaterials with polymers aims to improve gas-barrier properties and resistance to moisture in packaging (Chadha et al., 2022). By contrast, active food packaging goes beyond passive protection, actively interacting with the food. This interaction may involve the release of beneficial compounds, such as antimicrobial or antioxidant agents, or the removal of detrimental factors, such as O_2 or water vapor (de Azeredo et al., 2011). Smart or intelligent packaging—considered a promising field in active packaging coating—involves the development of nano(bio)sensors for assessing food quality, nano(bio)switches for the controlled release of preservatives, and nanocoatings with antimicrobial, antifungal, antioxidant, and barrier properties (Vasile, 2018).

Conventional materials utilized in food packaging include metal, ceramic (glass), and paper (cardboard). Although these materials remain in use, nanocoatings offer appealing options for food packaging. Enhancing food packaging through the integration of functional nanomaterials with polymer materials aims to elevate the mechanical and physical attributes of the packaging. This includes improving gas-barrier properties, resilience to temperature and humidity fluctuations, mechanical strength, and flexibility (Primožič et al., 2021). Paper remains a prevalent material in food packaging, reflecting its strength and barrier properties, especially in terms of water and grease resistance. A common practice involves coating paper with hydrophobic materials, such as paraffin wax and polyethylene, or using polysaccharides, such as starches, to enhance barrier properties against O_2, water, water vapor, and aroma profiles (Butkinaree et al., 2008). As another degradation factor, the presence of O_2 in packaging poses a threat to the shelf life of food because it creates a conducive environment for the growth of microorganisms. Bajpai et al. (2018) developed biodegradable and cost-effective films based on PVA/chitosan polymers. This formulation led to a significant reduction in O_2 and moisture permeability, by 25.6% and 10.2%, respectively. The findings suggested that these films could triple the preservation time of cherries compared to conventional packaging.

Various types of food coatings, such as thin layers or films, can be employed to envelop food, creating a barrier to mass transfer (Primožič et al., 2021). Edible nanocoatings, with a thickness of approximately 5 nm, find application in meat, fruit, vegetables, cheese, fast food, bakery goods, and confectionery items, acting as barriers against gases and moisture. Furthermore, they contribute to the flavor, color, enzymes, antioxidants, anti-browning, and extended shelf life of produced goods (Nile et al., 2020). Coatings comprising gelatin infused with nanocrystals of chitosan, along with nano-SiO_2, alginate nanolaminate, cellulose, chitosan or nano-SiO_2 have proven to be highly effective in the preservation of the quality of fresh foods during prolonged storage (Bajpai et al., 2018). Sarwar et al. (2018) formulated antimicrobial food packaging using PVA, nanocellulose, and Ag nanocomposite films, which demonstrated potent antibacterial efficacy against both (methicillin-resistant) *S. aureus* and *E. coli* (DH5-alpha). The largest inhibition zone, observed with 0.5 g Ag NPs and 12 wt% nanocellulose against DH5-alpha, measured 14 ± 0.70 mm. Qi et al. (2004) also developed chitosan nanocoatings that possess potential antibacterial properties due to their distinctive characteristics. Their study evaluated the antibacterial effectiveness of chitosan NPs against *E. coli*, *Salmonella enterica* serotype *choleraesuis*, *S. enterica* serovar *typhimurium*, and *S. aureus* by determining the minimum inhibitory concentration and minimum bactericidal concentration. The findings indicated that chitosan nanocoatings could impede the growth of a variety of tested bacteria (Qi et al., 2004). In Figure 5, an example of the development and mechanisms of active nanocomposites can be seen in a food packaging context.

It was previously emphasized that reinforcement agents are required to improve the mechanical and strength properties of packaging materials. Montmorillonite clay stands out as one of the most commonly utilized nanoreinforcements and nanofillers (Vasile, 2018). Utilizing clay as a potential filler has proven effective in enhancing the properties of thermoplastic starch, especially in the production of starch/clay nanocomposite films achieved through the dispersion of montmorillonite NPs. The incorporation of small amounts of sodium-montmorillonite has demonstrated improvements in both tensile strength and elongation at break, accompanied by an increase in decomposition temperature and a decrease in the relative diffusion coefficient of water vapor. Nanoclay composites are utilized in both flexible and rigid food packaging, owing to their outstanding barrier properties (Vasile, 2018). Recent research suggests that the incorporation of starch into nanocomposite films or coatings on paper food packaging improves barrier properties (Chavan et al., 2022). In another study, nanocomposite films were fabricated using whey protein isolate by blending the polymer with three distinct types of nanoclay at a 5% concentration. The incorporation of nanoclays led to a significant reduction in WVPs, enhancing the properties of the films (Oymacı & Altınkaya, 2009). Table 3 illustrates representative examples of biopolymer-based nanocomposites and their improved material properties.

Figure 5. Development and mechanisms of active nanocomposites and antimicrobial agents: A schematic representation
Source: Modified and adopted from Nath et al. (2022)

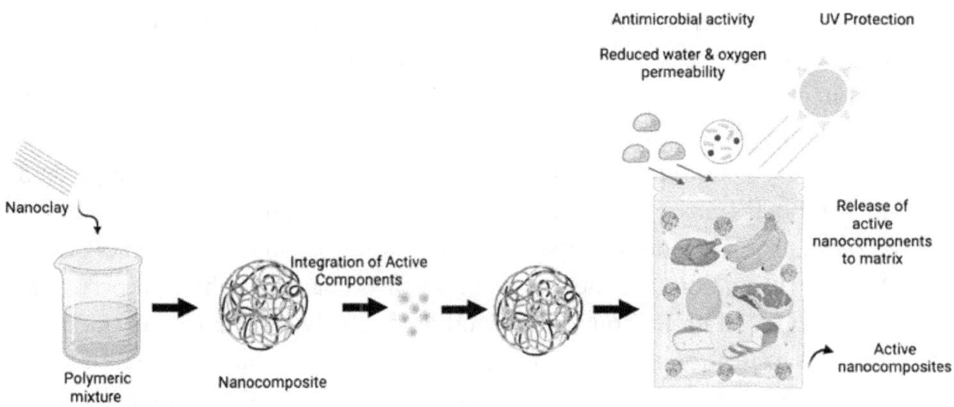

Presently, the bionanocomposites garnering the most attention for packaging applications include starch and its derivatives, such as PBS, PLA, and aliphatic polyester. As a result, bio-based packaging can serve the dual purposes of enhancing both packaging quality and incorporating active elements (Kuswandi, 2017). Bionanocomposites are employed in packaging various food items, including processed meats, cheese, confectionery, cereals, boil-in-the-bag foods, and items produced through extrusion-coating processes. This includes paperboard coatings for fruit juice and dairy products, as well as bottles for carbonated drinks manufactured through co-extrusion processes.

Notably, PET bottles incorporating nanocomposite and nylon bottles with nanoclays exhibit enhanced barrier characteristics (Vasile, 2018). Bionanocomposites are also being used in a number of food sectors. EcoSynthetix—an internationally operating biomaterials company—has innovated a novel technology named EcoSphere Biolatex®. Their studies have revolved around a natural binder derived from renewable sources like CS and potato starch (Bumbudsanpharoke et al., 2015). Nano-starch—frequently mentioned in terms of biodegradable food packaging—is finding application in various emerging areas. With a specific surface area approximately 400 times larger than natural starch granules, nano-starch demonstrates enhanced performance compared to traditional starch and is comparable to polyvinyl acetate adhesive. Notably, it has been employed in the production of McDonald's hamburger clamshells in the United States, replacing conventional adhesives (Bumbudsanpharoke & Ko, 2015).

Active packaging methods include the deployment of releasing systems, absorbers, and various mechanisms for preserving and enhancing the quality and safety of fruits and vegetables (Jafarzadeh et al., 2021). In this context, additives are incorporated into coating materials in minute quantities, imparting various property modifications. Nanotechnology-based functional coatings utilize nanomaterials, such as TiO_2, Ag, ZnO, proteins, polysaccharides, spice and herb extracts, vegetable and essential oils, bacteriocins (i.e., pediocins, nisin), and organic acids, particularly in antimicrobial coatings. These nanomaterial-containing coatings offer superior performance and processing characteristics compared to conventional coatings, including increased indentation resistance, high elasticity, rapid drying, no expansion upon water contact, and high WVP (Vasile, 2018). Antimicrobial films play a crucial role in reducing spoilage and hindering the development of pathogenic microorganisms. The antimicrobial functions are derived from the antimicrobial agents within the nanomaterial that are immobilized onto the film (Kuswandi & Moradi, 2019). The effective utilization of nanopackaging featuring antimicrobial properties has provided a degree of protection for our food products. In comparison to traditional food packaging, it not only functions as a passive barrier but also facilitates the release of antioxidant and antimicrobial compounds through direct interaction with

Table 3. Recent applications of nanocomposites in food packaging using bio-based materials

Nanotechnological Application	Composition	Observations and Improvements in Nanocomposites	Reference
Nanofilm	Apple Peel Pectin + 3% Montmorillonite Clay	Enhanced structural, thermal, and barrier properties, improved water vapor and oxygen barrier	Mangiacapra (2006)
	Chitosan + Clay + Glycerol	Enhanced thermal stability, improved tensile strength, modulus, and water resistance	Kusmono and Abdurrahim (2019)
	Waxy Maize Starch + Starch Nanocrystal + Glycerol	Enhanced mechanical properties, Reinforcing effect of waxy maize starch nanocrystals	Angellier et al. (2006)
	Cellulose Acetate+Alginate + Carrageenan	Antibacterial activity, better homogenization, biodegradable, cost-effective packaging	Rajeswari et al. 2020
Nanofilm	Pea Starch Nanocrystal + Polyvinyl Alcohol	Improved light transmittance, tensile strength, elongation break and lower moisture uptake	Chen et al. (2008)
Nanocomposite	Poly-caprolactone+Organo Nanoclay + Glycerol Monooleate + Chitosan	Enhanced antimicrobial and biodegradable properties	Cesur et al. (2018)
Nano-reinforced packaging	Plum + Zn-TiO2 nanoparticles+	Complete decomposition within 12 days of storage	Mann and Sooch (2023)
Nanofilm	Chitosan + Monolaurin	Increased film solubility, effective antibacterial activity	Lotfi et al. (2018)
Nanofilm	Corn Starch+Orange Peel Oil+Zein	Smooth and compact morphology, good oxidation resistance, bio-active edible packaging	Wang et al. (2019)
Nanofilm	Gelatin + Polyvinyl Alcohol + Bacterial Cellulose	Reduced water vapor transmission rate, eco-friendly packaging	Haghighi et al. (2021)
Nano coating film	Whey Protein isolate+Celluloe nanofiber+Titanium Dioxide	Strong antibacterial activity, enhancement in water resistance and physicomechanical properties, improved morohology	Alizadeh-Sani et al. (2018)

the food (Bajpai et al., 2018). Damm et al. (2008) investigated the antimicrobial effectiveness of polyamide 6/Ag-nano- and microcomposites against *E. coli*, varying the filler content. Polyamide 6 incorporated with 0.06 wt% Ag NPs demonstrated the complete eradication of bacteria within 24 hours. Chitosan has also been used due to its antimicrobial activity. Cellulose–chitosan films with incorporated monolaurin were formulated and subjected to comprehensive characterization, including microscopic, water solubility, structural, and antimicrobial assessments against *L. monocytogenes*. The research extended to evaluating the antimicrobial efficacy of these films on ultrafiltered cheese during a 14-day storage period at 4°C. The results indicated a significant increase in the diameter of the inhibition zone, underscoring the heightened antimicrobial activity attributed to the addition of monolaurin (Lotfi et al., 2018). Ge et al. (2012) introduced an innovative food packaging material, achieving this by fixing glucose oxidase within an electrospun nanofibrous membrane composed of PVA, chitosan, and tea extract. The experimental samples exhibited a noteworthy suppression of microbial growth and reproduction when subjected to a 1% O_2 concentration, thereby resulting in a significant extension of the lifespan of the food product. These findings underscore the prospective applicability of such a membrane in the domain of food preservation (Ge et al., 2012). A gelatin-based nanocomposite containing chitosan nanofiber and ZnO NPs (G/CHINF/ZnONPs) was tested as active packaging for chicken fillets and cheese. The study revealed strong antibacterial properties in the tested nanocomposites, particularly against foodborne pathogenic bacteria. In comparison to the neat gelatin film used for packaging, the G/CHINF/ZnONPs packaging exhibited a 1.4-fold increase in antibacterial effectiveness, with significantly lower bacterial counts of *E. coli* (5.6 ± 0.3 log CFU g^{-1}) and *P. aeruginosa* (5.6 ± 0.1 log CFU g^{-1}) after a storage period of 12 days (Amjadi et al., 2019). Nanocomposites possess the capacity to significantly enhance the barrier properties of the polymer matrix (Table 3).

Approximately 400 companies globally are currently engaged in exploring potential applications of nanotechnology in the food and food packaging industry. A diverse array of both branded and unbranded products, originating from various sources and countries, are readily accessible through online markets (Bumbudsanpharoke & Ko, 2015). Table 1 provides several instances, illustrating how nanotechnology is contributing to the sustainable and biodegradable preservation of products, ensuring prolonged freshness in an environmentally conscious manner. The infiltration of O_2 into food packaging can diminish the shelf life of products through diverse degradation reactions, including rancidity, the proliferation of aerobic microorganisms, browning, the depletion of vitamins, and the loss of essential flavor compounds. To mitigate these adverse impacts, it is crucial to eliminate or minimize the presence of O_2 in the package to levels deemed acceptable for a specific product (Mihindukulasuriya & Lim, 2014). Antioxidants act as scavengers for radicals, capturing free radicals and aiding in the postponement or prevention of oxidation. In recent years, there has been considerable research on active packaging incorporating antioxidant compounds. A recent investigation evaluated the antioxidant potential of green tea extract for preserving meat, highlighting its possible application in active films (Baghi et al., 2022). Antioxidants used in food packaging must satisfy specific criteria, including cost-effectiveness, non-toxicity, high activity at low concentrations, excellent permeability, and stable performance without compromising food quality (Choe et al., 2009). Zhang et al. (2021) conducted a study utilizing a rapid, straightforward, and environmentally friendly method to produce polyfunctional starch/tea polyphenols nanofibrous films (STNFs) through a one-step temperature-assisted electrospinning process. The study extensively examined the impact of incorporating tea polyphenols on the antioxidant activity of STNFs, revealing that the addition of tea polyphenols enhanced the antioxidant activity of STNFs, while maintaining optimal mechanical properties, as confirmed by tensile testing. Moalla et al. (2021) developed active food packaging films using chitosan, enhancing them with *Artemisia campestris* hydroalcoholic extract, aqueous extract, and essential oil. The study investigated the impact of incorporating *A. campestris* on the physical, mechanical, thermal, and antioxidant characteristics of the films. The addition of *A. campestris* active compounds, particularly extracts, significantly enhanced the antioxidant and UV–Vis barrier properties of chitosan films. Therefore, chitosan films enriched with *A. campestris* antioxidant compounds hold potential as alternatives for food packaging. The investigation successfully crafted functional films based on CMC incorporating Chinese chives root extract (CRE) at different concentrations (1%, 3%, and 5% w/w). The DPPH and ABTS radical scavenging abilities of CMC-CRE films increased by 58% and 82%, respectively. Overall, the film that resulted from incorporating CRE into CMC exhibited notable antioxidant and antimicrobial activities, suggesting its potential as a biocomposite food packaging material for the food sector (Riaz et al., 2020).

Packaging materials prevent prolonged exposure to sensory elements. However, consumers often rely on the sell-by dates established by producers based on idealized assumptions about storage or transportation conditions. For instance, the sell-by date on a carton of milk may suggest it is good for two weeks, but this may change if the milk is exposed to suboptimal temperatures during transportation or storage. The distinctive chemical and electro-optical properties of nanoscale particles provide potential solutions to address this issue (Duncan, 2011). Intelligent packaging systems utilize communication functions to enhance decision-making processes directed at preserving food quality, extending shelf life, and improving overall food safety (Schaefer & Cheung, 2018). The primary culprits in food spoilage—excess moisture and O_2—often necessitate the destruction of a package for assays to measure the vapor or gas content. The ongoing ability to non-invasively monitor the gas content in a package's headspace would offer a continuous assessment of the safety and quality of the enclosed food, extending beyond the production facility. Various gas-sensing methods that leverage nanotechnology have been developed to address this need (Duncan, 2011). Nanotechnology, specifically nanosensors embedded in plastic packaging, serves to identify the gases released during food spoilage, signaling consumers through color changes. The literature reports the development of sensors capable of detecting *E. coli* contamination in packaged foods. Furthermore, nanotechnology has the potential to minimize packaging waste in processed foods and contribute to the preservation of fresh foods, thus prolonging their shelf life (Adeyeye, 2019).

In smart packaging technology, in order to ensure the absence of O_2, a colorimetric oxygen nanosensor activated by UVA light has been developed. This sensor utilizes TiO_2 NPs to sensitize the reduction of methylene blue by triethanolamine within a polymer-encapsulated matrix. When exposed to UV light, the nanosensor undergoes a bleaching process, turning colorless, and upon exposure to O_2, it transitions back to a blue color. The rate of color change corresponds to the level of O_2 exposure (Kuswandi & Moradi, 2019). The change in color is also used as an indicator of the freshness of the product. Changes in polyaniline's total color consistently aligned with the levels of total volatile amines and the patterns of microbial growth in fish samples. These responses facilitated the continuous monitoring of fish spoilage, both under constant temperatures and temperature fluctuations, in real time (Kuswandi, 2017). In addition, smart packaging radiofrequency identification (RFID) functions have also been used as an active tag in electronic information-based systems. Using radio frequencies, data is transferred from a tag attached to the packaging so it can automatically trace and identify objects, including food products. Smart RFID presents improvements over barcodes and manual tracking systems, offering a longer reading range and robust operation under extreme conditions, such as high temperatures and pressures, with detection distances exceeding 100 m. Nanotechnology introduces nanosensors for cost-effective nano-enabled RFID tags that are small, flexible, and printable on food labels, enhancing tag versatility for low-cost production (Kuswandi & Moradi, 2019).

FUTURE PROSPECTS FOR BIODEGRADABLE NANOCOATINGS

In the last 10 years, there has been a remarkable focus on nano-sized structures, primarily because of their promising prospects across various industrial sectors. It is expected that nanobiotechnology holds significant potential to outperform traditional strategies in the food industry (Ahmad et al., 2023). While nanomaterials offer significant promise as a viable option for food packaging, there are certain research gaps that require careful consideration. Their remarkable physicochemical attributes, bioavailability, and chemical reactivity can potentially lead to a significant level of cytotoxicity (Garcia et al., 2018).

Safety considerations linked to the utilization of nanocomposite materials for food-contact applications are primarily driven by their distinct physical and chemical attributes, which can result in different reactivity toward biological environments than their macro-scale counterparts. The potential hazards to human health linked to nano packaging materials, such as nanocomposite coatings, hinge on two key factors—the possibility of NPs migrating into the food and the toxicological repercussions of these NPs on human well-being (Sahoo et al., 2021).

While NPs present on the packaging material's surface pose no harm to humans, concerns arise regarding their potential translocation and integration into the food, potentially affecting human health. The existing literature delves into the pathways, absorption, and distribution of NPs in the human body, with a particular focus on their cytotoxic and genotoxic effects (Nile et al., 2020). Numerous animal experiments have demonstrated the swift translocation of NPs from the point of entry to vital organs, including the brain, spleen, liver, kidney, and muscles, facilitated by the bloodstream. Moreover, recent findings indicate that NPs have the ability to traverse the blood–brain barrier, potentially influencing brain functions (Teow et al., 2011).

The behavior of nanomaterials concerning their migration and interactions with biological systems require careful examination. Incomplete cytotoxicity data, and the lack of comprehensive risk assessments and management studies hinder the widespread adoption of nanomaterials in the food technology industry. Additionally, the acceptance and approval of these materials by consumers will play a pivotal role in developing a substantial market for nanotechnology-based food applications. To establish a healthy environment, regulatory measures and accrediting institutions should set forth guidelines to foster consumer collaboration. Therefore, ensuring the commercialization of nanomaterials in the food packaging industry necessitates prioritizing consumer satisfaction, food safety, and sustainable manufacturing approaches (Ahmad et al., 2023). Insufficient consumer awareness, along with the absence of government guidelines, policies, and detection methods for assessing nanotechnology risks, highlights the need for a better comprehension of nanomaterial-based toxicity characterization and regulatory processes. The toxicity tends to increase as the size of metal-based NPs decreases, emphasizing the significant information gap regarding the effect of NPs on human health and the environment, particularly concerning the potential for NP-induced toxicity (Nile et al., 2020; Schrand et al., 2011).

CONCLUSION

In conclusion, biodegradable active biopolymers typically exhibit limited physical and thermal properties, such as water-vapor barrier properties. Therefore, modifying their structure or employing other techniques can be advantageous in controlling this attribute. These endeavors hold the potential to yield biodegradable active biopolymers possessing favorable barrier and mechanical characteristics. Simultaneously, greater focus should be directed toward reducing their costs and mitigating competition with food resources by enhancing the utilization of second- and third-generation raw materials.

In the realm of nanothechnology, despite ongoing research efforts in the field, there remains a significant gap in the comprehensive scientific exploration of naturally occurring nanosystems and their associated benefits. There is an urgent need for established and standardized test protocols to thoroughly investigate the effects of NPs on living cells, thereby facilitating a comprehensive risk assessment of human exposure to NPs. Understanding of the toxicology of NPs remains limited due to the lack of validated test methods and the inherent variability in reported data.

REFERENCES

Abbas, M., Buntinx, M., Deferme, W., Reddy, N., & Peeters, R. (2021). Oxygen Gas and UV Barrier Properties of Nano-ZnO-Coated PET and PHBHHx Materials Fabricated by Ultrasonic Spray-Coating Technique. *Nanomaterials (Basel, Switzerland)*, *11*(2), 449. doi:10.3390/nano11020449 PMID:33578826

Abdul Khalil, H. P. S., Banerjee, A., Saurabh, C. K., Tye, Y. Y., Suriani, A. B., Mohamed, A., Karim, A. A., Rizal, S., & Paridah, M. T. (2018). Biodegradable films for fruits and vegetables packaging application: Preparation and properties. *Food Engineering Reviews*, *10*(3), 139–153. doi:10.1007/s12393-018-9180-3

Achachlouei, B. F., & Zahedi, Y. (2018). Fabrication and characterization of CMC-based nanocomposites reinforced with sodium montmorillonite and TiO2 nanomaterials. *Carbohydrate Polymers*, *199*, 415–425. doi:10.1016/j.carbpol.2018.07.031 PMID:30143147

Adeyeye, S. A. O. (2019). Food packaging and nanotechnology: Safeguarding consumer health and safety. *Nutrition & Food Science*, *49*(6), 1164–1179. doi:10.1108/NFS-01-2019-0020

Adeyeye, S. A. O., & Ashaolu, T. J. (2021). Applications of nano-materials in food packaging: A review. *Journal of Food Process Engineering*, *44*(7), e13708. doi:10.1111/jfpe.13708

Agustin, S., Wahyuni, E. T., Suparmo, S., Supriyadi, S., & Cahyanto, M. N. (2021). Incorporation of pectin during biosynthesis of bacterial cellulose by *Gluconacetobacter xylinus* InaCC B404: Possibility for producing green food packaging. *Biodiversitas (Surakarta)*, *22*(5). Advance online publication. doi:10.13057/biodiv/d220512

Ahmad, A., Qurashi, A., & Sheehan, D. (2023). Nano packaging–Progress and future perspectives for food safety, and sustainability. *Food Packaging and Shelf Life*, *35*, 100997. doi:10.1016/j.fpsl.2022.100997

Ahmed, J., Hiremath, N., & Jacob, H. (2017). Antimicrobial efficacies of essential oils/nanoparticles incorporated polylactide films against *L. monocytogenes* and *S. typhimurium* on contaminated cheese. *International Journal of Food Properties*, *20*(1), 53–67. doi:10.1080/10942912.2015.1131165

Ainali, N. M., Kalaronis, D., Evgenidou, E., Bikiaris, D. N., & Lambropoulou, D. A. (2021). Insights into biodegradable polymer-supported titanium dioxide photocatalysts for environmental remediation. *Macromol*, *1*(3), 201–233. doi:10.3390/macromol1030015

Al-Naamani, L., Dutta, J., & Dobretsov, S. (2018). Nanocomposite zinc oxide-chitosan coatings on polyethylene films for extending storage life of okra (*Abelmoschus esculentus*). *Nanomaterials (Basel, Switzerland)*, *8*(7), 479. doi:10.3390/nano8070479 PMID:29966220

Alinaqi, Z., Khezri, A., & Rezaeinia, H. (2021). Sustained release modeling of clove essential oil from the structure of starch-based bio-nanocomposite film reinforced by electrosprayed zein nanoparticles. *International Journal of Biological Macromolecules*, *173*, 193–202. doi:10.1016/j.ijbiomac.2021.01.118 PMID:33482206

Alizadeh-Sani, M., Khezerlou, A., & Ehsani, A. (2018). Fabrication and characterization of the bionanocomposite film based on whey protein biopolymer loaded with TiO2 nanoparticles, cellulose nanofibers and rosemary essential oil. *Industrial Crops and Products*, *124*, 300–315. doi:10.1016/j.indcrop.2018.08.001

Amjadi, S., Emaminia, S., Nazari, M., Davudian, S. H., Roufegarinejad, L., & Hamishehkar, H. (2019). Application of reinforced ZnO nanoparticle-incorporated gelatin bionanocomposite film with chitosan nanofiber for packaging of chicken fillet and cheese as food models. *Food and Bioprocess Technology*, *12*(7), 1205–1219. doi:10.1007/s11947-019-02286-y

Angellier, H., Molina-Boisseau, S., Dole, P., & Dufresne, A. (2006). Thermoplastic starch– waxy maize starch nanocrystals nanocomposites. *Biomacromolecules*, *7*(2), 531–539. doi:10.1021/bm050797s PMID:16471926

Arun, R., Shruthy, R., Preetha, R., & Sreejit, V. (2022). Biodegradable nano composite reinforced with cellulose nano fiber from coconut industry waste for replacing synthetic plastic food packaging. *Chemosphere*, *291*, 132786. doi:10.1016/j.chemosphere.2021.132786 PMID:34762882

Ashfaq, A., Khursheed, N., Fatima, S., Anjum, Z., & Younis, K. (2022). Application of nanotechnology in food packaging: Pros and Cons. *Journal of Agriculture and Food Research*, *7*, 100270. doi:10.1016/j.jafr.2022.100270

Ashori, A., & Bahrami, R. (2014). Modification of physico-mechanical properties of chitosan-tapioca starch blend films using nano graphene. *Polymer-Plastics Technology and Engineering*, *53*(3), 312–318. doi:10.1080/03602559.2013.866246

Assis, R. Q., Lopes, S. M., Costa, T. M. H., Flôres, S. H., & de Oliveira Rios, A. (2017). Active biodegradable cassava starch films incorporated lycopene nanocapsules. *Industrial Crops and Products*, *109*, 818–827. doi:10.1016/j.indcrop.2017.09.043

Azaza, Y. B., Hamdi, M., Charmette, C., Jridi, M., Li, S., Nasri, M., & Nasri, R. (2022). Development and characterization of active packaging films based on chitosan and sardinella protein isolate: Effects on the quality and the shelf life of shrimps. *Food Packaging and Shelf Life*, *31*, 100796. doi:10.1016/j.fpsl.2021.100796

Azevedo, V. M., Silva, E. K., Pereira, C. F. G., da Costa, J. M. G., & Borges, S. V. (2015). Whey protein isolate biodegradable films: Influence of the citric acid and montmorillonite clay nanoparticles on the physical properties. *Food Hydrocolloids*, *43*, 252–258. doi:10.1016/j.foodhyd.2014.05.027

Azizi-Lalabadi, M., Ehsani, A., Ghanbarzadeh, B., & Divband, B. (2020). Polyvinyl alcohol/gelatin nanocomposite containing ZnO, TiO_2 or ZnO/TiO_2 nanoparticles doped on 4A zeolite: Microbial and sensory qualities of packaged white shrimp during refrigeration. *International Journal of Food Microbiology*, *312*, 108375. doi:10.1016/j.ijfoodmicro.2019.108375 PMID:31669767

Babaee, M., Garavand, F., Rehman, A., Jafarazadeh, S., Amini, E., & Cacciotti, I. (2022). Biodegradability, physical, mechanical and antimicrobial attributes of starch nanocomposites containing chitosan nanoparticles. *International Journal of Biological Macromolecules*, *195*, 49–58. doi:10.1016/j.ijbiomac.2021.11.162 PMID:34856218

Baghi, F., Gharsallaoui, A., Dumas, E., & Ghnimi, S. (2022). Advancements in biodegradable active films for food packaging: Effects of nano/microcapsule incorporation. *Foods*, *11*(5), 760. doi:10.3390/foods11050760 PMID:35267394

Bajpai, V. K., Kamle, M., Shukla, S., Mahato, D. K., Chandra, P., Hwang, S. K., Kumar, P., Huh, Y. S., & Han, Y. K. (2018). Prospects of using nanotechnology for food preservation, safety, and security. *Yao Wu Shi Pin Fen Xi*, *26*(4), 1201–1214. doi:10.1016/j.jfda.2018.06.011 PMID:30249319

Barage, S., Lakkakula, J., Sharma, A., Roy, A., Alghamdi, S., Almehmadi, M., Hossain, M. J., Allahyani, M., & Abdulaziz, O. (2022). Nanomaterial in food packaging: A comprehensive review. *Journal of Nanomaterials*, *2022*, 1–12. doi:10.1155/2022/6053922

Boey, J. Y., Mohamad, L., Khok, Y. S., Tay, G. S., & Baidurah, S. (2021). A review of the applications and biodegradation of polyhydroxyalkanoates and poly (lactic acid) and its composites. *Polymers*, *13*(10), 1544. doi:10.3390/polym13101544 PMID:34065779

Bumbudsanpharoke, N., Choi, J., & Ko, S. (2015). Applications of nanomaterials in food packaging. *Journal of Nanoscience and Nanotechnology*, *15*(9), 6357–6372. doi:10.1166/jnn.2015.10847 PMID:26716190

Bumbudsanpharoke, N., & Ko, S. (2015). Nano-food packaging: An overview of market, migration research, and safety regulations. *Journal of Food Science*, *80*(5), R910–R923. doi:10.1111/1750-3841.12861 PMID:25881665

Butkinaree, S., Jinkarn, T., & Yoksan, R. (2008). Effects of biodegradable coating on barrier properties of paperboard food packaging. *J. Met. Mater. Miner*, *18*, 219–222.

Cazón, P., Velazquez, G., Ramírez, J. A., & Vázquez, M. (2017). Polysaccharide-based films and coatings for food packaging: A review. *Food Hydrocolloids*, *68*, 136–148. doi:10.1016/j.foodhyd.2016.09.009

Chadha, U., Bhardwaj, P., Selvaraj, S. K., Arasu, K., Praveena, S., Pavan, A., Khanna, M., Singh, P., Singh, S., Chakravorty, A., Badoni, B., Banavoth, M., Sonar, P., & Paramasivam, V. (2022). Current trends and future perspectives of nanomaterials in food packaging application. *Journal of Nanomaterials*, *2022*, 1–32. doi:10.1155/2022/2745416

Chandio, A. D., Channa, I. A., Rizwan, M., Akram, S., Javed, M. S., Siyal, S. H., Saleem, M., Makhdoom, M., Ashfaq, T., Khan, S., Hussain, S., Albaqami, M., & Alotabi, R. G. (2021). Polyvinyl alcohol and nano-clay based solution processed packaging coatings. *Coatings*, *11*(8), 942. doi:10.3390/coatings11080942

Chaudhary, P., Fatima, F., & Kumar, A. (2020). Relevance of nanomaterials in food packaging and its advanced future prospects. *Journal of Inorganic and Organometallic Polymers and Materials*, *30*(12), 5180–5192. doi:10.1007/s10904-020-01674-8 PMID:32837459

Chausali, N., Saxena, J., & Prasad, R. (2022). Recent trends in nanotechnology applications of bio-based packaging. *Journal of Agriculture and Food Research*, *7*, 100257. doi:10.1016/j.jafr.2021.100257

Chavan, P., Sinhmar, A., Sharma, S., Dufresne, A., Thory, R., Kaur, M., Sandhu, K. S., Nehra, M., & Nain, V. (2022). Nanocomposite starch films: A new approach for biodegradable packaging materials. *Stärke*, *74*(5-6), 2100302. doi:10.1002/star.202100302

Chavan Qi, L., Xu, Z., Jiang, X., Hu, C., & Zou, X. (2004). Preparation and antibacterial activity of chitosan nanoparticles. *Carbohydrate Research*, *339*(16), 2693–2700. doi:10.1016/j.carres.2004.09.007 PMID:15519328

Choe, E., & Min, D. B. (2009). Mechanisms of antioxidants in the oxidation of foods. *Comprehensive Reviews in Food Science and Food Safety*, *8*(4), 345–358. doi:10.1111/j.1541-4337.2009.00085.x

Damm, C., Münstedt, H., & Rösch, A. (2008). The antimicrobial efficacy of polyamide 6/silver-nano-and microcomposites. *Materials Chemistry and Physics*, *108*(1), 61–66. doi:10.1016/j.matchemphys.2007.09.002

de Azeredo, H. M. (2009). Nanocomposites for food packaging applications. *Food Research International*, *42*(9), 1240–1253. doi:10.1016/j.foodres.2009.03.019

de Azeredo, H. M. C., Mattoso, L. C., & McHugh, T. H. (2011). Nanocomposites in food packaging–a review. *Advances in Diverse Industrial Applications of Nanocomposites*, 57-78. doi:10.1016/j.foodres.2009.03.019

de Jesús Salas-Méndez, E., Vicente, A., Pinheiro, A. C., Ballesteros, L. F., Silva, P., Rodríguez-García, R., ... de Rodríguez, D. J. (2019). Application of edible nanolaminate coatings with antimicrobial extract of *Flourensia cernua* to extend the shelf-life of tomato (*Solanum lycopersicum L.*) fruit. *Postharvest Biology and Technology*, *150*, 19–27. doi:10.1016/j.postharvbio.2018.12.008

de Lima Barizão, C., Crepaldi, M. I., Oscar de Oliveira, S., de Oliveira, A. C., Martins, A. F., Garcia, P. S., & Bonafé, E. G. (2020). Biodegradable films based on commercial κ-carrageenan and cassava starch to achieve low production costs. *International Journal of Biological Macromolecules*, *165*, 582–590. doi:10.1016/j.ijbiomac.2020.09.150 PMID:32991902

Di, Y., Iannace, S., Di Maio, E., & Nicolais, L. (2003). Nanocomposites by melt intercalation based on polycaprolactone and organoclay. *Journal of Polymer Science. Part B, Polymer Physics*, *41*(7), 670–678. doi:10.1002/polb.10420

Duncan, T. V. (2011). Applications of nanotechnology in food packaging and food safety: Barrier materials, antimicrobials and sensors. *Journal of Colloid and Interface Science*, *363*(1), 1–24. doi:10.1016/j.jcis.2011.07.017 PMID:21824625

Dwivany, F. M., Aprilyandi, A. N., Suendo, V., & Sukriandi, N. (2020). Carrageenan edible coating application prolongs Cavendish banana shelf life. *International Journal of Food Sciences, 2020*, 1–11. Advance online publication. doi:10.1155/2020/8861610 PMID:32908863

Egodage, D. P., Jayalath, H. T. S., Samarasekara, A. M. P. B., Amarasinghe, D. A. S., Madushani, S. P. A., & Senerath, S. M. N. S. (2017, May). Novel antimicrobial nano coated polypropylene based materials for food packaging systems. In *2017 Moratuwa Engineering Research Conference (MERCon)* (pp. 88-92). IEEE. 10.1109/MERCon.2017.7980462

El-Naggar, M. E., Hasanin, M., & Hashem, A. H. (2022). Eco-Friendly Synthesis of Superhydrophobic Antimicrobial Film Based on Cellulose Acetate/Polycaprolactone Loaded with the Green Biosynthesized Copper Nanoparticles for Food Packaging Application. *Journal of Polymers and the Environment, 30*(5), 1820–1832. doi:10.1007/s10924-021-02318-9

Elsabee, M. Z., & Abdou, E. S. (2013). Chitosan based edible films and coatings: A review. *Materials Science and Engineering C, 33*(4), 1819–1841. doi:10.1016/j.msec.2013.01.010 PMID:23498203

Farris, S., Unalan, I. U., Introzzi, L., Fuentes-Alventosa, J. M., & Cozzolino, C. A. (2014). Pullulan-based films and coatings for food packaging: Present applications, emerging opportunities, and future challenges. *Journal of Applied Polymer Science, 131*(13), app.40539. Advance online publication. doi:10.1002/app.40539

Ferrer, A., Pal, L., & Hubbe, M. (2017). Nanocellulose in packaging: Advances in barrier layer technologies. *Industrial Crops and Products, 95*, 574–582. doi:10.1016/j.indcrop.2016.11.012

Ferri, M., Papchenko, K., Degli Esposti, M., Tondi, G., De Angelis, M. G., Morselli, D., & Fabbri, P. (2023). Fully Biobased Polyhydroxyalkanoate/Tannin Films as Multifunctional Materials for Smart Food Packaging Applications. *ACS Applied Materials & Interfaces, 15*(23), 28594–28605. Advance online publication. doi:10.1021/acsami.3c04611 PMID:37265075

Garcia, C. V., Shin, G. H., & Kim, J. T. (2018). Metal oxide-based nanocomposites in food packaging: Applications, migration, and regulations. *Trends in Food Science & Technology, 82*, 21–31. doi:10.1016/j.tifs.2018.09.021

Ge, L., Zhao, Y. S., Mo, T., Li, J. R., & Li, P. (2012). *Immobilization of glucose oxidase in electrospun nanofibrous membranes for food preservation*. doi:10.1016/j.foodcont.2012.01.022

Goudarzi, V., Shahabi-Ghahfarrokhi, I., & Babaei-Ghazvini, A. (2017). Preparation of ecofriendly UV-protective food packaging material by starch/TiO2 bio-nanocomposite: Characterization. *International Journal of Biological Macromolecules*, *95*, 306–313. doi:10.1016/j.ijbiomac.2016.11.065 PMID:27884670

Hablot, E., Bordes, P., Pollet, E., & Avérous, L. (2008). Thermal and thermo-mechanical degradation of poly (3-hydroxybutyrate)-based multiphase systems. *Polymer Degradation & Stability*, *93*(2), 413–421. doi:10.1016/j.polymdegradstab.2007.11.018

Han, J. H., Lee, D. S., Min, S. C., & Chung, M. S. (2012). Eco-design of food and beverage packaging. In *Emerging Food Packaging Technologies* (pp. 361–379). Woodhead Publishing. doi:10.1533/9780857095664.4.361

Hegyesi, N., Zhang, Y., Kohári, A., Polyák, P., Sui, X., & Pukánszky, B. (2019). Enzymatic degradation of PLA/cellulose nanocrystal composites. *Industrial Crops and Products*, *141*, 111799. doi:10.1016/j.indcrop.2019.111799

Hu, X., Lu, C., Tang, H., Pouri, H., Joulin, E., & Zhang, J. (2022). Active Food Packaging Made of Biopolymer-Based Composites. *Materials (Basel)*, *16*(1), 279. doi:10.3390/ma16010279 PMID:36614617

Ibrahim, N. I., Shahar, F. S., Sultan, M. T. H., Shah, A. U. M., Safri, S. N. A., & Mat Yazik, M. H. (2021). Overview of bioplastic introduction and its applications in product packaging. *Coatings*, *11*(11), 1423. doi:10.3390/coatings11111423

Jafarzadeh, S., Nafchi, A. M., Salehabadi, A., Oladzad-Abbasabadi, N., & Jafari, S. M. (2021). Application of bio-nanocomposite films and edible coatings for extending the shelf life of fresh fruits and vegetables. *Advances in Colloid and Interface Science*, *291*, 102405. doi:10.1016/j.cis.2021.102405 PMID:33819726

Jang, E. J., Padhan, B., Patel, M., Pandey, J. K., Xu, B., & Patel, R. (2023). Antibacterial and biodegradable food packaging film from bacterial cellulose. *Food Control*, *109902*, 109902. Advance online publication. doi:10.1016/j.foodcont.2023.109902

Janjarasskul, T., and Suppakul, P. (2016). Active and intelligent packaging: the indication of quality and safety. *Crit. Rev. Food Sci. Nutr.* doi:10.1080/10408398.2016.1225278

Jayakumar, A., Heera, K. V., Sumi, T. S., Joseph, M., Mathew, S., Praveen, G., ... Radhakrishnan, E. K. (2019). Starch-PVA composite films with zinc-oxide nanoparticles and phytochemicals as intelligent pH sensing wraps for food packaging application. *International Journal of Biological Macromolecules*, *136*, 395–403. doi:10.1016/j.ijbiomac.2019.06.018 PMID:31173829

Jiang, W., Zhao, P., Song, W., Wang, M., & Yu, D. G. (2022). Electrospun zein/polyoxyethylene core-sheath ultrathin fibers and their antibacterial food packaging applications. *Biomolecules, 12*(8), 1110. doi:10.3390/biom12081110 PMID:36009003

Ju, J., Xie, Y., Guo, Y., Cheng, Y., Qian, H., & Yao, W. (2019). Application of edible coating with essential oil in food preservation. *Critical Reviews in Food Science and Nutrition, 59*(15), 2467–2480. doi:10.1080/10408398.2018.1456402 PMID:29580066

Kanmani, P., & Rhim, J. W. (2014). Development and characterization of carrageenan/grapefruit seed extract composite films for active packaging. *International Journal of Biological Macromolecules, 68*, 258–266. doi:10.1016/j.ijbiomac.2014.05.011 PMID:24832986

Khalid, S., Yu, L., Feng, M., Meng, L., Bai, Y., Ali, A., Liu, H., & Chen, L. (2018). Development and characterization of biodegradable antimicrobial packaging films based on polycaprolactone, starch and pomegranate rind hybrids. *Food Packaging and Shelf Life, 18*, 71–79. doi:10.1016/j.fpsl.2018.08.008

Khaneghah, A. M., Hashemi, S. M. B., & Limbo, S. (2018). Antimicrobial agents and packaging systems in antimicrobial active food packaging: An overview of approaches and interactions. *Food and Bioproducts Processing, 111*, 1–19. doi:10.1016/j.fbp.2018.05.001

Kim, H. Y., Park, S. S., & Lim, S. T. (2015). Preparation, characterization and utilization of starch nanoparticles. *Colloids and Surfaces. B, Biointerfaces, 126*, 607–620. doi:10.1016/j.colsurfb.2014.11.011 PMID:25435170

Kraśniewska, K., Pobiega, K., & Gniewosz, M. (2019). Pullulan–biopolymer with potential for use as food packaging. *International Journal of Food Engineering, 15*(9), 20190030. doi:10.1515/ijfe-2019-0030

Kraśniewska, K., Ścibisz, I., Gniewosz, M., Mitek, M., Pobiega, K., & Cendrowski, A. (2017). Effect of pullulan coating on postharvest quality and shelf-life of highbush blueberry (*Vaccinium corymbosum L.*). *Materials (Basel), 10*(8), 965. doi:10.3390/ma10080965 PMID:28820473

Kumari, S. V. G., Pakshirajan, K., & Pugazhenthi, G. (2022). Recent advances and future prospects of cellulose, starch, chitosan, polylactic acid and polyhydroxyalkanoates for sustainable food packaging applications. *International Journal of Biological Macromolecules, 221*, 163–182. Advance online publication. doi:10.1016/j.ijbiomac.2022.08.203 PMID:36067847

Kuswandi, B. (2017). Environmental friendly food nano-packaging. *Environmental Chemistry Letters*, *15*(2), 205–221. doi:10.1007/s10311-017-0613-7

Kuswandi, B., & Moradi, M. (2019). Improvement of food packaging based on functional nanomaterial. *Nanotechnology: Applications in energy, drug and food*, 309-344. doi:10.1007/978-3-319-99602-8_16

Lhamo, P., & Mahanty, B. (2023). Tuning Material Properties of Microbially Synthesized Poly (3-hydroxybutyrate-co-hydroxyvalerate) and Their Composites for Various Applications. *Journal of Polymers and the Environment*, *31*(11), 1–21. doi:10.1007/s10924-023-02908-9

Lin, D., Yang, Y., Wang, J., Yan, W., Wu, Z., Chen, H., Zhang, Q., Wu, D., Qin, W., & Tu, Z. (2020). Preparation and characterization of TiO2-Ag loaded fish gelatin-chitosan antibacterial composite film for food packaging. *International Journal of Biological Macromolecules*, *154*, 123–133. doi:10.1016/j.ijbiomac.2020.03.070 PMID:32171840

Liu, Y., Zhang, X., Li, C., Qin, Y., Xiao, L., & Liu, J. (2020). Comparison of the structural, physical and functional properties of κ-carrageenan films incorporated with pomegranate flesh and peel extracts. *International Journal of Biological Macromolecules*, *147*, 1076–1088. doi:10.1016/j.ijbiomac.2019.10.075 PMID:31739038

Lotfi, M., Tajik, H., Moradi, M., Forough, M., Divsalar, E., & Kuswandi, B. (2018). Nanostructured chitosan/monolaurin film: Preparation, characterization and antimicrobial activity against *Listeria monocytogenes* on ultrafiltered white cheese. *Lebensmittel-Wissenschaft + Technologie*, *92*, 576–583. doi:10.1016/j.lwt.2018.03.020

Majid, I., Nayik, G. A., Dar, S. M., & Nanda, V. (2018). Novel food packaging technologies: Innovations and future prospective. *Journal of the Saudi Society of Agricultural Sciences*, *17*(4), 454–462. doi:10.1016/j.jssas.2016.11.003

Manikandan, V., & Min, S. C. (2023). Roles of polysaccharides-based nanomaterials in food preservation and extension of shelf-life of food products: A review. *International Journal of Biological Macromolecules*, *126381*, 126381. Advance online publication. doi:10.1016/j.ijbiomac.2023.126381 PMID:37595723

Meereboer, K. W., Misra, M., & Mohanty, A. K. (2020). Review of recent advances in the biodegradability of polyhydroxyalkanoate (PHA) bioplastics and their composites. *Green Chemistry*, *22*(17), 5519–5558. doi:10.1039/D0GC01647K

Mesgari, M., Aalami, A. H., Sathyapalan, T., & Sahebkar, A. (2022). A comprehensive review of the development of carbohydrate macromolecules and copper oxide nanocomposite films in food nanopackaging. *Bioinorganic Chemistry and Applications, 2022*, 1–28. Advance online publication. doi:10.1155/2022/7557825 PMID:35287316

Mihindukulasuriya, S. D. F., & Lim, L. T. (2014). Nanotechnology development in food packaging: A review. *Trends in Food Science & Technology, 40*(2), 149–167. doi:10.1016/j.tifs.2014.09.009

Mirpoor, S. F., Corrado, I., Di Girolamo, R., Dal Poggetto, G., Panzella, L., Borselleca, E., Pezzella, C., & Giosafatto, C. V. L. (2023). Manufacture of active multilayer films made of functionalized pectin coated by polyhydroxyalkanoates: A fully renewable approach to active food packaging. *Polymer, 126136*. Advance online publication. doi:10.1016/j.polymer.2023.126136

Mitelut, A. C., Tanase, E. E., Popa, V. I., & Popa, M. E. (2015). Sustainable alternative for food packaging: Chitosan biopolymer-a review. *AgroLife Scientific Journal, 4*(2).

Moalla, S., Ammar, I., Fauconnier, M. L., Danthine, S., Blecker, C., Besbes, S., & Attia, H. (2021). Development and characterization of chitosan films carrying Artemisia campestris antioxidants for potential use as active food packaging materials. *International Journal of Biological Macromolecules, 183*, 254–266. doi:10.1016/j.ijbiomac.2021.04.113 PMID:33892038

Mohammad, Z. H., & Ahmad, F. (2024). Nanocoating and its application as antimicrobials in the food industry: A review. *International Journal of Biological Macromolecules, 254*, 127906. doi:10.1016/j.ijbiomac.2023.127906 PMID:37935295

Müller, K., Bugnicourt, E., Latorre, M., Jorda, M., Echegoyen Sanz, Y., Lagaron, J. M., Miesbauer, O., Bianchin, A., Hankin, S., Bölz, U., Pérez, G., Jesdinszki, M., Lindner, M., Scheuerer, Z., Castelló, S., & Schmid, M. (2017). Review on the processing and properties of polymer nanocomposites and nanocoatings and their applications in the packaging, automotive and solar energy fields. *Nanomaterials (Basel, Switzerland), 7*(4), 74. doi:10.3390/nano7040074 PMID:28362331

Nath, D., Santhosh, R., Pal, K., & Sarkar, P. (2022). Nanoclay-based active food packaging systems: A review. *Food Packaging and Shelf Life, 31*, 100803. doi:10.1016/j.fpsl.2021.100803

Nazrin, A., Sapuan, S. M., Zuhri, M. Y. M., Ilyas, R. A., Syafiq, R. S. F. K. S., & Sherwani, S. F. K. (2020). Nanocellulose reinforced thermoplastic starch (TPS), polylactic acid (PLA), and polybutylene succinate (PBS) for food packaging applications. *Frontiers in Chemistry*, *8*, 213. doi:10.3389/fchem.2020.00213 PMID:32351928

Nešić, A., Cabrera-Barjas, G., Dimitrijević-Branković, S., Davidović, S., Radovanović, N., & Delattre, C. (2019). Prospect of polysaccharide-based materials as advanced food packaging. *Molecules (Basel, Switzerland)*, *25*(1), 135. doi:10.3390/molecules25010135 PMID:31905753

Nile, S. H., Baskar, V., Selvaraj, D., Nile, A., Xiao, J., & Kai, G. (2020). Nanotechnologies in food science: Applications, recent trends, and future perspectives. *Nano-Micro Letters*, *12*(1), 45. doi:10.1007/s40820-020-0383-9 PMID:34138283

Nilsen-Nygaard, J., Fernández, E. N., Radusin, T., Rotabakk, B. T., Sarfraz, J., Sharmin, N., Sivertsvik, M., Sone, I., & Pettersen, M. K. (2021). Current status of biobased and biodegradable food packaging materials: Impact on food quality and effect of innovative processing technologies. *Comprehensive Reviews in Food Science and Food Safety*, *20*(2), 1333–1380. doi:10.1111/1541-4337.12715 PMID:33547765

Noshirvani, N., Ghanbarzadeh, B., Gardrat, C., Rezaei, M. R., Hashemi, M., Le Coz, C., & Coma, V. (2017). Cinnamon and ginger essential oils to improve antifungal, physical and mechanical properties of chitosan-carboxymethyl cellulose films. *Food Hydrocolloids*, *70*, 36–45. doi:10.1016/j.foodhyd.2017.03.015

Nouri, A., Yaraki, M. T., Lajevardi, A., Rahimi, T., Tanzifi, M., & Ghorbanpour, M. (2020). An investigation of the role of fabrication process in the physicochemical properties of κ-carrageenan-based films incorporated with *Zataria multiflora* extract and nanoclay. *Food Packaging and Shelf Life*, *23*, 100435. doi:10.1016/j.fpsl.2019.100435

Oymaci, P., & Altinkaya, S. A. (2016). Improvement of barrier and mechanical properties of whey protein isolate based food packaging films by incorporation of zein nanoparticles as a novel bionanocomposite. *Food Hydrocolloids*, *54*, 1–9. doi:10.1016/j.foodhyd.2015.08.030

Pan, H., Jiang, B., Chen, J., & Jin, Z. (2014). Assessment of the physical, mechanical, and moisture-retention properties of pullulan-based ternary coblended films. *Carbohydrate Polymers*, *112*, 94–101. doi:10.1016/j.carbpol.2014.05.044 PMID:25129721

Perdani, C. G., & Gunawan, S. (2021, November). A short review: Nanocellulose for smart biodegradable packaging in the food industry. *IOP Conference Series. Earth and Environmental Science, 924*(1), 012032. doi:10.1088/1755-1315/924/1/012032

Pereda, M., Amica, G., & Marcovich, N. E. (2012). Development and characterization of edible chitosan/olive oil emulsion films. *Carbohydrate Polymers, 87*(2), 1318–1325. doi:10.1016/j.carbpol.2011.09.019

Perera, K. Y., Hopkins, M., Jaiswal, A. K., & Jaiswal, S. (2023). Nanoclays-containing bio-based packaging materials: Properties, applications, safety, and regulatory issues. *Journal of Nanostructure in Chemistry*, 1–23. doi:10.1007/s40097-023-00525-5 PMID:36747507

Perera, K. Y., Jaiswal, A. K., & Jaiswal, S. (2023). Biopolymer-Based Sustainable Food Packaging Materials: Challenges, Solutions, and Applications. *Foods, 12*(12), 2422. doi:10.3390/foods12122422 PMID:37372632

Phisalaphong, M., & Jatupaiboon, N. (2008). Biosynthesis and characterization of bacteria cellulose–chitosan film. *Carbohydrate Polymers, 74*(3), 482–488. doi:10.1016/j.carbpol.2008.04.004

Pirozzi, A., Del Grosso, V., Ferrari, G., & Donsì, F. (2020). Edible coatings containing oregano essential oil nanoemulsion for improving postharvest quality and shelf life of tomatoes. *Foods, 9*(11), 1605. doi:10.3390/foods9111605 PMID:33158115

Pirozzi, A., Pataro, G., Donsì, F., & Ferrari, G. (2021). Edible coating and pulsed light to increase the shelf life of food products. *Food Engineering Reviews, 13*(3), 544–569. doi:10.1007/s12393-020-09245-w

Popović, S. Z., Lazić, V. L., Hromiš, N. M., Šuput, D. Z., & Bulut, S. N. (2018). Biopolymer packaging materials for food shelf-life prolongation. In *Biopolymers for food design* (pp. 223–277). Academic Press. doi:10.1016/B978-0-12-811449-0.00008-6

Primožič, M., Knez, Ž., & Leitgeb, M. (2021). (Bio) Nanotechnology in food science—Food packaging. *Nanomaterials (Basel, Switzerland), 11*(2), 292. doi:10.3390/nano11020292 PMID:33499415

Rad, F. H., Sharifan, A., Khodaiyan, F., & Shahabi-Ghahfarrokhi, I. (2016). Preparation and characterization of pullulan-soy protein concentrate blended film incorporated with *Zataria multiflora* and *Artemisia biennis* essential oils. *Jundishapur Journal of Natural Pharmaceutical Products*. Advance online publication. doi:10.17795/jjnpp37802

Rai, M., Ingle, A. P., Gupta, I., Pandit, R., Paralikar, P., Gade, A., Chaud, M. V., & dos Santos, C. A. (2019). Smart nanopackaging for the enhancement of food shelf life. *Environmental Chemistry Letters*, *17*(1), 277–290. doi:10.1007/s10311-018-0794-8

Regulation (EC) No. 1935/2004 of The European Parliament and of The Council of 27 October 2004 on Materials and Articles Intended to Come into Contact with Food and Repealing Directives 80/590/EEC and 89/109/EEC. Available online: https://www.interpack.com/cgi-bin/md_interpack/lib/pub/tt.cgi/Food_ Safety.html?oid=55076&lang=2&ticket=g_u_e_s_t

Rhim, J. W., Park, H. M., & Ha, C. S. (2013). Bio-nanocomposites for food packaging applications. *Progress in Polymer Science*, *38*(10-11), 1629–1652. doi:10.1016/j.progpolymsci.2013.05.008

Riaz, A., Lagnika, C., Luo, H., Nie, M., Dai, Z., Liu, C., Abdin, M., Hashim, M. M., Li, D., & Song, J. (2020). Effect of Chinese chives (*Allium tuberosum*) addition to carboxymethyl cellulose based food packaging films. *Carbohydrate Polymers*, *235*, 115944. doi:10.1016/j.carbpol.2020.115944 PMID:32122483

Rovera, C., Türe, H., Hedenqvist, M. S., & Farris, S. (2020). Water vapor barrier properties of wheat gluten/silica hybrid coatings on paperboard for food packaging applications. *Food Packaging and Shelf Life*, *26*, 100561. doi:10.1016/j.fpsl.2020.100561

Sá, N. M., Mattos, A. L., Silva, L. M., Brito, E. S., Rosa, M. F., & Azeredo, H. M. (2020). From cashew byproducts to biodegradable active materials: Bacterial cellulose-lignin-cellulose nanocrystal nanocomposite films. *International Journal of Biological Macromolecules*, *161*, 1337–1345. doi:10.1016/j.ijbiomac.2020.07.269 PMID:32777430

Saharudin, M. S., Ilyas, R. A., Awang, N., Hasbi, S., Shyha, I., & Inam, F. (2023). Advances in Sustainable Nanocomposites. *Sustainability (Basel)*, *15*(6), 5125. doi:10.3390/su15065125

Sahoo, M., Vishwakarma, S., Panigrahi, C., & Kumar, J. (2021). Nanotechnology: Current applications and future scope in food. *Food Frontiers*, *2*(1), 3–22. doi:10.1002/fft2.58

Saini, R. D. (2017). Biodegradable polymers. *Int. J. Appl. Chem*, *13*(2), 179–196. doi:10.5958/0974-4150.2018.00037.8

Samrot, A. V., Sean, T. C., Kudaiyappan, T., Bisyarah, U., Mirarmandi, A., Faradjeva, E., ... Kumar, S. S. (2020). Production, characterization and application of nanocarriers made of polysaccharides, proteins, bio-polyesters and other biopolymers: A review. *International Journal of Biological Macromolecules, 165*, 3088–3105. doi:10.1016/j.ijbiomac.2020.10.104 PMID:33098896

Sani, I. K., Pirsa, S., & Tağı, Ş. (2019). Preparation of chitosan/zinc oxide/*Melissa officinalis* essential oil nano-composite film and evaluation of physical, mechanical and antimicrobial properties by response surface method. *Polymer Testing, 79*, 106004. doi:10.1016/j.polymertesting.2019.106004

Sapper, M., Talens, P., & Chiralt, A. (2019). Improving functional properties of cassava starch-based films by incorporating xanthan, gellan, or pullulan gums. *International Journal of Polymer Science*. doi:10.1155/2019/5367164

Sartori, T., Feltre, G., Sobral, P. J. D. A., Cunha, R. L. D., & Menegalli, F. C. (2018). Properties of films produced from blends of pectin and gluten. *Food Packaging and Shelf Life, 18*, 221–229. doi:10.1016/j.fpsl.2018.11.007

Sarwar, M. S., Niazi, M. B. K., Jahan, Z., Ahmad, T., & Hussain, A. (2018). Preparation and characterization of PVA/nanocellulose/Ag nanocomposite films for antimicrobial food packaging. *Carbohydrate Polymers, 184*, 453–464. doi:10.1016/j.carbpol.2017.12.068 PMID:29352941

Schaefer, D., & Cheung, W. M. (2018). Smart packaging: Opportunities and challenges. *Procedia CIRP, 72*, 1022–1027. doi:10.1016/j.procir.2018.03.240

Schiessl, S., Kucukpinar, E., Cros, S., Miesbauer, O., Langowski, H. C., & Eisner, P. (2022). Nanocomposite coatings based on polyvinyl alcohol and montmorillonite for high-barrier food packaging. *Frontiers in Nutrition, 9*, 790157. doi:10.3389/fnut.2022.790157 PMID:35340548

Schrand, A. M., Rahman, M. F., Hussain, S. M., Schlager, J. J., Smith, D. A., & Syed, A. F. (2010). Metal-based nanoparticles and their toxicity assessment. *Wiley Interdisciplinary Reviews. Nanomedicine and Nanobiotechnology, 2*(5), 544–568. doi:10.1002/wnan.103 PMID:20681021

Shanmugam, R., Mayakrishnan, V., Kesavan, R., Shanmugam, K., Veeramani, S., & Ilangovan, R. (2022). Mechanical, barrier, adhesion and antibacterial properties of pullulan/graphene bio nanocomposite coating on spray coated nanocellulose film for food packaging applications. *Journal of Polymers and the Environment, 30*(5), 1–9. doi:10.1007/s10924-021-02311-2

Sharma, C., Dhiman, R., Rokana, N., & Panwar, H. (2017). Nanotechnology: An untapped resource for food packaging. *Frontiers in Microbiology*, *8*, 1735. doi:10.3389/fmicb.2017.01735 PMID:28955314

Shi, A. M., Wang, L. J., Li, D., & Adhikari, B. (2013). Characterization of starch films containing starch nanoparticles: Part 1: Physical and mechanical properties. *Carbohydrate Polymers*, *96*(2), 593–601. doi:10.1016/j.carbpol.2012.12.042 PMID:23768605

Shojaee-Aliabadi, S., Hosseini, H., Mohammadifar, M. A., Mohammadi, A., Ghasemlou, M., Ojagh, S. M., Hosseini, S. M., & Khaksar, R. (2013). Characterization of antioxidant-antimicrobial κ-carrageenan films containing *Satureja hortensis* essential oil. *International Journal of Biological Macromolecules*, *52*, 116–124. doi:10.1016/j.ijbiomac.2012.08.026 PMID:22959956

Sothornvit, R. (2019). Nanostructured materials for food packaging systems: New functional properties. *Current Opinion in Food Science*, *25*, 82–87. doi:10.1016/j.cofs.2019.03.001

Srinivasa, P. C., Ramesh, M. N., & Tharanathan, R. N. (2007). Effect of plasticizers and fatty acids on mechanical and permeability characteristics of chitosan films. *Food Hydrocolloids*, *21*(7), 1113–1122. doi:10.1016/j.foodhyd.2006.08.005

Swaroop, C., & Shukla, M. (2018). Nano-magnesium oxide reinforced polylactic acid biofilms for food packaging applications. *International Journal of Biological Macromolecules*, *113*, 729–736. doi:10.1016/j.ijbiomac.2018.02.156 PMID:29499267

Talebi, F., Misaghi, A., Khanjari, A., Kamkar, A., Gandomi, H., & Rezaeigolestani, M. (2018). Incorporation of spice essential oils into poly-lactic acid film matrix with the aim of extending microbiological and sensorial shelf life of ground beef. *Lebensmittel-Wissenschaft + Technologie*, *96*, 482–490. doi:10.1016/j.lwt.2018.05.067

Teow, Y., Asharani, P. V., Hande, M. P., & Valiyaveettil, S. (2011). Health impact and safety of engineered nanomaterials. *Chemical Communications*, *47*(25), 7025–7038. doi:10.1039/c0cc05271j PMID:21479319

Tonyali, B., McDaniel, A., Amamcharla, J., Trinetta, V., & Yucel, U. (2020). Release kinetics of cinnamaldehyde, eugenol, and thymol from sustainable and biodegradable active packaging films. *Food Packaging and Shelf Life*, *24*, 100484. doi:10.1016/j.fpsl.2020.100484

Vasile, C. (2018). Polymeric nanocomposites and nanocoatings for food packaging: A review. *Materials (Basel)*, *11*(10), 1834. doi:10.3390/ma11101834 PMID:30261658

Velickova, E., Winkelhausen, E., Kuzmanova, S., Alves, V. D., & Moldão-Martins, M. (2013). Impact of chitosan-beeswax edible coatings on the quality of fresh strawberries (Fragaria ananassa cv Camarosa) under commercial storage conditions. *Lebensmittel-Wissenschaft + Technologie*, *52*(2), 80–92. doi:10.1016/j.lwt.2013.02.004

Verma, C., Quraishi, M. A., Alfantazi, A., & Rhee, K. Y. (2023). *Biodegradable Synthetic Polymers in Sustainable Corrosion Protection: Present and Future Scenarios*. Advanced Industrial and Engineering Polymer Research., doi:10.1016/j.aiepr.2023.04.005

Versino, F., Lopez, O. V., Garcia, M. A., & Zaritzky, N. E. (2016). Starch-based films and food coatings: An overview. *Stärke*, *68*(11-12), 1026–1037. doi:10.1002/star.201600095

Vilarinho, F., Sanches Silva, A., Vaz, M. F., & Farinha, J. P. (2018). Nanocellulose in green food packaging. *Critical Reviews in Food Science and Nutrition*, *58*(9), 1526–1537. doi:10.1080/10408398.2016.1270254 PMID:28125279

Wang, Y., Zhang, R., Ahmed, S., Qin, W., & Liu, Y. (2019). Preparation and characterization of corn starch bio-active edible packaging films based on zein incorporated with orange-peel oil. *Antioxidants*, *8*(9), 391. doi:10.3390/antiox8090391 PMID:31514341

Wani, N. R., Dar, A. H., Dash, K. K., Pandey, V. K., Srivastava, S., Jan, S. Y., Deka, P., & Sabahi, N. (2023). Recent advances in the production of bionanomaterials for development of sustainable food packaging: A comprehensive review. *Environmental Research*, *116948*, 116948. Advance online publication. doi:10.1016/j.envres.2023.116948 PMID:37611789

Wardhono, E. Y., Pinem, M. P., Kustiningsih, I., Agustina, S., Oudet, F., Lefebvre, C., Clausse, D., Saleh, K., & Guénin, E. (2019). Cellulose nanocrystals to improve stability and functional properties of emulsified film based on chitosan nanoparticles and beeswax. *Nanomaterials (Basel, Switzerland)*, *9*(12), 1707. doi:10.3390/nano9121707 PMID:31795284

Wen, P., Zhu, D. H., Wu, H., Zong, M. H., Jing, Y. R., & Han, S. Y. (2016). Encapsulation of cinnamon essential oil in electrospun nanofibrous film for active food packaging. *Food Control*, *59*, 366–376. doi:10.1016/j.foodcont.2015.06.005

Wypij, M., Trzcińska-Wencel, J., Golińska, P., Avila-Quezada, G. D., Ingle, A. P., & Rai, M. (2023). The strategic applications of natural polymer nanocomposites in food packaging and agriculture: Chances, challenges, and consumers' perception. *Frontiers in Chemistry*, *10*, 1106230. doi:10.3389/fchem.2022.1106230 PMID:36704616

Xiao, Y., Liu, Y., Kang, S., Wang, K., & Xu, H. (2020). Development and evaluation of soy protein isolate-based antibacterial nanocomposite films containing cellulose nanocrystals and zinc oxide nanoparticles. *Food Hydrocolloids*, *106*, 105898. doi:10.1016/j.foodhyd.2020.105898

Xu, C., Chen, C., & Wu, D. (2018). The starch nanocrystal filled biodegradable poly (ε-caprolactone) composite membrane with highly improved properties. *Carbohydrate Polymers*, *182*, 115–122. doi:10.1016/j.carbpol.2017.11.001 PMID:29279105

Xu, H., Chai, Y., & Zhang, G. (2012). Synergistic effect of oleic acid and glycerol on zein film plasticization. *Journal of Agricultural and Food Chemistry*, *60*(40), 10075–10081. doi:10.1021/jf302940j PMID:22989052

Yadav, A., Mangaraj, S., Singh, R., Kumar, N., & Arora, S. (2018). *Biopolymers as packaging material in food and allied industry*. Academic Press.

Yu, Z., Dhital, R., Wang, W., Sun, L., Zeng, W., Mustapha, A., & Lin, M. (2019). Development of multifunctional nanocomposites containing cellulose nanofibrils and soy proteins as food packaging materials. *Food Packaging and Shelf Life*, *21*, 100366. doi:10.1016/j.fpsl.2019.100366

Zhang, D., Chen, L., Cai, J., Dong, Q., Din, Z. U., Hu, Z. Z., Wang, G.-Z., Ding, W.-P., He, J.-R., & Cheng, S. Y. (2021). Starch/tea polyphenols nanofibrous films for food packaging application: From facile construction to enhance mechanical, antioxidant and hydrophobic properties. *Food Chemistry*, *360*, 129922. doi:10.1016/j.foodchem.2021.129922 PMID:33965711

Zhang, D., Yu, G., Long, Z., Xiao, H., & Qian, L. (2016). Bio-wax latex-modified paper as antimicrobial and water-vapor-resistant packaging material. *Journal of Wood Chemistry and Technology*, *36*(3), 182–191. doi:10.1080/02773813.2015.1104545

Zhang, Y., Cui, L., Xu, H., Feng, X., Wang, B., Pukánszky, B., Mao, Z., & Sui, X. (2019). Poly (lactic acid)/cellulose nanocrystal composites via the Pickering emulsion approach: Rheological, thermal and mechanical properties. *International Journal of Biological Macromolecules*, *137*, 197–204. doi:10.1016/j.ijbiomac.2019.06.204 PMID:31255621

Zhao, Y., Li, B., Li, C., Xu, Y., Luo, Y., Liang, D., & Huang, C. (2021). Comprehensive review of polysaccharide-based materials in edible packaging: A sustainable approach. *Foods*, *10*(8), 1845. doi:10.3390/foods10081845 PMID:34441621

Zhou, J., Wen, Y., Marshall, M. R., Zhao, J., Gui, H., Yang, Y., Zeng, Z., Jones, D. L., & Zang, H. (2021). Microplastics as an emerging threat to plant and soil health in agroecosystems. *The Science of the Total Environment*, *787*, 147444. Advance online publication. doi:10.1016/j.scitotenv.2021.147444

Zhou, L., & Wang, Y. (2021). Physical and antimicrobial properties of zein and methyl cellulose composite films with plasticizers of oleic acid and polyethylene glycol. *Lebensmittel-Wissenschaft + Technologie*, *140*, 110811. doi:10.1016/j.lwt.2020.110811

ADDITIONAL READING

Correa-Pacheco, Z. N., Corona-Rangel, M. L., Bautista-Baños, S., & Ventura-Aguilar, R. I. (2021). Application of natural-based nanocoatings for extending the shelf life of green bell pepper fruit. *Journal of Food Science*, *86*(1), 95–102. doi:10.1111/1750-3841.15542 PMID:33258157

Hromiš, N. M., Lazić, V. L., Markov, S. L., Vaštag, Ž. G., Popović, S. Z., Šuput, D. Z., Džinić, N. R., Velićanski, A. S., & Popović, L. M. (2015). Optimization of chitosan biofilm properties by addition of caraway essential oil and beeswax. *Journal of Food Engineering*, *158*, 86–93. doi:10.1016/j.jfoodeng.2015.01.001

Kumar, L., Ramakanth, D., Akhila, K., & Gaikwad, K. K. (2022). Edible films and coatings for food packaging applications: A review. *Environmental Chemistry Letters*, *20*(1), 1–26. doi:10.1007/s10311-021-01339-z

Mohammad, Z. H., Ahmad, F., Ibrahim, S. A., & Zaidi, S. (2022). Application of nanotechnology in different aspects of the food industry. *Discover Food*, *2*(1), 12. doi:10.1007/s44187-022-00013-9

Nguyen, H. L., Tran, T. H., Hao, L. T., Jeon, H., Koo, J. M., Shin, G., Hwang, D. S., Hwang, S. Y., Parki, J., & Oh, D. X. (2021). Biorenewable, transparent, and oxygen/moisture barrier nanocellulose/nanochitin-based coating on polypropylene for food packaging applications. *Carbohydrate Polymers*, *271*, 118421. doi:10.1016/j.carbpol.2021.118421 PMID:34364562

Osorio-Madrazo, A., David, L., Trombotto, S., Lucas, J. M., Peniche-Covas, C., & Domard, A. (2010). Kinetics study of the solid-state acid hydrolysis of chitosan: Evolution of the crystallinity and macromolecular structure. *Biomacromolecules*, *11*(5), 1376–1386. doi:10.1021/bm1001685 PMID:20402497

Zhang, Y., Li, B., Zhang, Y., Xu, F., Zhu, K., Li, S., Tan, L., Wu, G., & Dong, W. (2019). Effect of degree of polymerization of amylopectin on the gelatinization properties of jackfruit seed starch. *Food Chemistry*, *289*, 152–159. doi:10.1016/j.foodchem.2019.03.033 PMID:30955597

Zhu, F. (2018). Relationships between amylopectin internal molecular structure and physicochemical properties of starch. *Trends in Food Science & Technology*, *78*, 234–242. doi:10.1016/j.tifs.2018.05.024

KEY TERMS AND DEFINITIONS

Active Packaging: Type of smart packaging designed to extend the shelf life of perishable products and improve their quality.

Antimicrobial Activtiy: Process of either killing or inhibiting disease-causing microbes, involving the prevention of bacterial growth and the inhibition of microbial colony formation.

Antoxidant Activity: The ability of a substance to neutralize or inhibit the damaging effects of reactive oxygen species (ROS) or free radicals.

Bactericidal: Having the ability to eliminate or utilized for the destruction of bacteria.

Bio-Based: Products which are sourced either entirely or partially from materials originating from living organisms.

Bio-Packaging: Biodegradable packaging that is commonly characterized as any type of packaging that undergoes natural disintegration and decomposition.

Bio-Plastic: Plastic derived from biological sources instead of petroleum, with many varieties being biodegradable.

Biodegradable Packaging: Packaging materials designed to break down naturally over time, typically through the action of microorganisms, into natural substances without causing harm to the environment.

Bionanomaterials: Materials derived from diverse biological elements, including plants, bacteria, peptides, fungi, nucleic acids, etc.

Cytotoxicity: The capacity of a substance to induce harm or destruction to living cells, especially within the realms of biological or medical investigations, where the evaluation of its effects on cell viability or function is conducted.

Deacetylation: Removal of acetyl from a compound typically through hydrolysis.

DPPH: Radical scavenging method that is a widely employed spectrophotometric method for assessing antioxidant capacity.

Electrospinning: A technique used to produce polymer, ceramic, metallic, and composite fibers from solutions, dispersions, or melts.

Enzymatic Hydrolysis: A catalytic breakdown of a chemical compound by reaction with water, such as the conversion of cellulosic materials into fermentable sugars by the addition of specific enzymes.

Migration: The transfer of substances, such as chemicals or components present in the packaging materials, into the food product.

Nanoparticles: Small particle with dimensions ranging from 1 to 100 nanometers in size.

Pickering Emulsion: An emulsion stabilized by solid particles, such as colloidal silica, that adsorb onto the interface between the water and oil phases.

Polymerization: Chemical process in which small molecules, known as monomers, react together to form a large chainlike or networked molecule referred to as a polymer.

Smart Packaging: A form of product packaging that employs interactive technology to enhance the overall customer experience.

Solvent Casting Method: A method for creating thermoplastic polymer samples involves immersing a mold into a polymer solution and removing the solvent to leave a polymer film attached to the mold.

Tear Strength: Capability of a material to endure failure in a direction perpendicular to the applied stress.

Tensile Strength: Maximum load that a material can withstand without fracturing when subjected to stretching forces.

Water Vapor Permeability: The capacity of a material to permit the passage of water vapor.

Chapter 6
Nanocoatings for Cosmetotextiles

Bancha Yingngam
https://orcid.org/0000-0001-7215-9123
Ubon Ratchathani University, Thailand

ABSTRACT

In recent years, nanotechnology has found significant applications in the textile industry, especially in the development of cosmetotextiles. This chapter explores the role of nanocoatings in enhancing the functional and sustainable attributes of these textiles. Integrating insights from materials science, sustainability, and cosmetic technology, the chapter offers a comprehensive overview of current nanocoating techniques, materials, and their environmental impacts. Furthermore, it critically evaluated challenges and opportunities in implementing sustainable nanocoatings within cosmetotextiles and proposed strategies for developing eco-friendly nanocoatings and performing lifecycle analyses. Consequently, nanocoatings are essential for improving efficient, sustainable cosmetotextiles and offer potential for future research and commercial use. Key directions for future work include the creation of biodegradable nanomaterials, conducting toxicological studies to ensure user safety, and adopting circular economy principles within the cosmetic textile industry.

INTRODUCTION

Nanotechnology has flourished over the past few decades, offering transformative solutions to long-standing challenges. Central to this progress is the development of nanocomposites and nanocoatings, which have found applications in various

DOI: 10.4018/979-8-3693-3136-1.ch006

sectors, including electronics, construction, medicine, food, cosmetics, fragrances, and, more recently, textiles (Dickson et al., 2024; Ma et al., 2023; Sun et al., 2024; Vats et al., 2023). As the fashion and textile industry addresses the urgent demands of sustainability and multifunctionality, nanocoatings have risen as a key solution, notably in the domain of cosmetic textiles, or "cosmetotextiles" (Ben Abdelkader et al., 2023). Cosmetotextiles combine the fields of textiles and cosmetics, enhancing traditional fabrics through scientific and technological advancements (Attia et al., 2023; Julaeha et al., 2023). However, a substantial challenge remains in the integration of active cosmetic agents into textiles. The key question is how to do this efficiently without undermining their long-term efficacy or the textile's intrinsic properties. Nanocoatings present a compelling answer. These coatings can create fabrics that gradually release cosmetic agents to the skin while maintaining their efficacy over prolonged periods (Janeni & Adassooriya, 2023). Nevertheless, with groundbreaking innovation comes great responsibility. As nanocoatings for cosmetotextiles become more commonplace, concern arises regarding their environmental footprint, biocompatibility, and potential long-term effects on human health. The production, use, and disposal of nanomaterials entail processes that could pose environmental threats. Issues range from the potential release of nanocarriers into water systems to the challenge related to their recycling or degradation. The uncharted territories of the environmental impact of nanocoatings necessitate urgent exploration (Goyal & Jerold, 2023). Moreover, direct human exposure to these nanocoatings through cosmetotextiles raises crucial inquiries about dermal absorption, potential allergenic reactions, and broader systemic impacts. Fundamentally, while nanocoatings in cosmetotextiles offer the possibility to redefine the human perception of clothing and personal care, they also highlight an immediate need to understand and navigate the trade-offs between innovation and sustainability. As global efforts to adopt sustainable practices increase, the imperative to ensure the eco-compatibility of nanocoatings becomes paramount (Janeni & Adassooriya, 2023).

The chapter seeks to explore the field of nanocoatings for cosmetotextiles, shedding light on both their transformative capacities and the challenges they present. The primary goal is to provide a roadmap for researchers, manufacturers, and stakeholders, guiding the evolution of cosmetotextiles toward a more sustainable direction. The specific aims of this chapter are as follows:

1. To provide an in-depth understanding of the diverse types of nanocoatings used in cosmetotextiles. This includes highlighting their functional capabilities, such as ultraviolet (UV) protection, moisture-wicking, and temperature regulation, as well as their cosmeceutical benefits, such as the release of vitamins or fragrances.

2. To explore the methods of applying nanocoatings to textiles, including dip-coating, layer-by-layer assembly, electrospraying, and other scalable techniques, while evaluating the pros and cons of each.
3. To examine the environmental impact and sustainability considerations of nanocoatings in cosmetotextiles. Discussions will encompass the use of biodegradable materials, lifecycle analysis, and potential environmental threats.
4. To assess the potential health and safety risks associated with the use of nanocoated cosmetotextiles.

BASIC CONCEPTS OF COSMETOTEXTILES AND NANOCOATINGS

Cosmetotextiles, also known as "cosmetic textiles" or "dermotextiles", represent an innovative intersection between textile engineering and dermatology (Goyal & Jerold, 2023). Serving as active interfaces, they interact with the skin to deliver a multitude of skincare benefits, thus enhancing their conventional roles of protection, comfort, and aesthetic allure with dermatologically rooted functionalities (Ben Abdelkader et al., 2023). These benefits span from moisturizing and anti-aging effects to UV protection and even therapeutic attributes such as anti-inflammatory and wound-healing properties (Lucia-Oana et al., 2023). Figure 1 concisely depicts the evolution of cosmetotextiles and the progress in textile technologies within the cosmetic field, detailing key milestones and highlighting promising future directions. This includes the rise of multifunctional and eco-friendly nanocoatings with the potential to reshape the industry. Since the inception of cosmetotextiles in the late 1980s, epitomized by Teijin Co. Ltd.'s innovative "Amino Jeans". The field has experienced profound development and transformation, marking numerous technological breakthroughs and significant milestones. The company ingeniously incorporated the amino acid arginine into the fabric, aiming at skin rejuvenation (Han et al., 2020). The 1990s marked a phase of steady progression, with companies integrating diverse substances into textiles to confer a variety of skin benefits (Elkalla et al., 2023). The advent of microencapsulation technologies in the early 2000s enabled the controlled and extended release of active ingredients, thereby amplifying the effectiveness of cosmetotextiles. Between the mid-2000s and early 2010s, the incorporation of a variety of cosmetic substances catalyzed the onset of widespread commercialization. Concurrently, advancements in nanotechnology marked the advent of nanocoatings, introducing functionalities such as UV protection and antibacterial properties (Shie et al., 2019). The latter part of the 2010s emphasized sustainability and a diminished environmental footprint. This shift was accompanied by intensified attention to safety and regulatory standards, exploring the long-term effects on both human health and the ecosystem (Yurtsever, 2019). The current decade is seeing a surge in demand for multifunctional, aesthetically appealing, and environmentally sustainable cosmetotextiles. Concurrent research is exploring the fields of advanced, sustainable nanocoatings. The sector anticipates innovations centered on multifunctionality, sustainability, and an enhanced user experience, with sustainable nanocoatings being perceived as potential pivotal innovations. The adoption of circular economy concepts, coupled with ongoing research into the long-term effects of nanomaterials, represents significant advancements. These innovations integrate aesthetics, functionality, and sustainability, addressing the continually evolving demands of consumers and environmental concerns (Ben Abdelkader et al., 2023).

While traditional cosmetotextiles have effectively served their intended purpose, nanocoated textiles offer an advanced and often superior alternative (Chen et al., 2023; Prabhakar et al., 2022). Table 1 compares traditional cosmetotextiles with their nanocoated counterparts in terms of functionality, durability, cosmetic agent release, environmental impact, cost, safety, health concerns, aesthetic appeal, and regulatory compliance. Traditional cosmetotextiles primarily derive their functionalities from the cosmetic ingredients they are infused with. These ingredients may offer benefits such as skin moisturization, anti-aging effects, or aroma release (Taieb & Mnejja, 2023). On the other hand, nanocoated textiles employ nanoparticle coatings to expand and diversify their functionalities, introducing features such as UV protection, moisture control, and antibacterial properties (Han et al., 2020). In terms of durability, conventional cosmetotextiles exhibit moderate durability. With regular washing and wear, their infused cosmetic benefits tend to fade. In contrast, nanocoated textiles offer markedly higher durability and resistance to wear. This durability arises from the small size of nanocarriers and the specialized techniques employed in their application, ensuring that the coating remains intact and effective even after numerous washes. Regarding cosmetic agent release, traditional textiles have a steady release mechanism, although it tends to wane as the product ages. Nanocoated textiles, with their intricate design and technology, provide a more controlled and often extended release of agents, optimizing both the product's utility and user experience (N. Zhang et al., 2023). From an environmental viewpoint, the footprint of traditional cosmetotextiles is primarily determined by the cosmetic substances utilized (Janeni & Adassooriya, 2023). However, nanocoated textiles offer a new perspective: the use of biodegradable nanocarriers presents a potentially more sustainable choice.

In the context of cost, the integration of advanced nanotechnology elevates the production expenses of nanocoated textiles relative to their traditional counterparts. This cost reflects the advanced materials and techniques employed, along with the enhanced benefits they bring (Allaka & Yepuri, 2023). Regarding safety and

Figure 1. Evolution of cosmetotextiles and advancements in textile technologies

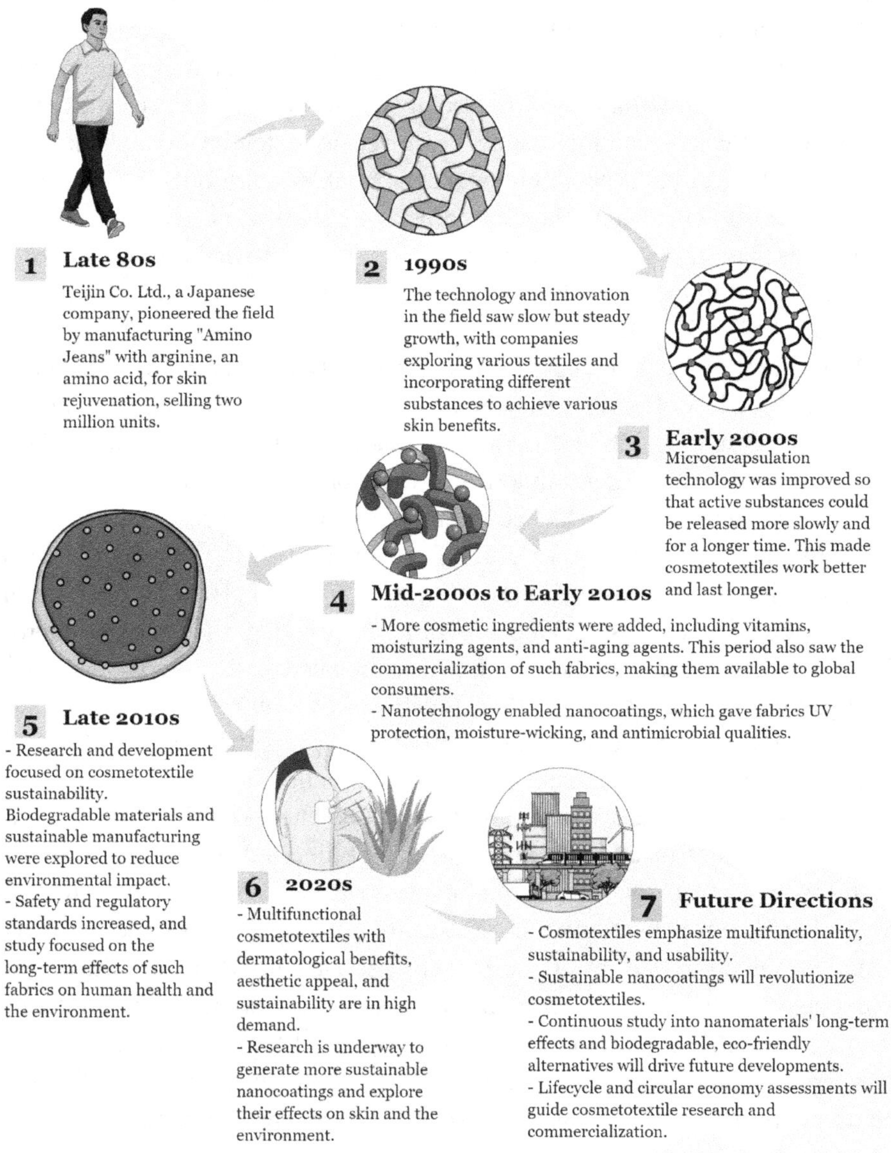

health, conventional cosmetotextiles are safe, but their safety depends on the specific ingredients infused. While many nanomaterials used in coatings are considered safe, there is a degree of uncertainty. Potential concerns related to nanoparticle exposure underscore the importance of continuous research and monitoring. Aesthetically, traditional cosmetotextiles might alter the fabric's texture due to the infused substances. However, nanocoated textiles, with their technological advantage, can heighten aesthetic appeal without compromising or altering the innate texture or feel of the fabric. Last, regulatory compliance remains pivotal for both types. Notably, nanocoated textiles demand added scrutiny owing to their use of nanomaterials (Attia et al., 2023).

TYPES OF NANOCOATING IN COSMETOTEXTILES

Nanocoatings in cosmetotextiles have the potential to significantly enhance the fabric's capabilities and functionalities. They offer a promising path to transform the way textiles interact with skin and their surrounding environment. The following types of coatings are commonly used in cosmetotextiles:

Hydrophobic and Hydrophilic Nanocoatings

Hydrophobic nanocoatings play a critical role in cosmetotextiles by providing vital water-repellent properties. They achieve this by lowering the surface energy of textiles, which prevents the adherence of water molecules to the fabric surface (Singh et al., 2023). This characteristic is invaluable for cosmetotextiles used in wet environments, helping wearers stay dry and comfortable while maintaining the aesthetic and functional properties of the textiles (Zheng et al., 2023). These coatings can be formulated from a range of materials, including silicon dioxide nanoparticles and fluoropolymers, and find extensive application in products such as moisture-wicking clothing, waterproof makeup capes, and water-resistant hairdressing gowns (He et al., 2023). Their advantages encompass enhanced comfort, prolonged textile longevity, aesthetic preservation, and reduced laundry frequency, which aid in water conservation (Atwah & Khan, 2023; Tanasa et al., 2023). In contrast, hydrophilic nanocoatings are designed to enhance the water-absorption capacity of cosmetotextiles. They increase the surface energy of textiles, enabling the fabric to attract and hold water molecules more efficiently. This feature is pivotal for cosmetotextiles intended for perspiration control or for delivering active cosmetic ingredients through the skin (Singh et al., 2023). Made from materials such as hydrogel nanoparticles and metallic oxide nanoparticles, these hydrophilic coatings are indispensable for producing moisturizing gloves and socks, sweat-absorbing sportswear, and cosmetic delivery textiles, such as facial masks enriched with skincare ingredients (Maton et al., 2023; Tahir et al., 2023). Employing these coatings yields numerous benefits, including superior moisture management, effective cosmetic ingredient delivery, increased user comfort, and potential enhancements in skin health and appearance. The choice between hydrophobic and hydrophilic nanocoatings demands careful consideration of

Table 1. Comparison of cosmetotextiles and nanocoated cosmetotextiles

Factors	Cosmetotextiles	Nanocoated Cosmetotextiles	References
Functionality	Basic functionalities based on infused ingredients.	Provide enhanced functionalities using nanoparticle coatings, enabling a broader range of benefits.	Puiu et al. (2023)
Durability	Exhibit moderate durability; effectiveness may fade with repeated washing due to the loss of infused ingredients.	Demonstrate high durability; coatings are resistant to wear and frequent wash, retaining effectiveness over time.	de Dio Ferreira et al. (2023)
Cosmetic agent release	Feature steady release of cosmetic agents but might diminish over time with wash and wear.	Allow for the controlled and prolonged release of agents, maintaining consistent benefits over extended periods.	Singh et al. (2023)
Environmental impact	The impact is dependent on the cosmetic substances used and their biodegradability; it can vary widely.	It has the potential for reduced environmental impact, especially when formulated with biodegradable nanoparticles and sustainable practices.	Goyal and Jerold (2023)
Cost	Typically, less expensive due to simpler technologies and materials used.	Often more costly due to the incorporation of advanced nanotechnologies and specialized materials.	Janeni and Adassooriya (2023)
Safety and health concerns	Generally, considered safe, but dependent on the nature of the infused ingredients and their concentrations; some may cause skin irritation in sensitive individuals.	Potential concerns related to nanoparticle exposure and interaction with biological systems; extensive research and testing are typically conducted to ensure safety., although many are considered safe.	Attia et al. (2023)
Aesthetic appeal	Possess traditional appearance; the infusion of ingredients may alter fabric feel and drape.	Can be engineered to enhance aesthetic appeal without changing the fabric texture, allowing for innovative design possibilities.	Vats et al. (2023)
Regulatory compliance	Adherence to standard textile and cosmetic regulations; varying regulations depending on the region.	Subject to additional regulatory scrutiny due to the use of nanomaterials; compliance with nanomaterial-specific regulations is crucial.	Tanasa et al. (2023)

the intended use, desired characteristics, and performance of the cosmetotextiles by manufacturers. It is also crucial to consider the application method, durability, safety, and substandability of the nanocoatings (Singh et al., 2023). Adherence to relevant regulations and standards is essential to ensuring the safety and efficacy of nanocoated cosmetotextiles. Looking ahead, the continuous efforts in research and development in nanocoatings promise the introduction of innovative materials and application methods designed to enhance the properties and performance of cosmetotextiles. The future might see the emergence of multifunctional nanocoatings, combining a range of properties such as hydrophobicity, hydrophilicity, antimicrobial traits, and UV protection to meet the diversifying needs of consumers in the cosmetotextiles industry (Zheng et al., 2023).

Antibacterial Nanocoatings

Antibacterial nanocoatings, among the most notable types, have secured a pivotal role, especially in healthcare and sportswear applications. These coatings primarily use silver or zinc oxide nanoparticles, celebrated for their antibacterial properties (Suellen Ferro de Oliveira & Kekhasharú Tavaria, 2023). For example, in healthcare settings, items such as wound dressings and hospital gowns may be coated with silver nanoparticles to reduce bacterial infection risks (Puiu et al., 2023). Similarly, in sportswear, antibacterial nanocoatings prove invaluable; zinc oxide nanoparticles are frequently incorporated into moisture-wicking socks and odor-resistant shirts to inhibit bacterial growth, ensuring hygiene and diminishing odors (N. Zhang et al., 2023). The mechanism by which zinc oxide nanoparticles inhibit bacterial growth on cosmetotextiles is multifaceted, involving three primary actions. First, the nanoparticles release zinc ions (Zn^{2+}), which disrupt bacterial cell membranes and interfere with their metabolic processes, leading to cell death and inhibiting bacterial growth. This mechanism is effective against a broad spectrum of bacteria (Hamk et al., 2023). Second, the nanoparticles generate reactive oxygen species (ROS), such as hydroxyl radicals and hydrogen peroxide, causing oxidative stress in bacterial cells. This oxidative damage impairs cellular functions and viability and is particularly effective due to the weak antioxidative defenses of bacteria (Abdelghany et al., 2023). Last, the nanoparticles physically interact with bacterial cells, attaching to their surface and penetrating them, causing physical damage and disrupting vital cellular processes. This combination of zinc ion release, ROS generation, and physical interaction makes zinc oxide nanoparticles a potent antibacterial agent, enhancing the antibacterial properties of cosmetotextiles (Abdelghany et al., 2023).

UV Protective Nanocoatings

The advent of UV-protective nanocoatings in the cosmetotextile industry marks a significant technological advancement. These nanocoatings, often comprising nanoparticles such as zinc oxide nanorods and titanium dioxide, are meticulously engineered to provide fabrics with enhanced protection against harmful UV radiation. Notably, these coatings are not only limited to UV protection but also incorporate additional benefits, such as moisture-wicking, anti-odor, or skin-nourishing properties. The integration of these multifunctional features into a single coating system represents a remarkable innovation in textile treatment and consumer product development (İlhan et al., 2023). The nanoparticles used in these coatings function through the absorption, reflection, or scattering of UV rays, thereby offering superior protection against sun exposure. Zinc oxide and titanium dioxide, in particular, are celebrated for their high refractive indices and UV-absorbing capabilities (de Dio Ferreira et al., 2023; Khan et al., 2023). This combination effectively shields the skin from a broad spectrum of UV radiation, making these nanocoatings especially valuable in textiles used for outdoor activities such as hiking, swimming, and sunbathing (Janeni & Adassooriya, 2023). An additional benefit of using nanoparticles in these coatings is their ability to maintain the inherent properties of the textiles, such as breathability and comfort. Despite the incorporation of nanoparticles, the nanoscale size ensures that the textile's flexibility and air permeability are not compromised. This feature is particularly crucial for maintaining the wearability and aesthetic appeal of textile products. Furthermore, some of these nanoparticles have been found to imbue textiles with skin-friendly properties, such as enhancing moisture retention and providing a barrier against pollutants. The current application of UV-protective nanocoatings is predominantly seen in sportwear and outdoor clothing, where protection from sun exposure is a critical requirement. However, the potential for these coatings extends beyond traditional uses. The ongoing research and development in this field are geared toward expanding their application to everyday wear, medical textiles, and even upholstery, offering a wide spectrum of protection and enhanced functionality. The continuous evolution in nanotechnology promises further advancements in the efficiency, durability, and ecological sustainability of these nanocoatings, making them an exciting area of growth in the textile industry (Khan et al., 2023).

Deodorizing Nanocoatings

Deodorizing nanocoatings play a pivotal role in cosmetotextile development, targeting malodor mitigation and enhancing the user experience. Designed to counteract or neutralize unpleasant odors, these coatings are particularly apt for cosmetotextiles frequently in contact with the skin and prone to sweat and body odors (Ristić et al., 2022). Various nanomaterials, such as silver nanoparticles, zinc oxide nanoparticles, and activated carbon nanoparticles, form the basis of these deodorizing coatings. Silver and zinc oxide nanoparticles exhibit antimicrobial properties, curbing the growth of odor-causing bacteria on textile surfaces and subsequently preventing malodor (Malhotra & Mandal, 2019). Meanwhile, activated carbon nanoparticles adsorb and trap odor molecules, efficiently removing them from the surrounding environment (Wang et al., 2020). Deodorizing nanocoatings are versatile in cosmetotextiles and tailored to unique needs and preferences. For instance, antimicrobial undergarments are meticulously designed to counteract odor by tackling the responsible bacteria, thereby ensuring hygiene and freshness. Odor-resistant sportswear lessens the effects of sweat-related odors, improving comfort during physical activities. Everyday clothing, such as shirts and trousers, benefits by maintaining freshness throughout the day with these coatings. Additionally, specialized cosmetic apparel with extended skin contact employs these nanocoatings to enhance deodorizing properties, thereby boosting wearer comfort and confidence. Incorporating deodorizing nanocoatings into cosmetotextiles boosts user comfort through odor control, prolongs textile life by staving off bacterial growth, and averts degradation, ensuring a consistent aesthetic appeal and freshness. Furthermore, effective odor control diminishes the need for regular washing, indicating a move toward environmental conservation by curtailing water and energy use (Bai et al., 2023). The integration of these advantages accentuates the value of deodorizing nanocoatings in optimizing the user experience, endorsing sustainability, and broadening the possibilities of cosmetic textile applications.

Cosmetic-Releasing Nanocoatings

Cosmetic-releasing nanocoatings are a revolutionary advancement in the field of cosmetotextiles, seamlessly blending fashion with skincare and therapeutic benefits. These coatings are typically ingeniously infused with microcapsules or nanocapsules filled with various cosmetic agents, such as vitamins, fragrances, or moisturizers (Kłosowska et al., 2023; Tansaoui et al., 2023). When textiles treated with these coatings come into contact with the skin, either mechanical friction or body heat triggers the release of the encapsulated agents, delivering their benefits directly to the skin (Elkalla et al., 2023). This innovative approach ensures a sustained and controlled release of cosmetic substances, improving the skin's appearance and feel without the need for separate applications of creams or lotions.

Temperature-Regulating Nanocoatings

Temperature-regulating nanocoatings represent a groundbreaking advancement in textile technology. They utilize materials such as graphene or aerogel in nanomaterial form to impart adaptive thermal properties to textiles (Jayan et al., 2021). Graphene, with its exceptional thermal conductivity, can distribute heat evenly across the fabric, while aerogel, known for its insulation properties, helps maintain a stable, comfortable temperature. The use of these materials in nanocoatings is a testament to the innovative strides being made in fabric treatment and design (Pakdel & Wang, 2023). Graphene's application in temperature-regulating nanocoatings is particularly noteworthy. It is a material that not only distributes heat uniformly but also responds dynamically to environmental changes. In colder environments, graphene-infused fabrics can help retain body warmth, offering a significant benefit in sportswear and outerwear, providing both comfort and protection against the cold (Allaka & Yepuri, 2023). Conversely, aerogel's role in these coatings is centered around its unparalleled insulating capabilities. When integrated into garments, aerogel particles can effectively deflect heat, preventing overheating and ensuring that the wearer remains cool and comfortable. This aspect is particularly valuable in designing apparel for warmer climates or for activities that involve high levels of physical exertion, where maintaining a stable body temperature is crucial for comfort and performance (Sheng et al., 2023). The dual functionality of temperature-regulating nanocoatings has opened up new possibilities in textile manufacturing. Currently, their application is predominantly seen in outdoor and athletic apparel, where adaptability to environmental conditions is essential. Looking forward, the potential uses of these innovative coatings extend to everyday wear, professional attire, and even specialized clothing for extreme environments (Pakdel & Wang, 2023).

Multifunctional Nanocoatings

Multifunctional nanocoatings in cosmetotextiles epitomize the fusion of diverse specialized coatings, providing a multifaceted solution tailored to specific lifestyle requirements. These composite coatings can combine UV protection, moisture-wicking, temperature regulation, and antibacterial attributes into a single, high-performance textile (N. Zhang et al., 2023). For instance, titanium dioxide nanoparticles can be incorporated for UV protection, whereas silver nanoparticles confer antibacterial benefits, all within the same fabric. This results in a garment that is not only shielded from damaging UV rays but also resilient against odor and bacterial growth, making it exceptionally suitable for outdoor activities with extended sun and sweat exposure (Zheng et al., 2023).

Biodegradable Nanocoatings

Biodegradable nanocoatings for cosmetotextiles represent a pivotal advancement in the quest for sustainability, merging functionality with eco-friendly considerations. Developed from materials such as chitosan, cellulose, or other plant-derived polymers, these coatings bestow a range of desirable properties—from moisture-wicking to UV protection—while being engineered to naturally decompose in the environment (Han et al., 2020; Janeni & Adassooriya, 2023). In contrast to traditional coatings, which may encompass synthetic polymers or metals, biodegradable nanocoatings breakdown into benign byproducts. This diminishes their ecological impact, positioning them as a thoughtful selection for eco-aware consumers (Peng et al., 2023).

NANOCOATING TECHNIQUES

A myriad of techniques exist for imparting these nanoscale coatings, each endowed with its own set of merits, drawbacks, and ideal contexts. Enumerated below are some predominant methods employed in this domain.

Dip-Coating

Dip-coating stands as a quintessential method for administering coatings, nanocoatings included, to textiles, among other substrates (Figure 2). At its core, the procedure entails submerging the textile into a liquid concoction brimming with the chosen nanoparticles—taking titanium dioxide for UV shielding or silver nanoparticles bestowing antibacterial properties (Lee et al., 2023). After ensuring thorough saturation, the textile is removed and subjected to a drying phase, typically engaging in air drying, curing, or thermal treatments. This drying phase enables nanoparticles to be infused onto textile fibers with the desired functional attributes (Pradhan et al., 2023). The allure of dip-coating lies in its simplicity, economic feasibility, and adaptability for scale-up, making it a preferred choice for both academic and industrial applications. However, it is not devoid of challenges. Factors such as the nanoparticle solution density, submersion duration, and drying technique influence the consistency and thickness of the resultant coating. Any mismanagement of these variables might spawn patchy nanoparticle spread, potentially jeopardizing the end product's efficacy. To illustrate, in textiles geared for UV defense, an inconsistent coat could spawn regions with diminished protective efficacy. Moreover, while its simplicity is commendable, dip-coating might falter for tasks demanding intricate, multilayered coatings or elaborate designs. Nevertheless, in the panorama of textile coatings, dip-coating retains its stature as an elementary yet indispensable tool, paving the way for refined methodologies and ventures (Ghazal et al., 2023).

Layer-By-Layer Assembly

Layer-by-layer assembly stands as an advanced method crafted for creating multilayered thin films or coatings on a wide range of substrates, including textiles. Distinct from rudimentary techniques such as dip-coating, this method boasts heightened customization and precision in the coating structure (Zhao et al., 2023). As depicted in Figure 3, the method unfolds with the sequential deposition of alternating layers of charged nanoparticles or polymers onto the designated substrate. The procedure commences by choosing an appropriate textile substrate, which is then rigorously cleaned to obliterate any contaminants. Thereafter, solutions replete with positively and negatively charged nanoparticles or polymers are constructed

Figure 2. Dip-coating procedure for cosmetotextiles embedded with nanoparticles

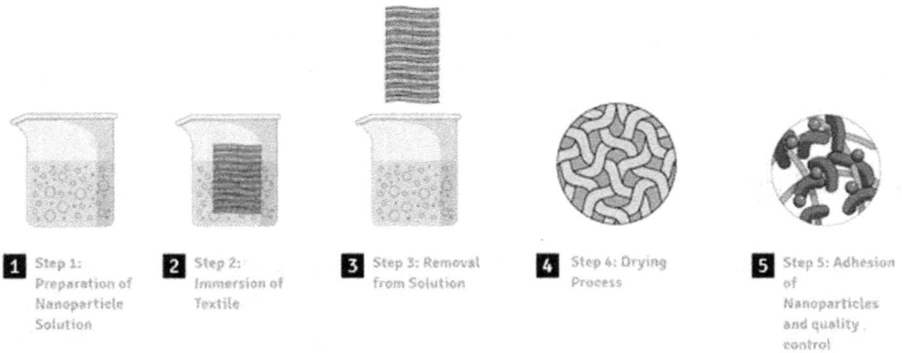

in preparation for the layering ritual. The substrate undergoes immersion in the solution, brimming with the inaugural layer of nanoparticles. After ensuring adequate adsorption, the substrate was rinsed and dried to remove excess particles. The same procedure is repeated for the subsequent layer, albeit with a solution bearing the opposite charge. Steps 3 through 6 oscillate until the coveted number of layers is manifested, with meticulous attention to securing full adherence and drying between layers. Upon achieving the desired multilayered build, a concluding curing or heat treatment is executed to reinforce the coating's stability and longevity. Consequently, comprehensive characterizations and assessments are undertaken to validate that the coating resonates with the stipulated specifications and desired attributes. This is followed by any finishing endeavors on the coated textile, such as tailoring or stitching (Escobar et al., 2023). This method facilitates the amalgamation of diverse functionalities into one holistic coating, a case in point being the unification of UV defense, antibacterial properties, and moisture management. Nonetheless, layer-by-layer assembly is not devoid of challenges. It is primarily characterized by its time-consuming nature, rendering it less amenable for vast-scale production. The application and rinsing of each layer must be approached independently, and based on the coating's intricacy, multiple layers might be needed. This escalates not only the fabrication duration but also the overall expenditure. Moreover, it demands pinpoint accuracy to sustain consistency in layer thickness and nanoparticle distribution, necessitating specialized apparatus and proficiency. Owing to these intricacies, layer-by-layer assembly is harnessed for niche applications where intricate customization is pivotal, such as medical equipment, innovative sensors, or elite textiles, rather than mainstream consumer commodities (Zhao et al., 2023).

Figure 3. Diagrammatic representation of the layer-by-layer assembly technique for cosmetotextiles

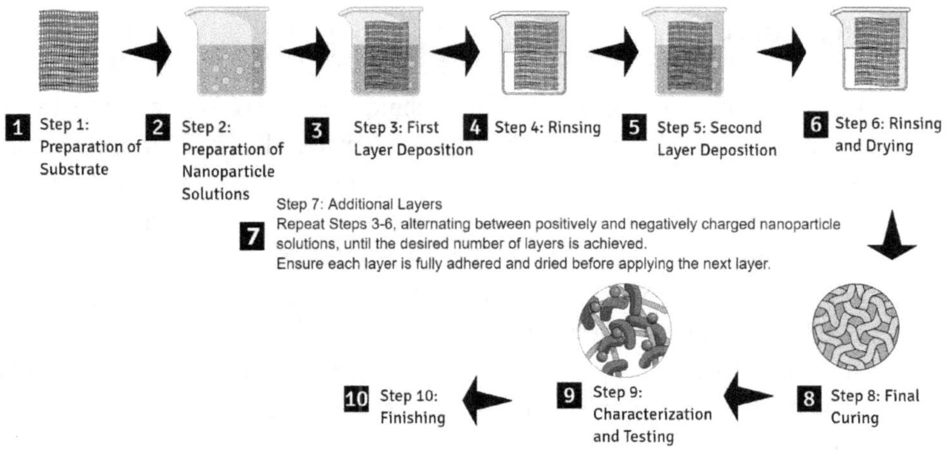

Electrospraying and Electrospinning

Electrospraying and electrospinning are advanced techniques employing electrical forces to generate finely uniform and meticulously controlled coatings on textile substrates. Central to both methods is the application of a high-voltage electric field to either a liquid polymer solution or a nanoparticle suspension, culminating in the material's deposition on the desired surface. Electrospraying is typically reserved for liquid-based nanoparticles. This results in the formation of tiny droplets sprayed onto the substrate, yielding a refined, even coating. This approach proves particularly beneficial in areas demanding stringent control over particle size and distribution, such as drug delivery mechanisms or sophisticated filtration membranes (Jha et al., 2023). Conversely, electrospinning refers to the creation of fibrous mats or coatings stemming from polymer solutions. Here, the electrical force elicits a liquid jet from the solution, which, upon reaching the substrate, crystallizes into ultrafine fibers. This method is a go-to for crafting nanofibrous frameworks in tissue engineering, high-calibre filters, or composite materials fortified with superior mechanical properties (Kiran et al., 2023). While electrospraying and electrospinning offer unparalleled precision and control for creating sophisticated material characteristics, they come with challenges. These methods require specialized equipment, including high-voltage power sources and precision-focused deposition setups, which can be costly and prohibitive for smaller-scale or budget-constrained projects. Additionally, optimizing the desired properties often involves extensive refinement, including adjusting voltages, calibrating solution concentrations, and modifying deposition times. Despite these hurdles, the high level of precision and control they provide makes them indispensable for advanced material development (Jha et al., 2023).

Sol-Gel Method

The sol-gel method stands as a versatile technique extensively employed in the creation of nanocoatings, thin films, and even bulk materials. The procedure begins with a "sol", a colloidal suspension consisting of nanoparticles dispersed in a liquid medium. This sol is then introduced to the textile substrate via methods such as dip-coating, spin-coating, or spraying. After this application, the sol embarks on a gelation phase, during which it metamorphoses from a liquid-like state into a solid gel. This shift entails the formation of a three-dimensional lattice, securely entrapping the nanoparticles within the gel matrix (Raduly et al., 2023). Typically, this consolidated gel phase is exposed to a curing stage, frequently involving elevated temperatures, to expunge any lingering solvent and fortify the coating (Mahltig, 2023). The sol-gel technique yields remarkably stable, uniform, and durable coatings that seamlessly bond to a multitude of substrates. It also facilitates the integration of a diverse array of materials, encompassing ceramics, metals, and polymers (Shiya et al., 2023). Nonetheless, the sol-gel method is not devoid of limitations. A notable impediment is the recurrently needed high-temperature curing phase, which is potentially incompatible with textiles of a heat-sensitive nature, such as specific synthetic polymers or organic fibers. Such a constraint might circumscribe the assortment of materials amenable to effective coating via this approach. Additionally, the sol-gel coatings might demand distinct precursors and solvents, potentially escalating the cumulative expense of the coating procedure. However, despite these constraints, the sol-gel method continues to be a formidable asset for fabricating top-tier nanocoatings (Raduly et al., 2023). Its adaptability renders it valuable across a spectrum of sectors, spanning from aerospace to healthcare.

Spray Coating

Spray coating is a straightforward and scalable technique that is frequently used in both industrial and research settings for the application of nanocoatings to textiles. In this method, a solution or suspension containing nanocarriers is loaded into a spray device. This device then atomizes the liquid into fine droplets, directing them onto the textile surface. As the solvent in the droplets evaporates, it leaves behind a coating of nanoparticles adhered to the fabric. Spray coating is appreciated for its speed and simplicity. It can quickly cover large surface areas and is adaptable to various fabric types, including both woven and nonwoven textiles (Ma et al., 2023). It proves especially useful for applications that demand a quick turnaround or for coating irregularly shaped or extensive substrates that might not be suitable for dip-coating or other methods. However, spray coating does have limitations. The primary concern is its level of precision in terms of nanocarrier distribution and coating thickness, which might not match the accuracy of methods such as electrospinning or layer-by-layer assembly. This can be problematic for applications that need highly uniform or specialized coatings. Moreover, the spray process can lead to some material waste, as portions of the sprayed solution might not adhere to the textile, potentially dispersing into the surrounding environment. The technique may also necessitate optimization to ensure that the sprayed droplets possess the appropriate size and velocity for a given application, introducing added complexity (Y. Zhang et al., 2023). Nevertheless, spray coating continues to be a sought-after and adaptable method for applying nanocoatings across various applications.

APPLICATION OF NANOCOATINGS IN SELECTED COSMETOTEXTILES

The application of nanocoatings in cosmetotextiles is a growing area of research and industrial development, with profound implications for enhancing both the functional and sustainable attributes of textiles. This section explores the specific areas where nanocoatings have a considerable impact in the field of cosmetotextiles.

Beauty and Skincare Products

In the beauty and skincare industry, cosmetotextiles enhanced with specialized nanocoatings are redefining how consumers approach skin wellness. Nanocoatings that encapsulate cosmetic agents such as moisturizers, vitamins, or even collagen are applied to fabrics used in facial masks, gloves, and even everyday clothing (Ristić et al., 2022; Taieb & Mnejja, 2023). Upon contact with the skin, these microcapsules release their contents in a controlled manner, turning the textile into a delivery system for skincare ingredients (Tahir et al., 2023). This technology allows for prolonged exposure to beneficial compounds, offering a convenient and effective way to supplement traditional beauty regimens. Another significant development is the use of nanocoatings for sun protection in fabrics designed for beauty and skincare applications. By embedding nanoparticles such as titanium dioxide or zinc oxide into textile fibers, these specialized fabrics can provide superior UV protection without the need for frequent reapplication of sunscreen lotions (Zheng et al., 2023). This is particularly valuable in beachwear or other outdoor apparel that is explicitly designed for sunbathing and prolonged exposure to sunlight. The versatility of nanocoatings also enables the development of multifunctional cosmetotextiles that offer multiple benefits simultaneously. Imagine a pair of gloves that not only moisturize your hands but also contain antibacterial properties or a scarf that provides UV protection while releasing a subtle fragrance (Y. Zhang et al., 2023; Zhao et al., 2023; Zheng et al., 2023). This dual, or even triple, functionality is made possible through advanced nanocoating technologies, making these textiles an increasingly popular choice for consumers seeking both beauty and practical benefits from their skincare products.

Table 2 outlines several selected studies, providing detailed information about the textile and (nano- or micro) coating materials used, the method of coating applied, and the corresponding key findings and potential applications. These studies cover a broad range of cosmetic agents, such as *Lavandula angustifolia* and *Citrus aurantifolia* essential oils, polyphenol-rich extracts, α-tocopherol, menthol, and *Cymbopogon winterianus* (citronella oil), each displaying unique beneficial properties and applications in the field of cosmetotextiles. For instance, one study by the author and colleagues concentrated on encapsulating menthol—a natural coolant limited by its low aqueous solubility and high volatility—in lipid-core nanocapsules, employing a nanoprecipitation method. This approach aimed to investigate menthol's viability in skin treatments. The resultant optimized nanocapsules, thoroughly characterized for various physicochemical properties and safety, revealed the potential to amplify the aqueous solubility and thermal stability of menthol. These nanocapsules could serve as substantial alternatives for cosmetotextiles, owing to their rapid dermal absorption properties (Yingngam, Chiangsom, et al., 2019). Additionally, Yingngam, Kacha, et al. (2019) explored a method involving oil-in-water emulsification and spray drying to encapsulate citronella oil in acacia gum microcapsules. The utilization of a central composite design was crucial in optimizing the encapsulation conditions, culminating in microcapsules that exhibited sustained volatility and diminished irritation properties of citronella oil. Scanning electron microscopy confirmed that these microcapsules could be easily impregnated onto nonwoven fabric, rendering them apt candidates for developing citronella essential oil-based cosmetotextiles.

Medical and Healthcare Textiles

In the medical and healthcare sectors, the application of nanocoatings on textiles has far-reaching implications for both patient care and medical staff protection. Antibacterial and antiviral nanocoatings, for instance, are being increasingly applied to hospital gowns, bed linens, and curtains to reduce the spread of infections. These coatings can utilize a range of nanoparticles, such as silver or copper, which possess natural antimicrobial properties, to create safer environments within healthcare facilities (Venkatraman et al., 2023). Moreover, wound dressings with nanocoatings releasing antiseptics or growth factors can accelerate the healing process and reduce the risk of infection (Venkatraman et al., 2023). Another avenue for application in

Table 2. Summary of selected studies illustrating various cosmetic agents encapsulated using different nanocoating materials and methods for application onto textile materials

Cosmetic Agent	Nanocoating Material	Method of Nanocoating	Textile Material	Key Findings	Application	Reference
Lavandula angustifolia Mill.	Polyurethane shell capsules	Exhaust method	Cotton	Environmentally friendly shells; spherical structures; different encapsulation efficiencies; consistency with recipes confirmed by thermogravimetric analysis	Aromatherapeutic (anxiety treatment) textiles	Özsevinç and Alkan (2023)
Citrus aurantifolia (Christm.) Swingle essential oil	Ethyl cellulose microcapsules	Microcapsules were prepared by a simple coacervation process and were immobilized onto the cotton fabric using a tad dry method with citric acid as a binder	Cotton fabric	High encapsulation efficiency and oil content; effective antibacterial activity; durable against washing cycles; governed by first-order kinetic mechanism	Antibacterial functional textile	Julaeha et al. (2023)
Polyphenol-rich extracts	Electrospun polyvinyl alcohol/chitosan fibrous patch	Electrospinning	Electrospun nanofibers	- Uniform, smooth-surfaced nanofibers (100–300 nm dia.) - High drug loading efficiency (89.5–97.7%) - Good water-absorbing properties - Sustained release of polyphenol-rich extracts - Potent bacterial inhibition of *Propionibacterium acnes* - Pronounced clinical therapeutic efficacy in treating mild-to-moderate facial acne - Improved biocompatibility or tolerability compared to benzoyl peroxide	Acne therapy Cosmetotextile	Tang et al. (2021)
α-Tocopherol	Chitosan nanospheres	Emulsion formation	Cotton fabric	- Successful encapsulation and improved stability - Good antibacterial activity - 64% release in dispersed form, 41% once applied on fabric due to cross-linking of nanospheres	Skincare cosmetotextile reduces skin dryness and offers antibacterial and antioxidant performance	Raza et al. (2020)

continued on following page

Table 2. Continued

Cosmetic Agent	Nanocoating Material	Method of Nanocoating	Textile Material	Key Findings	Application	Reference
Menthol	Polycaprolactone nanocapsules	Nanoprecipitation	Cotton fabric	- Successful encapsulation with optimized physicochemical features - Rapid ex vivo skin penetration and safe topical application - Enhanced aqueous solubility and improved thermal stability of menthol	Cosmetotextiles with fast dermal absorption properties	Yingngam, Chiangsom, et al. (2019)
Cymbopogon winterianus Jowitt ex Bor	Arabic gum microcapsules	Oil-in-water emulsification and spray drying method	Nonwoven fabric	- Successful encapsulation with prolonged volatility and reduced irritation - Easy impregnation onto nonwoven fabric	Citronella essential oil-based cosmetic textiles	Yingngam, Kacha, et al. (2019)

healthcare textiles is temperature-regulating nanocoatings. Such coatings, often incorporating materials such as aerogels or phase-change materials, can help maintain a constant microclimate between the textile and skin. This is particularly beneficial for burn victims, neonatal care, or patients suffering from thermoregulation issues, providing them with added comfort and potentially speeding up recovery times (Jayan et al., 2021). The integration of multifunctional nanocoatings into healthcare textiles is an emerging trend that combines several beneficial features into a single textile product. For example, a medical gown could be designed to be not only antibacterial but also flame-retardant and fluid-repellent (Samui et al., 2022). Alternatively, a wound dressing might incorporate both antibacterial agents and growth factors to simultaneously fight infection and promote tissue regeneration (Venkatraman et al., 2023). These multifunctional textiles, enhanced by advanced nanocoatings, represent a significant step forward in elevating the standards of care and safety in medical and healthcare settings.

Sportswear and Athletic Gear

The application of nanocoatings in cosmetotextiles has shown significant promise, particularly in sportswear and athletic gear. Advanced nanocoatings imbued with moisture-wicking properties can efficiently transport sweat away from the skin to the fabric's outer surface, enhancing the athlete's comfort and performance (McCann, 2023). Additionally, some nanocoatings can release microcapsules filled with essential vitamins or cooling agents upon friction or body heat, providing not only comfort but also skin benefits during physical activities. UV-protective nanocoatings are another game changer for outdoor sports apparel. Utilizing nanoparticles such as titanium dioxide or zinc oxide, these coatings offer superior protection against harmful ultraviolet rays, reducing the risk of skin cancer and sunburn. Such a feature is especially valuable for sports such as long-distance running, swimming, and hiking, where athletes are exposed to the sun for extended periods of time. Antibacterial nanocoatings, on the other hand, can inhibit the growth of odor-causing bacteria, maintaining the freshness of sportswear and contributing to better hygiene for athletes. The advent of multifunctional nanocoatings has also opened doors to combining multiple benefits into a single garment. Imagine a high-performance running shirt that not only wicks away moisture but also offers UV protection and releases essential nutrients to nourish the skin, all thanks to carefully engineered nanocoating (Maity et al., 2023). These multifunctional sportswear options not only enhance athletic performance but also offer health benefits, making them a highly attractive choice for health-conscious consumers.

LIFECYCLE ANALYSIS FOR SUSTAINABLE COSMETOTEXTILES

Lifecycle analysis for nanocoatings in cosmetotextiles offers a complex evaluation that quantifies the environmental, social, and economic impacts of these coatings from cradle to grave. Figure 4 demonstrates an overview of the comprehensive lifecycle assessment for nanocoatings in cosmetotextiles, detailing each phase from raw material acquisition to final impact assessment. The first stage, "raw material acquisition", emphasizes the environmental footprint of extracting or synthesizing nanoparticles such as titanium dioxide, silver, or graphene. This encompasses carbon emissions from mining operations, water usage, and the ecological impact on local biodiversity. The subsequent stage, "material processing and nanocoating production", investigates the processes utilized to refine the raw materials and produce the nanocoatings. The key metrics analyzed include energy consumption, waste production, and chemical emissions. In some instances, the use of toxic solvents or reagents may be involved, making their handling and disposal of paramount importance. During the "manufacturing" phase, the application of these nanocoatings to textiles is explored. Methods such as dip-coating, electrospraying, or layer-by-layer assembly each present unique environmental considerations. For instance, dip-coating necessitates large solution volumes, thereby raising queries about solvent recovery and wastewater treatment. The "usage" phase focuses on both the functional lifespan and efficacy of the nanocoated textile products, considering factors such as wash frequency, which can influence water consumption and detergent pollution. Moreover, if the nanocoating provides sun protection, moisture-wicking, or antimicrobial properties, then it could extend the textile's longevity, thus reducing the need for frequent replacements and consequently diminishing its environmental toll. Nonetheless, the potential health implications of prolonged skin contact with nanoparticles should be explored. The end-of-life phase addresses the disposal, recycling, or biodegradation of nanocoated textiles. Conventional disposal methods such as landfilling and incineration may lead to adverse environmental effects, particularly if nanoparticles contaminate the soil or water. Recycling is potentially complicated by the durability of the nanocoating and the mixed materials common in textiles. Biodegradable nanocoatings are attracting increasing attention for their promise to mitigate environmental repercussions at this junction (Goyal & Jerold, 2023). Last, the "life cycle impact assessment" quantitatively and interpretatively examines the data collected during the lifecycle analysis. This assesses the overarching environmental implications, factoring in aspects such as carbon footprint, eutrophication potential, and human toxicity. Such metrics guide manufacturers, policymakers, and consumers in making well-informed decisions regarding the sustainability and desirability of nanocoated cosmetotextiles.

Figure 4. Lifecycle analysis stages for sustainable cosmetotextiles

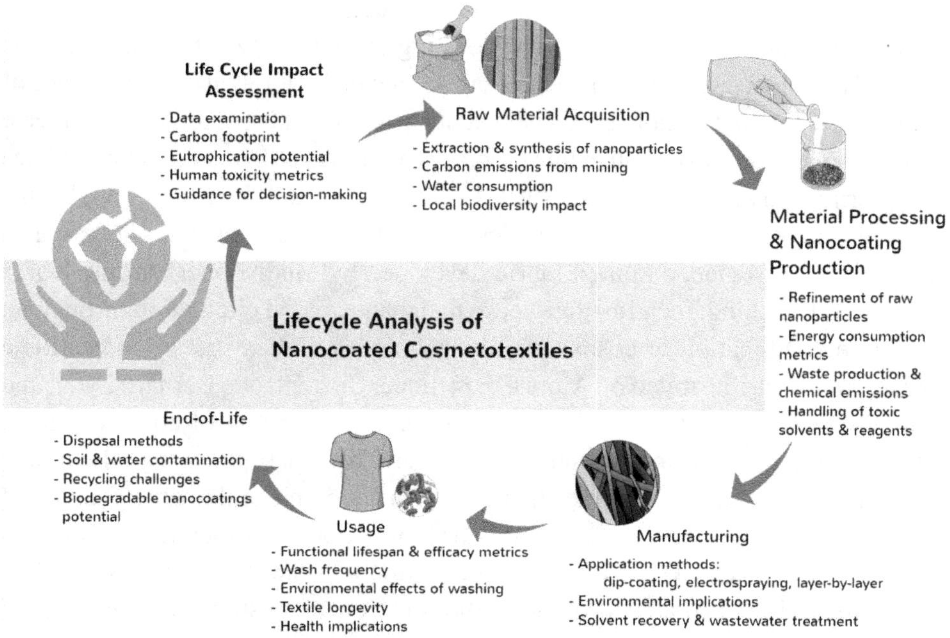

CURRENT MARKET TRENDS AND FUTURE DIRECTIONS

The cosmetotextile market is multifaceted, segmented by type—woven, nonwoven, and knitted; application—skincare, haircare, and body care; end-user—men, women, and unisex; and product category—apparel, footwear, and accessories. Several growth drivers propel the market forward: technological advancements, notably the incorporation of nanotechnology; escalating consumer demand driven by a rising interest in personal care and wellness; and a growing preference for sustainability and eco-friendly products (Goyal & Jerold, 2023; Y. Zhang et al., 2023; Zheng et al., 2023). On the regional spectrum, the Asia-Pacific region, dominated by China and India, takes the lead, attributed to its vast consumer base and manufacturing capabilities. Europe and North America follow, focusing on premium products and innovation, respectively. Latin America, the Middle East, and Africa emerged as markets with increasing demand and growing consumer awareness (https://www.technavio.com/report/cosmetics-market-industry-analysis). The rising appetite for functional textiles has galvanized innovation in nanocoatings. Multifunctional fabrics, enhanced with nanocarrier coatings, offer attributes that surpass traditional textile properties. These products are gaining a niche in the market, as they offer dual benefits—comfort and cosmetic treatment—without requiring the user to apply multiple products.

The Lens Patent Search (https://www.lens.org/) (accessed October 14, 2023) was deployed to investigate prevailing patents, revealing innovations at the confluence of cosmetic and textile technologies. The patents listed in Table 3 exemplify a diverse array of applications, from enhancing blood circulation to utilizing cationic particles to delivering active skin ingredients via textiles. The broad spectrum of applicants from various institutions and companies attests to the vast interest and reach in this field. For instance, a knitted fabric developed by Mediwinner and the Korea Dyeing and Finishing Tech Institute seeks to enhance blood circulation through an innovative combination of cosmetotextile techniques. Endeavors around cationic microparticles, as illustrated by Alpol Cosmetique, Univ Claude Bernard Lyon, and Centre Nat Rech Scient, show the potential multifunctionality of cosmetotextiles in incorporating active ingredients for skin benefits. Each patent entails unique approaches, materials, and utilities that reflect the dynamic and evolving nature of cosmetotextile technology. Ranging from the fabrication of cosmetotextiles imbued with nano/micro capsules to the conceptualization of wipes that administer active skin compounds, these patents epitomize the comprehensive strategy being adopted in this sector. This extensive range of patented innovations underlines the increasing importance and relevance of cosmetotextiles in the contemporary textile and cosmetic industries, promising potential advancements and new frontiers in the development and application of these synergistic technologies.

The future of nanocoatings in cosmetotextiles is ripe for transformative innovations that extend beyond just functional enhancements. Figure 5 summarizes the future directions for nanocoated cosmetotextiles from the author's perspective. The next big leap envisions a fusion of smart technology with nanocoatings, creating textiles that adapt dynamically to environmental conditions or individual needs, such as sportswear that releases cooling agents with rising temperatures or bedsheets that sense skin moisture and release hydrating substances. This would involve a convergence of textile engineering, nanotechnology, and IoT devices, with examples such as self-healing fabrics and health-monitoring wearable textiles. Additionally, evolving technology trends such as personalization, sustainability, and globalization are shaping the sector. Advances in artificial intelligence and data analytics may lead to textile products that adapt to individual preferences and environmental conditions. Sustainability could promote circular lifecycle models for textiles, emphasizing responsible sourcing, efficient production, and end-of-life recyclability or biodegradability. Furthermore, sustainability has transitioned from a niche concern to a mainstream expectation, leading to significant developments in eco-friendly nanocoatings. Biobased nanomaterials, such as those derived from algae or plant extracts, are increasingly replacing synthetic materials in line with the global push

Table 3. Selected patents based on cosmetotextiles, sourced from the lens patent search

Title	Identifiers	Family	Filed	Published	Applicants
Knitted fabric for improving blood circulation with cosmetotextile	• 091-164-285-663-125 • KR 20220063776 A	2s/2ex	Nov 9, 2020	May 18, 2022	Mediwinner, Korea Dyeing & Finishing Tech Institute
Cationic microparticles and use in the field of cosmetotextiles	• 017-950-371-249-945 • EP 2673413 B1	6s/6ex	Feb 9, 2012	Apr 11, 2018	Alpol Cosmetique, Univ Claude Bernard Lyon, Centre Nat Rech Scient
Cosmetotextile containing nano/micro capsule and manufacturing method thereof	• 159-393-851-495-176 • KR 20180010650 A	2s/2ex	Jul 22, 2016	Jan 31, 2018	Jeoun Il Dyeing and Finishing Co Ltd., Korea Dyeing & Finishing Tech Institute
Wipe for application of at least one active principle on the skin	• 064-027-080-137-818 • US 9668947 B2	20s/20ex	Jun 26, 2012	Jun 6, 2017	Tonani Alberto, Novello Andrea, Favero Erika, Favero Valerio, Zanolo S P A, Turati Idrofilo S P A
Nonwoven cosmetotextile for the medical-cosmetic sector	• 043-001-860-285-11X • ES 1174615 U	2s/2ex	Dec 21, 2016	Jan 23, 2017	Asoc De Investig De La Ind Textil (Aitex)

toward greener manufacturing processes (Y. Zhang et al., 2023). Additionally, manufacturers are exploring biodegradable nanocoatings that decompose naturally, thus reducing their environmental impact. For instance, chitosan sourced from crustacean shells is gaining traction in textile coatings, offering a biodegradable and antimicrobial solution (Ghosh et al., 2023). These innovations appeal to a growing segment of consumers who are willing to pay a premium for sustainable attributes. Consequently, they are beginning to influence regulatory perspectives on textile manufacturing and nanocarrier use, leading to stricter industry standards for quality and safety.

CONCLUSION

The application of nanocoatings for cosmetotextiles has carved new avenues for enhancing both the functionality and sustainability of textile products. Nanocoatings have demonstrated potential in bestowing textiles with antimicrobial, UV-protective, water-resistant, and cosmetic-releasing properties. Throughout this chapter, the author has explored the wide spectrum of materials suitable for formulating these nanocoatings, from metal nanoparticles and polymers to natural substances. Such nanomaterials not only extend the versatility of textile products but also provide

Figure 5. Directions for nanocoated cosmetotextiles: A multifaceted evolution driven by sustainability, smart technology integration, and personalization

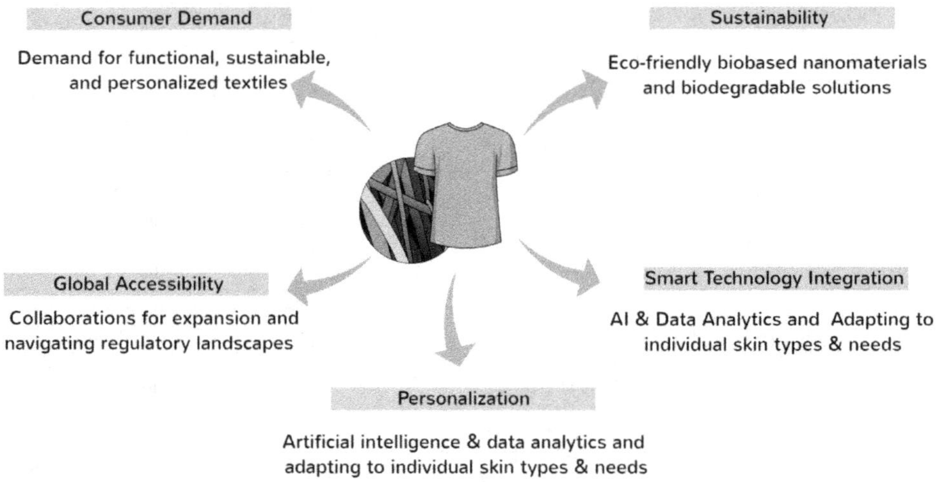

a way toward sustainability by minimizing the reliance on noxious chemicals and energy-intensive processes. In discussing applications, the author elucidated the synthesis methods and accentuated scalable techniques such as layer-by-layer assembly, dip-coating, and sol-gel processes. These methods present feasible strategies for industry-level production of cosmetotextiles furnished with nanocoatings. The emphasis on eco-friendly alternatives underlines the push for sustainable development in nanocomposites and nanocoatings. The exploration also touched upon the potential of biodegradable or biobased nanomaterials to mitigate environmental consequences, meeting both consumer expectations and regulatory mandates for greener products.

ABBREVIATION

ROS Reactive oxygen species

UV Ultraviolet

REFERENCES

Abdelghany, T. M., Al-Rajhi, A. M. H., Yahya, R., Bakri, M. M., Al Abboud, M. A., Yahya, R., Qanash, H., Bazaid, A. S., & Salem, S. S. (2023). Phytofabrication of zinc oxide nanoparticles with advanced characterization and its antioxidant, anticancer, and antimicrobial activity against pathogenic microorganisms. *Biomass Conversion and Biorefinery*, *13*(1), 417–430. doi:10.1007/s13399-022-03412-1

Allaka, G., & Yepuri, V. (2023). (in press). Inexpensive hydrophobic and infrared reflective coatings on the cotton and silk fabrics using sol-gel dip coating technique. *Materials Today: Proceedings*. Advance online publication. doi:10.1016/j.matpr.2023.06.042

Attia, N. F., Zakria, A. M., Nour, M. A., Abd El-Ghany, N. A., & Elashery, S. E. A. (2023). Rational strategy for construction of multifunctional coatings for achieving high fire safety, antibacterial, UV protection and electrical conductivity functions of textile fabrics. *Materials Today Sustainability*, *23*, 100450. doi:10.1016/j.mtsust.2023.100450

Atwah, A. A., & Khan, M. A. (2023). Influence of microscopic features on the self-cleaning ability of textile fabrics. *Textile Research Journal*, *93*(1-2), 450–467. doi:10.1177/00405175211069881

Bai, W., Yu, H., Liu, L., Pakdel, E., Tang, B., Su, H., Hurren, C., Liu, L., Wang, J., & Wang, X. (2023). The adsorption kinetics and mechanism of odorous gases onto textile fibers. *RSC Sustainability*, *1*(2), 357–367. doi:10.1039/D2SU00086E

Ben Abdelkader, M., Azizi, N., Salaün, F., Majdoub, M., & Chevalier, Y. (2023). Cosmetotextiles. In *Particle Technology and Textiles: Review of Applications* (pp. 235–274). De Gruyter. doi:10.1515/9783110670776-012

Chen, S., Wang, L., Lin, X., Ni, P., Liu, H., & Li, S. (2023). Catechol derivative urushiol's reactivity and applications beyond traditional coating. *Industrial Crops and Products*, *197*, 116598. doi:10.1016/j.indcrop.2023.116598

de Dio Ferreira, D., Galvão, T. D., & Appoloni, C. R. (2023). Washing and release of titanium nanoparticles from UV protective textiles. In *Advances in Healthcare and Protective Textiles* (pp. 447–466). Elsevier. doi:10.1016/B978-0-323-91188-7.00010-8

Dickson, D., Bock, B. D., & Thome, J. R. (2024). Heat transfer of uncoated and nanostructure coated commercially micro-enhanced refrigeration tubes under pool boiling conditions. *Applied Thermal Engineering*, *236*, 121757. doi:10.1016/j.applthermaleng.2023.121757

Elkalla, E., Khizar, S., Tarhini, M., Lebaz, N., Zine, N., Jaffrezic-Renault, N., Errachid, A., & Elaissari, A. (2023). Core-shell micro/nanocapsules: From encapsulation to applications. *Journal of Microencapsulation*, *40*(3), 125–156. doi:10.1080/02652048.2023.2178538 PMID:36749629

Escobar, K., Garrido-Miranda, K. A., Pulido, R., Naveas, N., Manso-Silván, M., & Hernandez-Montelongo, J. (2023). Coatings of cyclodextrin/citric-acid biopolymer as drug delivery systems: A review. *Pharmaceutics*, *15*(1), 296. doi:10.3390/pharmaceutics15010296 PMID:36678924

Ghazal, H., Khaleed, N., & Abdelaziz, E. (2023). Significance Advantages, and disadvantages of nanotechnology in textile finishing. *Egyptian Journal of Chemistry*, *66*(12), 467–482. doi:10.21608/ejchem.2023.195121.7624

Ghosh, S., Sarkar, B., Thongmee, S., & Mostafavi, E. (2023). Hybrid antibacterial, antifungal, and antiviral smart coatings. In *Antiviral and antimicrobial smart coatings: Fundamentals and applications* (pp. 431–452). Elsevier. doi:10.1016/B978-0-323-99291-6.00016-5

Goyal, N., & Jerold, F. (2023). Biocosmetics: Technological advances and future outlook. *Environmental Science and Pollution Research International*, *30*(10), 25148–25169. doi:10.1007/s11356-021-17567-3 PMID:34825334

Hamk, M., Akçay, F. A., & Avcı, A. (2023). Green synthesis of zinc oxide nanoparticles using *Bacillus subtilis* ZBP4 and their antibacterial potential against foodborne pathogens. *Preparative Biochemistry & Biotechnology, 53*(3), 255-264. https://doi.org/.2022.2076243 doi:10.1080/10826068

Han, J., Liu, L., Fan, Z., Zhang, Z., Yang, S., & Tang, Y. (2020). Grafting cosmetic active ingredients for the functionalization of cosmetotextiles. *IOP Conference Series. Materials Science and Engineering*, *782*(2), 022026. doi:10.1088/1757-899X/782/2/022026

He, B., Hou, X., Liu, Y., Hu, J., Song, L., Tong, Z., . . . Zhang, Q. (2023). Design of fluorine-free waterborne fabric coating with robust hydrophobicity, water-resistant and breathability. *Separation and Purification Technology, 311*, 123308. https://doi.org/. 123308 doi:10.1016/j.seppur.2023

İlhan, İ., Esen, M., Karaaslan, M., Akyürek, B. Y., & Çelikel, Ö. (2023). Investigation of EMI and UV-IR shielding properties of Kevlar and polyester/elastane nanocomposite fabrics coated by ECR and evaporation methods. *Journal of the Textile Institute*, 1–14. Advance online publication. doi:10.1080/00405000.2023.2200538

Janeni, J., & Adassooriya, N. M. (2023). Development of eco-friendly smart textiles from nanomaterials. In *Advances in Smart Nanomaterials and their Applications* (pp. 325–335). Elsevier. doi:10.1016/B978-0-323-99546-7.00002-1

Jayan, J. S., Sethulekshmi, A., Venu, G., Saritha, A., & Joseph, K. (2021). Sustainable nanotextiles: emerging antibacterial fabrics. In *Antimicrobial Textiles from Natural Resources* (pp. 619–651). Elsevier. doi:10.1016/B978-0-12-821485-5.00007-X

Jha, C. B., Santhosh, S., Singh, C., Bose, S., Manna, K., Varshney, R., & Mathur, R. (2023). 9 - Electrospun nanofiber a smart drug carriers: production methods, problems, solutions, and applications. In N. Sharma & B. S. Butola (Eds.), *Fiber and Textile Engineering in Drug Delivery Systems* (pp. 285–306). Woodhead Publishing. doi:10.1016/B978-0-323-96117-2.00002-9

Julaeha, E., Pandiangan, N. S., Eddy, D. R., Permadi, N., Harja, A., Wahyudi, T., & Al-Anshori, J. (2023). Ethyl cellulose-based microcapsules of *Citrus aurantifolia* (Christm.) Swingle essential oil with an optimized emulsifier for antibacterial cosmetotextiles. *Polymer*, *283*, 126265. Advance online publication. doi:10.1016/j.polymer.2023.126265

Khan, M. Z., Taghavian, H., Fijalkowski, M., Militky, J., Tomkova, B., Venkataraman, M., & Adach, K. (2023). Effect of microwave power on bactericidal and UV protection properties of the ZnO nanorods grown cotton fabrics. *Colloids and Surfaces. A, Physicochemical and Engineering Aspects*, *664*, 131135. doi:10.1016/j.colsurfa.2023.131135

Kiran, S., Adeel, S., Abrar, S., Iqbal, S., Naz, S., & Amin, N. (2023). Textile applications of nanofibers and nanocomposites. In Handbook of: Nanofibers and Nanocomposites: Characteristics, Synthesis, and Applications in Textiles (pp. 149-167). Jenny Stanford Publishing. doi:10.1201/9781003432746-5

Kłosowska, A., Wawrzyńczak, A., & Feliczak-Guzik, A. (2023). Microencapsulation as a route for obtaining encapsulated flavors and fragrances. *Cosmetics*, *10*(1), 26. doi:10.3390/cosmetics10010026

Lee, M., Hwang, S. Y., Koo, J. M., Jeon, H., Kim, H. J., Oh, D. X., & Park, J. (2023). Chitosan coating in the form of polymer and nanowhiskers on clothing fabrics for improved particulate matter removal efficiency in face mask filters. *Journal of Natural Fibers*, *20*(1), 2187507. doi:10.1080/15440478.2023.2187507

Lucia-Oana, S., Mirela, B., & Alexandra-Gabriela, E. (2023). Physical and mechanical characterization of cosmetic textiles with anti-acne effect. *Annals of the University of Oradea. Fascicle of Textiles. Leatherwork*, *24*(1), 113–118.

Ma, J., Fan, J., Xia, Y., Kou, X., Ke, Q., & Zhao, Y. (2023). Preparation of aromatic β-cyclodextrin nano/microcapsules and corresponding aromatic textiles: A review. *Carbohydrate Polymers*, *308*, 120661. doi:10.1016/j.carbpol.2023.120661 PMID:36813345

Mahltig, B. (2023). 3 Finishing of textile materials. In *Particle Technology and Textiles: Review of Applications* (pp. 45–66). De Gruyter. doi:10.1515/9783110670776-003

Maity, S., Singha, K., & Pandit, P. (2023). 1 - Introduction to functional and technical textiles. In S. Maity, K. Singha, & P. Pandit (Eds.), *Functional and Technical Textiles* (pp. 1–30). Woodhead Publishing. doi:10.1016/B978-0-323-91593-9.00021-3

Malhotra, S. P. K., & Mandal, T. (2019). Zinc oxide nanostructure and its application as agricultural and industrial material. *Contamin. Agric. Environ. Health Risks Remed*, *1*, 216–226. doi:10.26832/AESA-2019-CAE-0171-016

Maton, M., Gabut, S., Neut, C., Odou, P., Sacareau, C., Pinon, A., Vialette, M., Gerber, G., Martel, B., & Blanchemain, N. (2023). Antiviral functionalization of a polypropylene nonwoven textile structure as a self-decontaminating layer for respiratory masks. *Biomaterials Science*, *11*(10), 3502–3511. doi:10.1039/D2BM01988D PMID:36988417

McCann, J. (2023). 13 - Identification of design requirements for smart clothes and wearable technology. In J. McCann & D. Bryson (Eds.), *Smart Clothes and Wearable Technology* (2nd ed., pp. 327–369). Woodhead Publishing. doi:10.1016/B978-0-12-819526-0.00004-7

MuthuKathija, M., Sheik Muhideen Badhusha, M., & Rama, V. (2023). Green synthesis of zinc oxide nanoparticles using Pisonia Alba leaf extract and its antibacterial activity. *Applied Surface Science Advances*, *15*, 100400. doi:10.1016/j.apsadv.2023.100400

Özsevinç, A., & Alkan, C. (2023). Polyurethane shell medicinal lavender release microcapsules for textile materials: An environmentally friendly preparation. *Industrial Crops and Products*, *192*, 116131. doi:10.1016/j.indcrop.2022.116131

Pakdel, E., & Wang, X. (2023). Thermoregulating textiles and fibrous materials for passive radiative cooling functionality. *Materials & Design, 231*, 112006. https://doi.org/.112006 doi:10.1016/j.matdes.2023

Peng, X., Umer, M., Pervez, M. N., Hasan, K. M. F., Habib, M. A., Islam, M. S., Lin, L., Xiong, X., Naddeo, V., & Cai, Y. (2023). Biopolymers-based microencapsulation technology for sustainable textiles development: A short review. *Case Studies in Chemical and Environmental Engineering*, *7*, 100349. doi:10.1016/j.cscee.2023.100349

Prabhakar, P., Sen, R. K., Patel, M., Shruti, Dwivedi, N., Singh, S., Kumar, P., Chouhan, M., Yadav, A. K., Mondal, D. P., Solanki, P. R., Srivastava, A. K., & Dhand, C. (2022). Development of copper impregnated bio-inspired hydrophobic antibacterial nanocoatings for textiles. *Colloids and Surfaces. B, Biointerfaces*, *220*, 112913. doi:10.1016/j.colsurfb.2022.112913 PMID:36306694

Pradhan, L., Maiti, S., Mallick, A., Shahid, M., More, S. P., & Adivarekar, R. V. (2023). *3 Coating-and lamination-based smart textiles: techniques, features, and challenges. In Smart and Functional Textiles*. De Gruyter. doi:10.1515/9783110759747-003

Puiu, R. A., Bîrcă, A. C., Grumezescu, V., Duta, L., Oprea, O. C., Holban, A. M., . . . Andronescu, E. (2023). Multifunctional polymeric biodegradable and biocompatible coatings based on silver nanoparticles: A comparative in vitro study on their cytotoxicity toward cancer and normal cell lines of cytostatic drugs versus essential-oil-loaded nanoparticles and on their antimicrobial and antibiofilm activities. *Pharmaceutics, 15*(7), 1882. https://doi.org/15071882 doi:10.3390/pharmaceutics

Raduly, F. M., Rădițoiu, V., Rădițoiu, A., Frone, A. N., Nicolae, C. A., Răut, I., Constantin, M., & Grapin, M. (2023). Multifunctional finishing of cotton fabric with curcumin derivatives coatings obtained by sol-gel method. *Gels (Basel, Switzerland)*, *9*(5), 369. doi:10.3390/gels9050369 PMID:37232961

Raza, Z. A., Abid, S., Azam, A., & Rehman, A. (2020). Synthesis of alpha-tocopherol encapsulated chitosan nanoassemblies and their impregnation on cellulosic fabric for potential antibacterial and antioxidant cosmetotextiles. *Cellulose (London, England)*, *27*(3), 1717–1731. doi:10.1007/s10570-019-02862-7

Ristić, N., Marković-Nikolić, D., Zdravković, A., Mičić, I., & Ristić, I. (2022). Biofunctional textile materials: Cosmetic textiles. *Advanced Technologies*, *11*(1), 63–75. doi:10.5937/savteh2201063R

Samui, A. B., Hande, P. E., & Mondal, M. I. H. (2022). 25 - Care, maintenance and disposability of medical and protective textile products. In M. I. H. Mondal (Ed.), *Protective Textiles from Natural Resources* (pp. 793–837). Woodhead Publishing. doi:10.1016/B978-0-323-90477-3.00026-2

Sheng, Z., Liu, Z., Hou, Y., Jiang, H., Li, Y., Li, G., & Zhang, X. (2023). The rising aerogel fibers: Status, challenges, and opportunities. *Advancement of Science*, *10*(9), 2205762. doi:10.1002/advs.202205762 PMID:36658735

Shie, M.-Y., Shen, Y.-F., Astuti, S. D., Lee, A. K.-X., Lin, S.-H., Dwijaksara, N. L. B., & Chen, Y.-W. (2019). Review of polymeric materials in 4D printing biomedical applications. *Polymers*, *11*(11), 1864. doi:10.3390/polym11111864 PMID:31726652

Shiya, L., Xiaoke, J., Wei, T., Yanqing, L., Chengyan, Z., & Hongxia, Z. (2023). Research status and development trend of vitamin E loaded cosmeto-textiles. *Advanced Textile Technology*, *31*(1), 293. doi:10.19398/j.att.202206002

Singh, P., Ali, S. W., & Kale, R. D. (2023). Antimicrobial nanomaterials as advanced coatings for self-sanitizing of textile clothing and personal protective equipment. *ACS Omega*, *8*(9), 8159–8171. doi:10.1021/acsomega.2c06343 PMID:36910928

Suellen Ferro de Oliveira, C., & Kekhasharú Tavaria, F. (2023). The impact of bioactive textiles on human skin microbiota. *European Journal of Pharmaceutics and Biopharmaceutics*, *188*, 66–77. doi:10.1016/j.ejpb.2023.05.004 PMID:37182552

Sun, X., Gu, S., Wang, L., Wang, H., Xiong, S., Yin, X., & Yang, S. (2024). Multifunctional liquid-like magnetic nanofluids mediated coating with anticorrosion and self-healing performance. *Journal of Colloid and Interface Science*, *654*, 25–35. doi:10.1016/j.jcis.2023.09.182 PMID:37832232

Tahir, R., Albargi, H. B., Ahmad, A., Qadir, M. B., Khaliq, Z., Nazir, A., ... Jalalah, M. (2023). Development of sustainable hydrophilic *Azadirachta indica* loaded PVA nanomembranes for cosmetic facemask applications. *Membranes*, *13*(2), 156. doi:10.3390/membranes13020156 PMID:36837659

Taieb, A. H., & Mnejja, H. (2023). Design of microencapsulated nigelle sativa oil impregnated in textile structure. *Journal of Engineering Research and Reports*, *25*(5), 81–87. doi:10.9734/jerr/2023/v25i5913

Tanasa, F., Teaca, C.-A., Nechifor, M., Ignat, M., Duceac, I. A., & Ignat, L. (2023). Highly specialized textiles with antimicrobial functionality—Advances and challenges. *Textiles*, *3*(2), 219–245. doi:10.3390/textiles3020015

Tang, Y., Liu, L., Han, J., Zhang, Z., Yang, S., Li, S., Fan, Z., & Zhao, H. (2021). Fabrication and characterization of multiple herbal extracts-loaded nanofibrous patches for topical treatment of acne vulgaris. *Fibers and Polymers*, *22*(2), 323–333. doi:10.1007/s12221-021-0156-1

Tansaoui, H., Bouazizi, N., Behary, N., Campagne, C., El-Achari, A., & Vieillard, J. (2023). Assessing alternative pretreatment methods to promote essential oil fixation into cotton and polyethylene terephthalate Fiber: A comparative study. *Polymers*, *15*(6), 1362. doi:10.3390/polym15061362 PMID:36987143

Vats, M. K., Swami, M., Sheokand, B., Rawat, V., Negi, D. S., & Kumar, A. (2023). Chapter 12 - Antimicrobial textiles based on nanoparticles and composite, antiviral and antimicrobial coatings based on functionalized nanomaterials. In S. ul Islam, C. M. Hussain, & S. K. Shukla (Eds.), Antiviral and Antimicrobial Coatings Based on Functionalized Nanomaterials (pp. 389-420). Elsevier. doi:10.1016/B978-0-323-91783-4.00013-9

Venkatraman, P. D., Butler, J. A., & Britten, N. S. (2023). 2 - Advances in medical textiles. In S. Maity, K. Singha, & P. Pandit (Eds.), *Functional and Technical Textiles* (pp. 31–70). Woodhead Publishing. doi:10.1016/B978-0-323-91593-9.00003-1

Wang, J., Tang, B., Bai, W., Lu, X., & Wang, X. (2020). Deodorizing for fiber and fabric: Adsorption, catalysis, source control and masking. *Advances in Colloid and Interface Science*, *283*, 102243. doi:10.1016/j.cis.2020.102243 PMID:32854016

Yingngam, B., Chiangsom, A., Pharikarn, P., Vonganakasame, K., Kanoknitthiran, V., Rungseevijitprapa, W., & Prasitpuriprecha, C. (2019). Optimization of menthol-loaded nanocapsules for skin application using the response surface methodology. *Journal of Drug Delivery Science and Technology*, *53*, 101138. doi:10.1016/j.jddst.2019.101138

Yingngam, B., Kacha, W., Rungseevijitprapa, W., Sudta, P., Prasitpuriprecha, C., & Brantner, A. (2019). Response surface optimization of spray-dried citronella oil microcapsules with reduced volatility and irritation for cosmetic textile uses. *Powder Technology*, *355*, 372–385. doi:10.1016/j.powtec.2019.07.065

Yurtsever, M. (2019). Glitters as a source of primary microplastics: An approach to environmental responsibility and ethics. *Journal of Agricultural & Environmental Ethics*, *32*(3), 459–478. doi:10.1007/s10806-019-09785-0

Zhang, N., Wang, W., Zhou, M., Yu, Y., Wang, P., & Wang, Q. (2023). A controllable, universal, and natural supramolecular assembly coating strategy for multifunctionality textiles of antibacterial properties, UV resistance, antioxidant, and secondary reactivity. *Industrial Crops and Products*, *198*, 116637. doi:10.1016/j.indcrop.2023.116637

Zhang, Y., Xia, X., Ma, K., Xia, G., Wu, M., Cheung, Y. H., ... Farha, O. K. (2023). Functional textiles with smart properties: Their fabrications and sustainable applications. *Advanced Functional Materials, 33*(33), 2301607. doi:10.1002/adfm.202301607

Zhao, Y., Hu, J., Hu, X., Zhu, F., Su, J., & Han, J. (2023). A novel strategy for fabrication of antistatic and antibacterial fabric via layer-by-layer self-assembly. *Surface and Coatings Technology, 453*, 129143. doi:10.1016/j.surfcoat.2022.129143

Zheng, G., Cui, Y., Jiang, Z., Zhou, M., Wang, P., Yu, Y., & Wang, Q. (2023). Multifunctional composite coatings with hydrophobic, UV-resistant, antioxidative, and photothermal performance for healthcare. *Colloids and Surfaces. A, Physicochemical and Engineering Aspects, 667*, 131367. doi:10.1016/j.colsurfa.2023.131367 PMID:37025928

ADDITIONAL READING

Chen, D., Tang, L. W., Wang, Y. M., Tan, Y. Y., Fu, Y., Cai, W. H., Yu, Z. H., Sun, S., Zheng, J. Q., Cui, J. Q., Wang, G. S., Liu, Y., & Zhou, H. M. (2022). Speaking-induced charge-laden face masks with durable protectiveness and wearing breathability. *ACS Applied Materials & Interfaces, 14*(15), 17774–17782. doi:10.1021/acsami.2c01077 PMID:35362941

Han, H., Zhu, J., Wu, D. Q., Li, F. X., Wang, X. L., Yu, J. Y., & Qin, X. H. (2019). Inherent guanidine nanogels with durable antibacterial and bacterially antiadhesive properties. *Advanced Functional Materials, 29*(12), 1806594. Advance online publication. doi:10.1002/adfm.201806594

Hu, L. Y., Song, X. F., Zhao, X. F., Guo, F. W., Yang, F., & Xiao, P. (2021). A robust, hydrophobic CeO_2/NiCoCrAlY composite coating with excellent thermal stability and corrosion resistance prepared by air plasma spray. *Journal of Alloys and Compounds, 861*, 861. doi:10.1016/j.jallcom.2021.158623

Zhou, Z., Li, C. P., Wang, R. Y., Tao, H., Yang, D. L., Qin, S. Y., Meng, X., & Zhou, C. L. (2022). Crosslinking control of hydrophobic benzoxazine-based hybrid sol-gel coating for corrosion protection on aluminum alloys. *Progress in Organic Coatings, 171*, 171. doi:10.1016/j.porgcoat.2022.107059

Zhu, P. A., Chen, R. F., & Wang, L. Q. (2019). Topography-directed hot-water super-repellent surfaces. *Advancement of Science*, 6. PMID:31559129

Chapter 7
Properties and Characteristics of Nanocoatings for Medicinal Applications

John Tsado Mathew
https://orcid.org/0000-0001-9514-5330
Ibrahimm Badamasi Babangida Lapai, Nigeria

Abel Inobeme
Edo University, Uzairue, Nigeria

Abdulkadir Abdullahi
Federal University of Technology, Minna, Nigeria

Etsuyankpa Muhammad Bini
Federal University of Lafia, Nigeria

Elijah Yanda Shaba
Federal University of Technology, Minna, Nigeria

Monday Musah
Ibrahim BadamasiBabangida University, Lapai, Nigeria

Yakubu Azeh
Ibrahim Badamasi Babangida University, Lapai, Nigeria

Charles Oluwaseun Adetunji
Edo University, Uzairue, Nigeria

Aishetu Muhammad Ibrahim
Ibrahim Badamasi Babangida University, Lapai, Nigeria

Ezekiel Tanko
Federal University of Technology, Minna, Nigeria

Amos Mamman
Ibrahim Badamasi Babangida University, Lapai, Nigeria

ABSTRACT

Nanotechnology has emerged as a revolutionary field with diverse applications, including nanocoatings for medicinal purposes. Nanocoatings, with their unique properties and capabilities, have opened new avenues for drug delivery, medical

DOI: 10.4018/979-8-3693-3136-1.ch007

devices, and antimicrobial treatments. Nanocoatings can be engineered to encapsulate therapeutic agents, protecting them from degradation and facilitating targeted delivery to specific tissues or cells. The use of antimicrobial nanocoatings on medical surfaces has also shown promising results in preventing infections and combating antibiotic-resistant pathogens. While nanocoatings offer tremendous potential, concerns about potential toxicity, biocompatibility, and long-term effects require rigorous evaluation and testing. Furthermore, the study highlights ongoing research and future prospects in the field of medicinal nanocoatings. The development of multifunctional nanocoatings that combine drug delivery, sensing capabilities, and tissue regeneration holds promise for personalized medicine and advanced therapeutics.

INTRODUCTION

Nanotechnology has revolutionized various industries by manipulating materials at the nanoscale, typically at dimensions below 100 nanometers. One of the most promising and innovative applications of nanotechnology is the development of nanocoating, which are ultra-thin layers of nanomaterials applied to surfaces for a range of purposes. In recent years, nanocoating have gained significant attention and recognition for their remarkable potential in the field of medicine and healthcare (Ganeshkumar et al., 2023).

Nanotechnology, the science of manipulating matter at the nanoscale, has captivated the scientific community and various industries with its transformative potential. Among its numerous applications, the integration of nanotechnology into medicine is one of the most promising and groundbreaking advancements of the 21st century. Specifically, the development and utilization of nanocoating in the field of medicine have opened up new frontiers, offering solutions to longstanding challenges and ushering in a new era of healthcare innovation (Manivannan et al., 2024).

These coatings can be precisely engineered and applied to surfaces, creating surfaces with unique properties and functionalities. In medicine, nanocoating are poised to revolutionize multiple aspects, including drug delivery, medical device development, diagnostics, and more. This article will provide a comprehensive overview of the medicinal applications of nanocoating, emphasizing their profound impact on healthcare and patient outcomes (Mozafari et al., 2023).

Medicine has always been at the forefront of technological advancements, constantly seeking new ways to improve patient outcomes, minimize invasiveness, and enhance drug delivery. Nanocoating offer a versatile platform for achieving these goals. These coatings can be engineered to provide unique properties such as controlled drug release, improved biocompatibility, enhanced antimicrobial activity, and precise targeting within the body. As a result, nanocoating have opened up new avenues in diagnostics, treatment, and medical device development (Alotaibi& Federico, 2017).

This article will delve into the medicinal applications of nanocoating, highlighting their contributions to various facets of healthcare. We will explore how nanocoating are being utilized to improve drug delivery systems, enhance the biocompatibility of medical implants, combat infections, and enable advanced diagnostic techniques (Figure 1). Additionally, we will discuss the challenges and ethical considerations associated with this cutting-edge technology (Chopra et al., 2023).

Figure 1.

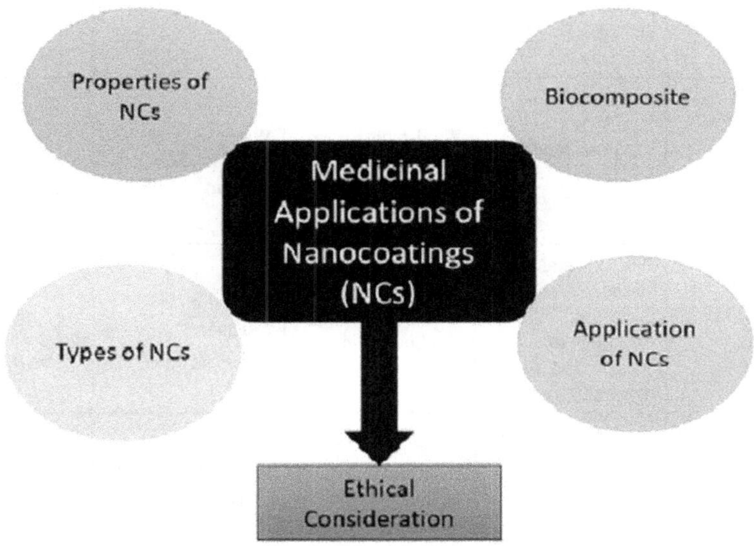

As we venture deeper into the realm of nanocoating in medicine, it becomes evident that these microscopic marvels are poised to transform the way we approach healthcare, offering innovative solutions to longstanding challenges and holding the promise of improved patient care and well-being (Butler et al., 2023).

TYPES OF NANOCOATING

Nanocoating are thin layers of nanomaterials applied to various substrates to enhance their properties and performance. These coatings are used across multiple industries, including medicine, electronics, automotive, and aerospace (Table 1).

Table 1. Types of nanocoating

Name	Description	Applications	Advantages	References
Anti-Reflective Nanocoating	Reduce light reflection and glare on surfaces, improving visibility and optical performance.	Eyeglasses, camera lenses, solar panels	Enhanced optical clarity, improved efficiency	Yao & He, 2014
Hydrophobic Nanocoating	Repel water and liquids, preventing moisture damage and making surfaces easy to clean.	Windshields, clothing, electronics	Water and stain resistance, self-cleaning	Yang et al., 2023
Anti-Corrosion Nanocoating	Protect metal surfaces from corrosion, extending their lifespan and durability.	Automotive parts, pipelines, bridges	Corrosion resistance, increased longevity	Al-Amiery et al., 2023
Anti-Bacterial Nanocoating	Inhibit the growth of bacteria and microbes on surfaces, promoting hygiene and cleanliness.	Healthcare facilities, food packaging	Reduced disease transmission, improved safety	Adetunji et al., 2023a
Self-Healing	Repairs minor damage automatically	Automotive paint, electronics	Extends product lifespan, reduces maintenance	Gopalakrishnan& Mishra, 2023
Thermal Barrier	Reduces heat transfer	Aerospace, building materials	Energy efficiency, temperature control	Lian et al., 2023
Anti-Scratch	Resists scratches and abrasion	Eyeglasses, phone screens, car bodies	Scratch resistance, durability	Becerra 2016
Conductive	Facilitates electrical conductivity	Electronics, sensors, coatings	Enables electronic functionality	Inobeme et al., 2022
UV-Blocking	Blocks ultraviolet (UV) radiation	Sunscreen, eyewear, coatings	Sun protection, prevents UV damage	Tanaka 2023

PROPERTIES AND CHARACTERISTICS OF NANOCOATING

Nanocoating exhibit a wide range of properties and characteristics that make them valuable for various applications. The specific properties and characteristics of nanocoating can vary depending on the materials used, the manufacturing techniques, and the intended applications (Farooq et al., 2022). Nanocoating have a broad range of uses across industries, from consumer electronics to healthcare to aerospace, due to their diverse and tailored properties (Table 2).

Table 2. Properties and characteristics of nanocoating

Property/Characteristic	Description	Examples	Applications	References
Thickness	Extremely thin, often nanometers in scale.	1-100 nm thickness	Optical coatings, electronics, corrosion control	Eessaa & El-Shamy, 2023
Durability	Resistant to wear, abrasion, and environmental factors.	High resistance to damage	Automotive, aerospace, marine	Karim et al., 2023
Surface Modification	Alters surface properties like hydrophobicity, hydrophilicity, etc.	Water-repellent, self-cleaning surfaces	Textiles, glass, electronics, medical devices	Pradhan & Grewal, 2023
Chemical Resistance	Protects against chemical degradation and corrosion.	Corrosion-resistant coatings	Chemical processing, infrastructure	Saxomiddin&Marg'uba, 2023
Self-Cleaning	Can repel dirt and contaminants, staying clean.	Easy-to-clean surfaces	Windows, solar panels, sanitary facilities	Mirabi 2020
Transparency	Often transparent, maintaining substrate appearance.	Clear coatings for optical clarity	Eyeglasses, displays, camera lenses	Huang et al., 2023
Adhesion	Strong adhesion to substrates for longevity.	Ensures coating remains in place	Automotive paints, protective coatings	Ramezani et al., 2023
Anti-Fouling	Prevents the attachment of biological organisms.	Resists bacterial and algal growth	Marine coatings, medical implants	Subbaiyan et al., 2023
Heat Resistance	Can withstand high temperatures without degradation.	High-temperature-resistant coatings	Aerospace, industrial equipment	Zhang et al., 2023
UV Protection	Blocks or absorbs ultraviolet (UV) radiation.	UV-blocking coatings	Sunscreen, UV protection for materials	Lv et al., 2023

NANOCOATING IN DRUG DELIVERY

Nanocoating in drug delivery are a specialized application of nanotechnology in the pharmaceutical field (figure 2). These coatings involve the use of nanoscale materials or structures to modify the surface properties of drug particles or delivery systems. Nanocoating in drug delivery can enhance the efficacy, stability, and targeted delivery of drugs (Adetunji et al., 2022a). There are some main aspects and benefits of nanocoating in drug delivery:

Figure 2.

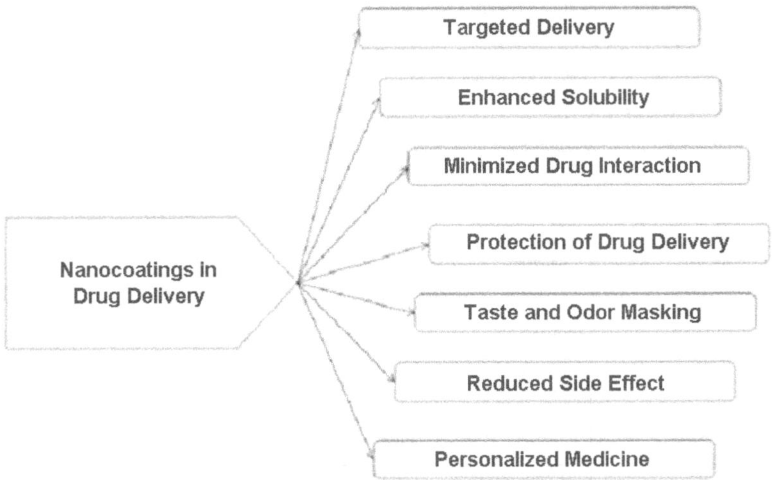

Improved Solubility: Many drugs have poor solubility in water or biological fluids, which can limit their absorption and therapeutic effectiveness. Nanocoating can be designed to improve the solubility of such drugs, allowing for better bioavailability (Bhalani et al. 2022).

Sustained Release: Nanocoating can be engineered to provide controlled and sustained release of drugs over an extended period. This can enhance patient compliance and reduce the frequency of dosing (Simos et al., 2021).

Targeted Delivery: Nanocoating can be designed to enable targeted drug delivery to specific cells, tissues, or organs. This minimizes systemic exposure and reduces side effects while maximizing the therapeutic effect at the target site (Adetunji et al., 2022b).

Protection of Drug Molecules: Nanocoating can protect drug molecules from degradation due to environmental factors such as moisture, light, or oxygen. This increases the shelf life of pharmaceutical formulations (Angelopoulou et al., 2022).

Taste and Odor Masking: Some drugs have unpleasant taste or odor, which can lead to poor patient compliance, especially in pediatric and geriatric populations. Nanocoating can be used to mask these sensory attributes (Vishvakarma et al., 2023).

Combination Therapies: Nanocoating can facilitate the combination of multiple drugs into a single dosage form. This is particularly useful for treating complex medical conditions that require the simultaneous administration of multiple drugs (Adetunji et al., 2023c).

Reduced Side Effects: By targeting drug delivery to specific sites, nanocoating can minimize the exposure of healthy tissues to the drug, reducing the likelihood of side effects (Afzal et al., 2022).

Enhanced Bioavailability: Nanocoating can improve the bioavailability of poorly absorbed drugs by increasing their dissolution rate and permeability through biological barriers.

Personalized Medicine: Nanocoating can be tailored to individual patient needs, allowing for the customization of drug release profiles and dosages (Alghamdi et al., 2022).

Minimization of Drug-Drug Interactions: Coating drug particles can reduce the likelihood of drug-drug interactions, as each drug remains encapsulated within its nanocoated particle.

Common materials used for nanocoating in drug delivery include polymers (e.g., polymeric nanoparticles), lipids (e.g., lipid-based nanoparticles), and inorganic materials (e.g., silica nanoparticles). These materials can be designed to have specific properties that suit the intended drug delivery application (Visan & Cristescu, 2023).

However, it's important to note that the development and regulatory approval of nanocoated drug delivery systems can be complex due to concerns related to safety, stability, and manufacturing processes. Extensive research and testing are typically required to ensure the efficacy and safety of these systems before they can be used in clinical practice (Yusuf et al., 2023).

BIOCOMPATIBLE NANOCOATING FOR MEDICAL IMPLANTS

Biocompatible nanocoating play a crucial role in improving the performance and longevity of medical implants. These coatings are designed to enhance the biocompatibility of implant materials, reduce the risk of adverse reactions, and promote better integration with the host tissue (Table 3).

Common materials used for biocompatible nanocoating include biodegradable polymers, hydroxyapatite, titanium oxide, and various organic and inorganic nanoparticles. These materials are chosen for their compatibility with biological systems and their ability to adhere well to implant surfaces (Abraham &Venkatesan, 2023).

The development of biocompatible nanocoating for medical implants involves rigorous testing to ensure safety and efficacy. Researchers and manufacturers must consider factors like cytotoxicity, immunogenicity, stability, and long-term biocompatibility. Regulatory agencies, such as the FDA in the United States, have established guidelines for the approval of medical devices, including those with nanocoating, to ensure their safety and effectiveness for patient use (Radu & Drăgănescu, 2023).

Overall, biocompatible nanocoating have the potential to significantly improve the performance and safety of medical implants, leading to better patient outcomes and enhanced quality of life for individuals with implanted medical devices (Butler et al., 2023).

Table 3. Examples of biocompatible nanocoating for medical implants

Nanocoating Material	Application	Benefits	References
Hydroxyapatite	Orthopedic Implants	Enhances osseointegration and bone growth	Burdusel et al., 2023
Polymer Nanocoating	Cardiovascular Stents	Reduces friction, promotes biocompatibility	Li et al., 2023
Silver Nanoparticles	Catheters and Implantable Devices	Antibacterial properties reduce infection risks	Esposito et al., 2023
Drug-Eluting Nanocoating	Drug-Eluting Stents	Controlled drug release for tissue healing	Alshimaysawee et al., 2023

Nanocoating Combating Infections

Nanocoating are thin films or layers of nanomaterials that are applied to various surfaces to provide specific functionalities or properties. They have gained significant attention in recent years for their potential to combat infections and improve hygiene in various settings (Sahoo et al., 2022). There are some ways nanocoating are being used to combat infections:

Antimicrobial Nanocoating: One of the primary applications of nanocoating is in the development of antimicrobial surfaces. These coatings are designed to inhibit the growth and spread of bacteria, viruses, and other microorganisms on surfaces. Silver nanoparticles are the commonly used such coatings due to their strong antimicrobial properties.

Self-cleaning Surfaces: Nanocoating can be engineered to be self-cleaning, which means they repel dirt, water, and contaminants. This property helps keep surfaces clean and minimizes the buildup of pathogens. Self-cleaning surfaces are especially useful in healthcare settings and high-traffic areas (Fu & Gray, 2021).

Drug Delivery Systems: Nanocoating can be used as carriers for antimicrobial drugs or vaccines. They can release drugs gradually over time, allowing for a sustained antimicrobial effect. This approach is particularly valuable in the development of wound dressings and medical implants (Paladini & Pollini, 2019; Abimbola et al., 2023).

Antiviral Coatings: In light of the COVID-19 pandemic, there has been a surge in research focused on developing nanocoating that can effectively deactivate viruses, including SARS-CoV-2. These coatings often use materials like copper or graphene to disrupt the viral structure and prevent infection (Ghosal 2023).

Air Purification: Nanocoating are used in air purification systems to trap and deactivate airborne pathogens. These coatings can be applied to filters or surfaces within ventilation systems to reduce the risk of infection transmission in indoor environments (Mahmoudi et al., 2023).

Food Packaging: Nanocoating are employed in food packaging materials to extend the shelf life of food products by preventing the growth of bacteria and fungi. This helps reduce foodborne illnesses and food waste (Adetunji et al., 2022a).

Dental and Orthopedic Implants: Nanocoating can improve the biocompatibility of dental and orthopedic implants while also providing antimicrobial properties. This reduces the risk of post-surgical infections and enhances the longevity of implants (Butler et al., 2023).

Textiles and Fabrics: Nanocoating can be applied to textiles to make them antimicrobial and stain-resistant. This is particularly important for healthcare workers' uniforms, sports clothing, and home textiles (Aguda & Lateef, 2022).

Water Purification: Nanocoating are used in water treatment systems to remove contaminants and pathogens from drinking water, making them safe for consumption (Mathew et al., 2023).

Touchscreens and High-Touch Surfaces: In public places, such as airports and hospitals, nanocoating can be applied to touchscreen interfaces and frequently touched surfaces like door handles to reduce the transmission of infections (Yong &Calautit, 2023).

While nanocoating offer significant promise in combatting infections, there are also challenges, including the potential for the development of resistance to antimicrobial agents and concerns about the environmental impact of nanoparticles.

Ongoing research and development are essential to address these issues and maximize the benefits of nanocoating in infection control and public health (Muteeb 2023).

NANOCOATING IN DIAGNOSTICS

Nanocoating play a significant role in the field of diagnostics by enhancing the performance of diagnostic devices and improving their accuracy and sensitivity. These coatings are ultra-thin layers of nanomaterials, typically ranging from a few nanometers to several hundred nanometers in thickness, which are applied to the surfaces of diagnostic tools or substrates. They offer various advantages in diagnostics, including improved biocompatibility, enhanced signal detection, and increased durability (Thwala et al., 2023). Nanocoating are utilized in diagnostics in the following ways:

Biocompatibility: Nanocoating can be used to modify the surface properties of diagnostic devices to make them more biocompatible. This is particularly important in applications like biosensors and lab-on-a-chip devices where the interaction between the diagnostic tool and biological samples needs to be minimized. Nanocoating can reduce the risk of sample contamination and improve the reliability of test results (Curulli 2023).

Signal Enhancement: Nanocoating can be engineered to enhance the sensitivity and specificity of diagnostic assays. For example, quantum dots, which are nanoscale semiconductor particles, can be used as labels in immunoassays and DNA tests. Their unique optical properties allow for highly sensitive detection of specific molecules, enabling the detection of diseases at an earlier stage (Adetunji et al., 2023b).

Antibacterial Coatings: In diagnostics, maintaining a sterile environment is crucial. Nanocoating can be designed with antimicrobial properties to prevent bacterial or fungal contamination on diagnostic surfaces. These coatings help ensure the accuracy of test results and reduce the risk of false positives or negatives (Sahoo et al., 2022).

Anti-fouling Coatings: In diagnostic devices such as microfluidic chips, the accumulation of biomolecules or debris on the surface can interfere with the flow of samples and reagents. Nanocoating with anti-fouling properties can minimize the adhesion of contaminants and improve the overall efficiency of diagnostic processes (Mumtaz et al., 2023).

Wear Resistance: Diagnostic tools often undergo repeated use and cleaning. Nanocoating can provide wear-resistant layers that protect the underlying materials from degradation, ensuring the longevity of the device and the consistency of test results (Krella, 2023).

Enhanced Imaging: In medical imaging diagnostics, nanocoating can be applied to imaging agents or contrast agents to improve their properties. For example, superparamagnetic nanoparticles can be coated to enhance their stability and targeting capabilities in magnetic resonance imaging or drug delivery (Difonzo et al., 2022).

Surface Functionalization: Nanocoating can be used to functionalize surfaces with specific biomolecules or ligands. This is especially useful in diagnostics that rely on capturing specific molecules or interactions between molecules, such as in protein or DNA microarrays (Díez-Pascual, 2022).

Disease-Specific Coatings: Nanocoating can be engineered to respond selectively to certain biomarkers associated with specific diseases. When these coatings are exposed to the target biomarkers, they produce a detectable signal, enabling the diagnosis of diseases like cancer, diabetes, or infectious diseases (Thwala et al., 2023).

Nanocoating have revolutionized the field of diagnostics by offering precise control over surface properties and enabling the development of highly sensitive and reliable diagnostic devices. Their versatility and tunability make them invaluable in the quest for faster, more accurate, and less invasive diagnostic techniques (Zambonino et al., 2023).

NANOCOATING FOR CANCER TREATMENT

Nanocoating hold great promise in the field of cancer treatment and therapy. These ultra-thin films or layers of nanoscale materials can be applied to various surfaces, including medical devices, drug delivery systems, and even directly to cancer cells, to enhance their effectiveness in cancer treatment (Mosleh-Shirazi et al., 2022). There are some ways nanocoating are being used in cancer treatment:

Drug Delivery: Nanocoating can be used to create nanoparticles that can carry cancer-fighting drugs to the tumor site. These nanoparticles are designed to release the drugs slowly and selectively, minimizing damage to healthy tissues while maximizing the therapeutic effect on cancer cells (Adetunji et al., 2022c).

Targeted Therapy: Nanocoating can be designed to target specific cancer cells or tissues. By attaching molecules or ligands to the nanocoating that bind specifically to cancer cells, the coatings can be directed to deliver drugs or therapies directly to the tumor, reducing side effects (Adetunji et al., 2022d).

Photothermal Therapy: Some nanocoating can be engineered to absorb specific wavelengths of light, effectively converting light energy into heat. This property can be used in photothermal therapy, where nanoparticles coated with materials like gold or carbon are delivered to the tumor site and then heated with laser light, killing cancer cells (Adetunji et al., 2021a).

Radiotherapy Enhancement: Nanocoating can be used to enhance the effectiveness of radiation therapy. Radiosensitizing nanoparticles can be coated with materials that increase their absorption of radiation, allowing for lower radiation doses while achieving the same therapeutic effect (Arif et al., 2023).

Imaging: Nanocoating can be used to improve cancer imaging techniques. For instance, nanoparticles coated with imaging agents can be designed to accumulate in tumors, making them more visible in imaging scans such as magnetic resonance imaging or CT scans (Liu & Grodzinski, 2021).

Biocompatible Coatings: Nanocoating can be applied to medical devices, such as catheters or implants, to make them more biocompatible and less likely to trigger an immune response. This is important in cancer treatment as these devices are often used in various procedures (Ramezani & Ripin, 2023).

Protection of Healthy Cells: In some cases, nanocoating can be used to protect healthy cells from the toxic effects of cancer treatments like chemotherapy. These coatings can be designed to shield normal cells while allowing cancer cells to be targeted (Yao et al., 2022).

Enhanced Drug Stability: Nanocoating can also help improve the stability and shelf life of cancer drugs, ensuring that they remain effective over time (Adetunji et al., 2021b).

While nanocoating have shown great promise in cancer treatment, it's essential to consider issues like biocompatibility, potential toxicity, and long-term effects when developing and using these technologies. Extensive research and testing are ongoing to ensure the safety and efficacy of nanocoating-based cancer treatments. Additionally, regulatory approval is required before many of these therapies can be used in clinical practice (Dessale et al., 2022).

NANOSTRUCTURED COATINGS IN COMMERCIALLY AVAILABLE MEDICINES

Nanostructured coatings play a pivotal role in enhancing the efficacy, safety, and targeted delivery of pharmaceuticals, contributing to the advancement of commercially available medicines. This article explores the impact of nanostructured coatings on pharmaceuticals, focusing on their applications and benefits. One key area where nanostructured coatings have demonstrated significant impact is in improving drug delivery systems. These coatings enable the controlled release of pharmaceutical agents, allowing for sustained therapeutic effects while minimizing side effects. By encapsulating drugs in nano-sized carriers, such as liposomes or polymer nanoparticles, coatings protect the drugs from degradation and facilitate their targeted delivery to specific cells or tissues. This targeted drug delivery minimizes the exposure of healthy tissues to the drug, enhancing its overall safety profile (Malik et al., 2023a).

Furthermore, nanostructured coatings contribute to the development of novel formulations with improved bioavailability. Many drugs face challenges related to low solubility, limiting their absorption and effectiveness. Nanostructured coatings can enhance the solubility of poorly water-soluble drugs, increasing their bioavailability and therapeutic impact. This improvement is particularly crucial for drugs with narrow therapeutic windows, where small changes in dosage can have significant effects on efficacy and safety. In the context of commercially available medicines, nanostructured coatings also play a crucial role in addressing challenges related to stability and shelf life. These coatings protect pharmaceutical formulations from environmental factors such as light, humidity, and oxidation, contributing to the preservation of drug integrity over time. This increased stability not only extends the shelf life of medicines but also ensures that patients receive pharmaceuticals with consistent quality and efficacy (Mathew et al., 2018, Bhalani et al., 2022).

Another significant application of nanostructured coatings in commercially available medicines is in the prevention of microbial contamination. Antimicrobial properties can be imparted to coatings through the incorporation of nanoparticles with inherent antimicrobial activity or the addition of antimicrobial agents. This is particularly relevant in the case of medical devices, where nanostructured coatings can prevent infections and enhance the overall safety of the device. Despite these notable benefits, challenges persist in the widespread adoption of nanostructured coatings in pharmaceuticals. Concerns related to toxicity, regulatory hurdles, and the scalability of manufacturing processes require careful consideration. Researchers and industry stakeholders continue to address these challenges through rigorous testing, collaboration, and advancements in manufacturing technologies (Jose *et al.*, 2023).

In addition, nanostructured coatings have emerged as a transformative technology in the realm of commercially available medicines. Their applications in drug delivery, stability enhancement, and antimicrobial protection contribute to the development of safer, more effective pharmaceutical products. As research in nanotechnology progresses, we can expect further innovations and refinements in nanostructured coatings, paving the way for the next generation of advanced and targeted therapeutic interventions (Abubakar et al., 2015).

CHALLENGES AND ETHICAL CONSIDERATIONS

The integration of nanocoating in diagnostics, like any technological advancement, presents both challenges and ethical considerations that need to be carefully addressed. There are challenges and ethical considerations associated with nanocoating in diagnostics:

- Toxicity and Safety: Some nanomaterials used in nanocoating may have unknown or poorly understood toxicological properties. Ensuring the safety of both patients and healthcare professionals when using diagnostic devices with nanocoating is a significant challenge (Joseph et al., 2023).
- Regulation and Standardization: Developing regulations and standards for nanocoating in diagnostics can be challenging due to the rapid evolution of nanotechnology. Ensuring that these coatings meet safety and performance criteria is essential to prevent potential harm and maintain diagnostic accuracy (Malik et al., 2023b).
- Cost: Integrating nanocoating into diagnostic devices can increase production costs. This can limit access to advanced diagnostics, particularly in resource-constrained healthcare settings (Sheta & El-Sheikh, 2022).
- Environmental Impact: The production and disposal of nanocoating may have environmental consequences. It is crucial to consider the life cycle of these materials and their potential impact on ecosystems (Mahesha et al., 2023).
- Long-term Stability: Ensuring the long-term stability of nanocoating, especially in diagnostic devices that are used repeatedly, is a technical challenge. Over time, coatings may degrade or lose their effectiveness, leading to unreliable test results (Ganeshkumar et al., 2023).
- Interference with Diagnostic Results: Poorly designed nanocoating can interfere with diagnostic results. For example, they may lead to false positives or negatives in certain tests, which can have serious consequences for patient care (Naresh & Lee, 2021).
- Informed Consent: Patients have the right to be informed about the use of nanocoating in diagnostic tests and any potential risks associated with them. Obtaining informed consent becomes crucial, especially when patients may not fully understand the technology involved (Butler et al., 2023).
- Equity and Access: The introduction of advanced diagnostic technologies with nanocoating should not exacerbate healthcare disparities. Ensuring equitable access to these technologies is an ethical imperative (Goisauf & Cano-Abadía, 2022).
- Privacy: Advanced diagnostics often involve the collection and analysis of sensitive patient data. Ethical considerations include how this data is stored, shared, and protected to maintain patient privacy (Chiruvella & Guddati, 2021).
- Beneficence and Non-maleficence: Healthcare providers must balance the benefits of using nanocoating in diagnostics (improved accuracy, early disease detection) with the potential risks (toxicity, misdiagnosis). Ethical decision-making should prioritize patient well-being (Thwala et al., 2023).

- Transparency and Accountability: The development and use of nanocoating in diagnostics should be transparent and accountable. This includes disclosing research findings, and potential conflicts of interest, and ensuring that diagnostic devices meet established safety and performance standards (Solanki et al., 2023).
- Environmental Responsibility: Ethical considerations extend to the environmental impact of nanocoating. Companies and researchers should strive to minimize the negative environmental consequences associated with the production and disposal of these materials (Miteu et al., 2023).
- Dual Use: Nanocoating developed for diagnostics may also have applications in other fields, including defense or surveillance. Ethical discussions should address the potential dual-use nature of these technologies and the need for responsible use (Bordin et al., 2020).
- International Collaboration: Ethical considerations in nanocoating and diagnostics often have global implications. International collaboration is essential to ensure that best practices, regulations, and ethical standards are upheld consistently across borders (Sundeep et al. 2023).
- Incorporating these challenges and ethical considerations into the development, regulation, and implementation of nanocoating in diagnostics is crucial for realizing the full potential of these technologies while minimizing risks and ensuring equitable and responsible healthcare practices. Researchers, healthcare providers, policymakers, and ethicists must work together to navigate these complex issues (Wasti et al., 2023).

CONCLUSION

Nanocoating represent a remarkable fusion of nanotechnology and medicine, offering unprecedented opportunities to revolutionize healthcare. From enhancing drug delivery and biocompatibility to combating infections and enabling precise diagnostics, these ultra-thin layers of nanomaterials hold the promise of improving patient outcomes, reducing healthcare costs, and advancing our understanding of diseases. As nanocoating continue to evolve, it is essential to address the associated challenges and ethical considerations to ensure their responsible and beneficial integration into healthcare systems worldwide. By doing so, we can unlock the full potential of nanocoating in medicine and shape a brighter.

REFERENCES

Abimbola, O. F., Adetunji, C. O., Dare, B. J., Mathew, J. T., Inobeme, A., Ghazanfar, S., Olaniyan, O. T., Ajiboye, M. D., & Dauda, W. (2023). Packaging of Animal-Based Food Products: The Need for Intelligent Freshness Sensors. In Sensing and Artificial Intelligence Solutions for Food Manufacturing. CRC Press.

Abraham, A. M., & Venkatesan, S. (2023). A review on application of biomaterials for medical and dental implants. *Proceedings of the Institution of Mechanical Engineers, Part L: Journal of Materials: Design and Applications, 237*(2), 249-273. 10.1177/14644207221121981

Abubakar, I., Mann, A., & Mathew, J. T. (2015). Evaluation of Phytochemical, Anti-nutritional and Antioxidant Potentials of Flower and Seed Methanol Extracts of *Senna alata* L. Grown in Nigeria. *American Journal of Applied Chemistry, 3*(3), 93–100. doi:10.11648/j.ajac.20150303.12

Adetunji, C. O., Mathew, J. T., Singh, K. R. B., Inobeme, A., Olaniyan, O., Vanya, N., & Singh, J. (2021). Conducting Polymer-Based Microbial Fuel Cells. Conducting Polymers for Advanced Energy Applications, 18, 337-344.

Adetunji, C. O., Mathew, J. T., Singh, K. R. B., Inobeme, A., Vanya, N., Olaniyan, O., Singh, J., & Singh, R. P. (2022). Zero-dimensional carbon nanomaterials for targeted drug delivery and photothermal therapy applications: In Zero-dimensional Carbon Nanomaterials: Fundamentals and applications. IOP Publishing.

Adetunji, C. O., Ogundolie, F. A., Mathew, J. T., Inobeme, A., Olotu, T., Adetunji, J. B., Dauda, W. P., Aborode, A. T., Ajayi, O. O., Ghazanfar, S., Aliyu, A., Imoisi, S. E., Moses-Oke, R. O., Egbuna, C., Uzualu, E. J., Ajenifujah-Solebo, S. O., Akinbo, O., Adetuyi, B. O., & Edetalehn, O. I. (2023). Chapter Twenty-Six - Patenting protocols, toxicity, risk assessments, and policy issues of nanomaterials with diverse applications in food, biomedical, and other relevant sectors. In Evaporative Coolers for the Postharvest Management of Fruits and Vegetables. Academic Press.

Adetunji, C. O., Ogundolie, F. A., Mathew, J. T., Inobeme, A., Olotu, T., Olaniyan, O. T., Ghazanfar, S., Ijabadeniyi, O. A., Ajiboye, M. D., Ajayi, O. O., Dauda, W., & Adetunji, J. B. (2022). Nanotube platforms for effective drug delivery applications. In Novel Platforms for Drug Delivery Applications. Elsevier Ltd.

Adetunji, C. O., Ogundolie, F. A., Mathew, J. T., Inobeme, A., Olotu, T., Olaniyan, O. T., Ijabadeniyi, O. A., Ajiboye, M. D., Ajayi, O. O., Dauda, W., Ghazanfar, S., & Adetunji, J. B. (2022). Graphene-based nanomaterials for targeted drug delivery and tissue engineering. In Novel Platforms for Drug Delivery Applications. Elsevier Ltd.

Adetunji, C. O., Ogundolie, F. A., Olaniyan, O., Mathew, J. T., Inobeme, A., Olotu, T., Ghazanfar, S., Ijabadeniyi, O. A., Ajiboye, M. D., Ajayi, O. O., Dauda, W. P., & Adetunji, J. B. (2022). Nanobiomaterials for Food Packaging Sensor Applications. In Bio- and Nano-sensing Technologies for Food Processing and Packaging. The Royal Society of Chemistry.

Adetunji, C. O., Olaniyan, O. T., Shariati, M. A., Hefft, D. I., Olugbenga, S. M., Rebezov, M., Ajayi, O. O., Anichkina, O., Bodunrinde, R. E., Adetunji, R. B., Adeniyi, M. J., Inobeme, A., & Mathew, J. T. (2023). Medicinal Properties of Sennaalata (L.) Roxb. (Syn. Cassia alata L.) and Its Biological Activities. In Phytochemical Composition and Pharmacy of Medicinal Plants. Apple Academic Press, Inc.

Adetunji, C. O., Olaniyan, O. T., Shariati, M. A., Hefft, D. I., Olugbenga, S. M., Rebezov, M., Ajayi, O. O., Baibalinova, G., Bodunrinde, R. E., Adetunji, R. B., Adeniyi, M. J., Inobeme, A., Mathew, J. T., Elufisan, T. O., & Oyedara, O. O. (2023). Medicinal and Pharmacological Attributes of AcalyphahispidaBurm. F. In Phytochemical Composition and Pharmacy of Medicinal Plants. Apple Academic Press, Inc.

Adetunji, C. O., Olugbemi, T. O., Osikemekha, A. A., Inobeme, A., & Mathew, J. T. (2021). *Environmental Impact of Polyurethane Chemistry*. American Chemical Society.

Afzal, O., Altamimi, A. S. A., Nadeem, M. S., Alzarea, S. I., Almalki, W. H., Tariq, A., Mubeen, B., Murtaza, B. N., Iftikhar, S., Riaz, N., & Kazmi, I. (2022). Nanoparticles in Drug Delivery: From History to Therapeutic Applications. *Nanomaterials (Basel, Switzerland), 12*(24), 4494. doi:10.3390/nano12244494 PMID:36558344

Aguda, O. N., & Lateef, A. (2022). Recent advances in functionalization of nanotextiles: A strategy to combat harmful microorganisms and emerging pathogens in the 21st century. *Heliyon, 8*(6), e09761. doi:10.1016/j.heliyon.2022.e09761 PMID:35789866

Al-Amiery, A. A., Yousif, E., Isahak, W. N. R. W., & Al-Azzawi, W. K. (2023). A review of inorganic corrosion inhibitors: Types, mechanisms, and applications. *Tribology in Industry, 44*(2), 313. doi:10.24874/ti.1456.03.23.06

Alghamdi, M. A., Fallica, A. N., Virzì, N., Kesharwani, P., Pittalà, V., & Greish, K. (2022). The Promise of Nanotechnology in Personalized Medicine. *Journal of Personalized Medicine, 12*(5), 673. doi:10.3390/jpm12050673 PMID:35629095

Alotaibi, Y. K., & Federico, F. (2017). The impact of health information technology on patient safety. *Saudi Medical Journal, 38*(12), 1173–1180. doi:10.15537/smj.2017.12.20631 PMID:29209664

Alshimaysawee, S., Fadhel Obaid, R., Al-Gazally, M. E., Alexis Ramírez-Coronel, A., & Bathaei, M. S. (2023). Recent Advancements in Metallic Drug-Eluting Implants. *Pharmaceutics*, *15*(1), 223. doi:10.3390/pharmaceutics15010223 PMID:36678852

Angelopoulou, P., Giaouris, E., & Gardikis, K. (2022). Applications and Prospects of Nanotechnology in Food and Cosmetics Preservation. *Nanomaterials (Basel, Switzerland)*, *12*(7), 1196. doi:10.3390/nano12071196 PMID:35407315

Arif, M., Nawaz, A. F., Ullah Khan, S., Mueen, H., Rashid, F., Hemeg, H. A., & Rauf, A. (2023). Nanotechnology-based radiation therapy to cure cancer and the challenges in its clinical applications. *Heliyon*, *9*(6), e17252. doi:10.1016/j.heliyon.2023.e17252 PMID:37389057

Becerra, L. (2016). *CMF design: The fundamental principles of color, material, and finish design*. Frame Publishers.

Bhalani, D. V., Nutan, B., Kumar, A., & Singh Chandel, A. K. (2022). Bioavailability Enhancement Techniques for Poorly Aqueous Soluble Drugs and Therapeutics. *Biomedicines*, *10*(9), 2055. doi:10.3390/biomedicines10092055 PMID:36140156

Bhalani, D. V., Nutan, B., Kumar, A., & Singh Chandel, A. K. (2022). Bioavailability Enhancement Techniques for Poorly Aqueous Soluble Drugs and Therapeutics. *Biomedicines*, *10*(9), 2055. doi:10.3390/biomedicines10092055 PMID:36140156

Bordin, G., Hristova, M., & Luque-Perez, E. (2020). *Horizon 2020-funded security research projects with dual-use potential: An overview (2014-2018)*. Publications Office of the European Union.

Burdusel, A. C., Neacsu, I. A., Birca, A. C., Chircov, C., Grumezescu, A. M., Holban, A. M., Curutiu, C., Ditu, L. M., Stan, M., & Andronescu, E. (2023). Microwave-Assisted Hydrothermal Treatment of Multifunctional Substituted Hydroxyapatite with Prospective Applications in Bone Regeneration. *Journal of Functional Biomaterials*, *14*(7), 378. doi:10.3390/jfb14070378 PMID:37504872

Butler, J., Handy, R. D., Upton, M., & Besinis, A. (2023). Review of Antimicrobial Nanocoating in Medicine and Dentistry: Mechanisms of Action, Biocompatibility Performance, Safety, and Benefits Compared to Antibiotics. *ACS Nano*, *17*(8), 7064–7092. doi:10.1021/acsnano.2c12488 PMID:37027838

Chiruvella, V., & Guddati, A. K. (2021). Ethical Issues in Patient Data Ownership. *Interactive Journal of Medical Research*, *10*(2), e22269. doi:10.2196/22269 PMID:34018968

Chopra, H., Mohanta, Y. K., Rauta, P. R., Ahmed, R., Mahanta, S., Mishra, P. K., Panda, P., Rabaan, A. A., Alshehri, A. A., Othman, B., Alshahrani, M. A., Alqahtani, A. S., & Basha, A. L. (2023). An Insight into Advances in Developing Nanotechnology Based Therapeutics, Drug Delivery, Diagnostics and Vaccines: Multidimensional Applications in Tuberculosis Disease Management. *Pharmaceuticals (Basel, Switzerland), 16*(4), 581. doi:10.3390/ph16040581 PMID:37111338

Curulli, A. (2023). Functional Nanomaterials Enhancing Electrochemical Biosensors as Smart Tools for Detecting Infectious Viral Diseases. *Molecules (Basel, Switzerland), 28*(9), 3777. doi:10.3390/molecules28093777 PMID:37175186

Dessale, M., Mengistu, G., & Mengist, H. M. (2022). Nanotechnology: A Promising Approach for Cancer Diagnosis, Therapeutics and Theragnosis. *International Journal of Nanomedicine, 17*, 3735–3749. doi:10.2147/IJN.S378074 PMID:36051353

Díez-Pascual, A. M. (2022). Surface Engineering of Nanomaterials with Polymers, Biomolecules, and Small Ligands for Nanomedicine. *Materials (Basel), 15*(9), 3251. doi:10.3390/ma15093251 PMID:35591584

Difonzo, M., Fliedel, L., Mignet, N., Andrieux, K., & Alhareth, K. (2022). How Could Nanomedicine Improve the Safety of Contrast Agents for Magnetic Resonance Imaging during Pregnancy? *Sci, 4*(1), 11. doi:10.3390/sci4010011

Eessaa, A. K., & El-Shamy, A. M. (2023). Review on fabrication, characterization, and applications of porous anodic aluminum oxide films with tunable pore sizes for emerging technologies. *Microelectronic Engineering, 279*, 112061. doi:10.1016/j.mee.2023.112061

Esposito, M. M., Glazer, J. R., & Turku, S. (2023). The Use of 3D Printing and Nanotechnologies to Prevent and Inhibit Biofilms on Medical Devices. *Hygiene, 3*(3), 325–338. doi:10.3390/hygiene3030024

Farooq, S. A., Raina, A., Mohan, S., Arvind Singh, R., Jayalakshmi, S., & Irfan Ul Haq, M. (2022). Nanostructured Coatings: Review on Processing Techniques, Corrosion Behaviour, and Tribological Performance. *Nanomaterials (Basel, Switzerland), 12*(8), 1323. doi:10.3390/nano12081323 PMID:35458032

Fu, H., & Gray, K. A. (2021). The key to maximizing the benefits of antimicrobial and self-cleaning coatings is to fully determine their risks. *Current Opinion in Chemical Engineering, 34*, 100761. doi:10.1016/j.coche.2021.100761 PMID:36569284

Ganeshkumar, S., Kumar, A., Maniraj, J., Babu, Y. S., Ansu, A. K., Goyal, A., ... Hassan, A. M. (2023). Exploring the potential of nanotechnology: A assessment of nano-scale multi-layered-composite coatings for cutting tool performance. *Arabian Journal of Chemistry*, *16*(10), 105173. doi:10.1016/j.arabjc.2023.105173

Ghosal, K. (2023). Tackling COVID-19 Using Antiviral Nanocoating's-Recent Progress and Future Challenges. *Particle & Particle Systems Characterization*, *40*(1), 2200154. doi:10.1002/ppsc.202200154 PMID:36711425

Goisauf, M., & Cano-Abadía, M. (2022). Ethics of AI in Radiology: A Review of Ethical and Societal Implications. *Frontiers in Big Data*, *5*, 850383. doi:10.3389/fdata.2022.850383 PMID:35910490

Gopalakrishnan, K., & Mishra, P. (2023). Self-Healing Polymer a Dynamic Solution in Food Industry: A Comprehensive Review. *Food Biophysics*, 1–17. doi:10.1007/s11483-023-09780-z

Huang, P. H., Laakso, M., Edinger, P., Hartwig, O., Duesberg, G. S., Lai, L. L., Mayer, J., Nyman, J., Errando-Herranz, C., Stemme, G., Gylfason, K. B., & Niklaus, F. (2023). Three-dimensional printing of silica glass with sub-micrometer resolution. *Nature Communications*, *14*(1), 3305. doi:10.1038/s41467-023-38996-3 PMID:37280208

Inobeme, A., Nayak, V., Mathew, J. T., Okonkwo, S., Ekwoba, L., Ajai, A. I., Bernard, E., Inobeme, J., Agbugui, M. M., & Singh, K. R. B. (2022). Chemometric approach in environmental pollution analysis: A critical review. *Journal of Environmental Management*, *309*(114653), 1–18. doi:10.1016/j.jenvman.2022.114653 PMID:35176568

Jose, A., Gizdavic-Nikolaidis, M., & Swift, S. (2023). Antimicrobial Coatings: Reviewing Options for Healthcare Applications. *Applied Microbiology*, *3*(1), 145–174. doi:10.3390/applmicrobiol3010012

Joseph, T. M., Kar Mahapatra, D., Esmaeili, A., Piszczyk, Ł., Hasanin, M., Kattali, M., Haponiuk, J., & Thomas, S. (2023). Nanoparticles: Taking a Unique Position in Medicine. *Nanomaterials (Basel, Switzerland)*, *13*(3), 574. doi:10.3390/nano13030574 PMID:36770535

Karim, M. A., Abdullah, M. Z., Deifalla, A. F., Azab, M., & Waqar, A. (2023). An assessment of the processing parameters and application of fiber-reinforced polymers (FRPs) in the petroleum and natural gas industries: A review. *Results in Engineering*, *18*, 101091. doi:10.1016/j.rineng.2023.101091

Krella, A. K. (2023). Degradation and Protection of Materials from Cavitation Erosion: A Review. *Materials (Basel), 16*(5), 2058. doi:10.3390/ma16052058 PMID:36903173

Li, D., Dai, D., Xiong, G., Lan, S., & Zhang, C. (2023). Composite Nanocoating of Biomedical Magnesium Alloy Implants: Advantages, Mechanisms, and Design Strategies. *Advanced Science (Weinheim, Baden-Wurttemberg, Germany), 10*(18), 2300658. doi:10.1002/advs.202300658

Lian, R., Ou, M., Guan, H., Cui, J., Piao, J., Feng, T., Ren, J., Wang, Y., Wang, Y., Liu, L., Chen, X., & Jiao, C. (2023). Facile fabrication of multifunctional energy-saving building materials with excellent thermal insulation, robust mechanical property, and ultrahigh flame retardancy. *Energy, 277*, 127773. doi:10.1016/j.energy.2023.127773

Liu, C. H., & Grodzinski, P. (2021). Nanotechnology for Cancer Imaging: Advances, Challenges, and Clinical Opportunities. *Radiology. Imaging Cancer, 3*(3), e200052. doi:10.1148/rycan.2021200052 PMID:34047667

Lv, S., Liang, S., Zuo, J., Zhang, S., Wang, J., & Wei, D. (2023). Lignin-based anti-UV functional materials: Recent advances in preparation and application. *Iranian Polymer Journal, 32*(11), 1–21. doi:10.1007/s13726-023-01218-0

Mahesha, K. N., Guddaraddi, A., Reddy, H., & GV, S. K. (2023). Unleashing the Potential of Nanotechnology in Agriculture. *Advanced Innovative Technologies in Agricultural Engineering for Sustainable Agriculture*, 97.

Mahmoudi, A., Tavakoly Sany, S. B., Ahari Salmasi, M., Bakhshi, A., Bustan, A., heydari, S., Rezayi, M., & Gheybi, F. (2023). Application of nanotechnology in air purifiers as a viable approach to protect against Coronavirus. *IET Nanobiotechnology / IET, 17*(4), 289–301. doi:10.1049/nbt2.12132 PMID:37096564

Malik, S., Muhammad, K., & Waheed, Y. (2023). a. Emerging Applications of Nanotechnology in Healthcare and Medicine. *Molecules (Basel, Switzerland), 28*(18), 6624. doi:10.3390/molecules28186624 PMID:37764400

Malik, S., Muhammad, K., & Waheed, Y. (2023). b. Nanotechnology: A Revolution in Modern Industry. *Molecules (Basel, Switzerland), 28*(2), 661. doi:10.3390/molecules28020661 PMID:36677717

Manivannan, S., Praveen, R., Kim, K., & Ramaraj, R. (2024). Introduction: Nanomaterials for Sustainable Energy Applications. In Nanomaterials for Sustainable Energy Applications (pp. 1-24). CRC Press.

Mathew, J. T., Adetunji, C. O., Inobeme, A., Musah, M., Azeh, Y., Otori, A. A., Shaba, E. Y., Mamman, A., & Tanko, E. (2023). Removal of Heavy Metals Using Bio-remedial Techniques. Springer Nature Switzerland AG. doi:10.1007/978-3-031-24086-7_6

Mathew, J. T., Dauda, B. E. N., Mann, A., Ndamitso, M. M., Fadipe, A. L., & Shaba, E. Y. (2018). Assessment of Nutritional Properties of Fermented and Unfermented Seed of *Cissus populnae* From Niger State, Nigeria. *Assumption University-eJournal of Interdisciplinary Research, 3*(2), 70–77.

Mirabi, S. (2020). The use of self-cleaning materials and the study of their impact on the quality of the visual appearance of large industrial cities. *Journal of Architectural Research and Development, 4*(6). Advance online publication. doi:10.26689/jard.v4i6.1435

Miteu, G. D., Emmanuel, A. A., Addeh, I., Ojeokun, O., Olayinka, T., Godwin, J. S., ... Benneth, E. O. (2023). Nanoscience Technology as A Pivot For Sustainable Agriculture And Its One Health Approach Awareness. *Science in One Health*, 100020.

Mosleh-Shirazi, S., Abbasi, M., Moaddeli, M. R., Vaez, A., Shafiee, M., Kasaee, S. R., Amani, A. M., & Hatam, S. (2022). Nanotechnology Advances in the Detection and Treatment of Cancer: An Overview. *Nanotheranostics, 6*(4), 400–423. doi:10.7150/ntno.74613 PMID:36051855

Mozafari, N., Mozafari, N., Dehshahri, A., & Azadi, A. (2023). Knowledge Gaps in Generating Cell-Based Drug Delivery Systems and a Possible Meeting with Artificial Intelligence. *Molecular Pharmaceutics, 20*(8), 3757–3778. doi:10.1021/acs.molpharmaceut.3c00162 PMID:37428824

Mumtaz, Z., Rashid, Z., Ali, A., Arif, A., Ameen, F., AlTami, M. S., & Yousaf, M. Z. (2023). Prospects of Microfluidic Technology in Nucleic Acid Detection Approaches. *Biosensors (Basel), 13*(6), 584. doi:10.3390/bios13060584 PMID:37366949

Muteeb, G. (2023). Nanotechnology—A Light of Hope for Combating Antibiotic Resistance. *Microorganisms, 11*(6), 1489. doi:10.3390/microorganisms11061489 PMID:37374990

Naresh, V., & Lee, N. (2021). A Review on Biosensors and Recent Development of Nanostructured Materials-Enabled Biosensors. *Sensors (Basel), 21*(4), 1109. doi:10.3390/s21041109 PMID:33562639

Paladini, F., & Pollini, M. (2019). Antimicrobial Silver Nanoparticles for Wound Healing Application: Progress and Future Trends. *Materials (Basel), 12*(16), 2540. doi:10.3390/ma12162540 PMID:31404974

Pradhan, R., & Grewal, H. S. (2023). Flexible and durable fluorine-free superhydrophobic films through a sustainable approach. *Chemical Engineering Journal*, *474*, 145812. doi:10.1016/j.cej.2023.145812

Radu, R.-D., & Drăgănescu, D. (2023). Present and Future of ZrO2 Nanostructure as Reservoir for Drug Loading and Release. *Coatings*, *13*(7), 1273. doi:10.3390/coatings13071273

Ramezani, M., Mohd Ripin, Z., Pasang, T., & Jiang, C.-P. (2023). Surface Engineering of Metals: Techniques, Characterizations and Applications. *Metals*, *13*(7), 1299. doi:10.3390/met13071299

Sahoo, J., Sarkhel, S., Mukherjee, N., & Jaiswal, A. (2022). Nanomaterial-Based Antimicrobial Coating for Biomedical Implants: New Age Solution for Biofilm-Associated Infections. *ACS Omega*, *7*(50), 45962–45980. doi:10.1021/acsomega.2c06211 PMID:36570317

Saxomiddino'g'li, Z. S., & Marg'uba, Y. (2023). Anticorrosive Materials: Safeguarding the Future of Infrastructure. *Journal of Universal Science Research*, *1*(7), 71–76.

Sheta, S. M., & El-Sheikh, S. M. (2022). Nanomaterials and metal-organic frameworks for biosensing applications of mutations of the emerging viruses. *Analytical Biochemistry*, *648*, 114680. doi:10.1016/j.ab.2022.114680 PMID:35429447

Simos, Y. V., Spyrou, K., Patila, M., Karouta, N., Stamatis, H., Gournis, D., Dounousi, E., & Peschos, D. (2021). Trends of nanotechnology in type 2 diabetes mellitus treatment. *Asian Journal of Pharmaceutical Sciences*, *16*(1), 62–76. doi:10.1016/j.ajps.2020.05.001 PMID:33613730

Solanki, R., Shankar, A., Modi, U., & Patel, S. (2023). New insights from nanotechnology in SARS-CoV-2 detection, treatment strategy and prevention. *Materials Today. Chemistry*, *29*, 101478. doi:10.1016/j.mtchem.2023.101478 PMID:36950312

Subbaiyan, R., Ganesan, A., & Varadharajan, V. (2023). Bioprospecting and Exploration of the Natural Antifouling Approaches against Marine Foulers. *Journal of Pure & Applied Microbiology*, *17*(3), 1374–1390. doi:10.22207/JPAM.17.3.02

Sundeep, D., Varadharaj, E. K., Umadevi, K., & Jhansi, R. (2023). Role of Nanomaterials in Screenprinted Electrochemical Biosensors for Detection of COVID-19 and Post-Covid Syndromes. *ECS Advances*, *2*(1), 016502. doi:10.1149/2754-2734/acb832

Tanaka, Y. (2023). Photoprotective Ability of Sunscreens against Ultraviolet, Visible Light and Near-Infrared Radiation. *Optics and Photonics Journal*, *13*(6), 140–146. doi:10.4236/opj.2023.136012

Thwala, L. N., Ndlovu, S. C., Mpofu, K. T., Lugongolo, M. Y., & Mthunzi-Kufa, P. (2023). Nanotechnology-Based Diagnostics for Diseases Prevalent in Developing Countries: Current Advances in Point-of-Care Tests. *Nanomaterials (Basel, Switzerland)*, *13*(7), 1247. doi:10.3390/nano13071247 PMID:37049340

Visan, A. I., & Cristescu, R. (2023). Polysaccharide-Based Coatings as Drug Delivery Systems. *Pharmaceutics*, *15*(9), 2227. doi:10.3390/pharmaceutics15092227 PMID:37765196

Vishvakarma, V., Kaur, M., Nagpal, M., & Arora, S. (2023). Role of Nanotechnology in Taste Masking: Recent Updates. *Current Drug Research Reviews*, *15*(1), 1–14. doi:10.2174/2589977514666220526091259 PMID:35619251

Wasti, S., Lee, I. H., Kim, S., Lee, J. H., & Kim, H. (2023). Ethical and legal challenges in nanomedical innovations: A scoping review. *Frontiers in Genetics*, *14*, 1163392. doi:10.3389/fgene.2023.1163392 PMID:37252668

Yang, G., Zhang, Z., Li, C., & Hu, J. (2023). A facile approach to fabricate omniphobic and robust polyurethane coatings for anti-smudge, anti-ink. *Progress in Organic Coatings*, *179*, 107488. doi:10.1016/j.porgcoat.2023.107488

Yao, L., & He, J. (2014). Recent progress in antireflection and self-cleaning technology–From surface engineering to functional surfaces. *Progress in Materials Science*, *61*, 94–143. doi:10.1016/j.pmatsci.2013.12.003

Yao, Y., Zhou, Y., Liu, L., Xu, Y., Chen, Q., Wang, Y., Wu, S., Deng, Y., Zhang, J., & Shao, A. (2020). Nanoparticle-Based Drug Delivery in Cancer Therapy and Its Role in Overcoming Drug Resistance. *Frontiers in Molecular Biosciences*, *7*, 193. doi:10.3389/fmolb.2020.00193 PMID:32974385

Yong, L. X., & Calautit, J. K. (2023). A Comprehensive Review on the Integration of Antimicrobial Technologies onto Various Surfaces of the Built Environment. *Sustainability (Basel)*, *15*(4), 3394. doi:10.3390/su15043394

Yusuf, A., Almotairy, A. R. Z., Henidi, H., Alshehri, O. Y., & Aldughaim, M. S. (2023). Nanoparticles as Drug Delivery Systems: A Review of the Implication of Nanoparticles' Physicochemical Properties on Responses in Biological Systems. *Polymers*, *15*(7), 1596. doi:10.3390/polym15071596 PMID:37050210

Zambonino, M. C., Quizhpe, E. M., Mouheb, L., Rahman, A., Agathos, S. N., & Dahoumane, S. A. (2023). Biogenic Selenium Nanoparticles in Biomedical Sciences: Properties, Current Trends, Novel Opportunities and Emerging Challenges in TheranosticNanomedicine. *Nanomaterials (Basel, Switzerland)*, *13*(3), 424. doi:10.3390/nano13030424 PMID:36770385

Zhang, Y., Ma, X., Yang, M., Lan, D., Li, X., Gao, Z., & Ji, T. (2023). Ablation resistance of ethylene propylene diene monomer insulation materials reinforced with liquid hyperbranched polycarbosilane coated aramid fibers. *Polymer Degradation & Stability*, *217*, 110510. doi:10.1016/j.polymdegradstab.2023.110510

ADDITIONAL READING

Canney, P. A., Moore, M., Wilkinson, P. M., & James, R. D. (1984). Ovarian cancer antigen CA125: A prospective clinical assessment of its role as a tumour marker. *British Journal of Cancer*, *50*(6), 765–769. doi:10.1038/bjc.1984.254 PMID:6208925

Das, J., Ivanov, I., Sargent, E. H., & Kelley, S. O. (2016). DNA clutch probes for circulating tumor DNA analysis. *Journal of the American Chemical Society*, *138*(34), 11009–11016. doi:10.1021/jacs.6b05679 PMID:27513828

Skoulidis, F., Li, B. T., Dy, G. K., Price, T. J., Falchook, G. S., Wolf, J., Italiano, A., Schuler, M., Borghaei, H., Barlesi, F., Kato, T., Curioni-Fontecedro, A., Sacher, A., Spira, A., Ramalingam, S. S., Takahashi, T., Besse, B., Anderson, A., Ang, A., ... Govindan, R. (2021). Sotorasib for Lung Cancers with KRAS p. G12C Mutation. *The New England Journal of Medicine*, *384*(25), 2371–2381. doi:10.1056/NEJMoa2103695 PMID:34096690

Chapter 8
Healing and Health Perspectives of Nanocoatings for Medicinal Applications

Abhinay Thakur
https://orcid.org/0000-0002-7186-1639
Lovely Professional University, India

Ashish Kumar
https://orcid.org/0000-0002-2003-6209
Nalanda College of Engineering, Bihar Engineering University, India & Department of Science, Technology and Technical Education, Government of Bihar, India

ABSTRACT

Offering unprecedented opportunities to revolutionize healing and healthcare, this chapter delves into the cutting-edge advancements in nanocoatings tailored specifically for medicinal applications, presenting a comprehensive overview of their potential to enhance healing processes and improve health outcomes. A major focus of this chapter is on the multifaceted roles of nanocoatings in medicinal contexts. These include drug delivery systems, wound healing, antimicrobial coatings, and tissue engineering scaffolds, among others. Furthermore, it discusses the challenges and regulatory considerations associated with the integration of nanocoatings into clinical practice. By exploring the latest research findings, this chapter provides valuable insights into how nanocoatings are poised to transform the landscape of medicinal applications. It underscores their potential to revolutionize healing processes, mitigate health-related challenges, and pave the way for a healthier future.

DOI: 10.4018/979-8-3693-3136-1.ch008

INTRODUCTION

Nanotechnology, a pioneering scientific field, is at the forefront of manipulating and engineering materials and devices at the nanoscale, a dimension where structures with unprecedented properties and functions emerge. This profound manipulation of matter has unleashed a transformative era of innovation in the realm of healthcare, a domain previously unimagined in its possibilities (Erdonmez et al., 2014; Meda et al., 2022; Rastogi et al., 2022). One of the most striking and celebrated applications of nanotechnology in medicine is its role in drug delivery. At the heart of this innovation lies nanoparticles, minuscule in size but monumental in their impact. Their tiny dimensions, combined with an expansive surface area, have bestowed upon them invaluable characteristics in the world of medical science. These characteristics have revolutionized the way drugs are administered and targeted within the human body, heralding a new era of precision medicine. Nanoparticles serve as remarkable vehicles for drug delivery, ensuring the precise loading and subsequent targeted release of therapeutic compounds to specific disease sites within the body. This level of precision is a paradigm shift in healthcare (Chaudhary et al., 2023; Saini et al., 2023). Traditional drug delivery methods often result in the dispersion of medications throughout the body, causing collateral damage to healthy tissues and, consequently, a range of undesirable side effects. In stark contrast, nanotechnology's drug delivery approach, facilitated by nanoparticles, minimizes these unintended consequences. The implications of this innovation are far-reaching. In the field of cancer treatment, for instance, the use of nanoparticles in drug delivery allows for the precise targeting of chemotherapy agents to cancerous cells while sparing healthy tissues. The result is not just more effective treatment overall but also a notable decrease in the crippling side effects of chemotherapy that are often experienced, like nausea and loss of hair.

Beyond oncology, nanotechnology's precision drug delivery holds immense potential for managing chronic diseases, from diabetes to heart conditions. The ability to precisely administer medications, regulating their release over time, allows for more stable and effective therapies. This, in turn, enhances patient well-being and quality of life. In addition to its groundbreaking applications in precision drug delivery, nanotechnology has also catalysed transformative changes in the realms of diagnostic tools and regenerative medicine, further enhancing the scope of its impact on healthcare (Akbari-Alavijeh et al., 2020; Mohammadi et al., 2018; Zhao et al., 2023). Concurrently, the chapter emphasizes the multifaceted potential of nanocoating in medical applications. The influence of nanotechnology on diagnostics and medical imaging is particularly noteworthy. The development of nanoscale contrast agents and sensors has elevated healthcare professionals' ability to gain unparalleled insights into the human body. These cutting-edge tools empower early disease detection, continuous health monitoring, and more accurate diagnoses, ushering in a paradigm shift in how diseases are identified and managed. For instance, the usage of nanoscale contrast agents in magnetic resonance imaging (MRI) provides enhanced clarity in identifying subtle abnormalities, significantly improving the diagnostic process (Keshtmand et al., 2023).

Furthermore, the field of regenerative medicine has benefited immensely from nanotechnology's advancements. Scaffold materials meticulously engineered at the nanoscale offer an ideal environment for tissue growth and regeneration. This progress holds tremendous promise in various domains, such as organ transplantation, tissue engineering, and personalized medical solutions. The capacity to create precise and supportive microenvironments for cell growth and tissue development has opened new horizons for addressing organ shortages and offering tailored regenerative solutions. Nonetheless, the focal point of this discussion is the versatile scope of nanocoating within medicinal applications. These specialized coatings, crafted with precision, can be adapted to serve a myriad of specific medical purposes (Corbett et al., 2000; Pratim Das et al., 2021; H. Singh & Kaur, 2023). They encompass controlled drug release, wound healing enhancement, antimicrobial protection, and the development of scaffolds for tissue engineering. Each of these applications profoundly influences healthcare practices.

For instance, nanocoating tailored for controlled drug release ensure that medications are administered at a controlled rate over an extended period. This not only optimizes therapeutic outcomes but also mitigates unwanted side effects, thereby enhancing patient well-being. In the context of wound healing, nanocoating expedite the natural healing process while simultaneously preventing infections, leading to improved patient outcomes and shorter recovery times. Figure 1 provides an illustrative depiction of virus inactivation based on the insights gained from studying the impact of copper nanoparticles and copper alloys on the structural integrity of viral capsids and viral RNA when exposed to them.

Phan et al. (2022) evaluated the bioactivity of cobalt chromium (CoCr) L605 alloy specimens and stents meant for cardiac stent deployments when trimethylsilane (TMS) plasma nanocoating were altered using NH_3/O_2 (2:1 molar proportion) plasma post-treatment. A number of factors were taken into consideration when evaluating biocompatibility, such as covering adherence, corrosion behaviour, ion discharge, toxicity, and cell growth. Furthermore, hydrophilicity and surface chemistry were examined to understand how surface characteristics affect biocompatibility. Using a water contact angle of 48.5° and a surface characteristic of oxygen (39.39 at. percent), silicon (31.92 at. percent), carbon (24.12 at. percent), and nitrogen, the results showed that NH_3/O_2 post-treated TMS plasma nanocoating display hydrophilic features (2.77 at. percent). The tape confirms that these plasma nanocoating exhibit good adherence to the alloy substrate and efficiently adapt to the surface morphology of the surface.Moreover, the L605 stent coatings show no signs of flaking or breaking following dilatation. With no indications of pitting corrosion or mineral sorption, these nanocoating improve the resistance to corrosion of the CoCr L605 alloy by increasing its corrosion capability and decreasing its corrosion rates. Ion discharge studies showed that, in comparison to unprotected L605 control specimens, the amounts of cobalt, chromium, and nickel ions were lower in the plasma-nanocoated L605 specimens by 64–79 percent, 67–69 percent, and 57–72 percent, correspondingly. Furthermore, in accordance with ISO 10993-05 and 10993-12 criteria, the plasma nanocoating show no toxicity.The results of a seven-day cell culture investigation show that the plasma nanocoating interfaces show a 62 $\pm$ 7.3 percent drop in porcine coronary artery smooth muscle cell (PCASMC) density and a comparable density of porcine coronary artery endothelial cells (PCAECs) when comparing to untreated L605 interfaces. Scanning electron microscopy (SEM) pictures at high magnification were utilised to thoroughly inspect the stent interfaces for possible malfunctions such coating separation or cracking. SEM pictures for both plasma-nanocoated and untreated L605 stents are shown in Figure 2A. These pictures were taken after the stents were crimped, which involved shrinking their diameter to 0.75 mm for 30 seconds and then expanding it to 3 mm for 30 seconds.Relative to their untreated L605 counterparts, the interfaces of the plasma-nanocoated stents appeared noticeably cleaner and more uniform at low resolution, with no visible flaws. But at high resolution, it was clear that both stent surfaces had multiple slip lines, which indicated that following the crimped and stretching processes, there had been multiple displacements, both horizontally and vertically. Crucially, there were no signs of debonding or fracture in the plasma nanocoating. The post-dilated rough surface of both untreated and plasma-nanocoated L605 stents showed several deep holes and grooves caused by the dilation procedures upon closer inspection at very higher resolution. Surprisingly, the stents covered with plasma showed no signs of microcracking or delamination. A tape experiment was used to evaluate the coatings' adhesive strength while according to predetermined guidelines (ASTM D3359). Figure 2B shows SEM images of plasma-nanocoated L605 specimens following the tape test. The edges of the lattice designs showed no evidence of layer separation in the vicinity of the cuts; instead, they were entirely smooth. Chips created by deformations gathered on both sides of the channels as leftover materials from the plasma nanocoating procedure. The greatest rating, 5B, was given to the plasma nanocoating's adherence to its substrates, signifying excellent adhesion without coating ablation. This evaluation was conducted at two additional locations on the same sample to ensure consistency and reliability.

Furthermore, antimicrobial nanocoating play a pivotal role in safeguarding healthcare settings against the persistent challenge of hospital-acquired infections. Moreover, nanocoating engineered for tissue engineering scaffolds herald a new era in the domain of regenerative medicine. These coatings offer the ideal microenvironment for tissue growth and regeneration, with the potential to revolutionize organ transplantation and tissue engineering practices (Chauhan et al., 2020; Khemthong et al., 2023). These nanocoating, designed with precision and tailored to specific medical purposes, serve as the linchpin in realizing the transformative potential of these groundbreaking advancements in healthcare. This initial exploration sets the stage for a more detailed examination of the applications and future perspectives of nanocoating in the subsequent chapters, ultimately reshaping the landscape of healthcare.

The objectives of the chapter are to provide a clear roadmap for what the reader can expect to learn and gain from the forthcoming sections. These objectives give the reader a sense of direction and purpose, ensuring that they can focus on the aspects of nanocoating in medicine that are most relevant to their interests. The objectives of chapter may include delving into the fundamental principles of nanocoating, discussing the latest advancements and case studies to exemplify their real-world applications, exploring the regulatory considerations necessary for the safe integration of nanocoating in clinical practice, and providing insights into the future prospects and challenges that lie ahead in this evolving field.

NANOCOATING FOR DRUG DELIVERY

The application of nanocoating in drug delivery represents a pivotal frontier in the ever-evolving field of nanomedicine. These specialized coatings, typically presented in the form of nanoparticles or nanofilms, are at the forefront of a transformative approach to drug administration. They offer an unprecedented level of precision and control over the release of therapeutic agents, paving the way for highly targeted and efficient drug delivery (Kaya, Lgaz, et al., 2023; Kaya, Thakur, et al., 2023; Kumar & Thakur, 2020; D. Sharma et al., 2023; Thakur et al., 2021; Thakur, Kumar, Kaya, Marzouki, Zhang, et al., 2022; Thakur, Kumar, Kaya, Vo, & Sharma, 2022; Thakur, Savaş, Kumar, et al., 2023; Thakur & Kumar, 2023a). At the core of nanocoating-enabled drug delivery is the ability to achieve precise control over the release of therapeutic compounds (Christy et al., 2020; Negut et al., 2018). Nanoparticles, with their minuscule size and high surface area, are harnessed to encapsulate and transport drugs. The nanocoating on these particles are engineered to regulate the kinetics of drug release. This precise control allows for the gradual and sustained release of medications over an extended period, in contrast to conventional drug delivery methods, which often lead to rapid, uncontrolled release.The advantage of controlled drug release cannot be overstated. It provides a means to maintain therapeutic drug levels within the body, optimizing treatment efficacy while minimizing side effects.

Patients benefit from a consistent and stable drug supply, resulting in improved symptom management and overall well-being. This controlled release is particularly valuable in the management of chronic diseases, where sustained drug levels are crucial for effective treatment (Bai et al., 2022; Yu et al., 2021).Nanocoating enable the realization of targeted drug delivery, a concept that significantly enhances the precision of treatment. By functionalizing the surface of nanoparticles with specific ligands or antibodies, drug-loaded nanoparticles can be guided to precise disease sites within the body. This level of precision minimizes drug exposure to healthy tissues, reducing the incidence of side effects and collateral damage.

For instance, in the context of cancer therapy, drug-loaded nanoparticles coated with targeting molecules can be directed to cancer cells, sparing healthy cells. This approach is a game-changer in oncology, where the indiscriminate effects of chemotherapy often lead to debilitating side effects. Moreover, the elevated permeability and retention (EPR) effect, characteristic of several tumours, promotes the aggregation of nanoparticles in cancerous tissues, significantly enhancing the effectiveness of cancer treatments. While the utilization of nanocoating in drug delivery offers immense promise, it is not without its challenges. Ensuring the stability and scalability of nanocoating techniques is a pressing concern (Malshe et al., 2010; Qi et al., 2022). The production of nanoparticles with consistent coatings in large quantities is essential for widespread medical use. Additionally, the potential toxicity of nanoparticles must be thoroughly evaluated to ensure patient safety. Opportunities within this field are vast. As nanotechnology continues to advance, opportunities for more precise and patient-specific drug delivery are on the horizon. Personalized medicine, where treatments are tailored to an individual's unique biology, is becoming increasingly feasible with the aid of nanocoating. Additionally, the development of multifunctional nanoparticles that combine drug delivery with diagnostic capabilities is a burgeoning area of research, promising earlier disease detection and treatment assessment.

Controlled Release Systems

The advent of controlled-release drug delivery systems, with nanocoating at their core, marks a ground-breaking achievement in contemporary medicine. These sophisticated systems offer the ability to precisely regulate the gradual and sustained release of medications over extended periods, ushering in a new era of therapy optimization. At the heart of this innovation lies the remarkable capability of nanocoating to govern the kinetics of drug release. This advancement is particularly significant in the management of chronic diseases, where maintaining consistent therapeutic drug levels within the body is essential for effective treatment. For instance, in the context of diabetes, nanocoating have played a pivotal role in crafting insulin-releasing nanoparticles (Chen et al., 2022; Sun et al., 2021). These nanoparticles ensure that insulin is released in a controlled and prolonged manner, closely mirroring the body's natural insulin production. As a result, patients experience stable blood sugar levels over extended durations, a development that not only enhances the efficacy of insulin therapy but also significantly improves patient comfort and adherence to the treatment regimen.

Controlled-release drug delivery systems are engineered with nanocoating that enable precise control over drug release kinetics. These coatings serve as a protective barrier, encapsulating therapeutic agents and allowing for modulation of the release by adjusting nanocoating properties. This capability translates into predictable, controlled, and steady drug release, in contrast to conventional drug delivery methods that often yield erratic kinetics. The primary advantage of controlled-release drug delivery systems is their ability to optimize therapy while minimizing side effects. By gradually releasing the drug over extended periods, these systems ensure that therapeutic drug levels remain within the body's desired range (Khatoon et al., 2018; Merugu et al., 2023; Trivedi et al., 2021). This steadiness guarantees that the drug effectively manages the intended condition, be it regulating blood pressure, alleviating pain, or controlling blood sugar levels. Consistent drug presence in the body mitigates fluctuations that can lead to unwelcome side effects or periods of insufficient therapeutic coverage. The value of controlled-release drug delivery systems is particularly pronounced in the diagnosis of chronic diseases, where sustained and consistent drug levels are imperative. A notable illustration is diabetes management. In this context, nanocoating have evaluated a pivotal role in the emergence of insulin-releasing nanoparticles, which ensure insulin is released in a controlled and prolonged manner, closely resembling the body's natural insulin production (Adebayo et al., 2023). This innovation has profound implications for diabetes management, offering patients stable blood sugar levels over extended periods (Parveen et al., 2020; Sharma et al., 2023; Thakur, Kaya, Abousalem, & Kumar, 2022; Thakur, Kumar, Sharma, Ganjoo, & Assad, 2022; Thakur, Sharma, Ganjoo, Assad, & Kumar, 2022; Thakur & Kumar, 2022; Verma et al., 2023). Consequently, patients benefit from the convenience of less frequent injections and an improved ability to control their condition, ultimately enhancing their quality of life and therapeutic outcomes.

Targeted Drug Delivery

Nanocoating have introduced a transformative dimension to the concept of targeted drug delivery. Their remarkable capacity to functionalize the surface of nanoparticles has opened up new frontiers in medication administration, where specific ligands or antibodies can be attached to guide drug-loaded nanoparticles to precise disease sites within the body. This precision not only revolutionizes drug delivery but also minimizes drug exposure to healthy tissues, thereby reducing undesirable side effects. An exemplar of this breakthrough is observed in the realm of cancer therapy, where drug-loaded nanoparticles, coated with these targeting molecules, can be accurately directed to cancer cells while sparing their healthy counterparts (Wang et al., 2023; Xu, Ma, et al., 2023). This novel approach results in more efficient treatment regimens with significantly fewer systemic side effects. Moreover, the phenomenon of enhanced permeability and retention (EPR), inherent to many tumour tissues, further bolsters the effectiveness of cancer treatments by facilitating the selective accumulation of nanoparticles in cancerous regions.

The potential for precision in drug delivery, unleashed by nanocoating, is a game-changer in the field of medicine. These coatings, applied to the surface of nanoparticles, provide the foundation for attaching specific ligands or antibodies that can recognize and bind to particular cell receptors or molecular targets. This means that when drug-loaded nanoparticles are introduced into the body, they can be guided with remarkable accuracy to disease sites. In the context of cancer, for example, targeting molecules attached to the nanoparticles can interact with the unique markers on cancer cells, ensuring that the therapeutic payload is selectively delivered to malignant tissues. This precision substantially reduces the risk of collateral damage to healthy cells, one of the primary challenges of conventional chemotherapy. Reducing side effects is a paramount objective in drug delivery, and nanocoating play a central role in achieving this goal. The targeted drug delivery facilitated by nanocoating ensures that medications are directed specifically to the affected areas, bypassing healthy tissues that would otherwise suffer the consequences of systemic exposure (Prakash et al., 2018; Yaashikaa et al., 2023). This precise delivery strategy translates into fewer adverse effects, improved patient comfort, and enhanced treatment adherence. This means that cancer patients will have fewer of the crippling side effects of chemotherapy, like weariness, nausea, and loss of hair. The reduction in side effects contributes not only to the physical well-being of patients but also to their overall quality of life during treatment.

In the fight against cancer, the EPR effect further amplifies the effectiveness of treatments involving drug-loaded nanoparticles. Many tumor tissues exhibit leaky blood vessels and poor lymphatic drainage, which results in the accumulation of nanoparticles within these tissues. This phenomenon allows for a more prolonged and concentrated exposure of cancer cells to therapeutic agents. The enhanced drug concentration at the tumor site maximizes the effectiveness of treatments, ultimately leading to improved outcomes and the potential for reduced treatment durations. For instance, in an experiment, Otto et al. (2022) employed a layer-by-layer (LbL) nanocoating analysis to enhance the stability of emulsions containing griseofulvin. These emulsions were further processed into lyophilized forms and reconstituted emulsions, with each stage of this process incorporating a layer of whey protein and one of the following polysaccharides: amylopectin, hyaluronic acid, or alginic acid, for the encapsulation of the drug. Importantly, the materials used for the coating are biologically friendly, environmentally sustainable, and readily available. Photostability assessments revealed that the LbL nanocoating offered six hours of protection against light-induced degradation after topical application. Investigations on the uptake of griseofulvin in vitro revealed that it mostly collected in the stratum corneum, having very little transdermal distribution. For the therapy of superficial fungal diseases, this suggests that lyophilized particulates, LbL-nanocoated emulsions, and regenerated lyophilized emulsions might be a suitable topical drug delivery strategy.

Arriagada et al. (2020) focused on the creation and characterization of a hybrid nano antioxidant carrier based on a porous silica nanoplatform with rosmarinic acid antioxidants immobilized on its surface. These nanocarriers were characterised and their antioxidant interaction assessed using a range of methods, such as zeta potential, dynamic light scattering (DLS), differential scanning calorimetry (DSC), Fourier transform-infrared spectroscopy (FT-IR), N_2 adsorption-desorption measurements, and a 2,2-diphenyl-1-picrylhydrazyl (DPPH•) assay. Two biorelevant media were also used to evaluate the drug distribution characteristics. Their focus was on the creation and characterisation of nano-RA, an antioxidant substance made of rosmarinic acid and a mesoporous silica nanostructure. Finding the controlled discharge controlling the bioactive component distribution from this model drug delivery mechanism was the main goal. The nanosystem showed promise in protecting oxidation-sensitive compounds. They effectively bonded rosmarinic acid (RA), an ester generated from hydroxycinnamic acid, onto amino-functionalized mesoporous silica nanoparticles by using methods similar to those described in their earlier work. The amide bond that formed among the amino group on the exterior of the nanoparticles and the carboxylic acid of RA allowed for this binding to occur (refer to Figure 3a). According to earlier research on the reaction duration needed to successfully transplant a polyphenol utilising EDC/NHS coupling in polymeric matrix, six hours of reaction time was determined to be adequate for RA to adhere to the surface. With an effectiveness of 88 percent, this adhesion technique loaded 216.2 mg of RA per gramme of nanoparticles. They used the impregnation/solvent evaporation procedure to place MH (probably a medicine or bioactive molecule) into the nanoparticles after the grafting procedure, using bare nanoparticles (MSN) acting as a reference (refer to Figure 3b). It was found that the packing effectiveness for MSN/MH was 9.4% w/w and for nano-RA/MH it was 23.1 percent w/w. Given that MH was incorporated into the MSN's porosity and only a little amount is adsorbed on the surface, these figures were anticipated. When MH was present in nano-RA, it not only entered the pores but also reacted with the functional groups of RA, increasing the quantity of MH adsorbed in comparison to pristine MSN.

The findings suggested that rosmarinic acid was appropriately disposed of as the antioxidant capacity of the substance was maintained. The results of kinetic tests showed that rosmarinic acid was produced more in simulated intestinal fluid than in gastric fluid, and that anomalous non-Fickian release was responsible for this emission. These results demonstrate the antioxidant nanocarrier's potential for usage in medication delivery applications in the future.

NANOCOATING IN WOUND HEALING

Nanocoating have emerged as a transformative technology in the field of wound healing, ushering in a new era of healthcare innovation. These coatings offer a spectrum of benefits that significantly enhance the natural healing processes, provide potent antimicrobial properties, and have yielded promising results in the development of advanced wound dressings (Maťátková et al., 2022; Özen et al., 2023; Yazdani et al., 2018). The implications of this breakthrough technology are profound, revolutionizing wound care and ultimately improving patient outcomes. Nanocoating have a profound impact on the natural healing processes of the human body. Wound healing is a complex biological phenomenon involving various stages, and nanocoating play a pivotal role in optimizing these processes. For instance, nanocoated wound dressings can create a microenvironment that promotes cell proliferation, tissue regeneration, and angiogenesis. They assist in maintaining an ideal moisture balance, minimizing scarring, and expediting the overall healing process. These coatings effectively address the challenges of chronic wounds, non-healing ulcers, and post-surgical incisions by providing the necessary conditions for accelerated and efficient healing (Hossain et al., 2022; Islam et al., 2019).

Another critical dimension of nanocoating in wound healing is their potent antimicrobial properties. Infections are a significant concern in wound management, and nanocoating serve as a robust defense against microbial invaders. Nanoparticles can be loaded with antimicrobial agents such as silver, iodine, or antibiotics and then applied to wound dressings. These agents are released gradually, creating a hostile environment for pathogens, which significantly reduces the risk of infection. Antimicrobial nanocoating are instrumental in preventing complications, particularly in the case of chronic wounds, where infection can be a severe hindrance to healing (Hasan et al., 2023; Pathan et al., 2020). The ability to control infection risk through nanocoating not only improves patient outcomes but also reduces healthcare costs associated with prolonged treatments and hospital stays. Nanocoating have catalysed the development of advanced wound dressings. These dressings are far more than passive coverings; they are dynamic and actively contribute to the healing process. Through nanocoating technology, wound dressings can be imbued with properties such as controlled drug release, growth factor delivery, and exudate management. For instance, nanocoated dressings can deliver growth factors to stimulate tissue regeneration, control inflammation, or alleviate pain. They can also efficiently manage wound exudate by absorbing excessive fluids, maintaining a moist wound environment, and promoting the removal of dead tissue, all of which are vital for proper healing. These advanced dressings represent a significant leap forward in wound care and provide healthcare professionals with a diverse toolbox for tailored wound management.

Enhanced Healing Processes

Nanocoating have emerged as a pivotal catalyst in the field of wound healing, significantly expediting the natural processes that the human body undertakes to repair and close wounds. These coatings offer a range of mechanisms through which this enhancement is achieved, providing a profound impact on the speed and effectiveness of wound healing. Two key avenues through which nanocoating drive these improvements are the creation of an optimal wound microenvironment and the incorporation of growth factors and bioactive molecules to stimulate angiogenesis and collagen synthesis (Montes et al., 2019; Yang et al., 2022).One of the fundamental ways in which nanocoating accelerate wound healing is by establishing an optimal microenvironment at the wound site. Wounds require specific conditions to heal efficiently, including controlled moisture levels and protection from external contaminants. Nanocoated wound dressings play a pivotal role in maintaining these ideal conditions (Bashir, Lgaz, et al., 2020; Bashir, Thakur, et al., 2020; Kumar & Thakur, 2023; Thakur, Kaya, Abousalem, Sharma, et al., 2022; Thakur, Kumar, et al., 2023; Thakur, Savaş, & Kumar, 2023; Thakur & Kumar, 2021, 2023b). The nanocoating regulate the moisture content, ensuring the wound remains neither too dry nor too wet. This controlled moisture environment is essential for promoting cell adhesion, proliferation, and differentiation, all of which are integral to tissue regeneration.

In chronic wounds, such as diabetic ulcers, which are characterized by their propensity for delayed healing, nanocoated dressings can be particularly transformative. These chronic wounds often exhibit a challenging and unfavourable microenvironment. Nanocoating have the capability to counteract these unfavourable conditions by creating a conducive environment that accelerates the healing process. The result is a more efficient and effective healing trajectory for patients who might otherwise face prolonged and complicated recoveries. Nanocoating can further expedite wound healing by incorporating growth factors and bioactive molecules into the wound care regimen (Y. Singh et al., 2023; Swain et al., 2023). These factors play a critical role in stimulating the biological processes essential for wound closure. Growth factors can enhance angiogenesis, which is the formation of new blood vessels. This is particularly important because blood vessels supply the wound with the oxygen and nutrients necessary for tissue regeneration. By promoting angiogenesis, nanocoated wound dressings ensure that the wound site receives the essential resources required for rapid healing. Moreover, nanocoating can also encourage collagen synthesis. Collagen is a key component of the extracellular matrix and provides structural support to tissues. Increased collagen synthesis results in more robust and effective wound closure, which is crucial for minimizing the risk of complications such as excessive scarring. By introducing growth factors and bioactive molecules, nanocoated dressings actively contribute to the promotion of these vital biological processes, reducing healing times and optimizing the overall quality of wound repair.

Case Studies in Wound Dressings

The practical application of nanocoating in wound healing is exemplified in the development of advanced wound dressings. These dressings are equipped with nanocoating that serve multiple functions. For instance, nanocoating may act as a barrier to moisture, preventing excessive fluid loss from the wound and providing a moist environment conducive to healing. Simultaneously, the nanocoating can contain antimicrobial agents to stave off infections (Berini et al., 2022; Xu, Zhou, et al., 2023). A case study in this realm involves the use of silver nanoparticles incorporated into wound dressings. The controlled release of silver ions from the nanocoating serves as a potent antimicrobial agent, ensuring the wound remains free from infections, ultimately promoting faster healing. Furthermore, the combination of nanocoating with advanced materials, such as nanofibers or hydrogels, enhances the overall performance of wound dressings. Nanocoating enable these materials to be more effective in terms of biocompatibility, mechanical properties, and drug release kinetics. In an experiment, Radulescu et al. (2016) developed an antimicrobial coating (NanoAg) for wound dressings made of polyester-nylon material (WDs) with the objective of supressing the risk of exogenous wound-related infections. Using XRD, SEM, TEM, Selected Area Electron Diffraction (SAED), and InfraRed Microscopy (IRM), the NanoAg-coated WDs were thoroughly examined. The biological evaluation encompassed both in vivo bioavailability experiments of AgNPs (silver nanoparticles) in mouse models and in vitro experiments of the interaction to fibroblast cell cultures. The technique was submerging conventional WDs in a glucose and NaOH mixture to precipitate silver nanoparticles, then gradually adding $AgNO_3$ reagent. Utilizing strains of Pseudomonas aeruginosa and Staphylococcus aureus, qualitative and quantitative in vitro investigations were used to assess the antibacterial efficacy of NanoAg-WDs. The in vitro and in vivo studies showed that the evaluated nanoparticles utilized to coat WDs were biocompatible, permitting normal human cell development and showing a typical biodistribution in a mouse model, all without causing any harmful consequences. Increased inhibitory activity towards microbial colonisation, adhesion, and biofilm development was found by analysing the altered WDs. The results indicate that these coatings could be used to create altered wound dressings (WDs) that have antibacterial qualities. This could be used as a prophylactic against exogenous microbial contamination of wound tissue, colonisation, and the consequent formation of biofilms.

Paladini et al. (2016) crafted a novel antibacterial wound dressing using a silver deposition technique relied on the photochemical production of silver nanoparticles. The end products are all-natural devices with outstanding adherence of the silver coatings that don't require binders. The quantity and dispersion of the silver particles on the cotton fibres were evaluated by SEM and thermogravimetric analysis (TGA) on the cotton gauze that had been treated with silver. Through a range of experiments, comprising agar migration, bacterial counting, biofilm quantification, fluorescence microscopy, and SEM microscopy, the antibacterial efficacy of the silver-treated gauzes was verified. Furthermore, the 3-[4,5-dimethylthiazol-2-yl]-2,5-diphenyltetrazolium bromide colorimetric assay (MTT) and the extract method on fibroblasts and keratinocytes were used to assess the possible cytotoxicity of the silver coating. The emission of silver in various mediums was measured using inductively coupled plasma mass spectrometry (ICP-MS), and the findings were correlated with the biological characterisation. Every result was contrasted with gauze prepared with a greater proportion of silver as a positive control and plain gauze as a negative control. Furthermore, Aldoasri et al. (2017) investigated the application of TiO_2 nanoparticles dispersed in an aqueous colloidal suspension directly onto historic marble stone surfaces using a spray-coating technique, creating a nanometric film. In order to decide if TiO_2 nanoparticles are suitable for use on historic stone surfaces, the research first evaluates their qualities. They used EDX, SEM to investigate surface characteristics, coating uniformity, and alterations during artificial ageing. The efficiency of the coated surface was assessed in the study using photo-induced impacts and UV light exposure. Molecular structural alterations in treated specimens were examined using attenuated total reflection infrared spectroscopy (ATR-FTIR). Stereo microscopy was used to analyse the hydrophobic qualities of the protective film, and colorimetric tests and capillary water absorption were used to gauge how effective the treatments are in terms of appearances. According to the findings, TiO_2 nanoparticles have potential for use on historic stone surfaces since they improve durability, offer self-cleaning and protection.

Feng et al. (2019) focused on preparing catalase (CAT)/ Zeolitic imidazole framework-8 (ZIF-8) composites with a protective nanocoating, achieved through controlled self-assembly using silanes or coordination complexes, such as tannic acid (TA) and Fe^{3+}. Extensive research was conducted on the characteristics and acid-resistance of CAT/ZIF-8 composites coated with a protective nanocoating. CAT/ZIF-8 and CAT/ZIF-8 with a protective nanocoating had recovered activities of 70% and 65%, correspondingly. The acid resistance of CAT/ZIF-8 with protective nanocoating was shown to be superior to that of normal CAT/ZIF-8 composites. For example, only 20% of the initial activity (or about 12 U/mg) of the CAT/ZIF-8 composites was preserved after 60 minutes of treatment in a pH 3.0 phosphate buffer solution. But CAT/ZIF-8 with a protective nanocoating was still able to function at roughly half its initial level (10 U/mg). Furthermore, in comparison to traditional CAT/ZIF-8, the thermal stability and storage stability of CAT/ZIF-8 composites were greatly enhanced by the nanocoating. Significantly, after 7 cycles, CAT/ZIF-8 with protective nanocoating maintained 40% of its initial activity, but CAT/ZIF-8 alone kept 8%. This work provides an effective method for creating enzyme/ZIF-8 composites that have good recyclability and strong acid resistance.

In conclusion, the integration of nanocoating in wound healing represents a substantial leap in the field of medical materials and treatments. These coatings actively promote enhanced healing processes, provide critical antimicrobial properties, and have led to significant advancements in the development of advanced wound dressings. Their potential to transform the way we manage wounds, particularly chronic wounds and infections, holds great promise for improving patient outcomes and reducing healthcare costs.

NANOCOATING FOR ANTIMICROBIAL APPLICATIONS

Nanocoating have emerged as a potent and innovative weapon in the ongoing battle against infections. Designed with antimicrobial properties, these coatings serve as a critical tool to reduce the risk of infections in various settings. Nanocoating engineered with an array of antimicrobial mechanisms makes them highly effective in combating infections. These mechanisms can include the release of antimicrobial agents, disruption of pathogen cell membranes, interference with microbial growth and replication, or the prevention of biofilm formation. These multifaceted approaches make nanocoating adaptable to a wide range of applications and infection scenarios (Blum & Molian, 2009; Du et al., 2022; Wani et al., 2023). In healthcare settings, where infections can have severe consequences for vulnerable patients, nanocoating play a crucial role. They can be applied to various surfaces, such as medical equipment, bed linens, and even healthcare worker uniforms, effectively reducing the risk of contamination. By actively preventing the growth and spread of pathogens, nanocoating contribute to a safer healthcare environment and help mitigate the risk of hospital-acquired infections.

Beyond healthcare, nanocoating find utility in public health and hygiene applications. They can be integrated into air filtration systems, water purification systems, and surface coatings, making public spaces safer by reducing the concentration of harmful microorganisms (Pal et al., 2021; Ray et al., 2016). This is especially relevant in densely populated areas, where the risk of infectious disease transmission is heightened. Nanocoating also play a significant role in enhancing food safety by reducing bacterial contamination on food contact surfaces and packaging materials. The impact of nanocoating extends to everyday life through various consumer products. Antimicrobial coatings can be found in textiles, personal protective equipment, and home appliances. For example, clothing with nanocoating can resist bacterial and fungal growth, reducing odors and extending the lifespan of garments. Antimicrobial coatings in household items, such as cutting boards and kitchen appliances, contribute to a cleaner and safer living environment. Nanocoating also support environmental stewardship by reducing the need for harsh chemical disinfectants and antibiotics. By controlling microbial growth in a targeted and efficient manner, these coatings promote sustainable infection control practices.

Combating Infections

Infections caused by bacteria, viruses, and fungi remain a substantial public health concern in both healthcare settings and everyday environments. Antimicrobial nanocoating have emerged as a versatile and multi-faceted approach to combat these infections. These coatings actively inhibit the growth and proliferation of microorganisms, rendering them invaluable in environments where infection control is paramount. From healthcare facilities to personal protective equipment (PPE) and textiles, antimicrobial nanocoating play a pivotal role in reducing the risk of infections and safeguarding public health. Infections caused by several pathogens, such as viruses, bacteria and fungi, are a persistent challenge in healthcare settings(Garg et al., 2022; Govardhana Reddy et al., 2021; Kolahalam et al., 2019). Healthcare-associated infections can also have serious repercussions, including higher mortality rates as well as higher healthcare expenses. In everyday environments, infections can also pose a threat, impacting personal well-being and public health. Antimicrobial nanocoating provide a multi-faceted approach to combating infections. These coatings are meticulously engineered to actively inhibit the growth and proliferation of microorganisms. They accomplish this through various mechanisms, such as disrupting microbial cell membranes, interfering with metabolic processes, or preventing the formation of biofilms. This versatility allows antimicrobial nanocoating to address a wide range of pathogens, making them adaptable to different infection scenarios.

In healthcare facilities, where the risk of infections is particularly high, antimicrobial nanocoating are a critical tool. These coatings can be applied to high-touch surfaces, such as doorknobs, bed rails, medical equipment, and countertops. By actively inhibiting the growth of bacteria and viruses on these surfaces, nanocoating help to reduce the risk of hospital-acquired infections. The implications for patient safety and healthcare efficiency are profound. The use of nanocoating in healthcare settings not only protects vulnerable patients but also minimizes the strain on healthcare resources. Antimicrobial nanocoating extend their impact to personal protective equipment (PPE) and textiles. In the era of contagious diseases and pandemics, this application is especially relevant. Face masks and clothing can be treated with nanocoating to reduce the transmission of pathogens. Antimicrobial coatings on PPE materials not only protect individuals from external pathogens but also prevent the potential spread of pathogens from the wearer to others. This dual role enhances public safety in situations where infection control is vital. Antimicrobial nanocoating offer not only individual protection but also broader environmental and public health benefits. By reducing the reliance on chemical disinfectants and antibiotics, these coatings support environmental sustainability. They also contribute to overall public health by minimizing the spread of infections and decreasing the burden on healthcare systems. In an experiment, Magovac et al. (2021) focused on the development of an environmentally friendly layer-by-layer (LbL) deposition technique to create flame-retardant and antimicrobial cotton. The process involves applying 8, 10, and 12 layers of phytic acid (PA) and chitosan (CH)-urea to cotton, followed by immersion in a copper (II) sulfate ($CuSO_4$) solution. The study found that using 12 layers of PA/CH-urea + Cu^{2+} effectively prevented the cotton from catching fire during vertical flammability testing (VFT), achieving a limiting oxygen index (LOI) value of 26%. Further analysis using a microscale combustion calorimeter (MCC) revealed a more than 61% reduction in peak heat release rates (pHRR) and over 54% reduction in total heat release (THR) compared to untreated cotton. Water, methane, carbon dioxide, carbon monoxide, and aldehydes were released from the 12-layer-treated cotton, according to thermal gravimetric-infrared (TG-IR) study. Methane generation decreased with the introduction of Cu^{2+} ions. Furthermore, there was no levoglucosan present in the treated cotton. The inflated morphology of the burned residue following burning served as evidence of the treatment's intumescent properties. Additionally, the research evaluated the antibacterial qualities and found that Staphylococcus aureus and Klebsiella pneumoniae were completely decreased. This research successfully functionalized cotton with an eco-friendly, multi-functional nanocoating that serves as both a flame retardant and an antibacterial agent, achieved through LbL deposition.

Allehyani et al. (2022) uniquely engineered polyester fabrics to incorporate hydrazide groups, which were utilized as ligands for trapping silver ions and subsequently reducing them to silver nanoparticles (AgNPs). The incorporation of polydopamine (PDA) into this polyester system effectively imparted antibacterial and antioxidant attributes. PDA's phenolic hydroxyl groups facilitated the conversion of silver ions into AgNPs. Furthermore, AgNPs-loaded polyester hydrazide was created using glucose as a reducing agent. Analysis of the materials' characteristics included tensile strength testing, thermogravimetric analysis (TGA), SEM, EDX, ATR-FTIR, and SEM. The antibacterial activity of the cloth against *Staphylococcus aureus* and *Escherichia coli* was evaluated in a broth test. The outcomes showed that the PDA-AgNP-infused composite nanocoating has outstanding tensile strength as well as antibacterial and antioxidant characteristics. These results imply that polyester textiles coated with PDA-AgNPs are suitable for possible biomedical textile applications. Nyga et al. (2021) covalently immobilized organic photosensitizers (azure A and 5-(4-aminophenyl)-10,15,20-(triphenyl)porphyrin) onto glass surfaces to create a photoactive monolayer. Three main phases were engaged in the process: two stages of chemical alteration after the preliminary chemical bonding of 3-aminopropyltriethoxysilane (APTES). The resultant mixed monolayer (MIX TC APTES-glass) was characterised using XPS and Raman spectroscopies, and its photoactive characteristics were evaluated using UV-Vis spectroscopy with α-terpinene acting as a chemical trap. The photosensitizers were discovered to have high stability as they continued to produce reactive oxygen species (ROS) when exposed to light even after becoming immobile on the glass surface. Under broad-spectrum light, 5-(4-aminophenyl)-10,15,20-(triphenyl)porphyrin combined with azure A enhanced ROS photogeneration. When compared to untreated glass surfaces, this light-activated nanocoating demonstrated encouraging antibacterial action against *Escherichia coli*, lowering the amount of attached germs.

Strategies for Antimicrobial Nanocoating

Antimicrobial nanocoating are engineered with various strategies to deter the growth of microorganisms, providing an effective defense against infection. These strategies encompass the use of nanoparticles, antimicrobial polymers, and controlled release mechanisms, each with distinct advantages and versatile applications. Let's delve into these strategies in detail.One common approach employed by antimicrobial nanocoating is the use of nanoparticles, such as silver, copper, or zinc oxide nanoparticles. These nanoparticles release ions that prove toxic to microbes, creating a hostile environment for their survival. The released ions interfere with the microorganisms' cellular functions, disrupt their membranes, and ultimately prevent their reproduction (El-Ashram et al., 2021; Hao et al., 2024; Li et al., 2023). Silver, in particular, has long been recognized for its potent antimicrobial properties. By incorporating silver nanoparticles into coatings, various materials can be rendered antimicrobial, effectively reducing the risk of infection. The key advantage of nanoparticles is their small size, which allows them to penetrate microbial cells and execute their antimicrobial action within the cells. This approach is versatile and has found applications in various sectors, including medical devices, textiles, and everyday items, making them resistant to microbial contamination (Dastkhoon et al., 2020; Saemian et al., 2019).

Another strategy involves the integration of antimicrobial polymers into nanocoating. These polymers contain molecules with the ability to actively target and disrupt microbial cell membranes, leading to the death of the microorganisms. The versatility of antimicrobial polymers allows them to be blended into nanocoating and applied to a wide range of surfaces and materials, expanding their utility in diverse settings. Antimicrobial polymers offer the advantage of specificity, targeting microbial cells without affecting host cells. They are particularly valuable in scenarios where selective microbial elimination is necessary, such as in medical equipment or personal protective gear, where preventing infection is paramount. The controlled release of antimicrobial agents is a critical feature in antimicrobial nanocoating. These coatings can provide a sustained release of antimicrobial compounds over time, ensuring long-lasting protection against pathogens. This approach is particularly valuable in medical implants, where post-surgical infections are a significant concern (Malekmohammadi et al., 2022; P. Sharma et al., 2019). By gradually releasing antimicrobial agents, these nanocoating can reduce the risk of infections that may occur after surgical procedures.

Controlled release mechanisms also find application in textiles, where sustained protection against microbial contamination is desirable, especially in clothing worn in healthcare settings or public spaces. In summary, antimicrobial nanocoating employ diverse strategies to deter the growth of microorganisms. The use of nanoparticles, antimicrobial polymers, and controlled release mechanisms allows for versatile and effective protection against infections. These strategies have broad applications, from medical devices and implants to textiles and everyday items, ultimately contributing to a safer and healthier environment.

NANOCOATING IN TISSUE ENGINEERING

Nanocoating have emerged as integral players in the field of tissue engineering, where their primary role is to support tissue growth and regeneration. These coatings facilitate the creation of scaffolds and materials that provide a conducive environment for cells to thrive and tissues to regenerate. This pivotal role in scaffold design enhances biocompatibility and drives forward the applications of regenerative medicine. Tissue engineering stands at the forefront of regenerative medicine, offering innovative solutions to address tissue loss, damage, and degeneration (Ghosh et al., 2021). It seeks to bridge the gap between the limited supply of donor organs and the ever-growing demand for organ transplantation. Moreover, it holds promise for the development of personalized medical solutions, where tissues can be engineered to meet the unique needs of individual patients. At the core of tissue engineering lies the scaffold, a three-dimensional framework that provides mechanical support and a nurturing environment for cells. The scaffold's composition and properties significantly influence cell behaviour, tissue formation, and ultimately, the success of regenerative therapies. Here is where nanocoating make their entrance.

Nanocoating enhance scaffold design in several ways. They provide fine-tuned control over the surface properties of the scaffold, making it more compatible with the surrounding biological environment. Nanoscale modifications, such as surface roughness and functionalization, promote cell adhesion, proliferation, and differentiation, accelerating tissue regeneration. Biocompatibility is a critical factor in tissue engineering. Nanocoating offer the opportunity to tailor the scaffold's surface properties to encourage favourable interactions with the body's cells. By mimicking the extracellular matrix's nanoscale features, nanocoated scaffolds create a microenvironment that fosters cellular growth and tissue formation, ultimately leading to improved biocompatibility. Nanocoating are particularly beneficial in regenerative medicine applications. They are applied to scaffold materials used in the regeneration of various tissues, including bone, cartilage, skin, nerve, and even complex organs like the heart and liver. By providing the right cues to the cells, nanocoated scaffolds expedite the regenerative process. For example, a nanocoated bone scaffold can promote osteoblast adhesion and calcium mineralization, facilitating bone tissue regrowth in patients with fractures or bone defects. In an experiment, Huang et al. (2023) focused on the development of porous scaffolds utilizing poly(lactic acid) (PLA) as the primary substance, with customizable pore structures achieved through 3D printing technology. A thorough investigation was conducted into how the pore structure affected the mechanical characteristics of the PLA scaffold that was 3D printed. Furthermore, by applying a polydopamine (PDA) covering to the well-designed PLA scaffolds, they improved their hydrophilicity and bioactivity. The main goal of this work was to create a custom biomimetic layout for the internal architecture of patient-specific bone lesion locations. In order to do this, they started working on creating a 3D porous bone scaffold model, which provides the perfect setting for cell adherence and transmission inside the broken bone.They accomplished this by employing poly(lactic acid) (PLA) as the polymer matrix and utilizing advanced 3D printing technology for precise fabrication. Furthermore, we sought to augment the bioactivity of these bone scaffolds and explored various surface enhancements. Theychosen strategy involved the application of a polydopamine (PDA) coating to the PLA scaffolds (as shown in Figure 4a). To comprehensively assess the PDA/PLA porous scaffolds, they conducted detailed examinations of their chemical composition, morphology, microstructure, and wettability. These analytical investigations allowed them to gain valuable insights into the material's properties. Additionally, we evaluated the biocompatibility of these materials through in vitro degradation and in vitro mineralization studies. This helped them understand how well the scaffolds would perform within a biological environment. The surface roughness of PDA/PLA scaffolds was shown in SEM pictures. In addition, the PDA/PLA scaffolds demonstrated modifiable photothermal responsiveness when exposed to near-infrared (NIR) light. These PLA/PDA scaffolds that were 3D printed have a lot of promise as a substitute material for fixing bone defects because they can be modified and have photothermal properties.

In Figure 5, the temperature alteration profiles of PLA/PDA scaffolds, which were subjected to varying immersion durations and irradiated with a NIR laser emitter (0.2 W/cm^2), are presented, along with the critical temperature values. Notably, the pristine PLA specimens displayed no photothermal impact, while the other PDA/PLA specimens exhibited an equivalent pattern. This pattern involved a swift temperature rise over a relatively brief period, followed by a prolonged, steady increase in temperature. Upon deactivation of the laser, all the samples swiftly cooled down when exposed to the surrounding air. The scaffold's heating rate was further boosted using a higher polydopamine covering, increasing the critical heating temperature from 46.3°C to 57.6°C. One could attribute this effective photothermal impact to the uniform dispersion of PDA on the PLA scaffold. It is noteworthy that the NIR power could be efficiently adjusted to manage the temperature of the bone scaffold. Studies have demonstrated that cancer cells could be destroyed by holding them at 42°C for 15–60 minutes. Therefore, when employing photothermal therapy, researchers should exercise caution in determining the heating duration and temperature to avoid harming normal somatic cells. To sum up, this study's PLA/PDA composite scaffold has more biological activity attributable to the PDA coating, which could have exciting implications for photothermal therapy.

Kao et al. (2022) utilizing fused deposition modelling, Ravi demonstrated the creation of a magnetic scaffold composed of calcium silicate (CS), iron oxide nanoparticles (Fe_3O_4), and poly-ε-caprolactone (PCL). The scaffold contains integrated magnetic characteristics that function as internal magnetic generators. An extra magnetic stimulus in the form of an external static magnetic field was used. According to the findings, a 5 percent concentration of Fe_3O_4 offered the best compromise between compressive strength (9.6 ± 0.9 MPa) and deterioration rate (21.6 ± 1.9 percent over a four-week period). Additionally, the Fe_3O_4-containing scaffold enhanced Wharton's jelly mesenchymal stem cells (WJMSCs)' adherence and demonstrated better in vitro bioactivity. Moreover, the Fe_3O_4-incorporated scaffold showed improved ability to promote osteogenic-related protein production, alkaline phosphatase activities, and WJMSC propagation. The CS scaffold containing Fe_3O_4 enhanced cellular activities and enabled the concurrent production of collagen I and osteocalcin under the combined action of the static magnetic field. All of these results highlight the possibility that three-dimensionally constructed Fe3O4-containing CS/PCL scaffolds could interact with a static magnetic field to modify cell behaviour and improve the chances of therapeutic applications in the field of bone tissue engineering.

Despite the remarkable advancements in tissue engineering, several challenges persist, including vascularization, immune response, and the development of functional tissues. Nanocoating, however, offer innovative solutions to address these challenges. Coatings can incorporate growth factors to stimulate angiogenesis (the formation of new blood vessels), which is essential for providing nutrients to the regenerating tissue. Furthermore, nanocoating can be engineered to modulate immune response, reducing the risk of graft rejection.

Scaffold Design and Biocompatibility

In the domain of tissue engineering, scaffold design serves as the cornerstone, and nanocoating exhibit a vital role in optimizing this process. The primary objective is to construct a three-dimensional framework that closely mimics the extracellular matrix, offering both mechanical support and a nurturing environment for cells to adhere, proliferate, and differentiate. Nanocoating, applied to scaffold materials, significantly contribute to achieving the desired level of biocompatibility necessary for successful tissue engineering. Scaffolds in tissue engineering act as architectural frameworks that guide the regeneration of damaged or lost tissues. The scaffold's composition, structure, and surface properties profoundly influence cellular behaviour, ultimately shaping the regenerative outcome (El-Nawasany et al., 2023; Mehta et al., 2019; Vahdati et al., 2022). The goal is to create a scaffold that optimally supports and nurtures cells, encouraging them to rebuild tissues as naturally as possible.

Nanocoating are the artisans behind the enhancement of scaffold surfaces. By introducing nanoscale modifications, these coatings refine the scaffold's surface properties, making them more favourable for cell attachment and growth. One remarkable example of this is the control over nanoscale surface roughness, achieved through nanocoating. These roughened surfaces provide an increased number of sites for cell adhesion, promoting stronger interactions between cells and the scaffold. As cells adhere and spread across the scaffold's surface, they lay the foundation for tissue formation. Nanocoating offer an additional layer of sophistication in scaffold design through the introduction of controlled release mechanisms. By engineering nanocoating to release specific biochemical cues, such as growth factors, cytokines, or signalling molecules, scaffold materials can actively guide cell behaviour (Unni et al., 2022; Zaitseva et al., 2020). These cues serve as molecular signposts that influence cell proliferation, differentiation, and tissue regeneration. For instance, a nanocoated scaffold may release growth factors that stimulate nearby cells to differentiate into specialized tissue cells, like osteoblasts for bone formation or myocytes for muscle tissue.

Biocompatibility is a non-negotiable attribute for any tissue-engineered scaffold. The surface properties of the scaffold material, modulated by nanocoating, are fine-tuned to create a microenvironment that fosters cellular growth and tissue formation. In essence, nanocoated scaffolds actively optimize biocompatibility, ensuring that they integrate seamlessly into the body and promote regenerative processes. In conclusion, scaffold design is a fundamental aspect of tissue engineering, and nanocoating substantially contribute to this process. Nanocoating enhance scaffold surface properties, creating an environment conducive to cell adhesion, proliferation, and differentiation. By offering precise control over surface roughness and the release of biochemical cues, nanocoating are key drivers of successful tissue regeneration, promising advanced solutions for the restoration of damaged or lost tissues in regenerative medicine.

REGENERATIVE MEDICINE APPLICATIONS

The application of nanocoating in tissue engineering extends its transformative potential across various domains of regenerative medicine. Within this multifaceted field, one of the most promising applications is the use of nanocoated scaffolds in organ transplantation. Creating functional, lab-grown organs represents a significant challenge, yet nanocoated scaffolds offer a viable solution. By seeding patient-specific cells onto these meticulously prepared scaffolds, researchers can actively stimulate the growth of replacement organs. This process holds immense promise in addressing the perennial shortage of donor organs and revolutionizing the world of transplantation (Agrawal et al., 2022; Guryanov et al., 2022). The shortage of donor organs for transplantation is a pressing global issue. Nanocoated scaffolds are emerging as a beacon of hope in this landscape. These scaffolds, precisely engineered with nanoscale coatings, create an ideal microenvironment for cells to thrive and form functional tissues. By seeding these scaffolds with patient-specific cells, scientists can initiate the complex process of growing replacement organs in the laboratory. As the cells adhere, proliferate, and differentiate on the nanocoated scaffold, they gradually transform into a functional organ. This revolutionary approach promises a future where patients in need of organ transplants can receive lab-grown, personalized organs, reducing the reliance on donor organs and mitigating transplant waiting lists (Chan et al., 2021; Radulescu et al., 2016). In the domain of orthopaedics, nanocoated scaffolds provide a ray of hope for those suffering from bone injuries or degenerative conditions. These scaffolds are designed to be implanted directly into areas of bone loss, acting as a framework for new bone formation. With nanocoating, the surface properties of the scaffold can be tailored to promote cell adhesion and tissue regeneration. Additionally, nanocoating can incorporate growth factors and biochemical cues that actively accelerate bone growth. This approach is invaluable in the treatment of fractures, bone defects, and degenerative bone conditions.

Nanocoating are also making a significant impact in the field of dentistry, particularly in the realm of dental implants and materials. In implant dentistry, nanocoated surfaces are applied to dental implants to enhance osseointegration. Osseointegration is the process of stable anchoring of dental implants within the jawbone. By using nanocoating to modify the surface properties of dental implants, they can promote more robust bone integration, ensuring the long-term success of dental prosthetics. In summary, nanocoating are revolutionizing the landscape of regenerative medicine. They find application in organ transplantation, offering the potential to create lab-grown replacement organs. In orthopaedics, nanocoated scaffolds accelerate bone regeneration, benefiting individuals with bone injuries or degenerative conditions. In dentistry, nanocoating enhance osseointegration, ensuring the stability and functionality of dental implants. The versatile applications of nanocoating in regenerative medicine herald a future where personalized, lab-grown tissues and organs become commonplace, transforming the way we approach transplantation and tissue repair.

CHALLENGES AND FUTURE PROSPECTS

Nanocoating in medical applications present significant potential but encounter challenges. The assurance of stability and scalability in nanocoating techniques is pivotal for widespread adoption in the medical field. Equally critical is addressing concerns related to nanoparticle toxicity to ensure patient safety. Despite these challenges, advancements in nanotechnology pave the way for more accurate and patient-specific drug delivery. Furthermore, achieving stability and scalability is crucial for the successful implementation of nanocoating in drug delivery. Challenges encompass attaining uniform coatings, preserving integrity during storage, and developing scalable production methods. Additionally, a thorough evaluation of nanoparticle toxicity is essential. Ongoing research seeks to comprehend interactions with biological systems, mitigating risks while securing the benefits of nanocoating and prioritizing patient well-being.

On a positive note, opportunities in medical nanocoating are substantial. As nanotechnology progresses, the potential for precise and patient-specific drug delivery becomes clearer, allowing tailored treatments based on individual genetic and physiological characteristics. A particularly promising aspect is the development of multifunctional nanoparticles that integrate drug delivery with diagnostics, transforming disease detection and treatment assessment. This approach holds the potential for earlier disease detection, real-time monitoring, and personalized treatment adjustments. Looking ahead, antimicrobial nanocoating exhibit promising trends in infection control. Advances may lead to coatings with finely tuned antimicrobial properties, targeting specific pathogens while sparing beneficial microorganisms. Precision is crucial for maintaining a balanced microbiota. Another emerging trend is the integration of nanocoating into smart materials, responding dynamically to pathogens to enhance infection prevention. Smart functionality in healthcare settings and food packaging detects spoilage or contamination, releasing antimicrobial agents to prolong shelf life. Researchers actively pursue environmentally friendly coatings to ensure benefits without adverse environmental consequences.

CONCLUSION

In conclusion, nanocoating represent a transformative technology in the field of medicine, offering a multitude of applications that have the potential to revolutionize healthcare. Their impact extends across precision drug delivery, wound healing enhancement, infection control, and regenerative medicine, ushering in a new era of medical treatments and patient care. Nanocoating have redefined drug delivery by enabling precise targeting and controlled release of medications. This has the potential to significantly improve patient outcomes, reduce side effects, and enhance treatment efficacy. In conditions like cancer, where precision is paramount, nanocoating hold the promise of more effective therapies. The application of nanocoating in wound healing expedites the natural healing processes, accelerates tissue regeneration, and prevents infections. This translates to improved patient outcomes, reduced healthcare costs, and enhanced quality of life for those dealing with chronic wounds and injuries. Antimicrobial nanocoating offer innovative ways to combat infections, from hospitals to everyday environments. They actively inhibit the growth and proliferation of microorganisms, reducing the risk of infections. In healthcare settings, this is crucial for patient safety, and in everyday life, it adds an extra layer of protection against harmful pathogens.

Nanocoating play a pivotal role in regenerative medicine, offering advanced solutions for tissue engineering. From organ transplantation to orthopaedic applications and dentistry, nanocoated scaffolds enhance tissue growth and regeneration. This opens new avenues for personalized medical treatments and addresses critical healthcare challenges, such as the shortage of donor organs. However, these promising advancements are not without their challenges. Safety, regulation, standardization, cost-effectiveness, and sustainability are areas that require focused attention. Researchers, industry leaders, and regulatory bodies are actively working to address these challenges, ensuring that nanocoating in medicine meet stringent safety standards, are cost-effective, and align with sustainable practices. The integration of nanocoating into medical practices is already underway, and their impact is palpable. As the field of nanocoating continues to evolve, it will undoubtedly shape the future of medicine. The result will be healthcare that is safer, more effective, and patient-centric, offering innovative solutions to age-old healthcare challenges and ultimately improving the lives of patients around the world. The future of medicine is being reshaped by nanocoating, and it holds the promise of a healthier and more resilient world.

REFERENCES

Adebayo, V. A., Adewale, O. B., Anadozie, S. O., Osukoya, O. A., Obafemi, T. O., Adewumi, D. F., Idowu, O. T., Onasanya, A., & Ojo, A. A. (2023). GC-MS analysis of aqueous extract of Nymphaea lotus and ameliorative potential of its biosynthesized gold nanoparticles against cadmium-induced kidney damage in rats. *Heliyon*, *9*(6), e17124. doi:10.1016/j.heliyon.2023.e17124 PMID:37484424

Agrawal, S., Kumar, V., Kumar, S., & Shahi, S. K. (2022). Plant development and crop protection using phytonanotechnology: A new window for sustainable agriculture. *Chemosphere*, *299*, 134465. doi:10.1016/j.chemosphere.2022.134465 PMID:35367229

Akbari-Alavijeh, S., Shaddel, R., & Jafari, S. M. (2020). Encapsulation of food bioactives and nutraceuticals by various chitosan-based nanocarriers. *Food Hydrocolloids*, *105*, 105774. doi:10.1016/j.foodhyd.2020.105774

Aldoasri, M. A., Darwish, S. S., Adam, M. A., Elmarzugi, N. A., & Ahmed, S. M. (2017). Protecting of marble stone facades of historic buildings using multifunctional TiO2 nanocoating. *Sustainability (Basel)*, *9*(11), 1–15. doi:10.3390/su9112002

Allehyani, E. S., Almulaiky, Y. Q., Al-Harbi, S. A., & El-Shishtawy, R. M. (2022). In Situ Coating of Polydopamine-AgNPs on Polyester Fabrics Producing Antibacterial and Antioxidant Properties. *Polymers*, *14*(18), 3794. Advance online publication. doi:10.3390/polym14183794 PMID:36145939

Arriagada, F., Günther, G., & Morales, J. (2020). Nanoantioxidant–based silica particles as flavonoid carrier for drug delivery applications. *Pharmaceutics, 12*(4), 1–17. doi:10.3390/pharmaceutics12040302 PMID:32224905

Bai, B., Saranya, S., Dheepaasri, V., Muniasamy, S., Alharbi, N. S., Selvaraj, B., Undal, V. S., & Gnanamangai, B. M. (2022). Biosynthesized copper oxide nanoparticles (CuO NPs) enhances the anti-biofilm efficacy against K. pneumoniae and S. aureus. *Journal of King Saud University. Science, 34*(6), 102120. doi:10.1016/j.jksus.2022.102120

Bashir, S., Lgaz, H., Chung, I. M., & Kumar, A. (2020). Effective green corrosion inhibition of aluminium using analgin in acidic medium: An experimental and theoretical study. *Chemical Engineering Communications, 0*(0), 1–10. doi:10.1080/00986445.2020.1752680

Bashir, S., Thakur, A., Lgaz, H., Chung, I. M., & Kumar, A. (2020). Corrosion inhibition efficiency of bronopol on aluminium in 0.5 M HCl solution: Insights from experimental and quantum chemical studies. *Surfaces and Interfaces, 20*(April), 100542. doi:10.1016/j.surfin.2020.100542

Berini, F., Orlandi, V., Gornati, R., Bernardini, G., & Marinelli, F. (2022). Nanoantibiotics to fight multidrug resistant infections by Gram-positive bacteria: Hope or reality? *Biotechnology Advances, 57*, 107948. doi:10.1016/j.biotechadv.2022.107948 PMID:35337933

Blum, R., & Molian, P. (2009). Liquid-phase sintering of nanodiamond composite coatings on aluminum A319 using a focused laser beam. *Surface and Coatings Technology, 204*(1), 1–14. doi:10.1016/j.surfcoat.2009.06.025

Chan, Y., Wu, X. H., Chieng, B. W., Ibrahim, N. A., & Then, Y. Y. (2021). Superhydrophobic nanocoating as intervention against biofilm-associated bacterial infections. *Nanomaterials (Basel, Switzerland), 11*(4), 1–33. doi:10.3390/nano11041046 PMID:33921904

Chaudhary, P., Ahamad, L., Chaudhary, A., Kumar, G., Chen, W.-J., & Chen, S. (2023). Nanoparticle-mediated bioremediation as a powerful weapon in the removal of environmental pollutants. *Journal of Environmental Chemical Engineering, 11*(2), 109591. doi:10.1016/j.jece.2023.109591

Chauhan, D. S., Prasad, R., Srivastava, R., Jaggi, M., Chauhan, S. C., & Yallapu, M. M. (2020). Comprehensive Review on Current Interventions, Diagnostics, and Nanotechnology Perspectives against SARS-CoV-2. *Bioconjugate Chemistry, 31*(9), 2021–2045. doi:10.1021/acs.bioconjchem.0c00323 PMID:32680422

Chen, M. C., Koh, P. W., Ponnusamy, V. K., & Lee, S. L. (2022). Titanium dioxide and other nanomaterials based antimicrobial additives in functional paints and coatings [Review]. *Progress in Organic Coatings, 163*, 106660. doi:10.1016/j.porgcoat.2021.106660

Christy, P. N., Basha, S. K., Kumari, V. S., Bashir, A. K. H., Maaza, M., Kaviyarasu, K., Arasu, M. V., Al-Dhabi, N. A., & Ignacimuthu, S. (2020). Biopolymeric nanocomposite scaffolds for bone tissue engineering applications – A review. *Journal of Drug Delivery Science and Technology, 55*, 101452. doi:10.1016/j.jddst.2019.101452

Corbett, J., McKeown, P. A., Peggs, G. N., & Whatmore, R. (2000). Nanotechnology: International Developments and Emerging Products. *CIRP Annals, 49*(2), 523–545. doi:10.1016/S0007-8506(07)63454-4

Dastkhoon, M., Ghaedi, M., Asfaram, A., & Alipanahpour Dil, E. (2020). Comparative study of ability of sonochemistry combined ZnS:Ni nanoparticles-loaded activated carbon in reductive of organic pollutants from environmental water samples. *Polyhedron, 180*, 114341. doi:10.1016/j.poly.2019.114341

Du, C., Du, T., Chang, Z., Yin, C., & Cheng, Y. (2022). On the interface between biomaterials and two-dimensional materials for biomedical applications. *Advanced Drug Delivery Reviews, 186*, 114314. doi:10.1016/j.addr.2022.114314 PMID:35568105

El-Ashram, S., El-Samad, L. M., Basha, A. A., & El Wakil, A. (2021). Naturally-derived targeted therapy for wound healing: Beyond classical strategies. *Pharmacological Research, 170*, 105749. doi:10.1016/j.phrs.2021.105749 PMID:34214630

El-Nawasany, L. I., Sundookh, A., Kadoum, L. A., Yasin, M. A., AlSalem, H. S., Binkadem, M. S., Al-Goul, S. T., Zidan, N. S., & Shoueir, K. R. (2023). Ameliorating characteristics of magnetically sensitive TPU nanofibers-based food packaging film for long-life cheese preservation. *Food Bioscience, 53*, 102633. doi:10.1016/j.fbio.2023.102633

Erdonmez, D., Mosayyebi, S., Erkan, K., Salimi, K., Nagizade, N., Saglam, N., & Rzayev, Z. M. O. (2014). Nanofabrication and characterization of PVA–organofiller/Ag nanocoating on pMAD plasmids. *Applied Surface Science, 318*, 127–131. doi:10.1016/j.apsusc.2014.02.007

Feng, Y., Zhong, L., Bilal, M., Tan, Z., Hou, Y., Jia, S., & Cui, J. (2019). Enzymes@ZIF-8 nanocomposites with protection nanocoating: Stability and acid-resistant evaluation. *Polymers, 11*(1), 1–14. doi:10.3390/polym11010027 PMID:30960011

Garg, R., Rani, P., Garg, R., Khan, M. A., Khan, N. A., Khan, A. H., & Américo-Pinheiro, J. H. P. (2022). Biomedical and catalytic applications of agri-based biosynthesized silver nanoparticles. *Environmental Pollution*, *310*, 119830. doi:10.1016/j.envpol.2022.119830 PMID:35926739

Ghosh, T., Nakano, K., Mulchandani, N., & Katiyar, V. (2021). Curcumin loaded iron functionalized biopolymeric nanofibre reinforced edible nanocoating for improved shelf life of cut pineapples. *Food Packaging and Shelf Life*, *28*, 100658. doi:10.1016/j.fpsl.2021.100658

Govardhana Reddy, P. V., Rajendra Prasad Reddy, B., Venkata Krishna Reddy, M., Raghava Reddy, K., Shetti, N. P., Saleh, T. A., & Aminabhavi, T. M. (2021). A review on multicomponent reactions catalysed by zero-dimensional/one-dimensional titanium dioxide (TiO2) nanomaterials: Promising green methodologies in organic chemistry. *Journal of Environmental Management*, *279*, 111603. doi:10.1016/j.jenvman.2020.111603 PMID:33172705

Guryanov, I., Naumenko, E., & Fakhrullin, R. (2022). Hair surface engineering: Combining nanoarchitectonics with hair topical and beauty formulations. *Applied Surface Science Advances*, *7*, 100188. https://doi.org/https://doi.org/10.1016/j.apsadv.2021.100188

Hao, L., Wang, L., Ma, Y., Bottini, M., Li, L., Cheng, H.-B., Gao, S., Liang, X., & Zhang, J. (2024). Nanomaterials relieving hypoxia for enhanced tumor therapy. *Coordination Chemistry Reviews*, *499*, 215482. doi:10.1016/j.ccr.2023.215482

Hasan, K. M. F., Champramary, S., Al Hasan, K. M. N., Indic, B., Ahmed, T., Pervez, M. N., Horváth, P. G., Bak, M., Sándor, B., Hofmann, T., Tolvaj, L., Horváth, A., Kóczán, Z., Sipos, G., Alpár, T., & Bejó, L. (2023). Eco-friendly production of cellulosic fibers from Scots pine wood and sustainable nanosilver modification: A path toward sustainability. *Results in Engineering*, *19*, 101244. doi:10.1016/j.rineng.2023.101244

Hossain, N., Islam, M. A., Chowdhury, M. A., & Alam, A. (2022). Advances of nanoparticles employment in dental implant applications. *Applied Surface Science Advances*, *12*, 100341. https://doi.org/https://doi.org/10.1016/j.apsadv.2022.100341

Huang, Z., Li, J., Chen, X., Yang, Q., Zeng, X., Bai, R., & Wang, L. (2023). Photothermal Sensitive 3D Printed Biodegradable Polyester Scaffolds with Polydopamine Coating for Bone Tissue Engineering. *Polymers*, *15*(2), 381. Advance online publication. doi:10.3390/polym15020381 PMID:36679260

Islam, N. U., Amin, R., Shahid, M., & Amin, M. (2019). Gummy gold and silver nanoparticles of apricot (Prunus armeniaca) confer high stability and biological activity. *Arabian Journal of Chemistry*, *12*(8), 3977–3992. doi:10.1016/j.arabjc.2016.02.017

Kao, C. Y., Lin, T. L., Lin, Y. H., Lee, A. K. X., Ng, S. Y., Huang, T. H., & Hsu, T. T. (2022). Synergistic Effect of Static Magnetic Fields and 3D-Printed Iron-Oxide-Nanoparticle-Containing Calcium Silicate/Poly-ε-Caprolactone Scaffolds for Bone Tissue Engineering. *Cells*, *11*(24), 3967. Advance online publication. doi:10.3390/cells11243967 PMID:36552731

Kaya, S., Lgaz, H., Thakkur, A., Kumar, A., Özbakır, D., Karakuş, N., & Ben Ahmed, S. (2023). Molecular insights into the corrosion inhibition mechanism of omeprazole and tinidazole : A theoretical investigation. *Molecular Simulation*, *49*(17), 1–15. doi:10.1080/08927022.2023.2256888

Kaya, S., Thakur, A., & Kumar, A. (2023). The role of in Silico / DFT investigations in analyzing dye molecules for enhanced solar cell efficiency and reduced toxicity. *Journal of Molecular Graphics & Modelling*, *124*(June), 108536. doi:10.1016/j.jmgm.2023.108536 PMID:37300949

Keshtmand, Z., Khademian, E., Poorjafari Jafroodi, P., Abtahi, M. S., & Tavakkoli Yaraki, M. (2023). Green synthesis of selenium nanoparticles using Artemisia chamaemelifolia: toxicity effects through regulation of gene expression for cancer cells and bacteria. *Nano-Structures & Nano-Objects, 36*, 101049. https://doi.org/https://doi.org/10.1016/j.nanoso.2023.101049

Khatoon, Z., McTiernan, C. D., Suuronen, E. J., Mah, T.-F., & Alarcon, E. I. (2018). Bacterial biofilm formation on implantable devices and approaches to its treatment and prevention. *Heliyon*, *4*(12), e01067. doi:10.1016/j.heliyon.2018.e01067 PMID:30619958

Khemthong, P., Phanthasri, J., Youngjan, S., Wanmolee, W., Samun, Y., Sosa, N., Rungnim, C., Kraithong, W., Sangkhun, W., Parnthong, J., Butburee, T., Thanee, K., Nakajima, H., Supruangnet, R., Towiwat, P., Chanvorachote, P., & Sukrong, S. (2023). Effect of the ethanol-to-water ratio on the properties of silica–carbon core–shell materials for prolonged antibacterial activity of thymol. *Applied Surface Science*, *635*, 157716. doi:10.1016/j.apsusc.2023.157716

Kolahalam, L. A., Kasi Viswanath, I. V., Diwakar, B. S., Govindh, B., Reddy, V., & Murthy, Y. L. N. (2019). Review on nanomaterials: Synthesis and applications. *Materials Today: Proceedings*, *18*, 2182–2190. doi:10.1016/j.matpr.2019.07.371

Kumar, A., & Thakur, A. (2020). Encapsulated nanoparticles in organic polymers for corrosion inhibition. In *Corrosion Protection at the Nanoscale*. Elsevier Inc. doi:10.1016/B978-0-12-819359-4.00018-0

Kumar, A., & Thakur, A. (2023). Overview of the Properties, Applicability, and Recent Advancements of Some Natural Products Used as Potential Inhibitors in Various Corrosive Systems. In Handbook of Research on Corrosion Sciences and Engineering (pp. 275–277). doi:10.4018/978-1-6684-7689-5.ch010

Li, Q., Wang, X., Wang, X., Zheng, L., Chen, P., & Zhang, B. (2023). Novel insights into versatile nanomaterials integrated bioreceptors toward zearalenone ultrasensitive discrimination. *Food Chemistry*, *410*, 135435. doi:10.1016/j.foodchem.2023.135435 PMID:36641913

Magovac, E., Vončina, B., Budimir, A., Jordanov, I., Grunlan, J. C., & Bischof, S. (2021). Environmentally benign phytic acid-based nanocoating for multifunctional flame-retardant/antibacterial cotton. *Fibers (Basel, Switzerland)*, *9*(11), 69. Advance online publication. doi:10.3390/fib9110069

Malekmohammadi, S., Mohammed, R. U. R., Samadian, H., Zarebkohan, A., García-Fernández, A., Kokil, G. R., Sharifi, F., Esmaeili, J., Bhia, M., Razavi, M., Bodaghi, M., Kumeria, T., & Martínez-Máñez, R. (2022). Nonordered dendritic mesoporous silica nanoparticles as promising platforms for advanced methods of diagnosis and therapies. *Materials Today. Chemistry*, *26*, 101144. doi:10.1016/j.mtchem.2022.101144

Malshe, A. P., Rajurkar, K. P., Virwani, K. R., Taylor, C. R., Bourell, D. L., Levy, G., Sundaram, M. M., McGeough, J. A., Kalyanasundaram, V., & Samant, A. N. (2010). Tip-based nanomanufacturing by electrical, chemical, mechanical and thermal processes. *CIRP Annals*, *59*(2), 628–651. doi:10.1016/j.cirp.2010.05.006

Maťátková, O., Michailidu, J., Miškovská, A., Kolouchová, I., Masák, J., & Čejková, A. (2022). Antimicrobial properties and applications of metal nanoparticles biosynthesized by green methods. *Biotechnology Advances*, *58*, 107905. doi:10.1016/j.biotechadv.2022.107905 PMID:35031394

Meda, U. S., Vora, K., Athreya, Y., & Mandi, U. A. (2022). Titanium dioxide based heterogeneous and heterojunction photocatalysts for pollution control applications in the construction industry. *Process Safety and Environmental Protection*, *161*, 771–787. doi:10.1016/j.psep.2022.03.066

Mehta, P., Al-Kinani, A. A., Arshad, M. S., Singh, N., van der Merwe, S. M., Chang, M.-W., Alany, R. G., & Ahmad, Z. (2019). Engineering and Development of Chitosan-Based Nanocoating for Ocular Contact Lenses. *Journal of Pharmaceutical Sciences, 108*(4), 1540–1551. doi:10.1016/j.xphs.2018.11.036 PMID:30513319

Merugu, R., Sharma, S., & Rastogi, M. (2023). An overview of nanotechnological applications in biomedical engineering. *Materials Today: Proceedings, 76*, 412–417. doi:10.1016/j.matpr.2022.12.028

Mohammadi, M., Mousavi Shaegh, S. A., Alibolandi, M., Ebrahimzadeh, M. H., Tamayol, A., Jaafari, M. R., & Ramezani, M. (2018). Micro and nanotechnologies for bone regeneration: Recent advances and emerging designs. *Journal of Controlled Release, 274*, 35–55. doi:10.1016/j.jconrel.2018.01.032 PMID:29410062

Montes, C., Villaseñor, M. J., & Ríos, Á. (2019). Analytical control of nanodelivery lipid-based systems for encapsulation of nutraceuticals: Achievements and challenges. *Trends in Food Science & Technology, 90*, 47–62. doi:10.1016/j.tifs.2019.06.001

Negut, I., Grumezescu, V., Ficai, A., Grumezescu, A. M., Holban, A. M., Popescu, R. C., Savu, D., Vasile, B. S., & Socol, G. (2018). MAPLE deposition of Nigella sativa functionalized Fe3O4 nanoparticles for antimicrobial coatings. *Applied Surface Science, 455*, 513–521. doi:10.1016/j.apsusc.2018.05.202

Nyga, A., Czerwińska-Główka, D., Krzywiecki, M., Przystaś, W., Zabłocka-Godlewska, E., Student, S., Kwoka, M., Data, P., & Blacha-Grzechnik, A. (2021). Covalent immobilization of organic photosensitizers on the glass surface: Toward the formation of the light-activated antimicrobial nanocoating. *Materials (Basel), 14*(11), 3093. Advance online publication. doi:10.3390/ma14113093 PMID:34200077

Otto, D. P., Otto, A., & de Villiers, M. M. (2022). In Vitro Skin Delivery of Griseofulvin by Layer-by-Layer Nanocoated Emulsions Stabilized by Whey Protein and Polysaccharides. *Pharmaceutics, 14*(3), 554. Advance online publication. doi:10.3390/pharmaceutics14030554 PMID:35335930

Özen, İ., Demir, A., Bahtiyari, M. İ., Wang, X., Nilghaz, A., Wu, P., Shirvanimoghaddam, K., & Naebe, M. (2023). Multifaceted applications of thymol/carvacrol-containing polymeric fibrous structures. *Advanced Industrial and Engineering Polymer Research.* https://doi.org/https://doi.org/10.1016/j.aiepr.2023.09.001

Pal, A., Goswami, R., & Roy, D. N. (2021). A critical assessment on biochemical and molecular mechanisms of toxicity developed by emerging nanomaterials on important microbes. *Environmental Nanotechnology, Monitoring & Management, 16*, 100485. doi:10.1016/j.enmm.2021.100485

Paladini, F., Di Franco, C., Panico, A., Scamarcio, G., Sannino, A., & Pollini, M. (2016). In vitro assessment of the antibacterial potential of silver nano-coatings on cotton gauzes for prevention of wound infections. *Materials (Basel)*, *9*(6), 1–14. doi:10.3390/ma9060411 PMID:28773531

Parveen, G., Bashir, S., Thakur, A., Saha, S. K., Banerjee, P., & Kumar, A. (2020). Experimental and computational studies of imidazolium based ionic liquid 1-methyl- 3-propylimidazolium iodide on mild steel corrosion in acidic solution Experimental and computational studies of imidazolium based ionic liquid 1-methyl-3-propylimidazolium. *Materials Research Express*, *7*(1), 016510. doi:10.1088/2053-1591/ab5c6a

Pathan, S. I., Chundawat, N. S., Chauhan, N. P. S., & Singh, G. P. (2020). A review on synthetic approaches of heterocycles via insertion-cyclization reaction. *Synthetic Communications*, *50*(9), 1251–1285. doi:10.1080/00397911.2020.1712609

Phan, T., Jones, J. E., Chen, M., Bowles, D. K., Fay, W. P., & Yu, Q. (2022). A Biocompatibility Study of Plasma Nanocoating onto Cobalt Chromium L605 Alloy for Cardiovascular Stent Applications. *Materials (Basel)*, *15*(17), 5968. Advance online publication. doi:10.3390/ma15175968 PMID:36079346

Poggio, C., Colombo, M., Arciola, C. R., Greggi, T., Scribante, A., & Dagna, A. (2020). Copper-alloy surfaces and cleaning regimens against the spread of SARS-CoV-2 in dentistry and orthopedics. From fomites to anti-infective nanocoating. *Materials (Basel)*, *13*(15), 1–12. doi:10.3390/ma13153244 PMID:32707757

Prakash, J., Sun, S., Swart, H. C., & Gupta, R. K. (2018). Noble metals-TiO2 nanocomposites: From fundamental mechanisms to photocatalysis, surface enhanced Raman scattering and antibacterial applications. *Applied Materials Today*, *11*, 82–135. doi:10.1016/j.apmt.2018.02.002

Pratim Das, P., Aditya Bachchan, A., Sahu, R., & Chaudhary, V. (2021). Whole body vibration: Effects on human body and role of biomaterials in repairing fracture joints and tissues. *Materials Today: Proceedings*, *43*, 141–147. doi:10.1016/j.matpr.2020.11.250

Qi, Z., Xue, X., Xu, X., Zhou, H., Li, W., Yang, G., & Xie, P. (2022). Detoxified and antimicrobial-enhanced olive mill wastewater phenols capping ZnO nanoparticles incorporated with carboxymethyl cellulose for fresh strawberry preservation. *Postharvest Biology and Technology*, *188*, 111891. doi:10.1016/j.postharvbio.2022.111891

Radulescu, M., Andronescu, E., Dolete, G., Popescu, R. C., Fufă, O., Chifiriuc, M. C., Mogoantă, L., Bălşeanu, T. A., Mogoşanu, G. D., Grumezescu, A. M., & Holban, A. M. (2016). Silver nanocoating for reducing the exogenous microbial colonization of wound dressings. *Materials (Basel), 9*(5), 1–15. doi:10.3390/ma9050345 PMID:28773468

Rastogi, A., Singh, A., Naik, K., Mishra, A., Chaudhary, S., Manohar, R., & Singh Parmar, A. (2022). A systemic review on liquid crystals, nanoformulations and its application for detection and treatment of SARS – CoV- 2 (COVID – 19). *Journal of Molecular Liquids, 362*, 119795. doi:10.1016/j.molliq.2022.119795 PMID:35832289

Ray, S., Raychaudhuri, U., & Chakraborty, R. (2016). An overview of encapsulation of active compounds used in food products by drying technology. *Food Bioscience, 13*, 76–83. doi:10.1016/j.fbio.2015.12.009

Saemian, T., Gharagozlou, M., Hossaini Sadr, M., & Naghibi, S. (2019). Synthesis of CoFe2O4@Cu3(BTC)2 nanocomposite as a magnetic metal–organic framework. *Polyhedron, 174,* 114163. doi:10.1016/j.poly.2019.114163

Saini, P., Kamalesu, L., & Manikanika. (2023). Review on nanotechnology "Impact on the food services industry." *Materials Today: Proceedings*. https://doi.org/https://doi.org/10.1016/j.matpr.2023.04.377

Sharma & Thakur. (2023). Synthesis, Electrochemical, Morphological, Computational and Corrosion Inhibition Studies of 3-(5-Naphthalen-2-yl-[1,3,4]oxadiazol-2-yl)-pyridine against Mild Steel in 1 M HCl. *Asian Journal of Chemistry, 35*(5), 1079–1088. https://doi.org/https://doi.org/10.14233/ajchem.2023.27711

Sharma, D., Thakur, A., Kumar, M., Kumar, A., & Jakhar, K. (2023). A convenient synthesis, electrochemical profiling, and morphological studies of a pyridine-based 1, 3, 4-oxadiazole hybrid : A promising study for corrosion mitigation of mild steel in strongly acidic environment. *Inorganic Chemistry Communications, 158*(P1), 111554. doi:10.1016/j.inoche.2023.111554

Sharma, P., Pant, S., Dave, V., Tak, K., Sadhu, V., & Reddy, K. R. (2019). Green synthesis and characterization of copper nanoparticles by Tinospora cardifolia to produce nature-friendly copper nano-coated fabric and their antimicrobial evaluation. *Journal of Microbiological Methods, 160*, 107–116. doi:10.1016/j.mimet.2019.03.007 PMID:30871999

Singh, H., & Kaur, K. (2023). Role of nanotechnology in research fields: Medical sciences, military & tribology- A review on recent advancements, grand challenges and perspectives. *Materials Today: Proceedings*. https://doi.org/https://doi.org/10.1016/j.matpr.2023.02.061

Singh, Y., Kumar, U., Panigrahi, S., Balyan, P., Mehla, S., Sihag, P., Sagwal, V., Singh, K. P., White, J. C., & Dhankher, O. P. (2023). Nanoparticles as novel elicitors in plant tissue culture applications: Current status and future outlook. *Plant Physiology and Biochemistry, 203*, 108004. doi:10.1016/j.plaphy.2023.108004 PMID:37714027

Sun, J., Gao, L., Wang, L., & Sun, X. (2021). Recent advances in single-cell analysis: Encapsulation materials, analysis methods and integrative platform for microfluidic technology. *Talanta, 234*, 122671. doi:10.1016/j.talanta.2021.122671 PMID:34364472

Swain, S. S., Paidesetty, S. K., Padhy, R. N., & Hussain, T. (2023). Nano-technology platforms to increase the antibacterial drug suitability of essential oils: A drug prospective assessment. *OpenNano, 9*, 100115. https://doi.org/https://doi.org/10.1016/j.onano.2022.100115

Thakur, A., Savaş, K., Kumar, A., Kaya, S., & Kumar, A. (2023). Recent Trends in the Characterization and Application Progress of Nano - Modified Coatings in Corrosion Mitigation of Metals and Alloys. *Applied Sciences, 13*, 730. https://doi.org/https://doi.org/10.3390/app13020730

Thakur, A., Kaya, S., Abousalem, A. S., & Kumar, A. (2022). Experimental, DFT and MC simulation analysis of Vicia Sativa weed aerial extract as sustainable and eco-benign corrosion inhibitor for mild steel in acidic environment. *Sustainable Chemistry and Pharmacy, 29*(July), 100785. doi:10.1016/j.scp.2022.100785

Thakur, A., Kaya, S., Abousalem, A. S., Sharma, S., Ganjoo, R., Assad, H., & Kumar, A. (2022). Computational and experimental studies on the corrosion inhibition performance of an aerial extract of Cnicus Benedictus weed on the acidic corrosion of mild steel. *Process Safety and Environmental Protection, 161*, 801–818. doi:10.1016/j.psep.2022.03.082

Thakur, A., Kaya, S., & Kumar, A. (2021). Recent Innovations in Nano Container-Based Self-Healing Coatings in the Construction Industry. *Current Nanoscience, 18*(2), 203–216. doi:10.2174/1573413717666210216120741

Thakur, A., Kaya, S., & Kumar, A. (2023). Recent Trends in the Characterization and Application Progress of Nano-Modified Coatings in Corrosion Mitigation of Metals and Alloys. *Applied Sciences (Basel, Switzerland), 13*(2), 730. doi:10.3390/app13020730

Thakur, A., & Kumar, A. (2021). Sustainable Inhibitors for Corrosion Mitigation in Aggressive Corrosive Media: A Comprehensive Study. *Journal of Bio- and Tribo-Corrosion, 7*(2), 1–48. doi:10.1007/s40735-021-00501-y

Thakur, A., & Kumar, A. (2022). Recent advances on rapid detection and remediation of environmental pollutants utilizing nanomaterials-based (bio)sensors. *The Science of the Total Environment, 834*(January), 155219. doi:10.1016/j.scitotenv.2022.155219 PMID:35421493

Thakur, A., & Kumar, A. (2023a). Ecotoxicity Analysis and Risk Assessment of Nanomaterials for the Environmental Remediation. *Macromolecular Symposia, 410*(1), 1–23. doi:10.1002/masy.202100438

Thakur, A., & Kumar, A. (2023b). Recent trends in nanostructured carbon-based electrochemical sensors for the detection and remediation of persistent toxic substances in real- time analysis. *Materials Research Express, 10*(3), 034001. https://doi.org/ https://doi.org/10.1088/2053-1591/acbd1a. doi:10.1088/2053-1591/acbd1a

Thakur, A., Kumar, A., Kaya, S., Benhiba, F., Sharma, S., Ganjoo, R., & Assad, H. (2023). Electrochemical and computational investigations of the Thysanolaena latifolia leaves extract : An eco-benign solution for the corrosion mitigation of mild steel. *Results in Chemistry, 6*(September), 101147. doi:10.1016/j.rechem.2023.101147

Thakur, A., Kumar, A., Kaya, S., Marzouki, R., Zhang, F., & Guo, L. (2022). Recent Advancements in Surface Modification, Characterization and Functionalization for Enhancing the Biocompatibility and Corrosion Resistance of Biomedical Implants. *Coatings, 12*(10), 1459. doi:10.3390/coatings12101459

Thakur, A., Kumar, A., Kaya, S., Vo, D. V. N., & Sharma, A. (2022). Suppressing inhibitory compounds by nanomaterials for highly efficient biofuel production: A review. *Fuel, 312*(September), 122934. doi:10.1016/j.fuel.2021.122934

Thakur, A., Kumar, A., Sharma, S., Ganjoo, R., & Assad, H. (2022). Computational and experimental studies on the efficiency of Sonchus arvensis as green corrosion inhibitor for mild steel in 0. 5 M HCl solution. *Materials Today: Proceedings, 66*, 609–621. doi:10.1016/j.matpr.2022.06.479

Thakur, A., Sharma, S., Ganjoo, R., Assad, H., & Kumar, A. (2022). Anti-Corrosive Potential of the Sustainable Corrosion Inhibitors Based on Biomass Waste: A Review on Preceding and Perspective Research. *Journal of Physics: Conference Series, 2267*(1), 012079. doi:10.1088/1742-6596/2267/1/012079

Trivedi, M. U., Greczynski, G., Kanth, P. C., Pandey, M. K., Ivanov, I. G., Syväjärvi, M., & Yazdi, G. R. (2021). Study of Cucurbit[7]uril nanocoating on epitaxial graphene to design a versatile sensing platform. *Applied Surface Science, 563*, 150096. doi:10.1016/j.apsusc.2021.150096

Unni, R., Reshmy, R., Latha, M. S., Philip, E., Sindhu, R., Binod, P., Pandey, A., & Awasthi, M. K. (2022). Enhancement of mechanical and thermal properties of Ixora coccinea L. plant root derived nanocellulose using polyethylene glycol-glutaraldehyde system. *Chemosphere, 298*, 134324. doi:10.1016/j.chemosphere.2022.134324 PMID:35307393

Vahdati, S. N., Behboudi, H., Navasatli, S. A., Tavakoli, S., & Safavi, M. (2022). New insights into the inhibitory roles and mechanisms of D-amino acids in bacterial biofilms in medicine, industry, and agriculture. *Microbiological Research, 263*, 127107. doi:10.1016/j.micres.2022.127107 PMID:35843196

Verma, C., Thakur, A., Ganjoo, R., Sharma, S., & Assad, H. (2023). Coordination bonding and corrosion inhibition potential of nitrogen-rich heterocycles : Azoles and triazines as specific examples. *Coordination Chemistry Reviews, 488*(November), 215177. doi:10.1016/j.ccr.2023.215177

Wang, Y., Liu, K., Zhang, M., Xu, T., Du, H., Pang, B., & Si, C. (2023). Sustainable polysaccharide-based materials for intelligent packaging. *Carbohydrate Polymers, 313*, 120851. doi:10.1016/j.carbpol.2023.120851 PMID:37182951

Wani, M. Y., Ganie, N. A., Dar, K. A., Dar, S. Q., Khan, A. H., Khan, N. A., Zahmatkesh, S., Manzar, M. S., & Banerjee, R. (2023). Nanotechnology future in food using carbohydrate macromolecules: A state-of-the-art review. *International Journal of Biological Macromolecules, 239*, 124350. doi:10.1016/j.ijbiomac.2023.124350 PMID:37028631

Xu, C., Ma, J., Liu, Z., Wang, W., Liu, X., Qian, S., Chen, L., Gu, L., Sun, C., Hou, J., & Jiang, Z. (2023). Preparation of shell-core fiber-encapsulated Lactobacillus rhamnosus 1.0320 using coaxial electrospinning. *Food Chemistry, 402*, 134253. doi:10.1016/j.foodchem.2022.134253 PMID:36162172

Xu, C., Zhou, S., Song, H., Hu, H., Yang, Y., Zhang, X., Ma, S., Feng, X., Pan, Y., Gong, S., Fan, F., Chen, P., & Chu, Q. (2023). Green tea polyphenols-derived hybrid materials in manufacturing, environment, food and healthcare. *Nano Today, 52*, 101990. doi:10.1016/j.nantod.2023.101990

Yaashikaa, P. R., Kamalesh, R., Senthil Kumar, P., Saravanan, A., Vijayasri, K., & Rangasamy, G. (2023). Recent advances in edible coatings and their application in food packaging. *Food Research International, 173*, 113366. doi:10.1016/j.foodres.2023.113366 PMID:37803705

Yang, Z., Zhang, D., Wang, K., He, J., Li, J., & Huang, C. (2022). Investigating graphdiyne based materials for rechargeable batteries. *Nano Today, 46*, 101588. doi:10.1016/j.nantod.2022.101588

Yazdani, J., Ahmadian, E., Sharifi, S., Shahi, S., & Maleki Dizaj, S. (2018). A short view on nanohydroxyapatite as coating of dental implants. *Biomedicine and Pharmacotherapy, 105*, 553–557. doi:10.1016/j.biopha.2018.06.013 PMID:29886376

Yu, Y., Zhang, J., Wang, J., & Sun, B. (2021). MicroRNAs: The novel mediators for nutrient-modulating biological functions. *Trends in Food Science & Technology, 114*, 167–175. doi:10.1016/j.tifs.2021.05.028

Zaitseva, O., Khudyakov, A., Sergushkina, M., Solomina, O., & Polezhaeva, T. (2020). Pectins as a universal medicine. *Fitoterapia, 146*, 104676. doi:10.1016/j.fitote.2020.104676 PMID:32561422

Zhao, Y., Zhang, Y., Dong, H., Wu, W., Yang, X., & He, Q. (2023). Functional biopolymers for food packaging: Formation mechanism and performance improvement of chitosan-based composites. *Food Bioscience, 54*, 102927. doi:10.1016/j.fbio.2023.102927

ADDITIONAL READING

Afshari, S., Montazer, M., Harifi, T., & Mahmoudi Rad, M. (2019). A coloured polyester fabric with antimicrobial properties conferred by copper nanoparticles. *Coloration Technology, 135*(6), 427–438. doi:10.1111/cote.12430

Al Zahrani, N. A., El-Shishtawy, R. M., Elaasser, M. M., & Asiri, A. M. (2020). Synthesis of novel chalcone-based phenothiazine derivatives as antioxidant and anticancer agents. *Molecules (Basel, Switzerland), 25*(19), 4566. doi:10.3390/molecules25194566 PMID:33036301

Choukairi, Z., Hazzaz, T., José, M. F., & Fechtali, T. (2020). The cytotoxic activity of salvia officinalis L. and rosmarinus officinalis L. Leaves extracts on human glioblastoma cell line and their antioxidant effect. *Journal of Complementary & Integrative Medicine, 17*(4), 20180189. doi:10.1515/jcim-2018-0189 PMID:32229691

Salam, M. A., Obaid, A. Y., El-Shishtawy, R. M., & Mohamed, S. A. (2017). Synthesis of nanocomposites of polypyrrole/carbon nanotubes/silver nano particles and their application in water disinfection. *RSC Advances, 7*(27), 16878–16884. doi:10.1039/C7RA01033H

Zahrani, A. L. (2020). Recent developments of gallic acid derivatives and their hybrids in medicinal chemistry: A review. *European Journal of Medicinal Chemistry, 204*, 112609. doi:10.1016/j.ejmech.2020.112609 PMID:32731188

Chapter 9

Nanocoatings in Medicine: Revolutionizing Healthcare Through Precision and Potential

G. Boopathy
https://orcid.org/0000-0003-2515-7277
Vel Tech Rangarajan Dr. Sagunthala R&D Institute of Science and Technology, Chennai, India

K. Gurusami
St. Peter's Institute of Higher Education and Research, Chennai, India

K. R. Vijayakumar
https://orcid.org/0009-0003-0527-4254
Dr. MGR Educational and Research Institute, Chennai, India

Mohit Hemanth Kumar
Alliance University, India

ABSTRACT

The use of nanotechnology in healthcare and medicine is crucial, particularly for nanocoatings. The therapeutic potential of nanocoatings on medical implants and devices is examined in this chapter. It goes into the ingredients, production processes, and health benefits of nanocoatings. Device functionality and biocompatibility can be enhanced by nanocoatings. They are essential in addressing healthcare issues including cardiovascular stents and orthopedic implants. Additionally, nanocoatings are used in medication delivery systems, tissue engineering, and wound healing. This chapter offers important knowledge on the medical applications of nanocoatings to experts and researchers. Nanotechnology has the potential to alter the medical industry, which would be advantageous for patients.

DOI: 10.4018/979-8-3693-3136-1.ch009

INTRODUCTION

Research and study on nanotechnology have shown enormous promise for innovative advances in a number of sectors, including healthcare and medicine. One interesting area of research is the use of nanotechnology in the form of nanocoating, which are essentially functionalized thin films of nanomaterials applied to a variety of surfaces. These nanocoating are especially interesting for use in the medical industry because of their special qualities and characteristics. Considering this, the suggested academic book chapter seeks to explore and fully look at the wide range of therapeutic applications linked to nanocoating, with a focus on highlighting their notable benefits to the field of medicine and patient care.

The chapter will begin with an introduction to nanocoating, providing a comprehensive overview of their composition, fabrication methods, and properties that make them suitable for medical applications. It will also discuss the importance of surface modification in biomedical materials and devices, emphasizing the role of nanocoating in enhancing their performance and biocompatibility. One of the key focuses of the chapter will be on the use of nanocoating in medical devices and implants. This section will delve into how nanocoating can improve the mechanical properties of implants, reduce the risk of infections through antimicrobial properties, and promote tissue integration for enhanced biocompatibility. Additionally, we will explore the application of nanocoating in cardiovascular stents, orthopaedic implants, and other medical devices, highlighting their potential to address critical healthcare challenges.

Furthermore, the chapter will discuss the role of nanocoating in drug delivery systems. Nanocoating offer a promising avenue for targeted and controlled drug delivery, increasing the therapeutic efficacy of medications while minimizing side effects. We will examine various strategies for incorporating drugs into nanocoating and their release mechanisms, emphasizing recent advancements and potential future developments in this area. We will discuss the mechanisms behind these effects and highlight case studies highlighting the successful application of nanocoating in wound management and tissue engineering. Another essential aspect that the chapter will cover is the safety and toxicity considerations of nanocoating in medicinal applications. The potential cytotoxicity and biocompatibility challenges of nanocoating will be addressed, along with ongoing research efforts to ensure their safe use in medical settings.

Finally, the planned chapter would give an in-depth examination of the numerous and potential medical applications of nanocoating. Through this investigation, we hope to shed insight on nanotechnology's enormous potential for revolutionising modern medicine and improving patient outcomes. The chapter will be an invaluable resource for researchers, academics, and professionals looking to comprehend the most recent advances in the field and fully realise the promise of nanocoating in therapeutic applications.

BASICS OF NANOCOATING

Nanocoating are thin layers of material with dimensions ranging from 1 to 100 nanometers. At this scale, distinct physical and chemical properties emerge, distinguishing nanocoating from their countless counterparts. Nanocoating have a high surface area-to-volume ratio and precise interactions with biological entities at the nanoscale. Nanocoating have the capacity to alter the surface properties of a wide range of substances, altering factors such as a surface's ability to reject or retain liquids, the degree of adhesion between surfaces, and a surface's compatibility with living organisms. The capacity to functionalize nanocoating with specific molecules or compounds improves their activity and enables more focused applications in medicine.

Nanocoating enable the molecular personalization of medical therapies. This level of precision is critical in the implementation of personalised medicine, which allows therapies to be tailored to the unique needs and characteristics of each patient. The ability to cautiously change drug delivery procedures and medical tools deviates from universal methodologies. The use of nanocoating in medicine is more than just a technological advancement; it has practical implications for therapeutic outcomes. Nanocoating contribute to the decrease of unwanted effects, enhancement of treatment effectiveness, and overall improvement of patient experiences by enabling accurate drug administration and improving the compatibility of medical implants with living tissues. Thin coats of nanomaterials called nanocoating are applied to surfaces to improve their characteristics and protect them from the damaging effects of corrosion. The synthesis of these coatings is often achieved by employing a variety of processes, including as chemical vapor deposition, spin coatings, sputtering, plasma spraying, cold spraying, and electroplating (Parray et al., 2023).

This chapter aims to give a thorough analysis of the role of nanocoating in redefining the present state of modern medicine. From fundamental principles to complex applications, the chapter will investigate several elements, revealing the opportunities, challenges, and potential developments associated with the incorporation of nanocoating into healthcare approaches. As we proceed on the course of this study, the section will take us through the fundamental concepts of nanocoating, their numerous applications in medicine, and the complicated relationship between the fields of precision medicine and nanotechnology. Furthermore, it will rigorously investigate the challenges experienced in the implementation of nanocoating, providing profound understanding into the solutions and advancements that build the framework for a future period in which healthcare is differentiated by exceptional accuracy and potential.

Organic Nanocoating

Organic nanocoating are thin films made of carbon-based compounds that are applied at the nanoscale. They take advantage of the special properties of organic materials to offer certain functions including improved adhesion, flexibility, and compatibility with living things. Organic nanocoating' intrinsic flexibility allows them to adapt to a wide variety of underlying surfaces, making them suitable for a wide range of materials and forms. The innate biocompatibility of many of the organic chemicals employed in these coatings lowers the likelihood of unfavorable reactions occurring when they come into touch with biological systems. Furthermore, organic coatings can be engineered to have their properties customized, which adds versatility for a variety of uses, such as delivering drugs and surface property change.

Organic nanocoating are used in drug delivery applications because of their ability to coat and protect therapeutic molecules, allowing for regulated and targeted release. It can improve the compatibility of medical implants with biological systems, reducing the risk of inflammatory responses and further improving implant integration with surrounding tissues. These coatings are applied to medical devices and surfaces to adjust qualities such as wettability, adhesion, and resistance to microbial settlement, thereby contributing to improved performance and safety. Metallic surfaces can have their level of corrosion protection increased using advanced nanocomposite coatings. When nanofillers are incorporated into organic matrices, key characteristics required for the use of anticorrosive agents are improved (Harb et al., 2020). The impact of the organic coating on the magnetic properties of nanoparticles is thoroughly analyzed in this study. The paper looks at different ways to control the surface chemistry and architecture of assemblies of nanoparticles (Abdolrahimi et al., 2021).

Study presents an inquiry into the hydrophobic and thermal stability of composite nanocoating made of both organic and inorganic components. These nanocoating make use of several silane alkyl chain groups. For the films made using Si/ITMS and Si/ETMS, the water contact angle attained by the glass substrate was measured to be $154 \pm 2°$ anD $150.4 \pm 2°$, respectively. It was found that the superhydrophobic coatings were remarkably resistant to high temperatures, retaining their stability up to 200°C (Syafiq et al., 2018). The reduced size of nanomaterials causes them to exhibit unique properties. The density of states and electronic energy spectrum of nanoparticles change. The effect of organic and inorganic coatings on magnetic nanoparticles is investigated in this article. An extensive overview of theories explaining the unanticipated magnetism caused by nanoscale phenomena is discussed (Crespo et al., 2013).

At lower levels of nanoparticle loading, the properties of nanocomposite coatings are improved. Ion exchange or covalent reactions are the processes that transform clay. Several organic groups are used to modify anisotropic nanoparticles, or sepiolite. The evaluation of the general effectiveness and surface characteristics of nanocomposite coatings is carried out (Wouters et al., 2010). Achieving the endurance and firmness of organic nanocoating in living tissues can be a tough challenge, necessitating ongoing research aimed at improving their long-term stability. Scholars work to improve the efficiency of organic coatings for specific reasons such as customising treatment discharge behaviour or enhancing the impact of surface changes.

Inorganic Nanocoating

Inorganic nanocoating are thin layers of inorganic compounds that are deposited uniformly on a nanoscale. These coatings take advantage of the unique properties of inorganic compounds to achieve specialised functionality such as increased lifetime, conductivity, or exclusive optical properties. Inorganic nanocoating are frequently recognised for their outstanding hardness and endurance, making them suitable for applications requiring abrasion resistance and environmental concerns. Inorganic substances have the ability to exhibit different properties such as electrical conductivity, magnetic susceptibility, or optical transparency, which imparts tremendous value in a variety of technological applications. Many inorganic nanocoating have high thermal stability, making them appropriate for use in situations with extreme temperatures.

In the field of medical imaging techniques, inorganic nanocoating are crucial in the development of contrast agents. When used in imaging modalities such as MRI or CT scans, these coatings serve to increase the visibility of specific tissues or structures. Furthermore, medical implants benefit from the use of inorganic coatings such as hydroxyapatite, which not only improves biocompatibility but also allows for smooth integration with adjacent tissues. Furthermore, the use of inorganic nanocoating adds considerably to the creation of diagnostic tools with increased sensitivity, allowing for the early detection of diseases or specific diagnostics.

Bionic nanocoating based on metalized retinin have strong antiviral and bactericidal properties. Composites based on silver nanoparticles show stronger bactericidal effects, while composites based on copper nanoparticles show stronger antiviral effects. Scientists used cutting-edge instruments like scanning electron microscopy (SEM) and energy-dispersive X-ray spectroscopy (EDS) to understand nanomaterials. They looked into the effects on human cells and found that nanocoating can both kill bacteria and provide viral protection. Surfaces with the ability to repair and think are a fascinating prospect because to this combination of biomolecule and metal (Kryuchkov et al., 2022). The combination of stearic acid and SiO_2 nanoparticles that were improved by oleic acid, according to the data, had the highest water contact angle—158.6°—and excellent sliding properties. These findings suggest the existence of both superhydrophobicity and self-cleaning properties. The combination of SiO_2 aerogel and stearic acid also showed a similar water contact angle, but with better durability. Conversely, the combination with intact SiO_2 nanoparticles showed superhydrophobicity but lacked durability and self-cleaning capabilities, while stearic acid alone showed hydrophobicity (Fallah et al., 2018).

The assurance of biocompatibility of inorganic nanocoating is critical for their effective use in medical contexts, and current research focuses on this specific aspect. Researchers working to promote the widespread use of inorganic nanocoating in medical applications have been concentrating on the problems of developing reliable synthesis procedures and ensuring production scalability.

Hybrid Nanocoating

Hybrid nanocoating are thin layers that combine organic and inorganic components on the nanoscale. The goal of hybrid coatings is to synergistically enhance specific features or capabilities by combining the benefits of both material categories. Hybrid nanocoating exhibit versatility by combining the pliability and customizability of organic materials with the rigidity and robustness of inorganic materials. The combination of organic and inorganic constituents allows for the creation of coatings with customised qualities such as increased mechanical strength, improved compatibility with living organisms, or advanced optical capabilities. The mutual interaction of organic and inorganic materials frequently results in coatings that outperform the capabilities of their individual constituents, providing a multifaceted approach to meet a wide range of applications.

The process of combining hybrid organic/inorganic molecules with integrated organic ligands and an amorphous oxide covering is discussed. The goal of these mixed coatings is to provide long-lasting CO_2 electro reduction catalysts. Coatings with hybrid oxides strengthen copper catalysts for the electroreduction of CO_2. As a result, the copper surface is effectively contained by these coatings in a state that is resistant to reduction (Albertini et al., 2023). This paper investigates the combination of hybrid nanocoating with silver nanoparticle (AgNP)-adorned self-assembled antimicrobial peptide (AMP) amphiphiles to prevent implant infections. The use of antimicrobial coatings has enormous promise for preventing implant infections. The enhanced antibacterial activity of hybrid nanocoating made of self-assembling organic-inorganic amphiphiles has been demonstrated (Ye et al., 2022).

In order to combine the stability and targeting abilities of inorganic materials with the qualities of controlled release demonstrated by organic materials, hybrid nanocoating are used in drug delivery systems. In the context of medical implants, hybrid coatings have the potential to improve both mechanical strength or durability, attained through the insertion of inorganic components, and biocompatibility, attained through the introduction of organic components. Biomolecules can be included into hybrid coatings to provide surfaces that interact well with biological systems for applications in tissue engineering and diagnostics. An ongoing task is to maintain stability and functionality while achieving the ideal balance between organic and inorganic ingredients to maximise synergistic effects. Research and development must prioritise ensuring the biocompatibility and security of hybrid nanocoating in medical facilities.

The investigation of the characteristics that make nanocoating very desirable for various applications as shown in Table 1 represents the study of the optimal nanocoating properties.

Table 1. Properties of ideal nanocoating

S. No.	Property	Description
1	**Thickness and Uniformity**	Nanocoating are incredibly small, particularly when viewed at the nanoscale. In order to preserve the substrate's original characteristics, these coatings are thin. For stable operation, a homogeneous coating on the surface is essential.
2	**Adhesion**	Strong adherence to the substrate material is a necessary quality of ideal nanocoating for longevity and durability. Delamination and separation are avoided via efficient intermolecular interactions.
3	**Chemical and Mechanical Stability**	Nanocoating should remain stable in a variety of environments and uses. Mechanical stresses must be tolerated by the coating without causing it to lose its integrity.
4	**Biocompatibility**	For medical uses, nanocoating should be biocompatible to reduce negative reactions with living tissues.
5	**Tailorability**	The capacity to design and tailor nanocoating for particular uses is a crucial characteristic. This involves modifying the thickness, composition, and surface functionalities.
6	**Optical Transparency**	In certain applications, nanocoating that are ideal for optical coatings or coatings for electronic displays will retain optical transparency, which enables clear visibility and minimal interference.
7	**Electrical Conductivity or Insulation**	Nanocoating can conduct or insulate electricity, depending on the application. The best coatings have precise control over electrical characteristics.
8	**Anti-Microbial Properties**	In some cases, nanocoating with anti-microbial properties help reduce microbial settlement and improve safety and hygiene.
9	**Environmental Friendliness**	Ideal nanocoating prioritize eco-friendly practices, reduce harmful substances, and promote sustainability.
10	**Ease of Application**	Nanocoating should be easily applied with appropriate methods, ensuring efficiency and cost-effectiveness in large-scale manufacturing.

APPLICATIONS OF NANOCOATING IN MEDICINE

The use of nanocoating in diagnosis, treatment, and patient care has completely changed medicine. They facilitate accurate medication administration, advance medical implants, and improve diagnostic equipment. In addition to supporting tissue engineering, nanocoating collaborate well with cutting-edge technologies to provide personalized treatment. The primary uses for nanocoating are depicted in Figure 1.

Figure 1. Medical applications of nanocoatings

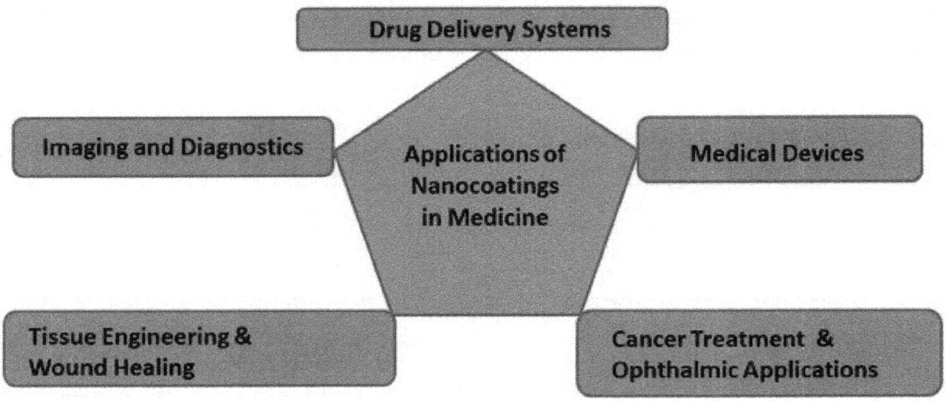

Drug Delivery Systems

The precise and controlled delivery of medicinal compounds made possible by the use of nanocoating in drug delivery systems has revolutionized the field of medicine. These novel technologies offer solutions to the problems associated with conventional drug delivery techniques, such as non-specific targeting and systemic side effects. In the field of medication distribution, two salient features stand out as being important:

1. Targeted Drug Delivery
2. Controlled Release

Targeted Drug Delivery

Targeted drug delivery refers to the precise distribution of pharmaceuticals to particular biological entities, anatomical structures, or organs in order to reduce exposure to non-targeted areas. The mechanism of action as follows:

- **Surface Modification**: Pharmaceutical compounds are specifically included into nanocoating so that they can be released in response to specific stimuli or conditions.
- **Active Targeting**: Surface-based functionalities of nanocoating have the capacity to deliberately target particular biomolecules or receptors linked to diseased cells with their actions.
- **Passive Targeting**: The enhanced permeability and retention (EPR) effect allows nanoparticles to collect within aberrant tissues, which is made possible by the nanoscale size.

By delivering medications directly to the desired site, targeted drug delivery reduces exposure to healthy tissues and, consequently, the possibility of undesirable reactions. This results in a decrease in side effects. Additionally, the concentration of therapeutic substances at the targeted site minimizes the required dosage while improving treatment effectiveness and efficacy of therapy.

Controlled Release

Controlled release, also known as sustained release, is the slow, controlled release of medicinal agents over a prolonged period of time. The mechanism of action as follows:

- **Encapsulation:** Pharmaceutical drugs can be encapsulated by nanocoating, creating a protective covering that controls the release kinetics.
- **Responsive Release:** The rate of release can be precisely adjusted in response to external stimuli like magnetic fields and light, as well as environmental variables like pH, temperature, and the presence of enzymes.

Reducing fluctuations and improving treatment outcomes are achieved by keeping medication concentrations in the bloodstream at steady levels. By reducing the frequency of drug dosage, the use of controlled release devices frequently improves patient adherence to treatment plans.

In the field of anti-inflammatory treatments, nanoparticle-mediated drug delivery platforms have found use, showing promise in enhancing anti-inflammatory outcomes while concurrently minimising undesirable effects (Wang et al., 2021). In the field of medicine, nanocoating have been used to create drug delivery systems that can selectively localise to specific areas, increase the effectiveness of therapy, and simultaneously reduce any negative side effects (Su et al., 2020). Drug administration and diagnostic procedures are both facilitated by nanomedicine. Additionally, there have been developments in the area of intelligent medication delivery devices that may react to outside stimuli (Aiacoboae et al., 2017). Due to their small size, compatibility with living things, and capacity to successfully maintain lipophilic pharmaceutical chemicals inside their core region, nanomicelles have shown promise as nanocarriers for the transportation of pharmacological drugs (Tawfik et al., 2020). Due to their small size, biological compatibility, and ability to safely contain liquid medicines within their nucleus, nanomicelles have shown promise as nanocarriers for the administration of pharmaceuticals. To measure nanoparticle size, TEM and SEM are used (Mirza et al., 2014).

Medical Devices

Nanocoating enhance the surface properties of medical devices, such as implants and prosthetics, reducing friction and obstructing microbial adherence while also improving biocompatibility. Nanocoating with antimicrobial properties help to prevent infections linked to medical devices, which lowers the risk of complications. Here are some key points regarding the use of nanocoating in medical devices:

- Medical equipment using nanocoating can have less friction. This enhances the longevity and functionality of mobile devices, which makes it significant.
- Antimicrobial qualities of nanocoating can stop infections. As an added layer of defence, they prevent microorganisms from adhering to device surfaces.
- Medical equipment with nanocoating have increased biocompatibility. By doing this, the possibility of negative reactions or immune system rejection is decreased.
- For continuous or focused drug release, some nanocoating improve drug delivery. For targeted therapy, these coatings can be included onto the surfaces of medical devices.
- Precise surface alterations are made possible by nanocoating, and this is essential for improved device integration with surrounding tissues, particularly for implants that need to be stable and functioning.
- Medical equipment that has nanocoating on it is more resilient to wear and tear, extending its lifespan and lowering the need for regular updates or replacements.
- Medical device surfaces that have certain nanocoating applied to them are less likely to create biofilm, which makes cleaning them easier and lowers the possibility of problems from microbial infection.

Currently, nanostructures are being created with the intention of using them in biomedical applications. Furthermore, biological components are being incorporated into biomedical technologies as they advance (Shaik et al., 2021). In the field of medicine, nanocoating have a wide range of uses, including the improvement of performance, the reduction of infection rates, and the support of biocompatibility on medical devices. Their use encompasses pharmaceutical administration, therapies, imaging, and therapeutics (Damodharan, 2021). Various processes are used in the field, including vapour deposition, plasma-assisted ion-beam-assisted techniques, chemical reduction, grafting, pulsed laser deposition (PLD), sputtering, extrusion, self-assembly, layer-by-layer (LbL) coating, dip coating, sol-gel coating, solution casting, and electrospinning. For nanoscale coating, the electrostatic LbL self-assembly method is used (Yilmaz et al., 2019). Medical devices such as dental implants, stents, and hip prosthesis have benefited from the application of nanocoating in order to improve operational effectiveness and reduce the spread of bacterial development. Both top-down and bottom-up approaches are used in the manufacturing of nanotechnology. Additionally, efforts have been made to incorporate silver nanoparticles into titanium nanotubes (Ahmed et al., 2016). In order to address issues with restenosis and thrombosis, nanocoating are now used in the creation of coatings for implantable medical devices (Arsiwala et al., 2013).

Imaging and Diagnostics

By improving the synthesis of contrast agents used in medical imaging modalities, inorganic nanocoating significantly improve the ability to distinguish between tissues and structures in procedures like MRI and CT scans. Nanocoating are utilised to create incredibly responsive biosensors that are used to locate biomarkers, enabling the early detection of diseases. The following are some main ways that inorganic nanocoating improve diagnostics and imaging in medicine:

- For medical imaging, inorganic nanocoating improve the production of contrast agents. They enhance the contrast between tissue and structure in imaging modalities.
- Highly sensitive biosensors for identifying illness biomarkers are made possible by nanocoating. This makes diagnosis easier and allows for prompt treatment.
- Biosensors with nanocoating can identify diseases early by focusing on particular biomarkers. Early identification lessens the effect of the disease and enhances treatment success.
- Personalized diagnostic procedures are made possible in precision medicine by nanocoating. They support the discovery of biomarkers for specialized cancer therapies.

- Nanocoating lessen negative effects and increase imaging accuracy. By using fewer contrast chemicals or radiation, they reduce the possibility of adverse consequences for patients.
- Different imaging applications can be catered for with nanocoating. Under some circumstances, they can permit imaging or react to influences.
- Combining nanotechnology with integration creates opportunities for multimodal imaging. Comprehensive diagnostics can be achieved by combining several approaches.

Numerous opportunities exist in the realm of nanotechnology for the application of clinical imaging techniques. In addition to acting as carriers for pharmacological compounds used in the treatment of cancer, nanoparticles can be used as effective contrast agents during imaging techniques. However, due to their intrinsic toxicity, the introduction of nanoparticles also brings a number of challenges (Prathna et al., 2021). Due to their optical properties and compatibility with living things, gold nanoparticles, also known as AuNMs, have bright futures in the field of biomedical imaging and diagnostics. To test the toxicity of these nanomaterials, in vitro cell culture techniques can be used. Additionally, Surface-enhanced Raman spectroscopy (SERS) can be used to improve molecular imaging. Additionally, research into the possible uses of gold nanoparticles in the fields of biological diagnostics and therapies is necessary. Further research must be done in order to determine the materials' long-term safety and to create effective plans for changing their surfaces (Sasidharan et al., 2015).

Nanotechnology increases the accuracy and precision of diagnostic techniques while lowering their cost. Nanomaterials make it possible to designate and portray desired cells and tissues intracellularly (Savaliya et al., 2015). There are several potential uses for nanocoating in the medical industry, particularly in the areas of imaging and diagnostics. This involves the use of nano diagnostics, which aims to improve sensitivity and make disease detection easier and early. Additionally, there are magnetic nanobeads that have been DNA-labeled and can recognise proteins and nucleic acids. Pathogens can also be measured using bioassays based on bioconjugate nanoparticles (Alharbi et al., 2014). In the field of diagnosing infectious diseases, nano diagnostic technologies have seen great success. The development of extremely complex and cutting-edge point-of-care nano diagnostics calls for coordinated and interdisciplinary efforts. With a specific focus on point-of-care nano diagnostics, recent trends in nano diagnostics for infectious disorders deserve special consideration. (Wang et al., 2017).

Tissue Engineering and Wound Healing

Nanocoating significantly contribute to the development of biomimetic scaffolds used in tissue engineering applications because they create an environment that is conducive to cellular growth and tissue regeneration. The replication of the inherent properties of the biological extracellular matrix is made possible by the manipulation of nanocoating, promoting advantageous cellular interactions. The creation of wound dressings intended to prevent infections and hasten the healing process of wounds benefits from the use of nanocoating containing antimicrobial agents. Responsive nanocoating can intelligently release therapeutic ingredients to effectively enhance and optimize the entire healing process by reacting to the unique conditions of the wound. The following are some salient features emphasizing their contributions:

- Nanocoating facilitate the use of biomimetic scaffolds in tissue engineering. It promote tissue regeneration and cell proliferation by imitating the extracellular matrix found in nature. The characteristics of biological tissues are replicated by nanocoating to promote advantageous cellular interactions.
- Nanoscale coatings encourage the continued growth of cells, which is crucial for tissue engineering. Regenerating healthy, functional tissues is the aim. Enhancing cell adhesion, growth, and differentiation all essential for effective tissue regeneration can be achieved by nanocoating.
- The use of nanocoating in wound healing is crucial. They encourage quicker healing and protect against infections.
- Certain nanocoating release therapeutic chemicals in response to wound situations. This treats the dynamic nature of wounds and improves the healing process.
- Anti-inflammatory nanocoating can speed up the healing of wounds. By controlling the inflammatory response, they aid in a faster and more effective healing process.
- Nanocoating can reduce the production of scar tissue by altering cellular activity. For tissue engineering applications that seek to minimize scarring, this is crucial.
- For particular tissue engineering and wound healing applications, nanocoating can be tailored. This makes it possible to create coverings that are tailored to the particular needs of various tissues and wounds.

To make drug delivery and cellular regeneration easier, nanocoating are useful in the field of tissue engineering, which includes diverse fields like bone, heart, brain, and skin tissue engineering (Ahmad et al., 2021). Exploration is being done into the creation of three-dimensional scaffolds that mimic the natural properties of the extracellular matrix. Additionally, researchers are looking into how to use nanomaterials to control cell adhesion and aid in the development of tissues. Nanomaterials' potential for use in tissue engineering is also being acknowledged. Additionally, damaged tissues can be replaced or restored using nanomaterials (Singla et al., 2019). The characteristics of tissue engineering scaffolds may be improved via nanotechnology. Substances have the ability to effectively restore a variety of organs at the nanoscale level (Abdollahiyan et al., 2021). Collagen films are created using the Langmuir-Blodgett process, while they are transferred using the Langmuir-Schaefer deposition method (Pastorino et al., 2014).

Potential uses for nanocoating exist in the field of wound healing. These coatings have the power to speed up and improve the healing process while also overcoming problems like infections, persistent wounds that won't heal, and the need to restore both cosmetic and functional qualities (Wang et al., 2021). The development of strong, high-functioning self-repairing coatings, the use of polymers with reversible covalent bonds, and the insertion of micro/nanocapsule containing curative agents are all extremely encouraging for future research applications in the automobile sector (Trentin et al., 2022). The application of therapeutic agents, the facilitation of wound closure, and the prevention of infection are three areas where nanocoating have the potential to be used in the field of wound healing (Mihai et al., 2019). Comparing them to common microorganisms, nanoparticles show special properties. Regulation of nanoparticles is possible by modification of protein and ion composition (Rajendran et al., 2018).

Cancer Treatment and Ophthalmic Applications

Nanocoating have the ability to enable the targeted delivery of chemotherapeutic drugs to tumour cells, thereby minimizing the harm done to healthy tissues. Theragnostic techniques are further supported by the incorporation of nanocoating with imaging agents, enabling the simultaneous diagnosis and therapy of cancer. The addition of nanocoating to contact lenses improves the comfort of the wearer, increases the amount of oxygen that can pass through the lenses, and reduces the possibility of bacterial adhesion. Ophthalmic implants with nanocoating allow for the controlled release of drugs for illnesses including glaucoma. Here are some key points highlighting their contributions:

- With the use of nanocoating, chemotherapeutic medications can be delivered to cancer cells specifically, minimizing harm to healthy tissues and lowering the negative effects of conventional chemotherapy.

- Drugs are delivered to tumour cells directly by nanocoating, minimizing damage to surrounding tissues, increasing the effectiveness of cancer treatments, and lowering systemic side effects.
- Imaging agents are incorporated into nanocoating to facilitate simultaneous diagnosis and therapy, give real-time data on treatment efficacy, and allow for necessary modifications.
- Because contact lenses cause less friction and discomfort, comfort has increased. Additionally, they have higher oxygen permeability to protect and preserve the health of the retina. Antimicrobial nanocoating reduce the chance of contact lens-related visual infections.
- Pharmaceuticals can be released gradually and under control through nanocoating on ophthalmic implants. This is essential for controlling intraocular pressure in glaucoma and related disorders.
- Ophthalmic equipment using antimicrobial nanocoating reduce the chance of infection, irritation, and problems.
- Ophthalmic device performance is increased, adverse response rates are reduced, and biocompatibility is enhanced by nanocoating.

Through the focused delivery of therapeutics to cancerous cells, nanomedicines, which also include nanocoating, demonstrate promising possibilities in the field of cancer treatment by reducing the negative consequences that may result from off-target toxicity (Tracey et al., 2021). Mechanisms generated from nanotechnology, such as carbon nanotubes, semiconductor nanocrystals, and nanostructured shells, have applications in the treatment of cancer (Taleuzzaman et al., 2021). In the areas of cancer theragnostic, nanovaccines, and gene therapy delivery platforms, it is common to see the use of nanoparticles that have achieved clinical approval. Additionally, it is conceivable to develop multifunctional nanomachines and nanorobots for targeted therapy by fusing individual nanoparticles with biomolecules. Rarely does the medication distribution to the tumour exceed 10% of the injected amount. Drug delivery systems that react to stimuli are currently beginning to appear (Loukanov et al., 2019).

Nanomaterials have the capacity to act as carriers for eye-specific medicinal drugs, improving their absorption and bioavailability and aiding in the diagnosis of disorders of the eye (Liu et al., 2021). In comparison to conventional treatment modalities, nanocarriers offer the potential for longer drug release and increased effectiveness in therapy, as demonstrated by the significant research done for the aim of ocular medication delivery (Singh et al., 2019). In the field of ophthalmology, the research investigates the potential use of nanoparticles in the treatment of anterior and posterior eye problems. Innovative diagnostic techniques and nanomedical equipment are used (Kamaleddin et al., 2017). In order to prevent bacterial infections, nanocoating have been applied to prosthetic eyes and silicone scleral buckles used in optical devices. Silver-containing antibacterial coatings created by RF co-sputtering have been used. These coatings are made of silver nanoclusters embedded in a silica matrix (Baino et al., 2017).

FUTURE PERSPECTIVES AND CHALLENGES

Emerging Trends

Examine the potential for modifying nanocoating to fit the unique characteristics of patients, resulting in the development of specialized and more effective therapeutic techniques. Think carefully about how nanocoating might be used with cutting-edge medical technology, such as robotics, 3D printing and artificial intelligence.

It is possible for nanocoating to improve the compatibility and functionality of robotic medical devices. As an example, coating robotic surgical instruments with nanomaterials can improve biocompatibility, reduce friction, and increase precision when doing surgery. The utilisation of robotic systems facilitates the methodical application of nanocoating, ensuring consistency and accuracy in implants and medical equipment. The application of nanocoating in the medical domain has the potential to augment the biocompatibility of 3D-printed objects. The performance and utility of 3D-printed tissues, organs, or implants may be improved by adding nanoparticles to coating solutions. The combination of nanocoating with 3D printing has the potential to create customized medical structures and devices for each patient while maximizing their biological and physical attributes.

Employing equipment like an MRI or PET-CT scan to diagnose at the nanoscale level. Using methods including chemotherapy, radiation, and photodynamic therapy, treatment at the nanoscale. Early illness diagnosis is made possible by nanotheranostics. Nanoscale therapy is offered by nanotheranostics (Suhag et al., 2020). The chapter looks at how nanobiotechnology is being used to treat cancer, neurological disorders, and cardiovascular diseases on an individual basis (Jain et al., 2011). Precipitation techniques used to create iron oxide nanoparticles. To create nanoparticles of solid pharmaceuticals, a milling procedure is used (McDonald et al., 2015).

Ethical and Regulatory Considerations

Examine the ethical issues associated with cutting-edge medical technologies and emphasize the importance of maintaining a delicate balance between ground-breaking innovations and the wellbeing of patients. Analyze the existing and evolving regulatory frameworks that govern the use of nanocoating in the medical field, and consider any potential adjustments that might be necessary to keep up with new discoveries.

Modern medical technologies like artificial intelligence and nanocoating require the collection and analysis of large amounts of patient data. Ensuring patient confidentiality and obtaining informed consent are critical to the use of personal health information. Maintaining patient autonomy and trust is essential to the moral application of these developments. Accurate information exchange about the use of data and the possible outcomes of customized treatments is crucial for securing informed consent.

Availability of advanced medical technologies can lead to disparities in healthcare. Ensuring that a diverse variety of people can benefit from novel treatments including nanocoating is a matter of ethical significance. It is possible to prevent the growing severity of already-existing healthcare disparities by addressing issues with affordability and accessibility. To serve a variety of patient populations, ethical considerations should direct the development and adoption of such technology.

Concerns about patient autonomy are raised by the growing use of artificial intelligence (AI) in the field of medical decision-making. When complex algorithms are used in decision-making, patients may find it difficult to understand and this could hinder their ability to make decisions about their care. Respecting patient autonomy requires that open lines of communication be established on the use of AI in medical decision making. It is the responsibility of healthcare providers to ensure that patients continue to be active partners in their care.

Unintended consequences arise from the initial lack of thorough understanding regarding the long-term implications of nanocoating and their interactions with the human body. The existence of ethical issues is apparent when questions about possible unexpected implications arise up. To successfully recognize and reduce any unexpected consequences that may result from the use of nanocoating on patients, it is necessary that frequent monitoring be carried out, continuous research attempts be sustained, and ethical responsibility be observed. Maintaining and sustaining trust requires honest and forthcoming communication about uncertainties.

As indicated by the existence of multiple nanomedicines now on the market and the continuing clinical trials for numerous of others, the use of nanotechnology in the domains of medicine and pharmaceutics has resulted in a new era in therapeutic techniques. (Santhini et al., 2020)It has been noted that the ethical issue, which is frequently disregarded, is a crucial concern for any fresh improvements. Additionally, if these technological advances are used improperly, illegally, or unethically, they could represent a risk (Wiwanitkit et al., 2017). The ethical and legal implications of nanomedicine are very important. Currently, the Food and Drug Administration (FDA) is working to create specific criteria for goods utilizing nanobiotechnology (Jain et al., 2017). Concerns about societal, legal, and ethical issues are raised by regenerative nanomedicine. The definitions, applications, dangers, laws, and availability must all be taken into consideration by healthcare professionals and researchers (Glenn et al., 2012).

Nanomaterials have particular qualities that make them suited for use in the medical industry. The use of nanomaterials has fascinating future applications in a number of fields, including imaging, therapy, drug delivery, and tissue engineering. Additionally, the combination of ultrasonic therapy with TiO_2 nanoparticles and the use of magnetic hyperthermia in the elimination of cancer cells present new research opportunities (Chen et al., 2013). The protection and wellbeing of people taking part in clinical trials including the study of nanoproducts in human beings is the main goal of the draft guidelines for nanomedicine (Marshall et al., 2011).

INNOVATIONS IN NANOCOATING DESIGN

Significant progress is being made in the design of smart nanocoating which will indicate in a new era of intelligent medical devices. These cutting-edge coatings have the ability to react dynamically to changing physiological conditions within the human body, maximizing the effectiveness of therapeutic interventions. The idea behind smart nanocoating involves a degree of adaptability that permits in-the-moment modifications, providing a promising path toward individualized and accurate medicinal treatments. A crucial part of this evolution is the investigation of multifunctional platforms inside nanocoating. A novel method for integrating healthcare technologies is the potential for nanocoating to serve two or more functions, such as acting as carriers for drug delivery and diagnostic tools at the same time. This confluence is a key step towards the future of nanotechnology in medicine since it not only streamlines medical operations but also holds the prospect of improved patient outcomes through a more clear and synergistic medical approach.

The creation of high-entropy nanocoating on the hard metal T12A is investigated in the study. It illustrates the possibility of avoiding intermetallic compound formation in these coatings (Popov et al., 2020). Nanocoating provide strength and protect against corrosion. The controlled synthesis of ceramic powders is made easier by the hydrothermal process (Singh et al., 2018). The top-down methodology includes the machining of microstructures in its scope. In order to introduce a nonsolvent, a solvent-mediated approach is used, which causes the coating of the dissolved substance (Bommana et al., 2019). Polymer nanocomposite coating design and production must be sustainable, according to experts in the industry. For the aim of optimization, the use of multiscale modelling and simulation techniques is essential (Xiao et al., 2011). The efficiency and security of nanomedicines are improved by nanotechnology. International harmonization is required for safety evaluation (Grewal et al., 2018).

CONCLUSION

This chapter offers a thorough examination of the revolutionary effects of nanocoating in the medical industry. The exposition begins with a perceptive introduction that lays the groundwork for understanding the rest of the narrative. The key components of nanocoating are thoroughly discussed, including their definitions, characteristics, types and fabrication methods. The chapter then proceeds to a full analysis of the various medical uses for nanocoating. Each application is extensively examined, demonstrating the attention to detail and adaptability that nanocoating offer to medical treatments. These applications range from drug delivery systems to medical equipment, from imaging and diagnostics to tissue engineering and wound healing. Notable portions focus on the use of nanocoating in ophthalmology and cancer treatment, highlighting their potential influence on important medical domains. The chapter offers a glimpse into the future by describing forthcoming trends, ethical and regulatory implications, and the ongoing advancements in nanocoating design. It acts as a guide for navigating the opportunities and difficulties of utilizing nanotechnology for revolutionary advances in healthcare.

REFERENCES

Abdollahiyan, P., Oroojalian, F., & Mokhtarzadeh, A. (2021). The triad of nanotechnology, cell signalling, and scaffold implantation for the successful repair of damaged organs: An overview on soft-tissue engineering. *Journal of Controlled Release*, *332*, 460–492. doi:10.1016/j.jconrel.2021.02.036 PMID:33675876

Abdolrahimi, M., Vasilakaki, M., Slimani, S., Ntallis, N., Varvaro, G., Laureti, S., Meneghini, C., Trohidou, K. N., Fiorani, D., & Peddis, D. (2021). Magnetism of nanoparticles: Effect of the organic coating. *Nanomaterials (Basel, Switzerland)*, *11*(7), 1787. doi:10.3390/nano11071787 PMID:34361173

Ahmad, A., Younas, M., Santulli, C., Kiyani, M. Z., Ali, R., Habbiba, O., & Zubair, M. (2021). Nanomedicine and tissue engineering. In *Nanomedicine Manufacturing and Applications* (pp. 261–277). Elsevier. doi:10.1016/B978-0-12-820773-4.00020-2

Ahmed, W., AlHannan, M., Yusuf, S., & Jackson, M. J. (2016). Nanocoating for medical devices. *Surgical Tools and Medical Devices*, 519-543.

Aiacoboae, A., Gheorghe, T., Lungu, I. I., Curutiu, C., Chifiriuc, M. C., Grumezescu, A. M., & Holban, A. M. (2017). Applications of nanoscale drugs carriers in the treatment of chronic diseases. In *Nanostructures for Novel Therapy* (pp. 37–55). Elsevier. doi:10.1016/B978-0-323-46142-9.00002-5

Albertini, P., Newton, M., Wang, M., Lecina, O. S., Green, P., Stoian, D., & Buonsanti, R. (2023). Hybrid Oxide Coatings Generate Stable Cu. *Catalysts*, CO2.

Alharbi, K. K., & Al-Sheikh, Y. A. (2014). Role and implications of nanodiagnostics in the changing trends of clinical diagnosis. *Saudi Journal of Biological Sciences*, *21*(2), 109–117. doi:10.1016/j.sjbs.2013.11.001 PMID:24600302

Arsiwala, A. M., Raval, A. J., & Patravale, V. B. (2013). Nanocoating on implantable medical devices. *Pharmaceutical Patent Analyst*, *2*(4), 499–512. doi:10.4155/ppa.13.30 PMID:24237126

Baino, F., Perero, S., Miola, M., & Ferraris, M. (2017). Antibacterial nanocoating for ocular applications. *Advances in Science and Technology (Owerri, Nigeria)*, *102*, 24–28. doi:10.4028/www.scientific.net/AST.102.24

Bommana, M. M., Kirthivasan, B., Gupta, S. S., Shikhar, A., & Shah, A. (2019). Nanostructures in pharma: elixir to oral medicine. In *Nanoparticles in Pharmacotherapy* (pp. 23–44). William Andrew Publishing. doi:10.1016/B978-0-12-816504-1.00001-6

Chen, S., Zhang, Q., Hou, Y., Zhang, J., & Liang, X. J. (2013). Nanomaterials in medicine and pharmaceuticals: Nanoscale materials developed with less toxicity and more efficacy. *European Journal of Nanomedicine*, *5*(2), 61–79. doi:10.1515/ejnm-2013-0003

Crespo, P., De la Presa, P., Marin, P., Multigner, M., Alonso, J. M., Rivero, G., ... Hernando, A. (2013). Magnetism in nanoparticles: Tuning properties with coatings. *Journal of Physics Condensed Matter*, *25*(48), 484006. doi:10.1088/0953-8984/25/48/484006 PMID:24201075

Damodharan, J. (2021). Nanomaterials in medicine–An overview. *Materials Today: Proceedings*, *37*, 383–385. doi:10.1016/j.matpr.2020.05.380

Fallah, M., Ghashghaee, M., Rabiee, A., & Ershad-Langroudi, A. (2018). Spray-deposition of an organic/inorganic blend for fabrication of a superhydrophobic surface: Effect of admixing with silica aerogel and modified silica nanoparticles. *Protection of Metals and Physical Chemistry of Surfaces*, *54*(5), 909–916. doi:10.1134/S2070205118050064

Glenn, L. M., & Boyce, J. S. (2012). Regenerative nanomedicine: ethical, legal, and social issues. *Nanotechnology in Regenerative Medicine: Methods and Protocols*, 303-316.

Grewal, A. S., Lather, V., Sharma, N., Singh, S., Narang, R. S., Narang, J. K., & Pandita, D. (2018). Recent updates on nanomedicine based products: Current scenario and future opportunities. *Applied Clinical Research, Clinical Trials and Regulatory Affairs*, *5*(2), 132–144. doi:10.2174/2213476X05666180611115135

Harb, S. V., Trentin, A., Uvida, M. C., & Hammer, P. (2020). Advanced organic nanocomposite coatings for effective corrosion protection. In *Corrosion Protection at the Nanoscale* (pp. 315–343). Elsevier. doi:10.1016/B978-0-12-819359-4.00017-9

Jain, K. K. (2011). Nanobiotechnology and personalized medicine. *Progress in Molecular Biology and Translational Science*, *104*, 325–354. doi:10.1016/B978-0-12-416020-0.00008-5 PMID:22093223

Jain, K. K., & Jain, K. K. (2017). Ethical and Regulatory Aspects of Nanomedicine. The Handbook of Nanomedicine, 605-620. doi:10.1007/978-1-4939-6966-1_18

Kamaleddin, M. A. (2017). Nano-ophthalmology: Applications and considerations. *Nanomedicine; Nanotechnology, Biology, and Medicine*, *13*(4), 1459–1472. doi:10.1016/j.nano.2017.02.007 PMID:28232288

Kryuchkov, M., Adamcik, J., & Katanaev, V. L. (2022). Bactericidal and Antiviral Bionic Metalized Nanocoating. *Nanomaterials (Basel, Switzerland)*, *12*(11), 1868. doi:10.3390/nano12111868 PMID:35683724

Liu, N., Wu, Q., Liu, Y., Li, J., Ji, P., & Fu, G. (2021). Application of Nanomaterials in the Treatment and Diagnosis of Ophthalmology Diseases. *Current Stem Cell Research & Therapy*, *16*(1), 95–103. doi:10.2174/1574888X15666200210104449 PMID:32039688

Loukanov, A., Nikolova, S., Filipov, C., & Nakabayashi, S. (2019). Nanomaterials for cancer medication: From individual nanoparticles toward nanomachines and nanorobots. *Pharmacia*, *66*(3), 147–156. doi:10.3897/pharmacia.66.e37739

Marshall, J. (2011). Draft guidelines for nanomedicine unveiled. *NATNews*.

McDonald, T. O., Siccardi, M., Moss, D., Liptrott, N., Giardiello, M., Rannard, S., & Owen, A. (2015). The application of nanotechnology to drug delivery in medicine. In *Nanoengineering* (pp. 173–223). Elsevier. doi:10.1016/B978-0-444-62747-6.00007-5

Mihai, M. M., Dima, M. B., Dima, B., & Holban, A. M. (2019). Nanomaterials for wound healing and infection control. *Materials (Basel)*, *12*(13), 2176. doi:10.3390/ma12132176 PMID:31284587

Mirza, A. Z., & Siddiqui, F. A. (2014). Nanomedicine and drug delivery: A mini review. *International Nano Letters*, *4*(1), 1–7. doi:10.1007/s40089-014-0094-7

Parray, M. R., Kumar, H., Singh, A., Tomar, B., & Shiva, S. (2023). Synthesis of nanomaterial coatings for various applications. In *Nanomaterials for Sustainable Tribology* (pp. 167–190). CRC Press. doi:10.1201/9781003306276-11

Pastorino, L., Dellacasa, E., Scaglione, S., Giulianelli, M., Sbrana, F., Vassalli, M., & Ruggiero, C. (2014). Oriented collagen nanocoating for tissue engineering. *Colloids and Surfaces. B, Biointerfaces*, *114*, 372–378. doi:10.1016/j.colsurfb.2013.10.026 PMID:24246194

Popov, V., Kostyuk, G., Tymofyeyev, O., Kostyk, K., & Naboka, O. (2020). Design of new nanocoating based on hard alloy. In *Design, Simulation, Manufacturing: The Innovation Exchange* (pp. 522–531). Springer International Publishing. doi:10.1007/978-3-030-50794-7_51

Prathna, T. C. (2021). Nanotechnology in Imaging Applications: An Overview. *Nanotechnology in Medicine*, 75-86.

Rajendran, N. K., Kumar, S. S. D., Houreld, N. N., & Abrahamse, H. (2018). A review on nanoparticle based treatment for wound healing. *Journal of Drug Delivery Science and Technology*, *44*, 421–430. doi:10.1016/j.jddst.2018.01.009

Santhini, O. E., & Kirupha, S. D. (2020). Nanomaterials and ethical issues. *Applications of nanomaterials in human health*, 275-286.

Sasidharan, A., & Monteiro-Riviere, N. A. (2015). Biomedical applications of gold nanomaterials: Opportunities and challenges. *Wiley Interdisciplinary Reviews. Nanomedicine and Nanobiotechnology*, *7*(6), 779–796. doi:10.1002/wnan.1341 PMID:25808787

Savaliya, R., Shah, D., Singh, R., Kumar, A., Shanker, R., Dhawan, A., & Singh, S. (2015). Nanotechnology in disease diagnostic techniques. *Current Drug Metabolism*, *16*(8), 645–661. doi:10.2174/1389200216666150625121546 PMID:26560320

Shaik, J., & Kumar, A. C. (2021). Nanostructures for biomedical devices. In *Nanoscale Processing* (pp. 299–326). Elsevier. doi:10.1016/B978-0-12-820569-3.00011-6

Singh, A., Dogra, T. S., Mandal, U. K., & Narang, R. K. (2019). Novel approaches for ocular drug delivery: A review. Int. J. Bio-. *Pharmaceutical Research*, *8*, 2722–2732.

Singh, A., Mittal, S., Mudgal, D., & Gupta, P. (2018). Design, development and application of nanocoating. *Nanomaterials and Their Applications*, 191-207.

Singla, R., Abidi, S. M., Dar, A. I., & Acharya, A. (2019). Nanomaterials as potential and versatile platform for next generation tissue engineering applications. *Journal of Biomedical Materials Research. Part B, Applied Biomaterials*, *107*(7), 2433–2449. doi:10.1002/jbm.b.34327 PMID:30690870

Su, S., & Kang, M, P. (. (2020). Recent advances in nanocarrier-assisted therapeutics delivery systems. *Pharmaceutics*, *12*(9), 837. doi:10.3390/pharmaceutics12090837 PMID:32882875

Suhag, D., Chauhan, M., Shakeel, A., & Das, S. (2020). Emerging trends in nanotheranostics. *Nanobiomedicine*, 335–368.

Syafiq, A., Pandey, A. K., Balakrishnan, V., Shahabuddin, S., & Abd Rahim, N. (2018). Organic-inorganic composite nanocoating with superhydrophobicity and thermal stability. *Pigment & Resin Technology*.

Taleuzzaman, M., Gupta, D. K., Kala, C., Gilani, S. J., & Beg, S. (2021). Nanotechnology-assisted medical devices in cancer treatment. In *Nanoformulation Strategies for Cancer Treatment* (pp. 195–205). Elsevier. doi:10.1016/B978-0-12-821095-6.00010-0

Tawfik, S. M., Azizov, S., Elmasry, M. R., Sharipov, M., & Lee, Y. I. (2020). Recent advances in nanomicelles delivery systems. *Nanomaterials (Basel, Switzerland)*, *11*(1), 70. doi:10.3390/nano11010070 PMID:33396938

Tracey, S. R., Smyth, P., Barelle, C. J., & Scott, C. J. (2021). Development of next generation nanomedicine-based approaches for the treatment of cancer: We've barely scratched the surface. *Biochemical Society Transactions*, *49*(5), 2253–2269. doi:10.1042/BST20210343 PMID:34709394

Trentin, A., Uvida, M. C., de Araújo Almeida, A., de Souza, T. A. C., & Hammer, P. (2022). Self-healing nanocoating. In *Nanotechnology in the Automotive Industry* (pp. 371–401). Elsevier. doi:10.1016/B978-0-323-90524-4.00018-9

Wang, H., Zhou, Y., Sun, Q., Zhou, C., Hu, S., Lenahan, C., Xu, W., Deng, Y., Li, G., & Tao, S. (2021). Update on nanoparticle-based drug delivery system for anti-inflammatory treatment. *Frontiers in Bioengineering and Biotechnology*, *9*, 630352. doi:10.3389/fbioe.2021.630352 PMID:33681167

Wang, M., Huang, X., Zheng, H., Tang, Y., Zeng, K., Shao, L., & Li, L. (2021). Nanomaterials applied in wound healing: Mechanisms, limitations and perspectives. *Journal of Controlled Release*, *337*, 236–247. doi:10.1016/j.jconrel.2021.07.017 PMID:34273419

Wang, Y., Yu, L., Kong, X., & Sun, L. (2017). Application of nanodiagnostics in point-of-care tests for infectious diseases. *International Journal of Nanomedicine*, *12*, 4789–4803. doi:10.2147/IJN.S137338 PMID:28740385

Wiwanitkit, V. (2017). Ethics in Nanomedicine: A Concern on New Nanotechnology. *Med One*, *2*(5).

Wouters, M., Rentrop, C., & Willemsen, P. (2010). Surface structuring and coating performance: Novel biocidefree nanocomposite coatings with anti-fouling and fouling-release properties. *Progress in Organic Coatings*, *68*(1-2), 4–11. doi:10.1016/j.porgcoat.2009.10.005

Xiao, J., & Huang, Y. (2011). Design of sustainable multifunctional nanocoating: A goal-driven multiscale systems approach. *Chinese Journal of Chemical Engineering*, *19*(4), 666–673. doi:10.1016/S1004-9541(11)60039-6

Ye, Z., Sang, T., Li, K., Fischer, N. G., Mutreja, I., Echeverría, C., & Aparicio, C. (2022). Hybrid nanocoating of self-assembled organic-inorganic amphiphiles for prevention of implant infections. *Acta Biomaterialia*, *140*, 338–349. doi:10.1016/j.actbio.2021.12.008 PMID:34896631

Yilmaz, O., & Yorgancioglu, A. (2019). Nanocoating: preparation, properties, and biomedical applications. In Polymeric Nanomaterials In Nanotherapeutics (pp. 299-331). Elsevier.

ADDITIONAL READING

Alghamdi, M. D., Nazreen, S., Ali, N. M., & Amna, T. (2022). ZnO Nanocomposites of Juniperus procera and Dodonaea viscosa Extracts as Antiproliferative and Antimicrobial Agents. *Nanomaterials (Basel, Switzerland)*, *12*(4), 664. doi:10.3390/nano12040664 PMID:35214995

Amna, T., Hassan, M. S., Yang, J., Khil, M. S., Song, K. D., Oh, J. D., & Hwang, I. (2014). Virgin olive oil blended polyurethane micro/nanofibers ornamented with copper oxide nanocrystals for biomedical applications. *International Journal of Nanomedicine*, *9*, 891–898. doi:10.2147/IJN.S54113 PMID:24611006

Hassan, M. S., Amna, T., Mishra, A., Yun, S. I., Kim, H. C., Kim, H. Y., & Khil, M. S. (2012). Fabrication, characterization and antibacterial effect of novel electrospun TiO_2 nanorods on a panel of pathogenic bacteria. *Journal of Biomedical Nanotechnology*, *8*(3), 394–404. doi:10.1166/jbn.2012.1393 PMID:22764408

Nguyen, L. T., Chau, J. K., Perry, N. A., de Boer, L., Zaat, S. A. J., & Vogel, H. J. (2010). Serum stabilities of short tryptophan- and arginine-rich antimicrobial peptide analogs. *PLoS One*, *5*(9), e12684. doi:10.1371/journal.pone.0012684 PMID:20844765

Walch, E. B. T., & Roos, C. (2020). Measurement of the mechanical properties of silver and enamel thick films using nanoindentation. *International Journal of Applied Glass Science*, *11*(1), 195–206. doi:10.1111/ijag.13875

Chapter 10
Sustainable Approach to Nanocomposites and Nanocoatings:
Research and Industrial Applications of Hydrophobic Nanocoatings

Sana Dahmani
Independent Researcher, Germany

ABSTRACT

The global hydrophobic coatings market is projected to reach $3.0 billion by 2030, growing at a CAGR of 4.2%. These innovative coatings are in high demand across sectors such as automotive, marine, construction, healthcare, textile, environmental management, and further industries thanks to their properties including and not limited to water-repellency, corrosion-resistance, and self-cleaning. Challenges faced by hydrophobic nanocoatings in durability, adhesion, cost-effectiveness, and sustainability are being addressed through research on wetting control and scalability. Hydrophobic coating-as-a-service business model offers specialized services for applying hydrophobic coatings on various surfaces, with the highest efficiency. Choosing providers with advanced expertise and high-quality materials is essential for long-lasting product protection.

DOI: 10.4018/979-8-3693-3136-1.ch010

INTRODUCTION

Nanotechnology was introduced in the 1980s and has found wide applications in various sectors, including aerospace, biomedical, cosmetics, transportation, and building construction. The concept dates back to the fourth century with the roman nanoparticle-based Lycurgus Cup which changes color according to the location of the light source it is exposed (Bayda et al., 2019). The global hydrophobic coatings market, initially valued at \$2.0 billion in 2020, is expected to experience substantial growth, reaching an estimated worth of \$3.0 billion by 2030, with a projected Compound Annual Growth Rate (CAGR) of 4.2% from 2021 to 2030 (Marpu, Prakhar and Eswara, 2022). This expansion underscores the increasing significance of hydrophobic coatings due to their exceptional water-repelling, corrosion-resistant and self-cleaning properties associated to it.

Hydrophobicity, or the propensity of water molecules to interact more with each other than with a given surface, can be assessed by employing drop shape analysis. In this technique, the contact angle is determined by examining the image of a sessile drop, where the point of contact between the liquid drop and the surface is observed. The contact angle (denoted as θ) characterizes the angle formed between the liquid surface and the tangent line to the outline of the contact area. A contact angle of around 90 degrees indicates that the surface is hydrophobic, while a contact angle of approximately 150 degrees signifies that the surface is superhydrophobic. Furthermore, on a superhydrophobic surface, the roll-off angle should be less than 5 degrees (Simpson et al., 2015).

Hydrophobic nanocoating, despite their promising applications, face challenges related to durability, adhesion, cost-effectiveness, and environmental sustainability. Researchers strive to address these challenges, focusing on heightened robustness, economic viability, and environmental conscientiousness. Emphasis is placed on precise manipulation of wetting characteristics and scalability in production methodologies. Hydrophobic nanocoatings show transformative potential in various industrial sectors. Additionally, the Hydrophobic Coating-as-a-Service model offers specialized coating services. These technologically advanced coatings, known for water and oil repellence, simplify maintenance and provide critical corrosion protection, significantly extending the lifespan of equipment. Specific formulations exhibit self-cleaning properties, reducing the need for recurrent maintenance. Applied in automotive, marine, construction, and industrial sectors, they act as protective shields, streamlining maintenance and ensuring the integrity of vessels and buildings. Engaging with reputable hydrophobic coating service companies is advisable for optimal solutions. The thin material films, serving as highly effective repellents, find versatility across automotive, aerospace, marine, construction, textiles, and other domains. In the automotive sector, they safeguard vehicles, contribute to improved fuel efficiency, and reduce drag. Similarly, in the aerospace sector, hydrophobic nanocoatings protect against ice and corrosion, improving fuel efficiency. These coatings also play a crucial role in safeguarding boats and buildings, enhancing energy efficiency and rendering textiles water and stain-resistant, with applications in clothing and medical devices.

The unique capabilities of hydrophobic nanocoatings offer tailored solutions, encompassing self-cleaning car surfaces, anti-icing coatings for aircraft, antifouling coatings for boats, water-resistant coatings for buildings, and stain-resistant coatings for fabrics. As a highly adaptable and promising technology, hydrophobic nanocoatings are poised to continue making substantial contributions across numerous sectors, advancing efficiency, and fortifying durability in diverse applications. Within nature, numerous surfaces, including lotus and rice leaves, butterfly wings, rose petals, gecko feet, spider silk, and fish scales, demonstrate outstanding instances of hydrophobicity or even super hydrophobicity with nonwetting behaviour in respect of water contact angle (WCA) respectively higher than 90 and 150 degrees. Hydrophobicity inspires biomimical industrial products with a large range of utilization and benefits including efficiency improvements, and providing alternative low-cost industrial solutions. The WCA is visualized in Figure 1.

The research request on Google Scholar namely "Hydrophobic Nanocoating" applied to the PRISMA 2009 Flow Diagram, included all types of articles including patents. The publications selected are all in English languages, collected until December 13.12.2023 and having in their title both "Hydrophobic" and "Waterproof" and "Nanocoating". The empty exported citations and the out-of-topic works were excluded as well. Among 984 found, one was not found which takes the raw data to 983 publications. 4 papers not written in english and 3 vacant references were excluded. Combined with the research query, the exclusion criteria led to 119 publication which were analysed.

The topics they cover are: energies including photovoltaic and wind, textile, surface engineering, electric power systems, fire control, optics and glass, aerospace, buidlings, healthcare, road pavement, materials, wood, food science and packaging, membranes, marine, oil-water separation and general uses. When going through those articles to analyze them, further areas of applications were encountered such as sports, pharmacology, plasma, pollution control etc. Thus, further 70 publications found during the research process were included in this manuscript. As a matter of fact, in this work, the citations were organised to provide a storytelling about the applications of hydrophobic nanocoating in several domains. The features of those materials such as antibacterial and waterproof properties etc. were evoked at the level of every field of application.

The research analysis results showed that the fields of application of hydrophobic nanocoating are numerous with the most relevant ones being: Surface Science,

Healthcare, Energies & Electricity, Hydraulics, Environmental Management & Sustainability and further applications such as optics and electromagnetic shielding. The healthcare applications include medical device, antibacterial and disinfection, the prevention of the spread of airborne viruses such as COVID-19; plasma nanocoating valuation in cardiovascular stent application and boosting functional performance of nonwovens etc. In terms of energies and hydraulics, those technologies showed an increasing efficiency of photovoltaic panels and wind-turbine blades in the renewable energies application field; membranes innovations for rapid and controlled water desalination and treatment; further icing properties, gradient refractive index core-shell coatings, solar cell performance through anti-dust coatings showing the performance of cleaning/self-cleaning hydrophobic nanocoating abilities. In the area of food and packaging field, innovations include polyvinylidenefluoride (PVDF)/thermoplastic polyurethane (TPU) nanofibers, hydrophobic paper coatings for improved food packaging, and edible nanocoatings for potential food-related applications. In terms of constructions, the coatings were found to be more effective at reducing the wettability of smooth surfaces than rough surfaces. A crucial further field of application is the textile industry. As a matter of fact, nanotechnology enhances textiles with properties including water repellence, antimicrobial action, and flame retardancy through nanocoatings. Innovations include hydrophobic MXene/black phosphorene nanocoatings, self-adhesive foam materials for various applications such as water resistance and draught sealing.

Antibacterial copper-based coatings, plasma-chemical coatings for textiles, and a transparent cellulose paper for multi-modal energy harvesting were widely used in this area. Finally, surface science including a range of subfields and a variety of nanocoatings, involves anti-corrosion and including graphene, CNTs and superhydrophobic nanocoating. This field encompasses the (1) Surface Chemistry of chemical interactions at material surfaces, (2) Surface Physics including electronic and magnetic behaviour at surfaces where hydrophobic nanocoating was used as magnetic shielding through reversible crumpling of 2D titanium carbide (MXene) nanocoatings for stretchable electromagnetic shielding and wearable wireless communication; (3) Surface Engineering through enhancing surface properties such as anticorrosion, (4) and Materials Science which aims at improving the material performance such as antifouling used in marine. Indeed, the antifouling coatings were found to be effective in preventing the adhesion of marine organisms to surfaces, such as ship hulls and offshore oil rigs.

Superhydrophobic coatings offer a multitude of practical uses, serving as a protective barrier that safeguards surfaces or materials against corrosion, contamination, and the harmful effects of various. The application of nanostructured transparent superhydrophobic coatings on glass surfaces results in diminished dust and water adhesion, ultimately enhancing the ease of cleaning. Certain formulations of these coatings have demonstrated effectiveness for several months, particularly in environments with higher frequencies of rainfall and elevated wind speeds (Mazumder et al., 2017). Furthermore, the investigation into wettability has been a catalyst for advancing self-cleaning technologies, meeting the increasing industry demand. Traditional cleaning methods involving mechanical and robotic cleaners are often impractical due to potential damage to glass surfaces, reduced electricity generation in heavy rain, and high equipment costs. In this review, we comprehensively discuss the application of transparent self-cleaning coatings, laying the foundation for future advancements in response to industry needs (Syafiq et al., 2022).

(Hooda et al., 2017) worked on a transparent superhydrophobic polystyrene/ZnO nanocoating with optimized roughness and contact angle was synthesized using dual-functionalized ZnO nanoparticles, demonstrating industrial potential. Another property of the hydrophobic nanocoating consists in wear resistance which pertain to the ability of a substance to withstand wear and tear (Yedra et al., 2014).

APPLICATIONS OF HYDROPHOBIC NANOCOATING

Healthcare

Hydrophobic nanocoatings enhance medical device safety and efficacy by reducing infection risks, improving biocompatibility, and extending device lifespan. Their versatile applications across implants, instruments, imaging, and drug delivery promise cost reduction and improved outcomes, signalling a transformative shift in the medical device industry. As technology progresses, these nanocoatings are poised to bring novel applications and advancements to medical technology.

Antimicrobial, self-healing, nanocoatings, biocompatible implant, medical imaging and corrosion protectionin Medicine: Biocompatible implants, designed for repairing or replacing damaged tissues and organs within the human body, play a crucial role in improving the health of individuals facing various medical conditions. However, challenges such as implant-related infections and device malfunction persist, highlighting the necessity for innovative solutions. Researchers are currently focusing on the development of next-generation biocompatible implants, utilizing advanced materials and technologies such antimicrobial features and self-healing capabilities. Various hydrophobic nanomaterials are being explored for potential use in biomedical implants. Titanium dioxide (TiO_2) stands out as a versatile material with excellent biocompatibility and surface modification capabilities. By engineering TiO_2, it can achieve super hydrophobicity, providing high resistance to bacteria and contaminants. Graphene oxide (GO), a thin layer of graphene, exhibits remarkable mechanical and electrical properties. Functionalizing GO enables the creation of hydrophobic surfaces resistant to wear and tear. Polymer nanofibers, such as polydimethylsiloxane (PDMS) and polyethylene glycol (PEG), can be structured into porous forms, promoting cell adhesion and facilitating sustained release of drugs or therapeutic agents. The utilization of those nanomaterials is explained with examples in Figures 2 and 3. Notably, nanocoatings represent a significant area of emphasis, offering a promising avenue to address these challenges and enhance the performance and safety of biomedical implants. Figure 4 shows the antimicrobial effect of nanomaterial on Hospital Acquired Infections (HAIs). Furthermore, Coated polymethyl methacrylate (PMMA) holds promise for effective antimicrobial nanocoatings in medical devices, including intraocular lenses (IOLs), dental implants, and bone cements. Antimicrobial agents such as silver nanoparticles and copper ions combat bacteria, fungi, and viruses on PMMA surfaces, preventing the proliferation of microorganisms such as Staphylococcus aureus, Escherichia coli, and Pseudomonas aeruginosa. Silver nanoparticle-coated PMMA IOLs specifically reduce the risk of posterior capsule opacification (PCO), a common post-IOL surgery complication.

These versatile nanocoatings offer infection mitigation across various medical applications, with PMMA bone cements reducing infection risks and silver nanoparticle-coated PMMA dental implants lowering the risk of peri-implantitis (Bellini, 2022). For instance, the study of (Wang et al., 2018) demonstrated that the best-coated PMMA slide achieved a remarkable transmittance of 94.5% and demonstrated strong antibacterial performance against both Staphylococcus aureus and Escherichia coli as well. Transmittance refers to the quantity of radiant energy that traverses the PMMA. Moreover, the coatings exhibited excellent adhesion to the PMMA substrata in (1) ultrasonic and (2) pencil hardness tests, presenting a promising approach for applications in optical devices and everyday plastic glass products. The ultrasonic hardness test (1) is a non-destructive method assessing material hardness by measuring sound wave velocity. Greater hardness results in faster sound wave propagation. The second (2) is a straightforward and budget-friendly technique for evaluating a coating's scratch resistance. It entails using a range of pencils with different hardness levels to draw on the coating, with the hardest pencil that doesn't cause scratches defining the coating's hardness.

In 2010, Vivek Polshettiwar's team developed DFNSc (Dendritic Fibrous Nanosilica) with customizable properties: particle sizes (50-1200 nm), high surface area (500-1200 m^2/g), and adjustable pore volume (0.32-2.18 cm^3/g). DFNS offers wide-ranging stability, unique morphology for exceptional accessibility, and sustainable synthesis through a low E-factor protocol. Explored for 12 years, DFNS finds applications in catalysis, solar energy, CO_2 capture, conversion, sensing, biomedicine, and energy storage (Polshettiwar, 2022). DFNS, or dendritic fibrous nano silica, is an emerging category of mesoporous nanoparticles attracting attention due to its unique features, including three-dimensional dendritic superstructures, sizable pore channels, and easily accessible internal surface areas (Shaban and Hasanzadeh, 2020). This material holds great promise for antimicrobial nanocoatings in medical applications, boasting attributes such as high surface area, substantial pore volume, tuneable pore size, impressive thermal and hydrothermal stability, and broad-spectrum antimicrobial properties. DFNS-based nanocoatings have proven highly effective, reducing bacterial populations on surfaces by up to 99.9%. This technology has the potential to prevent bacterial colonization on various medical devices, such as catheters, implants, and surgical instruments, thereby reducing the risk of infection (Bellini, 2022).

(Kurniawan et al., 2021) denoted that a large-scale antibacterial poly(vinyl chloride) (PVC) bed sheet coated with zinc oxide nanoparticles (ZnO-NP) was developed to combat contamination in hospital settings. The researchers employed a multi-layer coating technique, varying the number of coating cycles to control ZnO concentration. The results indicated that bed sheets subjected to 15 coating cycles exhibited the highest ZnO concentration (0.17 mg/cm^2) and demonstrated increased hydrophobicity (contact angle of 117° ± 0.76). Furthermore, the coated bed sheets displayed antibacterial properties against a range of bacteria, with a higher effectiveness against gram-positive compared to gram-negative bacteria. This nanocoating method could be extended to other medical devices at risk of infection transmission.

Implantable biomedical devices: These come into direct contact with tumor tissues benefit from biomaterial surfaces capable of inducing tumor cell death. Unfortunately, this particular antitumor feature is rare. A novel antitumor-cell nanocoating made of vanadium dioxide (VO2) was developed using customized reactive magnetron sputtering. This nanocoating demonstrates the ability to inhibit tumor growth; particularly in human cholangiocarcinoma cells which are cancers originating in the bile ducts constitute a category of malignancies. It achieves this by disrupting the mitochondrial electron transport chain, thus, increasing intracellular reactive oxygen species, destabilizing the mitochondrial membrane potential, and disrupting cell redox balance, ultimately leading to oxidative damage in the cancer cells (Li et al., 2018).

Another example of improvement of implant biocompatibilityconsists in silicone breast implants, commonly used for breast reconstruction but prone to failure due to capsular contracture. Recent research explores the use of biocompatible coatings to reduce this risk, focusing on the natural biopolymer silk fibroin (SF). SF, combined with recombinant elastin-like-peptide (ELP), is compared to the hydrophilic polymer polyethylene oxide (PEO). Characterization confirms uniform coatings that resist cracking. The SF/ELP blend, especially with 25 wt% ELP, shows improved crack resistance and reduces blood plasma protein adsorption by over 80%. Preliminary findings suggest non-cytotoxicity and support for fibroblast cell adhesion, growth, and proliferation, indicating the potential of SF/ELP coatings to mitigate breast implant failure (Souza et al., 2019; Li et al., 2023).

Hydrophobic antibacterial nanocoatings were effectively applied in textiles. Indeed, in the same superhydrophobic coating framework, (An et al., 2022) described a biocompatible, superhydrophobic coating with fluorination properties presented for medical wound dressings and; enhancing their antibacterial and haemostatic capabilities. To address the challenges of hemorrhage and microbial infections in wound healing, a biocompatible superhydrophobic coating is introduced for medical wound dressings. This coating-made from multifunctional antibacterial composite nanoparticles derived from acrylic acid (AA), 1H,1H,2H,2H-perfluorododecyl acrylate (PFDA), and poly (hexamethylene biguanide) hydrochloride (PHMB)-demonstrates antibacterial adhesion and bactericidal properties. In vitro and in vivo, haemostatic experiments confirm its effectiveness in promoting haemostasis and preventing adhesion. The best performance is achieved with 5.0 wt% fluorine content. This multifunctional fluoropolymer coating, offering antifouling, antibacterial, and haemostatic properties; holds significant promise for biomedical applications.

Several researches emphasized superhydrophobic antibacterial textile. Superhydrophobic coatings find extensive utility in various areas such as self-cleaning, anti-fogging, environmental remediation and antibacterial properties due to their exceptional non-wetting characteristics, often exhibiting water contact angles exceeding 150 degrees (Kumar and Nanda, 2019). As a matter of fact, (Xue et al., 2012) described cotton fibres coated with silver nanoparticles (Ag NPs) through a reduction process using a $[Ag(NH_3)_2]^+$ complex and glucose. The fibres were further modified with hexadecyltrimethoxysilane (HTMS) to create superhydrophobic cotton textiles. HTMS was applied to diverse materials including metals, plastics, glass, and ceramics, thus, constituting a robust and long-lasting hydrophobic layer through a siloxane bond. This hydrophobic layer enhances material properties by rendering it water and corrosion-resistant, improving the adhesion of coatings and adhesives; and making it non-stick and release-resistant, as liquids are unable to easily adhere to the hydrophobic surface.

The extensive range of applications for superhydrophobic coatings faces constraints due to the use of environmentally harmful substances and their limited sustainability in terms of material durability. A promising strategy to overcome these challenges involves adopting nature-inspired principles for designing and creating self-repairing coatings (Celik et al., 2023) introduced an innovative superhydrophobic coating both fluorine-free and biocompatible, drawing inspiration from the self-healing process seen in plant leaves' wax secretion. Comprising silica nanoparticles and carnauba wax, it rapidly self-heals with just one minute of moderate heating. Improved water repellency and thermal stability after healing are attributed to carnauba wax's low melting point, migrating to the surface of hydrophilic silica nanoparticles. The coating shows significant biocompatibility, with fibroblast L929 cell viability at around 90%. (An et al., 2022) developed a novel superhydrophobic coating with antibacterial and hemostatic properties, addressing blood loss and tissue adhesion in wound dressings. The composition includes acrylic acid, perfluorododecyl acrylate, and polyhexamethylene biguanide hydrochloride, showing promise for diverse biomedical applications, especially at a 5.0 wt% fluorine content.

In recent years, there has been significant interest in biomimetic superhydrophobic coatings for creating self-cleaning surfaces with potential environmental applications. Researchers have focused on using oxidized derivatives of carbon nanotubes (CNTs) and graphene to develop superhydrophobic surfaces. (Bains et al., 2020) study demonstrated an ionic liquid (IL)-functionalized multiwalled carbon nanotube (MWCNT) material created for hydrophobic coatings and characterized using various techniques. The synthesized ILs were assessed for their antibacterial properties against both Gram-positive and negative bacterial strains. Cytotoxicity tests on coated PVC material revealed no harm to viable cells. On the whole, the results suggest that the developed IL-1d@MWCNT-coated surface is an effective self-sterilizing solution for maintaining a bacteria-free surface, particularly in hospital biomedical devices.

Layer-by-layer (LbL) nanocoating, a versatile process, finds applications in bio motors, superhydrophobic surfaces, biosensors, implants, semiconductors, fibreoptics, and drug-delivery devices. This technique involves depositing oppositely charged polyelectrolytes, allowing integration with antimicrobial agents for controlled release against microorganisms. LbL coatings offer uniform coverage, customizable properties, making them promising for medical devices, healthcare cleanliness, reducing foodborne illnesses, and antimicrobial textiles. Inspired by mussels, an eco-friendly superhydrophobic gel nanocoating with transparency and stability is easily applied using a non-toxic spray. Notably, it exhibits rapid self-healing superhydrophobic properties when exposed to common organic solvents, suitable for diverse applications (Lichter et al., 2009; Keeney et al., 2015; Si et al., 2015).

During the COVID-19 pandemic, concerns regarding bio-contamination prompted the development of antimicrobial coatings to combat surface contamination and virus transmission. Incorporating non-migratory quaternary ammonium cations (QUATs) and positively charged silver nanoparticles in a polymer matrix creates coatings that repel water and oil, inactivate the coronavirus, and potentially inhibit virus replication. Super-hydrophobic nanocoatings, utilizing copper nanoparticles in a flexible polymer matrix, offer antibacterial and antiviral properties when spray-coated on surfaces like doors, knobs, wood, and fabrics. Addressing person-to-person transmission, nano-engineered facemasks, gloves, and PPE kits provide comfort, hydrophobic properties, and antimicrobial activity without compromising fabric texture. However, the efficacy of LbL nanocoating against viruses remains uncertain, necessitating regular material replacement. Recent studies show the creation of self-cleaning reusable face masks with hydrophobic silicon dioxide coatings over etched pores. Hydrophobic nanocoating also find applications in medical imaging, enhancing endoscopy, ultrasonography, and X-ray imaging, while providing multifaceted corrosion protection for medical devices. These coatings contribute to safer and more durable medical equipment, with the potential for broader adoption in advancing medical applications (Shirvanimoghaddam et al., 2021; Campos et al., 2020; Rai et al., 2021).

Pharma

Hydrophobic nanocoatings, ultra-thin layers with water- and oil-repellent properties, advance drug delivery protect drugs in the gastrointestinal tract. They enable targeted delivery, controlled release, and enhance medical device performance. Applications include drug-eluting stents and targeted nanoparticles for tumour treatment, revolutionizing pharmacology and reducing costs.

Medical Packaging

The creation of a biocompatible superhydrophobic coating is vital to reduce food spoilage, waste, and blood-related issues. In their study (Li et al., 2021) introduced an all-natural superhydrophobic material formed from lysozyme and cysteine, exhibiting strong interfacial stability and repelling water, viscous liquids, and biofluids like blood. The resulting phase-transited lysozyme (PTL) features a nanoscale roughness and amyloid-inspired adhesion, providing effective water and liquid repellence. This coating improves the water resistance of cellulose-based packaging materials without compromising their strength.

Textile

Hydrophobic nanocoatings play a vital role in textiles, offering versatility by enhancing water repellence, resisting stains, facilitating easy cleaning, and maintaining breathability. These coatings demonstrate remarkable durability and can be tailored for multifunctionality, encompassing antimicrobial properties, flame retardancy, and self-cleaning abilities, expanding their applications in diverse textile products. Innovative strategies have been explored to augment cotton fabric properties, such as (Abidi et al. 2012) utilizing nanoparticle vapor deposition for functionalization and (Shi et al., 2022) focusing on constructing a flame-retardant and hydrophobic coating through layer-by-layer assembly. (Wang et al., 2017) introduced a sustainable method for superhydrophobic materials using eco-friendly solvents, and (Joshi et al., 2012) effectively employed nanosilica and nanoclay for superhydrophobic cotton fabric. (Li et al., 2023) scaled up bioinspired nanocoatings for textiles, while (Li et al., 2010) demonstrated self-cleaning cellulosic materials. Lastly, (Obšel and Otřísal, 2022) experimentally confirmed that plasma chemical processes enable the application of nanocoatings with superhydrophobic and super oleophobic properties on suitable textile materials, showcasing the diverse potential of these coatings in textile innovation.

Hydrophobic nanocoating is a thin layer of nanomaterials applied to textiles, making them water-repellent by creating a barrier that causes liquids to bead up and roll off. It offers benefits like water repellence, stain resistance, easy cleaning, durability, breathability, and eco-friendliness. Commonly used in raincoats, sportswear, and various applications. In their study, (Rananavare and Lee, 2022) presented an eco-friendly approach to create hydrophobic cotton fabricsusing poly(glycidyl methacrylate) (PGMA) nanoparticles synthesized through a radical-induced dispersion polymerization method. The cotton fabric's surface is modified to enhance roughness and hydrophobicity by anchoring PGMA NPs through covalent bonds formed via the epoxy ring-opening reaction. Various characterizations confirm structural changes, resulting in increased stiffness and hydrophobicity (130° contact angle) of the fabric. Figure 5 shows a water repellent tissue.

Indeed, (Prabhakar et al., 2022) employed a novel approach to create hydrophobic antimicrobial nanocoatings on cotton fabric by combining mussel-inspired polydopamine (pDA) with graphene oxide (GO) and various copper compounds. This innovative bioinspired technique, using Cu(NO3)2 and CuNPs with GO, holds great promise in preventing the spread of bacterial and viral infections through contaminated clothing and has potential applications in healthcare settings, including PPE, gowns, face masks, and curtains. The rising consumer demand for durable, sustainable, and functional clothing has paved the way for the integration of nanomaterials into textiles. These nanomaterials offer features such as stain resistance, wrinkle-free properties, static elimination, and electrical conductivity while maintaining comfort and flexibility. Furthermore, nanomaterials enable the development of connected garments that can sense and respond to external stimuli through electrical, color, or physiological signals (Yetisen et al., 2016).

Superhydrophobic polyester (PET) fabrics were engineered by rising surface roughness and reducing surface energy through interactions involving natural polyphenols, ferrous sulfate heptahydrate, and hexadecyltrimethoxysilane (Chen et al., 2023). A durable and self-repairing superhydrophobic poly(ethylene terephthalate) (PET) fabric was created through a simple solution-dipping technique utilizing polydimethylsiloxane and octadecylamine (ODA). The surface roughness was achieved through the self-roughening process of ODA coating on PET fibers, eliminating the need for lithography or the addition of nanomaterials. Two primary approaches are employed in developing superhydrophobic PET textiles: plasma treatment and nanocoating. Plasma treatment involves exposing fibers to ionized gas, rendering them hydrophilic before coating them with a hydrophobic substance. Alternatively, nanocoating entails applying nanoscale materials like polymers or metal oxides to create the necessary microscopic roughness for superhydrophobic properties (Xue et al., 2016). Figure 6 shows the production process of superhydrophobic PET textiles.

Environmentally friendly LbL deposition was used to create flame-retardant and antimicrobial cotton. Cotton was coated with 12 bilayers of phytic acid (PA) and chitosan (CH)-urea, followed by immersion in copper (II) sulcate ($CuSO_4$) solution. The 12 bilayers with Cu^{2+} stopped flames during vertical flammability testing, achieving a limiting oxygen index (LOI) of 26%. Microscale combustion calorimeter (MCC) data demonstrated over 61% reduction in peak heat release rates (pHRR) and more than 54% reduction in total heat release (THR) compared to untreated cotton. Thermogravimetric analysis with infrared (TG-IR) revealed the release of water, methane, carbon dioxide, carbon monoxide, and aldehydes, with Cu^{2+} reducing methane production. The treated cotton exhibited intumescent behavior with a bubbled post-burn char structure and 100% reduction in Klebsiella pneumoniae and Staphylococcus aureus in antibacterial testing (Magovac et al, 2021). (Zhang, D., Williams, B.L., Liu, J. et al., 2021) developed an environmentally friendly fire-safe textile nanocoating system to enhance the flame retardancy and wash durability of natural cotton fabrics as well. This system involved a sandwich-like structure with interconnected networks formed through electrostatic interaction, hydrogen and chemical bonding, created via multi-layered coating and UV curing. Chemicals such as montmorillonite (MMT), ammonium polyphosphate (APP), acrylated polyurethane (PU), branched polyethylenimine (bPEI), and polydimethylsiloxane (PDMS) were used. The concentration of bPEI influenced wash durability, with higher concentrations providing better durability, but also affecting flame retardancy.

In one instance, a sustainable, superhydrophobic, and fire-resistant nanocoating called APP-PDA was successfully applied to cotton in alternating layers, confirmed through spectroscopic analysis. This treatment endowed the cotton with strong flame-retardant properties, reducing peak heat release rate by 59.6% and total heat release by 25.7%. The cotton maintained its superhydrophobicity even after dodecyl mercaptan treatment, making it promising for high-performance applications (Wang et al., 2019). Addressing polyester flammability, a four-bilayer (CH3/P-MAP) flame-retardant coating rendered the fabric self-extinguishing, reducing peak heat release rate by 42.6% and fire growth capacity by nearly 40%, proving highly effective with minimal bilayers (Petkovska and Mladenovic, 2023). Furthermore, a study explores multifunctional superhydrophobic nanoparticle coatings on cellulose-based materials through liquid flame spray, with various applications, including water-repellent clothing, self-cleaning textiles, and cost-effective lab-on-a-chip devices (TeisalaHannu, 2013).

Smart textiles, also known as electronic textiles (e-textiles), are fabric with integrated electronics like sensors and microcontrollers. They have diverse applications in healthcare, sports, military, and consumer electronics. These textiles are in early stages of development, offering potential to revolutionize various industries as technology advances. Scientists and engineers are particularly interested in using smart textiles for innovative wearable medical devices to enhance health monitoring and treatment options (Esfahani, 2021). Nanocoatings can help create smart textiles. They cover various coating methods such as sol-gel, cross-linking, plasma, and nanocomposite coatings, which are frequently employed in textile enhancement. Moreover, the sol-gel method can impart self-cleaning properties to textiles, particularly through photocatalysis using TiO2 nanoparticles. TiO2's photocatalytic activity, activated by UV light, generates negative electrons and positive holes that, when reacting with water and oxygen, produce active species like superoxide anions and hydroxyl radicals. These species can decompose contaminants on textile surfaces. Researchers have explored the application of TiO2 nanosols to cotton fabrics, resulting in improved properties, such as ultraviolet protection and antimicrobial activity against various pathogens (Pakdel et al., 2016; Daoud et al., 2004; Xin et al., 2004; Pakdel et al., 2018).

Conductive nanofibres and nanocoatings utilize nanosizefibers or fillers and apply specialized nanocoating techniques to create smart textiles. Nanotechnology offers the potential to improve the existing functionalities of textiles and develop smart textiles with novel properties or multifunctional capabilities (Shang and Zeng, 2013). Indeed, traditional textile treatments rely on wetting the textile structure for performance. However, the complexity of fiber surfaces, composition, and textile construction makes predicting wettability challenging. Water repellency is crucial for applications such as rainwear, upholstery, protective clothing, sportswear, and car interiors. New ultrahydrophobic materials using the "lotus effect" with water contact angles greater than 150°C are emerging. In textiles, hydrophobicity is typically determined by static water contact angles due to textile fabric's inherent roughness, making contact angle hysteresis measurement difficult (Ramaratnam et al., 2008). **Figure 7** is a photo of the lotus hydrophobic nanocoating happening in nature.

Energies

Fossil energy management: Hydrophobic nanocoatings in the fossil energy sector protect equipment from corrosion, reduce maintenance costs, enhance operational efficiency, and lower emissions. They safeguard metal surfaces in harsh conditions, minimize fluid transport friction, and mitigate pollutants in combustion chambers. Current applications include protecting pipelines, heat exchangers, and power plant components. Ongoing development holds promise for further sector enhancements.

Renewable energy management: Renewable energy sources, encompassing solar, wind, geothermal, hydroelectric, biomass, hydrogen power, and fuel cells, have emerged as compelling alternatives to address the depletion of traditional energy sources. Among them, photovoltaic (PV) systems stand out as the most prevalent and environmentally friendly choice (Rajput and Sudhakar, 2013).

Solar Energy: Solar energy is the predominant renewable energy source, leading to extensive utilization of solar panels with vast exposed surfaces. Recent data, as reported by the International Energy Agency (IEA), reveals a substantial surge in global solar photovoltaic (PV) generation, denoting a remarkable increase of 270 terawatt-hours in 2022, marking a 26% growth. This has propelled the total global solar PV capacity to nearly 1,300 terawatt-hours, significantly elevating the contribution of solar PV to approximately 4.5% of global electricity generation (IEA, 2023). Solar panels typically incorporate semiconductor materials, like silicon, enclosed within metal frameworks and safeguarded by glass encasements. Upon interaction with solar photons, these materials liberate electrons, initiating the generation of an electric charge, or photovoltaic (PV) charge. This phenomenon gives rise to direct current (DC), which is harnessed via the panel's internal wiring. Subsequently, an inverter transforms the DC electricity into alternating current (AC), the standard form for powering appliances (Nationalgrid, 2023). Photovoltaic (PV) cells are engineered through the introduction of boron and phosphorous doping into silicon, leading to the creation of "p-type" and "n-type" semiconductors. When sunlight impinges on the PN junction, electrons migrate from the n-type region to the p-type region through an external path, generating a flow of direct current, which can be stored in a battery (Mishra et al., 2019).

The efficiency of these solar panels hinges significantly on the exposure of photovoltaic (PV) cells to sunlight and their ability to absorb it. The performance of solar photovoltaic (PV) cells is contingent upon a multitude of factors which include the temperature of the cells, the spectral composition of sunlight, the installation angles, prevailing weather conditions, radiation density in the area, dust or soil accumulation, shadows cast on the panels, the effects of aging, the quality of diodes and wiring, and the inherent uncertainties associated with manufacturer ratings. A pivotal parameter influencing the efficiency of solar panels is the ability of their cover glass to either facilitate the absorption or reflection of incident radiation (Mondoc and Pop, 2010).

Accumulation of dust on solar systems leads to reduced light absorption, causing a decline in efficiency. Dust comprises various materials such as pollutants, cloth fibers, soil particles, vehicle exhaust particles, pedestrian-induced dust, water stains with salts, and bird droppings. This pervasive environmental dust buildup initiates with an initial layer, progressing to cover the entire surface. Over time, multiple layers accumulate, further diminishing the efficiency of solar systems (Mondoc and Pop, 2010; Karmouch, 2017). Solar power plants in hot desert and semiarid regions, chosen for abundant sunlight and minimal obstructions, face a significant challenge from dust accumulation. The prevalent dust hinders incoming radiation, necessitating frequent cleaning of PV solar panels. This task is crucial but poses challenges, especially for large-scale installations (Yilbas et al., 2017; Priyadarsini and Sharma, 2016). The solution lies in superhydrophobic antireflective coatings, which offer an effective means of creating self-cleaning, dust-free top glass for solar panels while simultaneously enhancing radiation absorption (Ennaceri et al., 2014 ;Mazumder et al., 2017).

The self-cleaning hydrophobic trend which research and industry tends to mimic is based on the ability of lotus leaves and rose petals to remove dust, has inspired them to reduce the surface of PV panels, a phenomenon often referred to as the "lotus effect" or "superhydrophobicity" (Mani and Pillai, 2010). Biomimetics, also known as biomimicry, is a multidisciplinary field leveraging nature's solutions to address human design challenges. Applied in the cleaning of photovoltaic systems, it manifests in examples such as photonic nanostructures mimicking floral epidermis for perovskite solar cells. Utilizing multifunctional nanocoatings, perovskite solar cells benefit from enhanced light trapping, water repellence, and protection against UV light and environmental moisture. These coatings achieve an impressive maximum power conversion efficiency of 24.61%, surpassing the reference cell at 21.01%, without requiring additional device optimizations. Moreover, the nanocoatings ensure stable operation under continuous exposure to air mass (AM) 1.5G (global) and UV light, even in real-world conditions (Vasilopoulou et al., 2022).

Photonic nanostructured TiO2 coatings enhance c-Si wafer solar conversion by about 30%, maintaining electrical performance. They offer broad absorption, high photocurrent density, and optical angular acceptance, promising efficiency gains in c-Si photovoltaics. Studies address environmental challenges, focusing on dust impact on PV module performance. Hydrophobic and superhydrophobic nanocoatings improve solar energy transmission, increasing power generation. Various materials, including modified silica-based films, are explored for dust repellency. Emphasizing durability, transparency, and self-cleaning, these studies propose eco-friendly solutions like recyclable eggshells for improved PV panel performance in dusty environments, fostering efficient and sustainable solar energy use (Santos, 2022; Ehsan et al., 2021; Aljdaeh et al., 2021; Tayel et al., 2022; Salehi and Eshaghi, 2022; Aldawoudet al., 2022; Cherupurakal et al., 2021; Chungsiriporn et al., 2020).

Hydrophobic silica nanoparticles serve as anti-reflective coatings in photovoltaic devices, crucial for enhancing power conversion efficiency. In optical systems, multifunctional coatings, such as anti-reflective coatings, play a revolutionary role. In PV devices, nano coatings on glass provide both hydrophobic and anti-reflective properties, featuring easy cleaning under extreme weather conditions. The project focuses on depositing anti-reflective hydrophobic coatings using silica nanoparticles, exemplified by KleanBoostTM, a superhydrophobic nano-textured coating for glass. The research explores its anti-reflective and anti-soiling properties, revealing a "dust herding" effect due to the hydrophobic surface, demonstrating the coating's anti-soiling and cleaning mechanisms (Nayashevsky et al., 2017).

Another example focuses on creating transparent and hydrophobic nanolayers on PET-SiOx and ETFE-SiOx substrates to enhance the protection of photovoltaic devices. It assesses the stability and chemical resistance of these coatings through accelerated aging tests, highlighting their improved stability and hydrophobic properties (Rossi et al., 2015). In addition, a breakthrough approach involving artemisinin (ART) is introduced to create a "perovskite/perovskite" bilayer structure that enhances the efficiency and reliability of perovskite photovoltaic cells. This bilayer structure offers improved electron blocking, high open-circuit voltage, and enhanced hydrophobicity, resulting in more reliable PV cells (Hou et al., 2020). As a further example, an innovative self-cleaning hydrophobic nanocoating developed in Egypt which consists of polydimethylsiloxane (PDMS) and silicon dioxide (SiO2), significantly improves the efficiency of solar panels by reducing dust accumulation and enhancing self-cleaning properties, resulting in a remarkable 30.7% increase in power output and efficiency (Bellini, 2022). Antireflective coatings (ARCs) applied to glass substrates protecting solar cells aim to improve light transmission and maintain surface cleanliness. These coatings typically use nanomaterials such as MgF2, SiO2, SnO3, TiO2, Al2O3, Nb2O5, and ZrO2. Efforts focus on enhancing ARCs' properties, including self-cleaning, durability, and resistance to abrasion. However, combining various nanomaterials has become necessary to meet the diverse demands of the solar cell industry effectively (Wu et al., 2023).

Wind energy: Hydrophobic nanocoatings offer diverse benefits in wind energy applications, including enhancing wind turbine efficiency by reducing blade friction, protecting against icing to maintain performance, and extending the overall lifespan by preventing corrosion and damage. These coatings are applied to wind turbine blades, creating anti-icing solutions and corrosion protection for various components. As hydrophobic nanocoating technology advances, its integration into wind energy systems is expected to grow, further improving efficiency, reliability, and cost-effectiveness in this renewable energy sector.

(Liu et al., 2023) developed MoS_2/ZnO/PDMS superhydrophobic nano-anti-icing coatings for wind turbine blades to combat blade icing issues in cold regions. These coatings displayed superhydrophobic characteristics with high contact angles and low sliding angles. They significantly reduced ice adhesion strength by 60.2%, and wind tunnel tests demonstrated decreased icing levels on the coated blade air foil surface compared to uncoated blades, particularly at low temperatures and wind speeds. Icing represents a frequent natural occurrence with significant consequences for wind turbines leading to increased attention on anti-icing coatings. (Feng et al, 2023) conducted a research work on hydrophobic nano-carbon coating. This latter was developed from corn-straw-biogas residue. Characterization analyses confirmed its suitable pore and crystal structure. This approach offers an effective method for creating anti-icing coatings on wind turbine blades and outdoor equipment.The study of (Qu, Sun and Feng et al., 2020) outlines a method for applying robust hydrophobic polypyrrole (PPy) coatings to wind turbine blades, yielding a water contact angle of 127.2° and exceptional anti-icing properties. Strong hydrophobic PPy-coated blades exhibited minimal icing, with significantly lower icing weight and thickness compared to uncoated blades.

Further renewable energies: Nanogenerators (NGs) have emerged as a versatile technology capable of tapping into this spectrum of energy reservoirs, including temperature fluctuations and sound waves; and converting them into electrical power. The practical application of NGs extends to energizing autonomous and environmentally friendly electronic devices, including sensors, wearable technology, and implantable medical devices. NGs can be broken down into three principal types which are piezoelectric, triboelectric and pyroelectric nanogenerators. (Su et al., 2023) presened a robust and self-powered superhydrophobic wearable piezoelectric sensor with superhydrophobic and antibacterial properties. The sensor features a flexible piezoelectric nanofiber membrane with carbon nanotube electrodes, protected by a conformal hydrophobic nanocoating. It can accurately monitor various human activities and maintains durable sensing performance even in harsh conditions. The sensor's long-term stability allows it to be integrated into shoe insoles for human gait monitoring, making it a promising technology for practical applications in tactile sensing, motion detection, and biomedical signal monitoring (Xiong et al., 2019) showed a versatile triboelectric nanogenerator (TENG) was developed by integrating foldability, transparency, and waterproof properties into an origami transparent cellulose electronic paper (CEP). The CEP-TENG utilizes a unique silver nanowires-bundle mesh (AgBM) embedded into ethyl cellulose paper, allowing it to harvest energy from various sources, including human motion, wind, and water. It can operate even in contact with acidic or alkaline water due to an added hydrophobic cellulose oleoyl ester nanocoating. The origami feature enables it to be reconfigured for maximum energy harvesting efficiency, making it suitable for self-powered applications like outdoor displays, windows, and wearable power sources.

Electrical Engineering and Electronics

Hydrophobic nanocoatings show significant promise in electronics, providing protection against water and moisture damage for devices in challenging environments like smartphones and wearables. Beyond safeguarding, these coatings enhance device performance by reducing corrosion and friction, leading to prolonged battery life and faster operation. They also increase reliability by shielding devices from dust, dirt, and contaminants, crucial for applications like medical devices and industrial control systems. In the developmental stages, hydrophobic nanocoatings have the potential to revolutionize the electronics industry. They improve electrical device performance, reduce corrosion and friction, contribute to cost reduction, and find applications in smartphones, wearables, outdoor electronics, PCBs, medical devices, and power grids. In summary, hydrophobic nanocoatings hold promise for enhancing the durability, reliability, and efficiency of electrical devices with an expanding role in various applications.

Agriculture and Food and Packaging

Agriculture: Hydrophobic nanocoatings benefit agriculture by enhancing crop yields, reducing pesticide use, and safeguarding against extreme weather. Applications include coating seeds, leaves, soil, and agricultural products to improve germination, nutrient uptake, water retention, and pest resistance. These coatings have the potential to transform agriculture, promoting water conservation, food safety, and environmental sustainability. In a study on 'Chandler' strawberries, hydrophobic nanocoatings with natural antimicrobials prolonged shelf life during simulated transportation vibrations. Limonene liposomes showed lower fungal growth and higher acidity and phenolic content on the 14th day of storage. Research suggests exploring liposome coatings with varying particle sizes for improved coating integrity during transportation (Dhital et al., 2017)

Food and Beverage and Packaging: Recent research focuses on superhydrophobic coatings made from edible, non-toxic, and eco-friendly materials to address this challenge in a sustainable manner (Ruzi et al., 2022). Hydrophobic nanocoatings offer a range of applications in the food and beverage industry. They enhance food safety by preventing bacterial and mold growth, extend shelf life by reducing spoilage, and maintain food quality. Additionally, these coatings are instrumental in reducing packaging waste and making food processing equipment easier to clean. As a result, they not only cut costs and enhance sustainability but also improve consumer satisfaction by providing safer, fresher, and better-tasting products. This technology is poised to revolutionize the industry, with potential for further applications as it continues to develop. (Ghosh et al., 2021) presents an inventive edible coating with silk nanodiscs (SND) in chitosan (CS), aiming to improve thermal, hydrophobic, optical, mechanical, and physicochemical properties for enhanced banana fruit storage. SND fabrication involves silk fibroin (SF) acid hydrolysis, resulting in disc-shaped nanostructures with heightened crystallinity (approx. 95.0%), confirmed by FESEM, FETEM, and XRD analyses. In a related study, (Ghosh et al., 2022) introduces nanochitosan (NCS) as an edible nanomodifier, enhancing starch (ST)/ guar gum (GG) biocomposites for effective edible food packaging on fresh cuts. NCS efficacy is evaluated through a thorough structure-property-performance analysis, showcasing potential for meeting strict standards in edible food packaging.

The packaging industry faces challenges, including ink migration, which is harmful to consumers. This study aims to reduce ink penetration into products using an eco-friendly superhydrophobic coating prepared with silicon dioxide. Duplex paper sheets were coated with this coating to seal printing material pores and act as a barrier to prevent ink leakage. The polymer offers the dual benefit of protection and reduced ink penetration, improving properties such as roughness, water absorption, and printing quality (Fahmy and Ibrahim, 2021). Superhydrophobic coatings are gaining prominence for food packaging due to their self-cleaning and anti-fouling properties. These coatings help prevent contamination and fouling of food packages, reducing food waste and enhancing the consumer experience.

A study introduced an efficient one-step method to create hydrophobic paper by incorporating chitin nanocrystals (ChNCs) and hexadecyltrimethoxysilane (HDTMS). The resulting 2% ChNCs/HDTMS coated paper exhibited high hydrophobicity with a water contact angle exceeding 130°, primarily due to the low surface energy provided by HDTMS. This modification significantly enhanced the paper's tensile strength and water vapor barrier compared to uncoated paper, thanks to the addition of ChNCs. The hydrophobic paper demonstrated excellent environmental resilience and mechanical stability and was biocompatible, with L929 cell survival exceeding 80%. Additionally, it effectively contained various liquids and could serve as a liquid-absorbing straw, making it a promising option for eco-friendly food packaging applications (He et al., 2022). The Nano and Advanced Materials Institute (NAMI) leads in materials technology with the introduction of innovative, eco-friendly nano-shield coatings designed as per- and polyfluoroalkyl substances (PFAS) alternatives in paper-based food packaging, enabling technology transfer for commercial deployment (NAMI, 2023).

Hydro Nanocoating, a superhydrophobic coating, addresses water-induced damage prevalent in the paper and cardboard industry, causing about 5% of the EU's product losses. This liquid coating, employing superhydrophobic silica nanoparticles, significantly improves water resistance, reducing water absorption by up to 30 times compared to untreated cardboard and 40% compared to existing paper protection alternatives. The coating extends the lifespan of paper, cardboard, and corrugated cardboard by up to threefold, preventing water-related damage and potential €9 million product losses. This technology is poised to impact the European goods packaging sector, especially for corrugated cardboard, safeguarding over 34% of European products and reducing supply chain losses by up to 90%. It is compatible with common inks, lowering production costs, enhancing gluing adhesion, and complying with EU regulations and food safety standards for various packaging types (EU, 2016).

Construction

The building industry is witnessing a growing adoption of hydrophobic nanocoating, primarily attributed to its advantageous properties such as moisture resistance, self-cleaning abilities, and multifunctionality. Hydrophobic nanocoatings offer key advantages in construction, providing protection against water, enhancing energy efficiency, and reducing maintenance costs. Superhydrophobic nanocoating offers numerous advantages, including reduced dirt accumulation, self-cleaning properties, and enhanced resistance to moisture and corrosion.Despite being in development, these coatings hold significant potential for revolutionizing the industry. Companies are actively working on applications like water damage protection on roofs, repelling elements on exterior walls, self-cleaning windows, and water-resistant flooring for high-traffic areas like kitchens and bathrooms. The ongoing advancement of hydrophobic nanocoating technology suggests that they will find an ever-expanding role in building and construction applications. Furthermore, those coatings exhibit anti-fungal properties which inhibit fungal growth on building materials, slowing down its progression. For instance, the study conducted by (MARILYN, 2019) involved the application of ZnO nanoparticles and various binders onto clay tiles, which were then positioned on an air conditioning ledge. The tiles underwent close monitoring to assess the effectiveness of ZnO nanoparticles in inhibiting fungal growth. However, a notable challenge was encountered in identifying a suitable location that promotes fungal development. (Moradi and Rezaei, 2022) worked on the creation of anti-corrosion and super-hydrophobic polypropylene/graphene oxide nanocomposite coatings on carbon steel demonstrates high effectiveness in preventing degradation of construction metals, including mild steel.(Chang, 2018) highlights the potential of nanoparticles, particularly superhydrophobic ones, for functional nanocoatings in construction and industrial settings, offering benefits like self-cleaning, improved durability, and moisture/corrosion resistance. Scanning electron microscopy (SEM) and transmission electron microscopy (TEM) analyze their morphology and dimensions, while a comprehensive database is created for these nanoparticles, including WCAs and material properties. The research explores chemical structures and functional groups contributing to hydrophobicity. Suitable adhesive binders are identified, with WCA testing revealing their hydrophobicity and simple tests gauging durability. Copper oxide (CuO) and multi-walled carbon nanotubes (MWCNT) exhibit top-performing WCAs exceeding 150°, with MWCNT bonded with water-based polyurethane (WPU) achieving a WCA of 136.5°, showing promise for building and industrial applications.

Water-repellent hydrophobic surfaces and anti-fungal nanocoatings are gaining traction in construction for their exceptional qualities. Hydrophobic nanocoatings exhibit impressive water-repelling and self-cleaning features, while anti-fungal coatings prevent fungal growth, leading to significant reductions in maintenance costs and time. Evaluation of hydrophobicity includes water contact angle measurements and penetration tests. Super hydrophobic nanostructured anti-stain surface coatings for smart green home applications know growing application of hydrophobic nanocoatings in various industries, emphasizing their potential for self-cleaning surfaces, energy efficiency, and cost reduction. The research of (Yiyang, 2019) involves testing different silanes and polymers for their hydrophobic properties on glass and kitchen tiles by measuring the water contact angle (WCA). By varying coating thickness, incorporating functionalized SiO2 spheres, and conducting tests in different conditions, the study aims to enhance the hydrophobicity of nanocoatings. Results show coatings achieving superhydrophobicity with WCAs exceeding 120° and highly hydrophilic coatings with WCAs less than 10°, demonstrating the significant effect of SiO2 spheres on hydrophobicity. According to (Raelyn, 2020), the water contact angle (WCA) of a surface is determined using the drop-coating method. A detailed analysis of the results has revealed that incorporating SiO2 spheres does enhance hydrophobicity.

(Bavithran et al., 2019) developed lightweight and cost-effective composite partition sheets using noncombustible glass fiber, waste nylon from fishing nets, and a polyester matrix. Coating these sheets with fumed silica nanopowders in polyester resin significantly improved mechanical properties. These sheets show promise for use in affordable building construction. (Wang et al., 2023) created a superhydrophobic coating for ancient bricks using a simple method with fluorinated B72, methyl-modified silica/titania, and hexamethyldisilazane. The coating inhibits microbial invasion and weathering without altering the brick's appearance. This approach preserves ancient structures without structural changes like cement, offering protection and conservation.

Hydrophobic polymer-based nano-coatings are used for building applications. They utilize nanotechnology for facade maintenance, particularly for self-cleaning purposes. They enhance building upkeep and reduce long-term costs. Instead of addressing maintenance concerns solely after construction, there's a growing emphasis on incorporating facade maintenance into the early design stages (Ying, 2018). Smart buildingsare a great field of application of hydrophobic nanocoating. (Zulfiqar et al., 2017) introduced a method for creating mechanically robust superhydrophobic surfaces on construction materials, which can recover their properties after extensive abrasion. Silica nanoparticles were sprayed on brick, marble, and glass surfaces, creating hydrophobic coatings. TiO2-nanocomposites, synthesized non-aqueously, demonstrated photocatalytic potential for preserving stone heritage. These nanoparticles, integrated into polymeric solutions, formed crack-free coatings on marble, reducing water absorption. Increased nanoparticle content enhanced contact angles, with the highest concentration showing significant photocatalytic activity. Accelerated ageing tests confirmed the durability and effectiveness of these TiO2-based nanocomposites in protecting historical stone surfaces (Gherardi et al., 2016).

In the framework of cultural heritage protection, (La Russa et al., 2016), studied three nano-TiO_2 coatings on calcarenite and Carrara marble. Paraloid B72 had poor performance, while the fluorinated polymer-TiO2 (AKP-TiO2) and acrylic water suspension (Fosbuild) with TiO2 demonstrated effective hydrophobic and photo-degrading effects. Fosbuild showed the best performance, recommended for conservation at 100 g/m2 for marble and 1600 g/m2 for calcarenite, emphasizing penetration and anchoring within the stone's pore structure.

Marine and Transportation

Naval and offshore: The naval and offshore industries face two major challenges: corrosion and marine organism attachment, both of which can harm system performance and necessitate costly maintenance. In the modern era, graphene-based nanocomposites are a vibrant area of research known for their capacity to significantly enhance various polymer characteristics. Graphene exhibits superhydrophobic behavior and antibacterial properties, offering the potential for advanced nanotechnology-driven coatings surpassing conventional naval paints in marine settings. This study focuses on investigating the hydrophobic attributes of a graphene-infused marine varnish, with concentrations ranging from 0.1% to 25% (w/w) (Gago et al., 2016).

Marine antifouling coatingshelp prevent fouling organisms from attaching to surfaces in water. It is used on ships, underwater structures, and marine equipment to improve efficiency, performance, and maintenance. They typically contain biocides or other chemicals that deter or inhibit the growth of fouling organisms. There are several marine antifouling coatings such as Polymer/graphene-derived nanocomposites. Following the ban on organo-tin antifouling coatings in 2003 by the International Maritime Organization, research shifted towards environmentally friendly alternatives, particularly silicone fouling release (FR) nanocoatings. These coatings possess effective fouling prevention and self-cleaning properties. Innovative polymer–graphene nanocomposites with reduced surface energy and rough surfaces are explored for their eco-friendly fouling release capabilities. Various graphene–inorganic nanofiller hybrids dispersed in polymeric resins are used to deter fouling organisms (Selim et al., 2017; Shimaa et al., 2023). (Selim et al., 2018) studies the impact of Ag@SiO2 core–shell nanofillers on silicone nano-paints, an ultrahydrophobic nanocomposite was developed for antifouling coatings.

The nanofillers, with controlled size and growth direction, were integrated into the silicone surface. Surface self-cleaning properties, durability, and antifouling performance were evaluated. A well-distributed 0.5 wt% of Ag@SiO2 core–shell exhibited superior self-cleaning with a water contact angle of 156° and surface free energy of 11.15 mN m−1.

Graphene-based nanocomposites are a promising avenue of research to enhance polymer properties, particularly in naval and offshore applications, where corrosion and marine life adherence are significant challenges. By introducing graphene into varnish coatings, this study investigated its impact on hydrophobicity and corrosion resistance in steel samples. Graphene doping ratios ranging from 0.1% to 25% (w/w) were explored. Results showed an increase in contact angle, reaching 143.6º, indicating near-superhydrophobic behavior ($\theta > 150^{o}$) (Selim et al., 2023). Furthermore, graphene-doped samples exhibited improved corrosion resistance, suggesting the potential for graphene-based marine coatings to offer superior hydrophobicity, environmental resistance, and biocide-free fouling control compared to existing naval paints based on fluoropolymers with contact angles around 95º (Gago et al., 2016).

Corrosion resistanceis a prominent characteristic of hydrophobic nanocoatings used in the maritime industry. Composite coatings, comprised of epoxy resin (EP), polydimethylsiloxane (PDMS), and SiO2 nanoparticles, were fabricated through liquid spraying and laser curing techniques. By varying the laser curing powers, the coatings' microstructure, hydrophobicity, corrosion resistance, and friction properties were examined. According to (Du et al., 2022), these coatings displayed surface micropores and clustered SiO2 nanoparticles. They exhibitedsuperhydrophobicity due to Si-(CH3)2 hydrophobic groups, which improved with higher laser curing powers, achieving a maximum contact angle of $161.90 \pm 1.85°$. Additionally, the coatings demonstrated exceptional self-cleaning capabilities and robust resistance to corrosion in artificial seawater. This study introduces an innovative approach to achieving highly durable, superhydrophobic coatings.

Aircraft: These coatings offer numerous benefits, including preventing ice formation on aircraft surfaces for safer takeoff and landing, enhancing aerodynamics to reduce fuel consumption and emissions, and protecting against corrosion and damage to extend aircraft lifespan. While hydrophobic nanocoatings are still in development, companies are actively working to introduce them in aviation. Practical applications range from preventing ice accumulation on windscreens to reducing drag and enhancing fuel efficiency on wings, to interior coatings that resist dirt, bacteria, and reduce cleaning costs, potentially revolutionizing aviation. On the other hand, ice accumulation poses a major threat to power systems, with superhydrophobic coatings showing promise for overcoming de-icing challenges and improving overhead transmission line resilience (Li et al., 2023).

Liu et al., 2016 developed a super-hydrophobic surface was created using nanocomposite materials consisting of silica nanoparticles and self-assembled monolayers of 1H,1H,2H,2H-perfluorooctyltriethoxysilane (POTS) through spin coating and chemical vapor deposition methods. The scanning electron microscope images indicate a porous structure in the silica nanoparticles, forming small-scale air pockets. With an average water contact angle of 163° and water droplets bouncing off, the surface exhibits super-hydrophobic properties. The durability of the coating was assessed using a pressurized pneumatic water impinging impact test rig, demonstrating that the obtained coatings are durable, though further improvements are necessary for aerospace applications.

Anti-icing effect on aircraft is another benefit of hydrophobic nanocoating. (Knausgård, 2012) thesis explores wetting properties, particularly related to icephobicity, of closed-cell structures using LAMMPS simulations. The simulations align with Wenzel and Cassie-Baxter theories, but the impact of tuning roughness scale differs from theoretical predictions on a small scale. Closed-cell surfaces did not show a comparative advantage over open-cell structures. Experimental testing on a nearly-superhydrophobic black silicon surface revealed that water condensation significantly compromises water repellency. While superhydrophobic surfaces exhibit potential for anti-icing applications, challenges remain, such as icephobicity deterioration from mechanical damage and reduced performance in humid conditions.

Automobile: The transportation industry is rapidly expanding globally, driven by population growth and increasing demand for efficient systems. However, challenges arise for manufacturers and researchers in the automobile sector, including the need for safe, economical, low-emission vehicles. To address these challenges, strategies such as replacing heavy metallic components with lightweight, high-strength composites and integrating nanoparticles into fuels, tires, lubricants, coatings, and paints have been employed. Titanium, known for its low density, high rigidity, corrosion resistance, and other qualities, along with epoxy resin, which offers improved mechanical, electrical, chemical properties, has found significant use in this context. Various applications of titanium and epoxy in the automobile industry, including lightweight parts, anti-fouling, anti-fogging, self-cleaning, and corrosion-resistant materials, showed to improve engine performance, and more, demonstrating their substantial contributions to automotive advancements (Akinlabi et al., 2019).

Optic: Hydrophobic nanocoatings find diverse applications in optics, including anti-reflective coatings to reduce glare, self-cleaning coatings for minimal maintenance, and protective coatings safeguarding optical surfaces from contaminants. Examples of their use today include camera lenses for improved image quality, eyeglass lenses for glare reduction and scratch resistance, medical optics for better imaging and hygiene, and solar panels to prevent dust buildup and enhance efficiency.

A self-cleaning glass method was developed by (Maharjan et al., 2020) it involves creating a self-cleaning hydrophobic nanocoating (SCHN) on glass surfaces through a scalable manufacturing process. Initially, planar glass surfaces are roughened using diamond micro-/nano-particle abrasives to form microscopic grooves. After cleaning, the surface is vapor-deposited with trichloro (1H,1H,2H,2H-perfluorooctyl) silane (TCPFOS) in controlled humidity conditions, forming a covalent bond with the substrate. The SCHN exhibits a low sliding angle of 14° for water droplets, allowing them to slide down without leaving a trail. The reduced adhesion of dirt to the surface enables the removal of dirt particles by sliding or rolling water droplets. The coating maintains high light transmission and withstands various environmental challenges, including abrasion, acid rain, saline exposure, alkali solution, and temperature cycling.

This technology has the potential to transform the optics industry, enhancing device performance, durability, and reducing maintenance costs (Víg et al., 2023) explored the effects of water droplets on solar cell performance, focusing on uneven illumination and the anti-reflective impact of water droplets. A C++ model within the RayTracer framework program is used to quantitatively analyze the phenomena, incorporating equations for light refraction, reflection, and Fresnel equations for drops of varying shapes, positions, and coverage ratios. Their study considers spectral components of sunlight and variations in incident angles, corroborating results with actual solar cell data. Their findings have broader applications, potentially shedding light on the optical and energetic consequences of water droplets on plant leaves or human skin.

In their study (Özpirin and Ebil, 2018), protective, self-cleaning, and super-hydrophobic polymeric thin films for electro-optical (EO) systems in aviation and military applications are developed using initiated chemical vapor deposition (iCVD). Poly (glycidyl methacrylate) (PGMA), poly (1H, 1H, 2H, 2H-Perfluorodecyl acrylate) (PPFDA) homopolymers, and P(GMA-PFDA) copolymer are employed for coating fabrication. Optical modeling and simulations assess the impact of film thickness and refractive index on performance, measured within the 400-1000 nm range. Coatings up to 1 μm exhibit no significant effect on optical performance. The P(GMA-PFDA) copolymer, incorporating a fluoro monomer, achieves optimal results with good mechanical properties and super-hydrophobicity. (Jaafar, 2022) work focuses on the development of a hybrid nanocoating for eyeglass lenses in the medical field. The nanocoating aims to prevent water droplet condensation on the lenses. Silicon rubber - RTV and Al2 O3 nanoparticles were used in various concentrations. The resulting hydrophobic hybrid nanocoating exhibits anti-dust properties and a 100% reduction in adhesion of both water and dust, improving eyeglass lens clarity and acuity. (Smitha et al., 2007) synthesized a hydrophobic silica-biopolymer hybrid coatings, through a sol–gel method using colloidal silica, gelatin, and vinyltrimethoxysilane, exhibit water repellence, adjustable hydrophobic properties, and high optical transmittance, suggesting potential applications in optical devices and transparent biocompatible coatings on substrates like leather.

Environmental Sciences

Hydrophobic nanocoatings offer diverse applications for a sustainable environment, including water purification, desalination, energy efficiency, renewable energy enhancement, and agriculture. Examples include more efficient water purification membranes, cost-effective desalination membranes, energy-efficient windows, self-cleaning solar panels, and soil coatings that reduce water evaporation and enhance nutrient retention for improved agriculture. These applications contribute to sustainability by reducing energy consumption, conserving resources, and minimizing pollution. Hydrophobic nanocoatings hold great promise for advancing environmental sustainability across various sectors.

Water treatment / water-oil separation: Hydrophobic nanocoatings play a pivotal role in various water-oil separation applications. They enable the creation of superhydrophobic surfaces with water contact angles exceeding 150 degrees, which repel water but attract oil. These coatings are used to develop membranes for efficient water-oil separation, where pores smaller than oil droplets but larger than water molecules ensure selective permeation. Hydrophobic nanocoatings are also applied to enhance existing materials such as filters, meshes, and sponges for improved water-oil separation. Notably, these nanocoatings find utility in oil spill cleanup, industrial wastewater treatment, and food processing, enabling the development of more effective materials and technologies for these applications. For instance, (Kim et al. 2019) developed a continuous ZIF-8/reduced graphene oxide (RGO) nanocoating by growing ZIF-8 on RGO-coated polyurethane foam. The surface modification of the foam involved immersing it in GO solution, followed by mild thermal treatment to convert the GO layer into an RGO layer. The enhanced adhesion between the polymer foam and ZIF-8 layer was achieved through the inserted RGO layer, facilitating preferential ZIF-8 layer growth. The resulting PU foam exhibited synergetic hydrophobic/oleophilic properties, enabling selective oil absorption and ultrafast hexane flux via vacuum filtration. (Wang et al., 2020) addressed oil-water separation using poly (methyl methacrylate) (PMMA) nanoparticles cured cotton fabric, providing hydrophobicity and efficient oil-water separation. (Tsai et al., 2021) developed electrospun nanofibrous membranes for gravity-driven separation of oil-rich emulsions, achieving high separation efficiency for various emulsions. (Zhu et al., 2022) coated polyurethane foams with chitosan and nano-montmorillonite via layer-by-layer assembly for emergency oil absorption, reducing flammability and improving thermal stability. (Li et al., 2023) employed polydopamine nanoparticles and polydimethylsiloxane to modify PVDF membranes for vacuum membrane distillation, achieving superhydrophobic properties and outstanding anti-wetting and anti-fouling performance.

Barthwal et al., 2022 devised a straightforward method for fabricating superhydrophobic and superoleophilic sponges using zinc-based metal-organic frameworks (MOF-5) to achieve oil-water separation and antibacterial features. The process involved rendering hydrophilic MOF-5 superhydrophobic with PFOTS, followed by a dip-coating process using modified hydrophobic MOF-5 (H-MOF-5) and polydimethylsiloxane (PDMS) to create the superhydrophobic sponge. This sponge exhibited a high-water contact angle (156°), efficient oil-water separation (over 95% efficiency after ten cycles), and significant antibacterial properties against Staphylococcus aureus. (Li et al. 2020) developed a superhydrophobic aerogel by coating kapok fibers with a MOF nanocoating (ZIF-8), resulting in a lotus leaf-like surface with superhydrophobic properties. This aerogel demonstrated exceptional oil and organic absorbent capabilities, outperforming ZIF-8* particles by 17-40 times and maintaining over 75.64% sorption capacity after 50 cycles. Finally, (Karimnezhad, 2017) created hydrophobic/oleophilic membranes for n-hexane removal from water, using para-aminobenzoate alumoxane, boehmite-epoxide, and Stearate Alumoxane nanoparticles. The resulting membranes were hydrophobic and oleophilic, maintaining their efficiency through multiple uses in a gravity-driven separation system, achieving 84% separation efficiency for a 20% n-hexane in water mixture.

Water desalination and membranes: Hydrophobic nanocoatings offer various advantages in water desalination applications, primarily through enhancing membrane performance, enabling novel desalination technologies, and reducing costs. They improve water flux and reduce fouling in desalination membranes, particularly in reverse osmosis (RO) systems, increasing water throughput and reducing fouling by significant percentages. Additionally, hydrophobic nanocoatings facilitate the development of self-cleaning membranes that repel contaminants, thereby diminishing the need for chemical cleaning. These coatings find application in solar-powered desalination devices, where they enhance efficiency by reducing energy requirements for water evaporation. This technology has the potential to elevate water desalination by improving water quality, decreasing energy consumption, prolonging membrane lifespan, and lowering maintenance costs, ensuring its widespread adoption in various desalination processes.

Hsieh et al., 2009 introduced a one-step coating technique for creating silica-based coatings that are exceptionally repellent to both water and oil, eliminating the need for additional low-surface energy treatments. The coatings are characterized by the stacking of spherical silica nanoparticles, resulting in primary surface roughness. An optimal fluorine-to-silicon ratio (F/Si: 2.13) in the silica coating is identified as critical for achieving the highest levels of water (168.1°) and oil repellence (165.2°). These repellent properties are attributed to the combined effects of silica sphere stacking and surface fluorination, resulting in a substantial area of contact with air for water and ethylene glycol droplets, thus ensuring stable water and oil repellence. This one-step nanocoating process holds significant potential for enhancing polymeric or flexible substrates with superhydrophobic and super oleophobic properties. To enhance nanofiltration membranes (NFMs), hydrophobic polymer microfiltration substrates were coated with tannic acid/diethylenetriamine (TA/DETA) to improve wettability for effective interfacial polymerization. Laser confocal microscopy visualized solution spreading, crucial for defect-free polyamide layers. Molecular simulation and performance evaluation showcased the impact of TA/DETA coatings. The resulting NFMs exhibited higher water permeation flux, excellent pressure resistance (up to 0.9 MPa), and solvent resistance due to stable hydrophobic substrates (Zhang et al., 2020).

Environmental-friendly superhydrophobic bioactive nanocoatings: Nowadays, the focus is on developing eco-friendly bioactive-superhydrophobic nanocoatings (bioactive-SHC) for diverse applications, extending the durability of coated structures and enhancing their properties. These coatings offer unique characteristics owing to their surface texture and chemistry, allowing control over liquid dewetting. Recent research is shifting toward hydrofluorocarbon-free SHC, given growing regulatory restrictions. Biocompatibility, bioactivity, and mechanical properties of coating materials remain important factors. Various biomaterials show promise in enhancing physical traits and bioactivity. This chapter delves into water repellency and recent advancements in bioactive-SHC, outlining coating methods and addressing key challenges like durability and scalability, as well as current and future requirements (Boutamart et al., 2022).

Wen et al., 2019developed an eco-friendly superhydrophobic nanocoating derived from recycled eggshells for energy-efficient applications. This study offers an eco-friendly method for crafting a superhydrophobic coating using recycled eggshells and ZnO, addressing ecological concerns associated with conventional coatings. The coating, modified with stearic acid (STA) and carboxymethyl cellulose (CMC), demonstrates resilience to UV irradiation and mechanical damage, delaying icing and providing self-cleaning properties. This sustainable and low-cost coating holds promise for practical applications. Hydrophobic nanocoatings are increasingly used in the building industry for their moisture resistance, self-cleaning properties, and environmental friendliness. This study investigates the impact of functionalized SiO2 spheres on these coatings, testing various silanes and polymers to enhance hydrophobicity. By achieving a Water Contact Angle (WCA) of over 150°, superhydrophobic coatings are found effective. The integration of SiO2 spheres is shown to enhance hydrophobicity, contributing to their suitability for green building applications (Lee Yi Ling Raelyn, 2020).

The widespread use of nanocoatings in various industries, especially in smart green home applications, is driven by their multifunctionality. Hydrophobic nanocoatings, essential for self-cleaning properties, are gaining popularity for applications on buildings, vehicles, and food products, leading to reduced energy, labor, and maintenance costs. This study explores the impact of functionalized SiO2 spheres on hydrophobic nanocoatings applied to different surfaces, such as glass and kitchen tiles. Various silanes and polymers are tested for their hydrophobicity using Water Contact Angle (WCA) measurements. Through extensive research and experiments, the study identifies superhydrophobic and hydrophilic coatings, with WCAs above 120 and below 10, respectively. The addition of functionalized SiO2 spheres significantly influences hydrophobicity, making this study a valuable reference for nanocoatings on glass and tiles (Yiyang, 2019).

Wood: Hydrophobic nanocoatings, in addition to their remarkable water-repelling capabilities, function as microscopic guardians for wood, offering a range of benefits such as dirt and grime repellency, an anti-microbial shield, sun protection, enhanced durability, and easier maintenance. To enhance wood resistance to weathering, Chinese fir (Cunninghamia lanceolata) underwent a two-step modification process: first, titanium dioxide (TiO2) nanocoatings were grown using a sol-gel process, and then low-surface free-energy treatment was applied with hydrolyzedhexadecyltrimethoxysilane (HDTMS). The resulting coatings exhibited dual-size roughness, covalently linked alkyl groups, and significantly improved hydrophobicity with a water contact angle (WCA) increase of up to 138%. The TiO2 coatings demonstrated strong UV absorption, enhancing the wood substrate's photostability. A similar two-step process, inspired by the lotus effect, involved growing nanofilms on Chinese fir wood with silica sol and subsequent hydrophobization using HDTMS (Wang et al., 2014). The resulting nanofilm coating displayed uniform spherical nanoparticles, covalently linked HDMTS, and elevated wood hydrophobicity with water contact angles of approximately 141° (longitudinal) and 150° (transverse) (Wang et al., 2013). Wood surfaces were coated with PVA/SiO2 nanofibers via roll electrospinning, followed by hydrophobization using OTS. Characterization confirmed OTS monolayers, and the coating's adhesive strength and scratch resistance were evaluated. (Kumar et al., 2016).

Fire control: Hydrophobic nanocoatings offer significant applications in fire management, spanning prevention, suppression, and safety. They can create fire-resistant materials, enhance firefighting gear, and improve fire safety in buildings. Specific examples include fire-resistant fabrics, coatings for various surfaces, more effective fire-fighting foams and gels, and fire-retardant insulation. These applications have the potential to transform fire management by reducing fire risks and minimizing damage. Additionally, hydrophobic nanocoatings can contribute to environmental protection by developing eco-friendly fire-fighting solutions and sustainable fire-resistant materials. (Teisala et al., 2010) introduced a continuous roll-to-roll process, Liquid Flame Spray (LFS), for depositing transparent titanium dioxide nanoparticle-based nanocoatings on pigment-coated paperboard at atmospheric conditions, achieving water contact angles over 160°, high hydrophobicity allowing water droplets to bounce off, and demonstrating scale-up potential for industrial applications at line speeds up to 150 m/min.Superhydrophobic, flame-retardant M-GOWR@MF sponge with efficient oil absorption, separation, and fire warning capabilities was developed using a simple two-step surface-modifying method, making it versatile for diverse applications (Mao et al., 2021).

Pollution control: Hydrophobic nanocoatings offer diverse applications in pollution management. They enhance air and water pollution control through more effective filters and membranes, contribute to soil remediation by encapsulating pollutants, and aid noise pollution control with sound-absorbing materials. Specific applications include self-cleaning air filters, advanced water filtration membranes, and improved soil remediation methods. As the technology evolves, hydrophobic nanocoatings have the potential to further revolutionize pollution management, offering more environmentally friendly and efficient solutions. Addressing contamination and dust accumulation on electrical insulators is crucial for grid reliability. Electric discharge in wet conditions poses a threat to the power grid. Hydrophobic and dustproof coatings are increasingly vital in the electricity industry. This study used silica nanoparticles in a silicon rubber matrix to coat ceramic insulators, enhancing their resistance to dust and moisture. Hydrophobic properties are critical for efficient power transmission. Nanocomposites were applied to 70 kN insulators and rigorously tested according to industry standards, including electrical, salt fog, and hydrophobicity tests. The nanocomposite with 3 wt % silica showed the best performance (De Santos et al., 2021; Zolriasatein, 2022).

OTHER APPLICATIONS

Sports

Hydrophobic nanocoatings have gained considerable toehold in the world of sports. Their integration brings forth several benefits, including reduced drag, improved grip, and protection against environmental elements. These advantages lead to enhanced sports equipment performance and durability. For instance, hydrophobic nanocoatings applied to swimsuits minimize water resistance, aiding swimmers' performance. Bicycles benefit from reduced air resistance, enhancing efficiency, while skis glide more smoothly. These coatings also bolster grip and control on sports equipment like tennis rackets, baseball bats, and golf balls. By reducing drag and extending equipment lifespans, hydrophobic nanocoatings stand to transform the sports industry, promising increased performance and efficiency across various disciplines. As technology evolves, their integration into additional sports applications appears highly probable, further elevating the industry's capabilities. Recently introduced nanocoatings with hydrophobic properties, designed to reduce surface friction, have made their commercial debut. These coatings offer potential innovations for various industries, creating smart surfaces with decreased surface tension. Their application to ski surfaces holds promise for enhancing ski performance (Flanagan, 2009).

Water-Repellent Footwear: Ruzafa-Silvestre et al., 2022 investigated functional nanocoatings for rendering footwear leather materials water-repellent using chemical plasma polymerization with hexamethyldisiloxane (HMDSO). The effects of monomers on the deposition time were evaluated and the resulting polysiloxane polymers displayed hydrophobic properties on leather, enhancing water resistance without altering the leather's visual appearance and physical properties. These plasma coating treatments offer a more sustainable and less polluting alternative to conventional chemical processing in the footwear industry, making them suitable for automated product-finishing processes.

Plasma or the "the fourth state of matter": Matter exists in three primary states: solid, liquid, and gas. There is a fourth state known as Plasma, which is generated by heating a gas. Plasma is essentially an ionized gas, created through the ionization process, where electrons are removed from atoms, resulting in both negatively and positively charged particles (Mott-Smith, 1971). For instance, (Huang et al., 2010) study aimed to create a hydrophobic nano-coating and examined various operational parameters during low-pressure plasma processing. Radio frequency power at 13.56 MHz was used to generate low-pressure plasma, and a gas mixture of difluoromethane (CH2F2) and argon (Ar) was employed. The deposition of CH2F2 plasma polymerized films increased with higher RF plasma power input and varied system gas pressure, resulting in enhanced hydrophobic surface properties. The study also explored the effects of CH2F2/Ar plasma on the films' surface characteristics as a function of the Ar content, finding that the optimal CH2F2:Ar ratio was 1.5:1, yielding superhydrophobic coatings with water contact angles exceeding 150°. Atomic force microscopy (AFM) analysis indicated that ion bombardment from CH2F2/Ar plasma increased surface roughness and effectively formed hydrophobic coatings on heat-sensitive materials.

Road Pavements

In this framework, (Nascimento et al., 2012) developed a superhydrophobic asphalt material with antifreeze properties using nanostructured surface modifiers. Coating the asphalt surface through a wet chemical process, the material exhibited excellent superhydrophobicity (163-degree contact angle) and maintained this effect after wear testing. This advancement holds promise for applications in road engineering. Moreover, Numerous difficulties have emerged concerning the use of traditional approaches, such as employing sodium chloride (NaCl) to prevent asphalt binder icing and improve anti-icing characteristics. The limitations linked to these methods encompass environmental incongruity, leading to the drying of nearby vegetation, corrosion of metallic components in vehicles, and structural harm to pavement element (Nejad et al., 2016).

Surface Engineering and Material

In a comprehensive exploration of diverse superhydrophobic coatings, researchers have developed innovative formulations and techniques to achieve desirable properties for various applications. (Shen et al., 2015) utilized ethyl acetoacetate as a chelating agent to create a nano-TiO2 coating on Type 316L stainless steel, exhibiting crack-free porous surfaces with uniform anatase TiO2 particles. (Karimnezhad, 2017) reported the fabrication of hydrophobic/oleophilic membranes for n-hexane removal using para-aminobenzoate alumoxane, boehmite-epoxide, and Stearate Alumoxane. (Shao et al., 2021) presented a self-cross linkable poly(perfluorohexylethyl acrylate-co-isocyanato ethyl methacrylate) nanocoating with dynamic water repellence and mechanical strength. (Wang et al., 2013) introduced a simple approach for superhydrophobic nanocoating using functional ZnO nanocrystals, achieving remarkable superhydrophobicity. (Xu et al., 2010) described a sol–gel dip-coating method for preparing superhydrophobic, transparent, and stable organic–inorganic composite nanocoatings. (Feng et al., 2013) employed electroless galvanic deposition to create superhydrophobic silver nanocoatings on copper substrates, exhibiting high thermal stability. (Syafiq et al., 2018) developed a highly transparent and stable superhydrophobic organic-inorganic composite nanocoating through a sol-gel dip-coating method. (Haapanen et al., 2018) explored large-scale fabrication of superhydrophobic surfaces using TiO2 nanoparticles in a roll-to-roll pilot scale process.

Fallah et al., 2018 achieved facile and low-cost fabrication of superhydrophobic and self-cleaning glass slides through spray-coating. (Chagas and Weibel, 2014) utilized a simple dipping process to fabricate poly(propylene) (PP) superhydrophobic surfaces coated with titanium dioxide nanoparticles. (Yu et al., 2020) developed a nano-SiO2/fluorinated polyacrylate polyurethane composite coating with good hydrophobicity and anti-aging properties. (Selim et al., 2023) prepared PDMS nanocomposite antifouling coatings with graphene oxide nanosheets and magnetite nanospheres, exhibiting superhydrophobicity and high fouling release efficiency. (Ganguly and Margel, 2023) discussed the preparation of superhydrophobic surfaces inspired by lotus leaves, categorizing materials into inorganic, organic, and hybrid types, with emphasis on polymer systems. (John et al., 2022) introduced a novel bilayer coating approach, combining hydroxy-terminated polydimethylsiloxane modified cardanol-based polyurethane with silane-grafted ZnO nanoparticles, resulting in a superhydrophobic coating with enhanced anticorrosion properties. (Srinivasan et al., 2008) acknowledged financial support for their study involving various contributors.

Cao et al., 2022 integrated nanocoatings and microstructure fabrication to create a robust superhydrophobic surface with rectangular-groove microstructures coated by nanocoatings. (Ye et al., 2017) proposed a cost-effective and environmentally friendly approach for superhydrophobic coatings involving hydrophobic functionalized silica nanoparticles and cross-linking agent incorporation. (Lin et al., 2013) employed a sol–gel dip-coating method to create a transparent superhydrophobic nanocoating with sub-100 nm roughness, demonstrating potential applications. (Wimalasiri et al., 2017) developed silica-based nonfluorinated superhydrophobic coatings for natural rubber surfaces, exhibiting high water contact angle and sliding angle values. (Zemajtis et al., 2023) achieved a superhydrophilic to superhydrophobic transformation of TiO2 nanoparticles doped zinc phosphate coating systems. (Wang et al., 2017) fabricated a superhydrophobic TiO2-oleic acid nanocomposite coating through a vacuum cold spray process. (Hsieh et al., 2009) demonstrated the fabrication of a stable superhydrophobic surface using nanoscale TiO(2) spheres. (Arukalam et al., 2019) evaluated the transition of a PDMS–ZnO nanocoating from superhydrophobic to hydrophobic state for anticorrosion and barrier properties after exposure to marine water. (Yazdani-Ahmadabadi et al., 2016) investigated photopolymericnanocoatings with varying hydrophobic nanosilica contents, showcasing changes in glass transition temperature and cross-link density.

Xi et al., 2020 achieved a multifunctional nanocoating by dip-coating a mixed suspension of acid-catalysed silica sol and hydrophobic silica nanoparticles, exhibiting high transmittance, superhydrophobicity, mechanical robustness, and stability. (Gutiérrez-Cano et al., 2018) used Raman spectra to assess the thermal degradation of graphite-based thin film hydrophobic coatings. (Kuznetsov et al., 2011) enhanced adsorptive passivation on iron (steel) surfaces using organic inhibitors, demonstrating improved efficiency and protection against atmospheric corrosion. (Zhang et al., 2019) fabricated a multifunctional super-hydrophobic polyethersulfone coating exhibiting super-hydrophobicity, self-cleaning, mechanical stability, and enhanced corrosion resistance. (Ammar et al., 2017) developed hybrid polymeric nanocomposite coatings with SiO2 nanoparticles, showcasing improved dispersion, hydrophobicity, and anticorrosion properties. (Ahmad et al., 2016) achieved hierarchical surfaces with enhanced mechanical and corrosion resistance through ultra shot penning of stainless steel, chemical vapor deposition, and hydrophobic coating. (Cai et al.,

2016) utilized hydrophobic R974 nanoparticles to create superhydrophobic films with high bonding strength, exhibiting self-cleaning properties and drag reduction in water. Reuna et al. (2023) achieved a superhydrophilic to superhydrophobic transformation of TiO2 nanoparticles doped zinc phosphate coating systems. (Selim et al., 2017) demonstrated elastomeric PDMS/SiO2-doped ZnO nanocomposites with self-cleaning and fouling release properties, exhibiting high hydrophobicity, physical characteristics, and stability under various conditions. (Koh and Lee, 2021) enhanced the hydrophobicity of insulating surfaces using hybrid hydrophobic MoS2 nanoparticles, demonstrating superhydrophobic self-cleaning behavior, good electrical resistance, UV-degradation resistance, and improved water contact angle.

(Marvast et al., 2023) explored Ce-treated steel surfaces modified with ZIF8 nanoparticles, showing increased resistance against corrosion and improved durability and stability under harsh conditions. In their recent review (Sharma et al., 2022) emphasized the significance of anti-corrosive coatings for sustainable and durable materials, particularly focusing on graphene-based super-hydrophobic surfaces for potential anti-wear applications. (Lee et al., 2020) focused on achieving conformal and ultrathin coatings of highly conductive PEDOT:PSS on hydrophobic surfaces for resistive-based pressure sensors, utilizing a solvent exchange process for enhanced coating properties. (Xing et al., 2020) improved friction and hydrophobic performances of contact interfaces using laser-induced periodic surface structures (LIPSS) and atomic layer deposition (ALD) MoS2 nano-coatings, regulating friction and wetting properties for enhanced performance. (Maeztu et al., 2017) proposed a multilayer sol-gel nanocoating on aluminum alloys for surfaces with dual hydrophobicity and corrosion resistance, optimizing nanocoatings using a hybrid sol-gel matrix with corrosion inhibitors and a simple sol-gel matrix for hydrophobic properties. (Ying et al., 2018) investigated hydrophobic mesoporous silica-coated metal-organic framework (MOF) MIL-101Cr for catalytic activity enhancement and improved stability in the oxidation of indene, demonstrating promising catalytic performance. (Sankar et al., 2016) explored nano-sized phosphates of rare earth materials, specifically LaPO4, for their unique hydrophobic properties without chemical modification, displaying notable hydrophobicity and inertness towards molten metals. (Feng et al., 2018) fabricated zinc nano-coatings with superhydrophobicity and enhanced antifouling performance on carbon steel, featuring a micro/nanostructure and a stearic acid layer for high water contact angles and improved oleophobicity. (Dolatzadeh et al., 2011) studied organosilane-treated SiO_2 nanoparticles dispersed in polyurethane coatings for their influence on electrochemical behavior, demonstrating improved corrosion performance in resultant nanocoatings.

Hydrophobic nanocoatings, applied as thin films on diverse surfaces, exhibit versatility in applications. They serve anti-wetting, flame retardant, anti-de-icing, anti-graffiti, and anti-microbial functions. Additionally, these coatings contribute to creating selective permeable membranes, ink-resistant prints, corrosion-resistant surfaces, and biofilm-resistant layers. They enhance the performance of materials like carbon nanotubes, offer electromagnetic shielding, and play a role in electrophoresis and electrospray applications. The coatings are instrumental in developing self-cleaning glass surfaces, and practical methods like spraying and dip coating make them adaptable for a wide range of surface engineering and chemistry applications.

CONCLUSION

The global hydrophobic coatings market showed a great expansion which reflects the increasing demand and diverse applications of hydrophobic coatings in industries such as automotive, marine, construction, and industrial settings, driven by their water-repelling, corrosion protection, and self-cleaning properties. Hydrophobic nanocoatings represent a highly promising technological advancement with a multitude of potential applications spanning various industries. Nevertheless, the broad implementation of these coatings faces several impediments, encompassing issues of durability, adhesion, cost-effectiveness, and environmental sustainability. Researchers are actively engaged in tackling these challenges to devise hydrophobic nanocoatings that are not only more robust but also economically viable and ecologically responsible. Additionally, challenges in achieving the desired degree of hydrophobicity, maintaining precise control over the coating's wetting characteristics, and the scalability of production processes necessitate thorough investigation. Despite these complexities, hydrophobic nanocoatings hold substantial potential to bring about transformative changes across diverse industrial sectors.

HydrophobicCoating-as-a-service represents a business model wherein companies provide specialized hydrophobic coating services to a diverse clientele. This comprehensive service encompasses cleaning, surface preparation, and precise application of hydrophobic coatings on a wide array of surfaces, including vehicles, watercraft, architectural structures, and industrial machinery. Hydrophobic coatings offer a multitude of advantages, including their remarkable water and oil repellent properties, which render surfaces more easily maintainable and cleaner. These coatings are instrumental in corrosion protection, effectively extending the longevity of various equipment and infrastructure. Some hydrophobic coatings also possess self-cleaning capabilities, repelling dirt and contaminants, thereby diminishing the need for frequent cleaning and maintenance. For industrial use, creating superhydrophobic or superamphiphobic surfaces must be cost-effective, reproducible, and scalable. Fabricating overhang structures for superamphiphobicity is more challenging than for superhydrophobic surfaces. Precise control of surface protrusions is crucial, particularly as liquid surface tension decreases. Durability requires resistance to UV, acids, bases, and solvents, depending on the coating's chemical composition. Long-term mechanical stability is a challenge, especially for surfaces with overhang structures, which show weaker resistance to shearing. Abrasion can alter both structure and surface chemistry, increasing adhesion and contact angle hysteresis (Paven et al., 2016).

Various materials in artificial superhydrophobic nanocoatings offer specific advantages. Inorganic options like carbon-based and silica-based materials demonstrate excellent chemical resistance, while metallic oxides (e.g., ZnO, TiO2, SnO2) exhibit diverse surface morphologies. Transparent high-polymer coatings provide simplicity, flexibility, and controllable thickness. Challenges include metallic oxide corrosion, the high cost of certain materials, and the toxicity of organic polymers, prompting ongoing research for solutions (Si and Guo, 2015). (Das et al., 2015) highlight the challenge of developing next-gen materials for oil/water separation. Stimuli-responsive polymeric materials are explored for innovative oil/water separation solutions. (Rajamohan et al., 2023) note the primary challenge of designing mechanically durable coatings in this field. Nayshevsky's work (2020) addresses anti-reflective, anti-soiling, and self-cleaning coatings, particularly relevant for photovoltaic cover glass applications. The study emphasizes the need for coatings to combat soiling exacerbated by natural dew, with fluorinated ethylene-propylene (FEP) polymer coatings showing promise (Nayshevsky, 2020).

The robustness of this study lies in its narrative presentation and emphasis on elucidating the hydrophobic nanocoating attributes pertinent to the specified domain of application. Due to the inadequacy of the initial search query in yielding exhaustive study cases and technologies within the targeted areas, additional references were incorporated. These supplementary sources serve to encompass the salient features derived from both overarching reviews and those dedicated to specific domains, thus enhancing the comprehensiveness of the discussion. This work has a scope on the techniques used without digging very much in the details to keep the study oriented towards applications rather than fabrication and testing technologies. However, resorting to Google Scholar as the only engine research and applying the search query excludes several innovations and papers.

The hydrophobic coating as a service is indispensable for a range of industries, encompassing the automotive sector, where it safeguards vehicles from environmental factors, facilitating effortless maintenance. In the marine industry, these coatings protect boats from saltwater corrosion and marine fouling. For the construction sector, hydrophobic coatings offer vital protection against water damage and corrosion of buildings. In the industrial domain, these coatings are instrumental in safeguarding various equipment from corrosion and other forms of damage. Hydrophobic coating service providers bring a wealth of expertise and experience to their clients, ensuring the safe and effective application of coatings on a diverse range of surfaces. Moreover, they offer convenience by performing on-site applications, saving customers valuable time and effort. These service providers prioritize quality, employing high-quality materials and state-of-the-art equipment to ensure precise coating application, resulting in a long-lasting protective layer. Therefore, when considering the application of hydrophobic coatings to surfaces, it is advisable to engage with reputable hydrophobic coating service companies to assess specific needs and receive tailored recommendations for optimal coating solutions.

REFERENCES

Abidi, N., Aminayi, P., Cabrales, L., & Hequet, E. (2012). Super-hydrophobic cotton fabric prepared using nanoparticles and molecular vapor deposition methods. *ACS Symposium Series, 1107*, 149–165. 10.1021/bk-2012-1107.ch008

Akinlabi, S., Ogbonna, O., Mashinini, P. M., Adediran, A., Fatoba, O., & Akinlabi, E. (2019). Titanium and Epoxy for Automobile Application. *RE:view*. Advance online publication. doi:10.15224/978-1-63248-170-2-09

Aldawoud, A., Aldawoud, A., Aryanfar, Y., Assad, E. H., Mamdouh, S., & Shubham, A. (2022). Performance Enhancement of Self-Cleaning Hydrophobic Nanocoated Photovoltaic Panels in a Dusty Environment. *Energies*, *14*(20), 6800. doi:10.3390/en14206800

Aldawoud, A., Aldawoud, A., Aryanfar, Y., El Haj Assad, M., Sharma, S., & Alayi, R. (2022). Reducing PV soiling and condensation using hydrophobic coating with brush and controllable curtains. *The International Journal of Low Carbon Technologies*, *17*, 919–930. doi:10.1093/ijlct/ctac056

Ammar, S., Ramesh, K., Ma, I. A. W., Farah, Z., Vengadaesvaran, B., Ramesh, S., & Arof, A. K. (2017). Studies on SiO2-hybrid polymeric nanocomposite coatings with superior corrosion protection and hydrophobicity. *Surface and Coatings Technology*, *324*, 536–545. doi:10.1016/j.surfcoat.2017.06.014

An, X., Li, W., Zhang, Y., Ding, J., Zhang, S., Hu, T., Li, S., Ren, Y., Liu, P., & Li, X. (2022). Antibacterial and superhydrophobic hemostatic properties of fluorinated random copolymer composite nanocoatings. *Progress in Organic Coatings*, *168*, 106869. doi:10.1016/j.porgcoat.2022.106869

An, X., Li, W., Zhang, Y., Ding, J., Zhang, S., Hu, T., Sen Li, Y. R., Liu, P., & Li, X. (2022). Antibacterial and superhydrophobic hemostatic properties of fluorinated random copolymer composite nanocoatings. *Progress in Organic Coatings*, *168*, 106869. doi:10.1016/j.porgcoat.2022.106869

Arukalam, I. O., & Li, Y. (2019). Anticorrosion and barrier properties appraisal of poly(dimethylsiloxane)–ZnO nanocoating transition from superhydrophobic to hydrophobic state. *Journal of Coatings Technology and Research, 16*(4), 1077–1088. doi:10.1007/s11998-018-00182-2

Bains, D., Singh, G., Bhinder, J., Agnihotri, P. K., & Singh, N. (2020). Ionic Liquid-Functionalized Multiwalled Carbon Nanotube-Based Hydrophobic Coatings for Robust Antibacterial Applicatio. *ACS Applied Bio Materials, 3*(4), 2092–2103. doi:10.1021/acsabm.9b01217 PMID:35025261

Bavithran, B., Nagarajan, V., & Vinod Kumar, K. (2018). Fire-resistant hydrophobic nanocoated composite partition sheets. *Journal of Thermoplastic Composite Materials.* Advance online publication. doi:10.1177/0892705718799814

Bayda, S., Adeel, M., Tuccinardi, T., Cordani, M., & Rizzolio, F. (2019). The History of Nanoscience and Nanotechnology: From Chemical-Physical Applications to Nanomedicine. *Molecules (Basel, Switzerland), 25*(1), 112. doi:10.3390/molecules25010112 PMID:31892180

Ben John, P. R. (2022). Design and fabrication of nano textured superhydrophobic and anti-corrosive silane-grafted ZnO/bio-based polyurethane bilayer coating. *Surface and Coatings Technology, 451*, 129036. doi:10.1016/j.surfcoat.2022.129036

Boutamart, M., Jhabli, O., Nouneh, K., Rafqah, S., Eid, A. I., Briche, S., & Agzenai Ben Salem, Y. (2022). Environmental-friendly superhydrophobic bioactive nanocoatings. In A. Barhoum, J. Jeevanandam, & M. K. Danquah (Eds.), *Fundamentals of Bionanomaterials* (pp. 77–110). Elsevier. doi:10.1016/B978-0-12-824147-9.00004-2

c, Z., Patel, F., Mastoi, N. R., &Saddiqa, A. (2016). Effect of sandblasting, annealing and hydrophobic treatment on the nano-mechanical and corrosion behaviour of n-TiO 2 -coated 316L stainless steel. *Current Science, 110*(3), 353–362. https://www.jstor.org/stable/24906780

Cai, C., Sang, N., Teng, S., Shen, Z., Guo, J., Zhao, X., & Guo, Z. (2016). Superhydrophobic surface fabricated by spraying hydrophobic R974 nanoparticles and the drag reduction in water. *Surface and Coatings Technology, 307*, 366–373. doi:10.1016/j.surfcoat.2016.09.009

Campos, E. V. R., Pereira, A. E. S., de Oliveira, J. L., Carvalho, L. B., Guilger-Casagrande, M., de Lima, R., & Fraceto, L. F. (2020). How can nanotechnology help to combat COVID-19? Opportunities and urgent need. *Journal of Nanobiotechnology, 18*(1), 125. doi:10.1186/s12951-020-00685-4 PMID:32891146

Cao, D., Ji, S., Zhao, J., Liu, Z., & Dai, H. (2022). Design and superhydrophobic performance analysis of the rectangular-groove microstructure coated by nanocoatings. *Materials Chemistry and Physics*, *291*, 126539. doi:10.1016/j.matchemphys.2022.126539

Celik, N., Sahin, F., Ozel, S. S., Sezer, G., Gunaltay, N., Ruzi, M., & Onses, M. (2023). Superhydrophobic coatings for food packaging applications: A review. *Food Packaging and Shelf Life*, *32*, 100823. doi:10.1016/j.fpsl.2022.100823

Chagas, G. R., & Weibel, D. E. (2014). Development of Poly(propylene) Superhydrophobic Surfaces by Functionalised TiO2 Nanoparticles: Effect of Solvents and Dipping Times. *Macromolecular Symposia*, *344*(1), 55–62. doi:10.1002/masy.201300218

Chen, C., & Peng, X. (2023). Constructing superhydrophobic sands layer with PTFE nanocoating for desert water storage and oil/water separation. *Colloids and Surfaces A: Physicochemical and Engineering Aspects, 657*(A), 130517.

Chen, X., Wang, J., Xie, A., Wang, B., Wu, J., Chen, G., & Xing, T. (2023). Fabrication of transparent self-healing superhydrophobic coatings with high water contact angle stability. *Langmuir*, *39*(44), 15817–15827. doi:10.1021/acs.langmuir.3c02508 PMID:37877472

Chen, C., & Peng, X. (2023). Constructing superhydrophobic sands layer with PTFE nanocoating for desert water storage and oil/water separation. *Colloids and Surfaces A: Physicochemical and Engineering Aspects, 657*(Part A), 130517. ISSN 0927-7757. doi:10.1016/j.colsurfa.2022.130517

Cherupurakal, N., Mozumder, M. S., Mourad, A.-H. I., & Lalwani, S. (2021). Recent advances in superhydrophobic polymers for antireflective self-cleaning solar panels. *Renewable & Sustainable Energy Reviews*, *151*, 111538. doi:10.1016/j.rser.2021.111538

Chungsiriporn, J., Pongyeela, P., & Chairerk, N. (2020). Sol-gel self-cleaning superhydrophobic nanocoating for glass surface of solar cell. *Songklanakarin Journal of Science and Technology*, *42*(4), 923–927.

Contreras, C. B., Chagas, G., Strumia, M. C., & Weibel, D. E. (2014). Permanent superhydrophobic polypropylene nanocomposite coatings by a simple one-step dipping process. *Applied Surface Science*, *307*, 234–240. doi:10.1016/j.apsusc.2014.04.019

Daoud, W. A., & Xin, J. H. (2004). Nucleation and growth of anatase crystallites on cotton fabrics at low temperatures. *Journal of the American Ceramic Society*, *87*(5), 953–955. doi:10.1111/j.1551-2916.2004.00953.x

Das, S., Kumar, S., Samal, S. K., Mohanty, S., & Nayak, S. K. (2018). A Review on Superhydrophobic Polymer Nanocoatings: Recent Development and Applications. *Industrial & Engineering Chemistry Research, 57*(8), 2727–2745. doi:10.1021/acs.iecr.7b04887

de Santos, H., & Sanz-Bobi, M. Á. (2021). Research on the pollution performance and degradation of superhydrophobic nano-coatings for toughened glass insulators. *Electric Power Systems Research, 191*, 106863. doi:10.1016/j.epsr.2020.106863

Dhital, R., Joshi, P., Becerra-Mora, N., Umagiliyage, A., Chai, T., Kohli, P., & Choudhary, R. (2017). Integrity of edible nano-coatings and its effects on the quality of strawberries subjected to simulated in-transit vibrations. *Lebensmittel-Wissenschaft + Technologie, 80*, 257–264. doi:10.1016/j.lwt.2017.02.033

Dolatzadeh, F., Moradian, S., & Jalili, M. M. (2011). Influence of various surface treated silica nanoparticles on the electrochemical properties of SiO2/polyurethane nanocoatings. *Corrosion Science, 53*(12), 4248–4257. doi:10.1016/j.corsci.2011.08.036

Du, J., Yuan, H., Xia, H., Kou, H., Zhang, Y., Xing, W., & Zhang, C. (2022). Improvement of superhydrophobicity and durability of EP+PDMS/SiO2 composite coatings by adjusting laser curing powers. *Materials Chemistry and Physics, 289*, 126428. doi:10.1016/j.matchemphys.2022.126428

Ehsan, R. M., Simon, S. P., Kumar, K. A., & Sriharsha, T. (2021). Effect of Soiling on Photovoltaic Modules and Its Mitigation Using Hydrophobic Nanocoatings. *IEEE Journal of Photovoltaics, 11*(3), 742–749. doi:10.1109/JPHOTOV.2021.3062023

Ennaceri, H., Alami, H. E., Brik, H., Mokssit, O., & Khaldoun, A. (2014). Lotus effect and superhydrophobic coatings for concentrated solar power systems (CSP). *Proceedings of International Conference on Composite Materials & Renewable Energy Applications*, 14.

Fahmy, H., & Ibrahim, S. (2021). The effect Ecofriendly Super hydrophobic Nano composites Coating and Penetration of Printing Inks on Paperboard Packaging Materials. *International Design Journal, 11*(1), 319–326. doi:10.21608/idj.2021.133025

Falahaty marvast, M., Gholambargani, M., Ramezanzadeh, M., & Ramezanzadeh, B. (2023). Enhancing Self-Healing and Adhesion Bonding Properties: Investigating the Promising Potential of Ce-ZIF8 MOF Hybrid Thin Films as pH-Responsive Nanocoatings. *Langmuir, 39*(27), 9579–9594. doi:10.1021/acs.langmuir.3c01363 PMID:37366573

Fallah, M., Ghashghaee, M., Rabiee, A., & Ershad-Langroudi, A. (2018). Spray-Deposition of an Organic/Inorganic Blend for Fabrication of a Superhydrophobic Surface: Effect of Admixing with Silica Aerogel and Modified Silica Nanoparticles. *Protection of Metals and Physical Chemistry of Surfaces*, *54*(5), 909–916. doi:10.1134/S2070205118050064

Feng, F., Wang, R., Yuan, W., & Li, Y. (2023). Study on anti-icing performance of biogas-residue nano-carbon coating for wind-turbine blade. *Coatings*, *13*(5), 814. doi:10.3390/coatings13050814

Feng, X., Shi, Y., Wang, Y., Yue, G. R., & Yang, W. (2013). Preparation of superhydrophobic silver nano coatings with feather-like structures by electroless galvanic deposition. *Chinese Science Bulletin*, *58*(16), 1887–1891. doi:10.1007/s11434-013-5755-9

Feng, Y., Chen, S., & Frank Cheng, Y. (2018). Stearic acid modified zinc nano-coatings with superhydrophobicity and enhanced antifouling performance. *Surface and Coatings Technology*, *340*, 55–65. doi:10.1016/j.surfcoat.2018.02.053

Flanagan, T. (2009). Alpine skiing technology: Faster, higher, stronger. *Sports Technology*, *2*(1-2), 5–6. doi:10.1080/19346182.2009.9648492

Gago, Miguel, León, & Ibarra. (2016). *Graphene-based hydrophobic varnish for marine applications*. Academic Press.

Ganguly, S., & Margel, S. (2023). Superhydrophobic nanoscale materials for surface coatings. In S. Thomas & J. S. George (Eds.), *Polymer-Based Nanoscale Materials for Surface Coatings* (pp. 479–500). Elsevier. doi:10.1016/B978-0-32-390778-1.00029-3

Gherardi, F., Colombo, A., Goidanich, S., &Toniolo, L. (2016). *Smart hydrophobic TiO2-nanocomposites for the protection of stone Cultural Heritage*. Academic Press.

Ghosh, T., & Katiyar, V. (2022). Nanochitosan functionalized hydrophobic starch/guar gum biocomposite for edible coating application with improved optical, thermal, mechanical, and surface property. *International Journal of Biological Macromolecules*, *211*, 116–127. doi:10.1016/j.ijbiomac.2022.05.079 PMID:35561853

Ghosh, T., Mondal, K., Giri, B. S., & Katiyar, V. (2021). Silk nanodisc based edible chitosan nanocomposite coating for fresh produces: A candidate with superior thermal, hydrophobic, optical, mechanical and food properties. *Food Chemistry*, *360*, 130048. doi:10.1016/j.foodchem.2021.130048 PMID:34034054

Gutiérrez-Cano, V., Rodríguez, F., González, J. A., & Díaz, V. (2018). Thermal degradation of hydrophobic graphite-based thin film nano-coatings observed by Raman spectroscopy. *Thin Solid Films*, *648*, 8–11. doi:10.1016/j.tsf.2017.12.023

Haapanen, J., Aromaa, M., Teisala, H., Juuti, P., Tuominen, M., Sillanpää, M., Stepien, M., Saarinen, J. J., Toivakka, M., Kuusipalo, J., & Mäkelä, J. (2018). On the limit of superhydrophobicity: Defining the minimum amount of TiO2 nanoparticle coating. *Materials Research Express*, *6*(3), 035004. Advance online publication. doi:10.1088/2053-1591/aaf2ee

Hooda, A., Goyat, M. S., Gupta, R., Prateek, M., Agrawal, M., & Biswas, A. (2017). Synthesis of nano-textured polystyrene/ZnO coatings with excellent transparency and superhydrophobicity. *Materials Chemistry and Physics*, *193*, 447–452. doi:10.1016/j.matchemphys.2017.03.011

Hou, Y., Wu, C., Yang, D., Wang, K., Ye, T., Brownlie, L., Wang, K., & Priya, S. (2020). Artemisinin (ART)-Induced "perovskite/perovskite" bilayer structured photovoltaics. *Nano Energy*, *78*, 105133. doi:10.1016/j.nanoen.2020.105133

Hsieh, C. T., Lai, M. H., & Cheng, Y. S. (2009). Fabrication and superhydrophobicity of fluorinated titanium dioxide nanocoatings. *Journal of Colloid and Interface Science*, *340*(2), 237–242. doi:10.1016/j.jcis.2009.08.029 PMID:19775698

Huang, C., Pan, C.-H., & Liu, C.-H. (2010). Deposition of hydrophobic nano-coatings with low-pressure radio frequency CH2F2/Ar plasma processing. *Thin Solid Films*, *518*(13), 3570–3574. doi:10.1016/j.tsf.2009.12.013

Jaafar, H. T., & Diaa, B. M. (2022). Hybrid Nanocoating Waterproof of Silicon Si and Nano-aluminum Oxide Al2O3 for Medical Lenses Applications. *Malaysian Journal of Medicine and Health Sciences*, *18*(SUPP11), 1–5.

Jiang. (2019). *Super hydrophobic nanostructured anti-stain surface coatings for smart green home applications.* ScholarBank@NUS Repository.

Joshi, M., Bhattacharyya, A., Agarwal, N., & Parmar, S. (2012). Nanostructured coatings for superhydrophobic textiles. *Bulletin of Materials Science*, *35*(6), 933–938. Advance online publication. doi:10.1007/s12034-012-0391-6

Juntima, Pongyeela, & Chairerk. (2020). Sol-gel self-cleaning superhydrophobic nanocoating for glass surface of solar cell. *Songklanakarin Journal of Science and Technology*, *42*(4), 923–927.

Karimnezhad, H. (2017). *Fabrication of Hydrophobic Membrane for the Separation of n-Hexane/Water Mixture Using Novel Oleophilic Nanoparticle and Kevlar Fabric, as a Superior Support*. Academic Press.

Karimnezhad, H. (2017). Fabrication of Hydrophobic Membrane for the Separation of n-Hexane/Water Mixture Using Novel Oleophilic Nanoparticle and Kevlar Fabric, as a Superior Support. *Journal of Water and Environmental Nanotechnology*, *2*(3), 145–156. doi:10.22090/jwent.2017.03.002

Karmouch, R., Hor, H. E. (2017). Solar cells performance reduction under the effect of dust in Jazan Region. *Journal of Fundamental and Renewable Energy Applications, 7*.

Keeney, M., Jiang, X. Y., Yamane, M., Lee, M., Goodman, S., & Yang, F. (2015). Nanocoating for biomolecule delivery using layer-by-layer self-assembly. *Journal of Materials Chemistry. B*, *3*(45), 8757–8770. doi:10.1039/C5TB00450K PMID:27099754

Kim, D. W., Eum, K., Kim, H., Kim, D., de Mello, M. D., Park, K., & Tsapatsis, M. (2019). Continuous ZIF-8/reduced graphene oxide nanocoating for ultrafast oil/water separation. *Chemical Engineering Journal*, *372*, 509–515. doi:10.1016/j.cej.2019.04.179

Knausgård, K. (2012). *Superhydrophobic Anti-Ice Nanocoatings*. Institutt for konstruksjonsteknikk.

Koh, E., & Lee, Y. T. (2021). Development of hybrid hydrophobic molybdenum disulfide (MoS2) nanoparticles for super water repellent self-cleaning. *Progress in Organic Coatings*, *153*, 106161. doi:10.1016/j.porgcoat.2021.106161

Kumar, A., & Nanda, D. (2019). Methods and fabrication techniques of superhydrophobic surfaces. In S. K. Samal, S. Mohanty, & S. K. Nayak (Eds.), *Superhydrophobic Polymer Coatings* (pp. 43–75). Elsevier. doi:10.1016/B978-0-12-816671-0.00004-7

Kumar, A., Ryparovà, P., Petrič, M., Tywoniak, J., & Hajek, P. (2017). Coating of wood by means of electrospun nanofibers based on PVA/SiO2 and its hydrophobization with octadecyltrichlorosilane (OTS). *Holzforschung*, *71*(3), 225–231. doi:10.1515/hf-2016-0108

Kurniawan, F. H., Chinavinijkul, P., & Nasongkla, N. (2021). Hydrophobic and antibacterial bed sheet using ZnO nanoparticles: A large-scale technique. *Journal of Drug Delivery Science and Technology*, *62*, 102339. doi:10.1016/j.jddst.2021.102339

Kuznetsov, Y. I. (2011). Progress in inhibition of metal corrosion and modification of safety nanolayers on metals. *Protection of Metals and Physical Chemistry of Surfaces*, *47*(7), 821–829. doi:10.1134/S2070205111070100

La Russa, M. F., Rovella, N., Alvarez de Buergo, M., Belfiore, C. M., Pezzino, A., Crisci, G. M., & Ruffolo, S. A. (2016). Nano-TiO2 coatings for cultural heritage protection: The role of the binder on hydrophobic and self-cleaning efficacy. *Progress in Organic Coatings*, *91*, 1–8. doi:10.1016/j.porgcoat.2015.11.011

Lee, J. J., Gandla, S., Lim, B., Kang, S., Kim, S., Lee, S., & Kim, S. (2020). Alcohol-based highly conductive polymer for conformal nanocoatings on hydrophobic surfaces toward a highly sensitive and stable pressure sensor. *NPG Asia Materials*, *12*(1), 65. doi:10.1038/s41427-020-00238-z

Lee. (2020). *Synthesis and applications of hydrophobic nanostructured coatings for green buildings*. ScholarBank@NUS Repository.

Li, B., Bai, J., He, J., Ding, C., Dai, X., Ci, W., Zhu, T., Liao, R., & Yuan, Y. (2023). A Review on Superhydrophobic Surface with Anti-Icing Properties in Overhead Transmission Lines. *Coatings*, *13*(2), 301. doi:10.3390/coatings13020301

Li, C., Liu, W., Mao, J., Hu, L., Yun, Y., & Li, B. (2023). Superhydrophobic PVDF membrane modified by dopamine self-polymerized nanoparticles for vacuum membrane distillation. *Separation and Purification Technology*, *304*, 122182. doi:10.1016/j.seppur.2022.122182

Li, J., Jiang, M., Zhou, H., Jin, P., Cheung, K. M. C., Chu, P. K., & Yeung, K. W. K. (2018). Vanadium Dioxide Nanocoating Induces Tumor Cell Death through Mitochondrial Electron Transport Chain Interruption. *Global Challenges (Hoboken, NJ)*, *3*(3), 1800058. doi:10.1002/gch2.201800058 PMID:31565366

Li, J., Tian, J., Gao, Y., Qin, R., Pi, H., Li, M., & Yang, P. (2021). All-natural superhydrophobic coating for packaging and blood-repelling materials. *Chemical Engineering Journal*, *410*, 128347. doi:10.1016/j.cej.2020.128347

Li, S., Wei, Y., & Huang, J. (2010). Facile fabrication of superhydrophobic cellulose materials by a nanocoating approach. *Chemistry Letters*, *39*(1), 20–21. doi:10.1246/cl.2010.20

Li, W., Shi, J., Zhao, Y., Huo, Q., Sun, Y., Wu, Y., Tian, Y., & Jiang, Z. (2020). Superhydrophobic Metal–Organic Framework Nanocoating Induced by Metal-Phenolic Networks for Oily Water Treatment. *ACS Sustainable Chemistry & Engineering*, *8*(4), 1831–1839. doi:10.1021/acssuschemeng.9b05937

Li, W., Yin, Z., Qi, L., Yu, B., & Xing, W. (2023). Scalable production of bioinspired MXene/black phosphorene nanocoatings for hydrophobic and fire-safe textiles with tunable electromagnetic interference and exceeding thermal management. *Chemical Engineering Journal*, *460*, 141870. doi:10.1016/j.cej.2023.141870

Lichter, J. A., Van Vliet, K. J., & Rubner, M. F. (2009). Design of Antibacterial Surfaces and Interfaces: Polyelectrolyte Multilayers as a Multifunctional Platform. *Macromolecules*, *42*(22), 8573–8586. doi:10.1021/ma901356s

Lin, J., Chen, H., Fei, T., & Zhang, J. (2013). Highly transparent superhydrophobic organic–inorganic nanocoating from the aggregation of silica nanoparticles. *Colloids and Surfaces. A, Physicochemical and Engineering Aspects*, *421*, 51–62. doi:10.1016/j.colsurfa.2012.12.049

Liu, B., Liu, Z., Li, Y., & Feng, F. (2023). A wind tunnel test of the anti-icing properties of MoS2/ZnO hydrophobic nano-coatings for wind turbine blades. *Coatings*, *13*(4), 686. doi:10.3390/coatings13040686

Liu, J., Janjua, Z. A., Roe, M., Xu, F., Turnbull, B., Choi, K. S., & Hou, X. (2016). Super-hydrophobic/icephobic coatings based on silica nanoparticles modified by self-assembled monolayers. *Nanomaterials (Basel, Switzerland)*, *6*(12), 232. doi:10.3390/nano6120232 PMID:28335360

Liu, S., Chen, X., Yu, M., Li, J., Liu, J., Xie, Z., Gao, F., & Liu, Y. (2022). Applications of Titanium Dioxide Nanostructure in Stomatology. *Molecules (Basel, Switzerland)*, *27*(12), 3881. doi:10.3390/molecules27123881 PMID:35745007

Maeztu, J. D., Rivero, P. J., Berlanga, C., Bastidas, D. M., Palacio, J. F., & Rodriguez, R. (2017). Effect of graphene oxide and fluorinated polymeric chains incorporated in a multilayered sol-gel nanocoating for the design of corrosion resistant and hydrophobic surfaces. *Applied Surface Science*, *419*, 138–149. doi:10.1016/j.apsusc.2017.05.043

Magovac, E., Vončina, B., Budimir, A., Jordanov, I., Grunlan, J. C., & Bischof, S. (2021). Environmentally Benign Phytic Acid-Based Nanocoating for Multifunctional Flame-Retardant/Antibacterial Cotton. *Fibers (Basel, Switzerland)*, *9*(11), 69. doi:10.3390/fib9110069

Maharjan, S., Liao, K.-S., Wang, A. J., Barton, K., Haldar, A., Alley, N. J., Byrne, H. J., & Curran, S. A. (2020). Self-cleaning hydrophobic nanocoating on glass: A scalable manufacturing process. *Materials Chemistry and Physics*, *239*, 122000. doi:10.1016/j.matchemphys.2019.122000

Mani, M., & Pillai, R. (2010). Impact of dust on solar photovoltaic (PV) performance: Research status, challenges, and recommendations. *Renewable & Sustainable Energy Reviews*, *14*(9), 3124–3131. doi:10.1016/j.rser.2010.07.065

Mao, M., Xu, H., Guo, K.-Y., Zhang, J.-W., Xia, Q.-Q., Zhang, G.-D., ... Tang, L.-C. (2020). Mechanically flexible, super-hydrophobic and flame-retardant hybrid nano-silica/graphene oxide wide ribbon decorated sponges for efficient oil/water separation and fire warning response. *Composites. Part A, Applied Science and Manufacturing*, *106191*. Advance online publication. doi:10.1016/j.compositesa.2020

Marilyn. (2019). *Water repellent hydrophobic surfaces and anti-fungal nanocoatings for building applications*. Academic Press.

Mazumder, M. K., Horenstein, M. N., Joglekar, N. R., Sayyah, A., Stark, J. W., Bernard, A. A. R., Garner, S. M., Yellowhair, J. E., Lin, H. Y., Eriksen, R. S., Griffin, A. C., Gao, Y., Centra, R. L., & Lloyd, A. H. (2017). Mitigation of dust impact on solar collectors by water-free cleaning with transparent electrodynamic films: Progress and challenges. *IEEE Journal of Photovoltaics*, *7*(5), 1342–1353. doi:10.1109/JPHOTOV.2017.2721462

Mishra, A., Bhatt, N., & Bajpai, A. K. (2019). Nanostructured superhydrophobic coatings for solar panel applications. In Nanomaterials-Based Coatings (pp. 397–424). doi:10.1016/B978-0-12-815884-5.00012-0

MokhlespourEsfahani, M. I. (2021). Smart textiles in healthcare: a summary of history, types, applications, challenges, and future trends. In Micro and Nano Technologies, Nanosensors and Nanodevices for Smart Multifunctional Textiles (pp. 93-107). Elsevier. doi:10.1016/B978-0-12-820777-2.00006-6

Mondoc, B., & Pop, F. (2010). Factors influencing the performance of photovoltaic power plant. *Proceedings of Third International Conference on Modern Power Systems (MPS),* 271-276.

Moradi, M., & Rezaei, M. (2022). Construction of highly anti-corrosion and super-hydrophobic polypropylene/graphene oxide nanocomposite coatings on carbon steel: Experimental, electrochemical and molecular dynamics studies. *Construction & Building Materials*, *317*, 126136. doi:10.1016/j.conbuildmat.2021.126136

Mott-Smith, H. D. (1971). History of 'plasmas'. *Nature*, *233*(5316), 219. doi:10.1038/233219a0 PMID:16063290

Nayashevsky, I., Xu, Q., Barahman, G., & Lyons, A.M. (2017). *Anti-reflective and anti-soiling properties of a KleanBoost™, a superhydrophobic nano-textured coating for solar glass*. Academic Press.

Nayshevsky, I. (2020). *Hydrophobic Fluoropolymer Nano-Coatings: Study of Anti-Reflective and Anti-Soiling Properties and Applications*. Academic Press.

Nejad, F. M., Tanzadeh, J., Soozangar, M., & Otadi, A. (2016). Performance Evaluation of the Impact of Modified Silica Nano-Materials on the Hydrophobicity of Hot-Mix Asphalt. *International Conference on Transportation and Development 2016*. 10.1061/9780784479926.101

Norazmi, F. S. (2019). *Synthesis and characterization of hydrophobic silica nanoparticles as anti-reflective coating for photovoltaic* [Masters thesis]]. Universiti Teknologi Malaysia, Faculty of Science.

Obšel, V., & Otřísal, P. (2022). Application of methods of preparation of plasmo-chemical nanocoatings designated for hydrophobic and oleophobic correction of surface working of textile materials for filtration protective suit. *Challenges to National Defence in Contemporary Geopolitical Situation*, *2022*(1), 275–287. doi:10.47459/cndcgs.2022.35

Özpirin, M., & Ebil, Ö. (2018). Transparent block copolymer thin films for protection of optical elements via chemical vapor deposition. *Thin Solid Films*, *674*, 1–5. doi:10.1016/j.tsf.2018.06.044

Pakdel, E., Fang, J., Fang, J., Sun, L., & Wang, X. (2018). Nanocoatings for Smart Textiles. In Smart Textiles (pp. 247–300). doi:10.1002/9781119460367.ch8

Pakdel, E., Wang, J., Allardyce, B. J., Rajkhowa, R., & Wang, X. (2016). Functionality of nano and 3D-microhierarchical TiO2 particles as coagulants for sericin extraction from the silk degumming wastewater. *Separation and Purification Technology*, *170*, 92–101. doi:10.1016/j.seppur.2016.06.025

Paven, M., Mammen, L., & Vollmer, D. (2016). Smart Materials for Advanced Environmental Applications. In P. Wang (Ed.), *RSC Smart Materials No. 20*. The Royal Society of Chemistry. doi:10.1039/9781782622192-00209

Petkovska, J., Mladenovic, N., Marković, D., Radoičić, M., Chiang, H.-C., Palen, B., Mirakovski, D., Dimova, V., Radetić, M., Grunlan, J. C., & Jordanov, I. (2023). Environmentally benign few-bilayer intumescent nanocoating for flame retardant enzyme/plasma modified polyester fabric. *Polymer Degradation & Stability*, *214*, 110406. doi:10.1016/j.polymdegradstab.2023.110406

Polshettiwar, V. (2022). Dendritic Fibrous Nanosilica: Discovery, Synthesis, Formation Mechanism, Catalysis, and CO2 Capture-Conversion. *Accounts of Chemical Research*, *55*(10), 1395–1410. doi:10.1021/acs.accounts.2c00031 PMID:35499964

Prabhakar, P., Sen, R. K., Patel, M., Shruti, Dwivedi, N., Singh, S., Kumar, P., Chouhan, M., Yadav, A. K., Mondal, D. P., Solanki, P. R., Srivastava, A. K., & Dhand, C. (2022). Development of copper impregnated bio-inspired hydrophobic antibacterial nanocoatings for textiles. *Colloids and Surfaces. B, Biointerfaces*, *220*, 112913. doi:10.1016/j.colsurfb.2022.112913 PMID:36306694

Priyadarsini, G. B., & Sharma, A. K. (2016). Design of multi-layer anti-reflection coating for terrestrial solar panel glass. *Bulletin of Materials Science*, 39.

Raelyn, L. Y. L. (2020). *Synthesis and applications of hydrophobic nanostructured coatings for green buildings*. Academic Press.

Rai, M., Bonde, S., Yadav, A., Bhowmik, A., Rathod, S., Ingle, P., & Gade, A. (2021). Nanotechnology as a Shield against COVID-19: Current Advancement and Limitations. *Viruses*, *13*(7), 1224. doi:10.3390/v13071224 PMID:34202815

Rajamohan, P. D. (2023). *A Review on Nanoparticles Based and Non-Nanoparticles Based Hydrophobic Nanocoatings and Its Applications*. Academic Press.

Rajput, D. S., & Sudhakar, K. (2013). Effect of dust on the performance of solar PV panel. *International Journal of Chemtech Research*, *5*, 1083–1086.

Ramaratnam, K., Iyer, S. K., Kinnan, M. K., Chumanov, G., Brown, P. J., & Luzinov, I. (2008). Ultrahydrophobic Textiles Using Nanoparticles: Lotus Approach. *Journal of Engineered Fibers and Fabrics*, *3*(4). doi:10.1177/155892500800300402

Rananavare, A. P., & Lee, J. (2022). Hydrophobic cotton fabric synthesized via dispersion polymerization from poly(glycidyl methacrylate) nanoparticles for self-cleaning applications. *Progress in Organic Coatings, 170*.

Reuna, J., Kanter, R., Kandelin, N., Kiuru, K., Koivuluoto, H., & Guina, M. (2023). Ice resistance of hydrophobic fluoropolymerized nanostructured alumina films for antireflective coatings. *Nano Express*, *3*(4), 044002. doi:10.1088/2632-959X/acad19

Rossi, G., Altavilla, C., Scarfato, P., Ciambelli, P., & Incarnato, L. (2015). Durability study of transparent and flexible nanolayer barrier for photovoltaic devices. *Polymer Degradation & Stability*, *112*, 160–166. doi:10.1016/j.polymdegradstab.2014.12.022

Ruzafa-Silvestre, C., Juan-Fernández, B., Carbonell-Blasco, M. P., Bañón-Gil, E., Orgilés-Calpena, E., & Arán-Ais, F. (2022). Organosilicon-Based Plasma Nanocoating on Crust Leather for Water-Repellent Footwear. *Materials (Basel)*, *15*(20), 7255. doi:10.3390/ma15207255 PMID:36295320

Ruzi, M., Celik, N., & Onses, M. (2022). Superhydrophobic coatings for food packaging applications: A review. *Food Packaging and Shelf Life*, *32*, 100823. doi:10.1016/j.fpsl.2022.100823

Sahoo, J., Sarkhel, S., Mukherjee, N., & Jaiswal, A. (2022). Title of the Article. *ACS Omega*, *7*(50), 45962–45980. doi:10.1021/acsomega.2c06211 PMID:36570317

Salehi, H., Eshaghi, A., Rezazadeh, M., & Zabolian, H. (2022). Antireflective and anti-dust modified silica-based thin film on solar cell cover glass. *Journal of Alloys and Compounds*, *892*, 162228. doi:10.1016/j.jallcom.2021.162228

Sankar, S., Nair, B. N., Suzuki, T., Anilkumar, G. M., Padmanabhan, M., Hareesh, U. N., & Warrier, K. G. (2016). Hydrophobic and Metallophobic Surfaces: Highly Stable Non-wetting Inorganic Surfaces Based on Lanthanum Phosphate Nanorods. *Scientific Reports*, *6*(1), 22732. doi:10.1038/srep22732 PMID:26955962

Selim, M. S., Fatthallah, N. A., Shenashen, M. A., Higazy, S. A., Madian, H. R., Selim, M. M., & El-Safty, S. A. (2023). Bioinspired Graphene Oxide–Magnetite Nanocomposite Coatings as Protective Superhydrophobic Antifouling Surfaces. *Langmuir*, *39*(6), 2333–2346. doi:10.1021/acs.langmuir.2c03061 PMID:36719844

Selim, M. S., Shenashen, M. A., El-Sockary, M. A., Fatthallah, N. A., Higazy, S. A., El-Saeed, A. M., El-Azabawy, O. E., & El-Safty, S. A. (2023). Polymer/graphene-derived nanocomposites as advanced marine antifouling coatings. In *Micro and Nano Technologies: Advances in Nanotechnology for Marine Antifouling*. Elsevier. doi:10.1016/B978-0-323-91762-9.00007-1

Selim, M. S., Shenashen, M. A., Elmarakbi, A., Fatthallah, N. A., Hasegawa, S., & El-Safty, S. A. (2017). Synthesis of ultrahydrophobic and thermally stable inorganic–organic nanocomposites for self-cleaning foul release coatings. *Chemical Engineering Journal*, *320*, 653–666. doi:10.1016/j.cej.2017.03.067

Shaban, M., & Hasanzadeh, M. (2020). Biomedical applications of dendritic fibrous nanosilica (DFNS): Recent progress and challenges. *RSC Advances*, *10*(61), 37116–37133. doi:10.1039/D0RA04388E PMID:35521236

Shao, J., Sheng, W., Wang, C., & Ye, Y. (2021). Solvent-free fabrication of tough self-crosslinkable short-fluorinated copolymer nanocoatings for ultradurable superhydrophobic fabrics. *Chemical Engineering Journal*, *416*, 128043. doi:10.1016/j.cej.2020.128043

Sharma, S., Kumar, A., Bashir, S., Ramesh, K., & Ramesh, S. (2022). A comprehensive review: Super hydrophobic graphene nanocomposite coatings for underwater and wet applications to enhance corrosion resistance. *FlatChem*, *31*, 100326. doi:10.1016/j.flatc.2021.100326

Shen, G. X., Du, R. G., Chen, Y. C., Lin, C. J., & Scantlebury, D. (2005). Study on hydrophobic nano-titanium dioxide coatings for improvement in corrosion resistance of type 316L stainless steel. *Corrosion*, *61*(10), 943–950. doi:10.5006/1.3280894

Shi, X.-H., Liu, Q.-Y., Li, X.-L., Du, A.-K., Niu, J.-W., Li, Y.-M., Li, Z., Wang, M., & Wang, D.-Y. (2022). Construction phosphorus/nitrogen-containing flame-retardant and hydrophobic coating toward cotton fabric via layer-by-layer assembly. *Polymer Degradation & Stability*, *197*, 109839. doi:10.1016/j.polymdegradstab.2022.109839

Shirvanimoghaddam, K., Akbari, M. K., Yadav, R., Al-Tamimi, A. K., & Naebe, M. (2021). Fight against COVID-19: The case of antiviral surfaces. *APL Materials*, *9*(3), 031112. doi:10.1063/5.0043009 PMID:33842101

Si, Y., & Guo, Z. (2015). Superhydrophobic nanocoatings: From materials to fabrications and to applications. *Nanoscale*, *7*(14), 5922–5946. doi:10.1039/C4NR07554D PMID:25766486

Si, Y., Zhu, H., Chen, L., Jiang, T., & Guo, Z. (2015). A multifunctional transparent superhydrophobic gel nanocoating with self-healing properties. *Chemical Communications*, *51*(94), 16794–16797. doi:10.1039/C5CC06977G PMID:26434728

Simpson, J. T., Hunter, S. R., & Aytug, T. (2015). Superhydrophobic materials and coatings: A review. *Reports on Progress in Physics*, *78*(8), 1–14. doi:10.1088/0034-4885/78/8/086501 PMID:26181655

Smitha, S., Shajesh, P., Mukundan, P., & Warrier, K. G. K. (2007). Antiwetting silica–gelatin nanohybrid and transparent nano coatings synthesised through an aqueous sol–gel process. *Journal of Sol-Gel Science and Technology*, *42*(2), 157–163. doi:10.1007/s10971-007-1535-9

Souza, J. C. M., Sordi, M. B., Kanazawa, M., Ravindran, S., Henriques, B., Silva, F. S., Aparicio, C., & Cooper, L. F. (2019). Nano-scale modification of titanium implant surfaces to enhance osseointegration. *Acta Biomaterialia*, *94*, 112–131. doi:10.1016/j.actbio.2019.05.045 PMID:31128320

Srinivasan, S., Praveen, V. K., Philip, R., & Ajayaghosh, A. (2008). Bioinspired Superhydrophobic Coatings of Carbon Nanotubes and Linear π Systems Based on the "Bottom-up" Self-Assembly Approach. *Angewandte Chemie International Edition*, *47*(31), 5750–5754. doi:10.1002/anie.200802097 PMID:18604864

Su, C., Huang, X., Zhang, L., Zhang, Y., Yu, Z., Chen, C., Ye, Y., & Guo, S. (2023). Robust superhydrophobic wearable piezoelectric nanogenerators for self-powered body motion sensors. *Nano Energy*, *107*, 108095. doi:10.1016/j.nanoen.2022.108095

Sumit, Jeon, & Lim. (2022). Superhydrophobic sponge decorated with hydrophobic MOF-5 nanocoating for efficient oil-water separation and antibacterial applications. *Sustainable Materials and Technologies, 33*, 468–477. doi:10.1016/j.cej.2017.03.052

Suyambulingam, G. R. T., Jeyasubramanian, K., Mariappan, V. K., Veluswamy, P., Ikeda, H., & Krishnamoorthy, K. (2017). Excellent floating and load bearing properties of superhydrophobic ZnO/copper stearate nanocoating. *Chemical Engineering Journal*, *320*, 468–477. doi:10.1016/j.cej.2017.03.052

Syafiq, A., Balakrishnan, V., Ali, M. S., Dhoble, S. J., Rahim, N. A., Omar, A., & Bakar, A. H. A. (2022). Application of transparent self-cleaning coating for photovoltaic panel: A review. *Current Opinion in Chemical Engineering*, *36*, 100801. doi:10.1016/j.coche.2022.100801

Syafiq, A., Pandey, A. K., Balakrishnan, V., Shahabuddin, S., & Abd Rahim, N. (2018). Organic-inorganic composite nanocoatings with superhydrophobicity and thermal stability. *Pigment & Resin Technology*.

Taghvaei, E., Moosavi, A., Nouri-Borujerdi, A., Daeian, M. A., & Vafaeinejad, S. (2017). Superhydrophobic surfaces with a dual-layer micro- and nanoparticle coating for drag reduction. *Energy*, *125*, 1–10. doi:10.1016/j.energy.2017.02.117

Tayel, S. A., Abu El-Maaty, A. E., Mostafa, E. M., & Elsaadawi, Y. F. (2022). Enhance the performance of photovoltaic solar panels by a self-cleaning and hydrophobic nanocoating. *Scientific Reports*, *12*(1), 21236. doi:10.1038/s41598-022-25667-4 PMID:36481954

Teisala, H., Tuominen, M., Aromaa, M., Mäkelä, J. M., Stepien, M., Saarinen, J. J., Toivakka, M., & Kuusipalo, J. (2010). Development of superhydrophobic coating on paperboard surface using the Liquid Flame Spray. *Surface and Coatings Technology*, *205*(2), 436–445. doi:10.1016/j.surfcoat.2010.07.003

Thomas, T. M., & Mahapatra, P. S. (2021). Condensation of Humid Air on Superhydrophobic Surfaces: Effect of Nanocoatings on a Hierarchical Interface. *Langmuir*, *37*(44), 12767–12780. doi:10.1021/acs.langmuir.1c01012 PMID:34714651

Tsai, Y.-T., Maggay, I. V., Venault, A., & Lin, Y.-F. (2021). Fluorine-free and hydrophobic/oleophilic PMMA/PDMS electrospun nanofibrous membranes for gravity-driven removal of water from oil-rich emulsions. *Separation and Purification Technology*, *279*, 119720. doi:10.1016/j.seppur.2021.119720

Vasilopoulou, M., da Silva, W. J., Soultati, A., Kim, H. P., Kim, B. S., Reo, Y., & Nazeeruddin, M. K. (2022). Photonic nanostructures mimicking floral epidermis for perovskite solar cells. *Cell Reports. Physical Science*, *3*(9), 101019. doi:10.1016/j.xcrp.2022.101019 PMID:36259071

Víg, P., Tóth, J., & Seres, I. (2023). Optical effects of water droplets on the performance of solar cell covered with a hydrophobic thin layer. *Solar Energy*, *251*, 366–381. doi:10.1016/j.solener.2023.01.028

Wang, H., Kumar, R., & Memon, H. (2020). Strongly Hydrophobic and Superoleophilic PMMA Based Nanocoated Cotton Fabrics. *Coatings*, *10*(10), 943. doi:10.3390/coatings10100943

Wang, J., Xue, Z., Yu, T., Liu, Z., & Mu, T. (2017). Activation of cellulose assisted by CO2 for the preparation of a superhydrophobic nanocoating. *Chemistry, an Asian Journal*, *12*(14), 1773–1779. doi:10.1002/asia.201700335 PMID:28471506

Wang, K., Bu, N., Zhen, Q., Liu, J., & Bashir, S. (2023). Modified nano-SiO2/TiO2 hybrid fluorinated B-72 as antimicrobial and hydrophobic coatings for the conservation of ancient bricks. *Construction & Building Materials*, *365*, 130090. doi:10.1016/j.conbuildmat.2022.130090

Wang, S., Du, X., Deng, S., Fu, X., Du, Z., Cheng, X., & Wang, H. (2019). A polydopamine-bridged hierarchical design for fabricating flame-retarded, superhydrophobic, and durable cotton fabric. *Cellulose (London, England)*, *26*(11), 7009–7023. doi:10.1007/s10570-019-02586-8

Wang, X., Chai, Y., & Liu, J. (2013). Formation of highly hydrophobic wood surfaces using silica nanoparticles modified with long-chain alkylsilane. *Holzforschung*, *67*(6), 667–672. Advance online publication. doi:10.1515/hf-2012-0153

Wang, X., Han, Z., Liu, Y., & Wang, Q. (2020). Micro-nano surface structure construction and hydrophobic modification to prepare efficient oil-water separation melamine formaldehyde foam. *Applied Surface Science*, *505*, 144577. doi:10.1016/j.apsusc.2019.144577

Wang, X., Liu, S., Chang, H., & Liu, J. (2014). Sol-Gel Deposition of TiO2 Nanocoatings on Wood Surfaces with Enhanced Hydrophobicity and Photostability. Wood and Fiber Science. *Journal of the Society of Wood Science and Technology*, *46*, 109–117.

Wang, Y., Li, B., Xu, C., & Nie, R. (2013). Superhydrophobic ZnONanocoatings for Different Substrates. *Asian Journal of Chemistry*, *25*(10), 5571–5572. doi:10.14233/ajchem.2013.OH21

Wang, Y., Wang, Y., Li, X., Li, J., Su, L., Zhang, X., & Du, X. (2018). Dendritic Silica Particles with Well-Dispersed Ag Nanoparticles for Robust Antireflective and Antibacterial Nanocoatings on Polymeric Glass. *ACS Sustainable Chemistry & Engineering*, *6*(11), 14071–14081. doi:10.1021/acssuschemeng.8b02617

Wang, Z., Chen, X., Gong, Y., Zhang, B., & Li, H. (2017). Superhydrophobic nanocoatings prepared by a novel vacuum cold spray process. *Surface and Coatings Technology*, *325*, 52–57. doi:10.1016/j.surfcoat.2017.06.044

Williams, A. G., Moore, E., Thomas, A., & Johnson, J. A. (2023). Graphene-Based Materials in Dental Applications: Antibacterial, Biocompatible, and Bone Regenerative Properties. *International Journal of Biomaterials*. doi:10.1155/2023/8803283

Wu, J., Tu, J., Yu, S., Wu, H., Xie, Y., Yang, Y., Xv, Z., & Zhang, Q. (2023). Hollow core-shell nanocoatings with gradient refractive index structure for enhanced photovoltaic performance. *Colloids and Surfaces. A, Physicochemical and Engineering Aspects*, *667*, 131424. doi:10.1016/j.colsurfa.2023.131424

Xi, R., Wang, Y., Li, X., Zhang, X., & Du, X. (2020). A facile strategy to form a three-dimensional network structure for mechanically robust superhydrophobic nanocoatings with enhanced transmittance. *Journal of Colloid and Interface Science*, *563*, 42–53. doi:10.1016/j.jcis.2019.12.049 PMID:31865047

Xi, R., Wang, Y., Li, X., Zhang, X., & Du, X. (2020). A facile strategy to form three-dimensional network structure for mechanically robust superhydrophobic nanocoatings with enhanced transmittance. *Journal of Colloid and Interface Science*, *563*, 42–53. doi:10.1016/j.jcis.2019.12.049 PMID:31865047

Xiang, Y., Si, Y., Xin, Y., & Guo, Z. (2017). One-step strategy to prepare utility ZnO–stearic acid (STA) superhydrophobic nanocoating. *Chemistry Letters*, *46*(9), 1393–1395. doi:10.1246/cl.170545

Xin, J. H., Daoud, W. A., & Kong, Y. Y. (2004). A new approach to UV-blocking treatment for cotton fabrics. *Textile Research Journal*, *74*(2), 97–100. doi:10.1177/004051750407400202

Xing, Y., Wu, Z., Yang, J., Wang, X., & Liu, L. (2020). LIPSS combined with ALD MoS2 nano-coatings for enhancing surface friction and hydrophobic performances. *Surface and Coatings Technology*, *385*, 125396. doi:10.1016/j.surfcoat.2020.125396

Xiong, J., Luo, H., Gao, D., Zhou, X., Cui, P., Thangavel, G., Parida, K., & Lee, P. S. (2019). Self-restoring, waterproof, tunable microstructural shape memory triboelectric nanogenerator for self-powered water temperature sensor. *Nano Energy*, *61*, 584–593. doi:10.1016/j.nanoen.2019.04.089

Xu, Q., Sun, Z., Feng, F., Li, Y., Tong, G., & Noor, R. S. (2020). Preparation and anti-icing of hydrophobic polypyrrole coatings on wind turbine blade. *International Journal of Rotating Machinery*, *8626457*, 1–9. Advance online publication. doi:10.1155/2020/8626457

Xu, Q. F., Wang, J. N., & Sanderson, K. D. (2010). Organic–Inorganic Composite Nanocoatings with Superhydrophobicity, Good Transparency, and Thermal Stability. *ACS Nano*, *4*(4), 2201–2209. doi:10.1021/nn901581j PMID:20302323

Xu, Y.-D., Zhu, Z.-Y., Xu, T.-Z., Abadikhah, H., Wang, J.-W., Xu, X., & Agathopoulos, S. (2018). Fabrication and characterization of robust hydrophobic lotus leaf-like surface on Si3N4 porous membrane via polymer-derived SiNCO inorganic nanoparticle modification. *Ceramics International*, *44*(14), 16443–16449. doi:10.1016/j.ceramint.2018.06.058

Xue, C.-H., Chen, J., Yin, W., Jia, S.-T., & Ma, J.-Z. (2012). Superhydrophobic conductive textiles with antibacterial property by coating fibers with silver nanoparticles. *Applied Surface Science*, *258*(7), 2468–2472. doi:10.1016/j.apsusc.2011.10.074

Xue, C.-H., Li, Y.-R., Hou, J.-L., Zhang, L., Ma, J.-Z., & Jia, S.-T. (2015). Self-roughened superhydrophobic coatings for continuous oil–water separation. *Journal of Materials Chemistry. A, Materials for Energy and Sustainability*, *3*(19), 10248–10253. doi:10.1039/C5TA01014D

Yazdani-Ahmadabadi, H., Rastegar, S., & Ranjbar, Z. (2016). Acidic hydrophobic nanosilica particle-induced mechanical dissipation and adhesion promotion of acrylate-based photopolymericnanocoatings. *Progress in Organic Coatings*, *100*, 173–177. doi:10.1016/j.porgcoat.2016.05.004

Ye, H., Zhu, L., Li, W., Liu, H., & Chen, H. (2017). Constructing Fluorine-Free and Cost-Effective Superhydrophobic Surface with Normal-Alcohol-Modified Hydrophobic SiO2 Nanoparticles. *ACS Applied Materials & Interfaces*, *9*(1), 858–867. doi:10.1021/acsami.6b12820 PMID:27957841

Yedra, Á., Vilar, N., González, M., & Manteca, C. (2014). *Advanced Materials and Nanomaterials Unit.* Fundación Centro Tecnológico de Componentes (CTC).

Yeong, Y. H., Davis, A., Steele, A., Loth, E., & Bayer, I. S. (2014). Spray deposition effects on superhydrophobicity and durability of nanocoatings. *Surface Innovations*, *2*(2), 70–78. doi:10.1680/si.13.00044

Yetisen, A. K., Qu, H., Manbachi, A., Butt, H., Dokmeci, M. R., Hinestroza, J. P., Skorobogatiy, M., Khademhosseini, A., & Yun, S. H. (2016). Nanotechnology in Textiles. *ACS Nano*, *10*(3), 3042–3068. doi:10.1021/acsnano.5b08176 PMID:26918485

Yilbas, B. S., Keles, O., & Toprakli, A. Y. (2017). Surface engineering towards self-cleaning applications: Laser textured silicon surface. *Procedia Engineering*, *184*, 716–724. doi:10.1016/j.proeng.2017.04.147

Ying. (2018). Hydrophobic polymer-based nano-coatings for building applications. Academic Press.

Yu, F., Gao, J., Liu, C., Chen, Y., Zhong, G., Hodges, C., Chen, M., & Zhang, H. (2020). Preparation and UV aging of nano-SiO2/fluorinated polyacrylate polyurethane hydrophobic composite coating. *Progress in Organic Coatings*, *141*, 105556. doi:10.1016/j.porgcoat.2020.105556

Zemajtis, F., Hasan, A. B. M. R., Yetik, O., Trtik, P., Pillai, K. M., & Sobolev, K. (2023). From superhydrophilicity to superhydrophobicity: High-resolution neutron imaging and modeling of water imbibition through porous surfaces treated with engineered nano-coatings. *Scientific Reports*, *13*(1), 11176. doi:10.1038/s41598-023-38324-1 PMID:37429919

Zhang, D., Williams, B. L., Liu, J., Hou, Z., Smith, A. T., Nam, S., Nasir, Z., Patel, H., Partyka, A., Becher, E. M., Lofink, B. J., Santos, V. H., Peng, X., & Sun, L. (2021). An environmentally-friendly sandwich-like structured nanocoating system for wash durable, flame retardant, and hydrophobic cotton fabrics. *Cellulose (London, England)*, *28*(16), 10277–10289. doi:10.1007/s10570-021-04177-y

Zhang, X., Wang, H., Zhang, X., Zhao, Z., & Zhu, Y. (2019). A multifunctional super-hydrophobic coating based on PDA modified MoS2 with anti-corrosion and wear resistance. *Colloids and Surfaces. A, Physicochemical and Engineering Aspects*, *568*, 239–247. doi:10.1016/j.colsurfa.2019.02.016

Zhu, G., Wang, J., Yuan, X., & Yuan, B. (2022). Hydrophobic and fire-safe polyurethane foam coated with chitosan and nano-montmorillonite via layer-by-layer assembly for emergency absorption of oil spill. *Materials Letters*, *316*, 132009. doi:10.1016/j.matlet.2022.132009

Zolriasatein, A., & Abdollahzadeh, S., & MotakefKazemi, N. (2023). An Examination of the Application of Polyurethane-Based Hydrophobic Nanocoatings Inspired by Natural Structures in Extracting Water from Air Humidity. *Journal of Water and Environmental Nanotechnology*, *8*(2), 151–163. doi:10.22090/jwent.2023.02.005

Zulfiqar, U., Awais, M., Hussain, S. Z., Hussain, I., Husain, S., & Subhani, T. (2017). Durable and self-healing superhydrophobic surfaces for building materials. *Materials Letters*, *192*, 56–59. doi:10.1016/j.matlet.2017.01.070

ADDITIONAL READING

Bellini, E. (2022). Hydrophobic Nanocoating to Reduce Soiling in Solar Panels. https://www.pv-magazine.com/2022/12/14/hydrophobic-nanocoating-to-reduce-soiling-in-solar-panels/

Chang, F. C. (2018, June 20). Superhydrophobic Nanoparticles-Based Hard Coatings for Building and Industrial Application. ScholarBank@NUS Repository.

EU. (2016). Light Management in Crystalline Silicon Solar Cells with Photonic Nanocoatings. https://cordis.europa.eu/project/id/744881/reporting

International Energy Agency. (2023). Renewables: Solar PV. https://www.iea.org/energy-system/renewables/solar-pv

Marpu, P., & Eswara, P. (2022). Hydrophobic Coatings Market Size, Share & Trends Analysis Report by Product (Silica-Based, Fluoropolymer-Based, Others), by Application (Aerospace &Defense, Automotive, Building & Construction, Others), by Region, and Segment Forecasts, 2022-2030. Allied Market Research. https://www.alliedmarketresearch.com/hydrophobic-coatings-market

NAMI. (2023). https://www.innovationhub.hk/articles/nami

Nationalgrid. (2023). How Does Solar Power Work? https://www.nationalgrid.com/stories/energy-explained/how-does-solar-power-work#:~:text=Solar%20panels%20are%20usually%20made,and%20produces%20an%20electric%20charge

Santos. (2022). Light Management in Crystalline Silicon Solar Cells with Photonic Nanocoatings. https://run.unl.pt/bitstream/10362/155362/1/Santos_2022.pdf

Teisala, H. (2013). Multifunctional Superhydrophobic Nanoparticle Coatings for Cellulose-based Substrates by Liquid Flame Spray. https://trepo.tuni.fi/handle/10024/114287

The engineer. (2015, November 16). Superhydrophobic coating is non-toxic and economical to produce. https://www.theengineer.co.uk/content/news/superhydrophobic-coating-is-non-toxic-and-economical-to-produce

Compilation of References

Abbasi, H., Antunes, M., & Velasco, J. I. (2020). Effects of Graphene Nanoplatelets and Cellular Structure on the Thermal Conductivity of Polysulfone Nanocomposite Foams. Polymers.

Abbas, M., Buntinx, M., Deferme, W., Reddy, N., & Peeters, R. (2021). Oxygen gas and uv barrier properties of nano-ZnO-coated pet and phbhhx materials fabricated by ultrasonic spray-coating technique. *Nanomaterials (Basel, Switzerland)*, *11*(2), 1–14. doi:10.3390/nano11020449 PMID:33578826

Abdeen, D. H., El Hachach, M., Koc, M., & Atieh, M. A. (2019). A Review on the Corrosion Behaviour of Nanocoatings on Metallic Substrates. Materials. doi:10.3390/ma12020210

Abdelghany, T. M., Al-Rajhi, A. M. H., Yahya, R., Bakri, M. M., Al Abboud, M. A., Yahya, R., Qanash, H., Bazaid, A. S., & Salem, S. S. (2023). Phytofabrication of zinc oxide nanoparticles with advanced characterization and its antioxidant, anticancer, and antimicrobial activity against pathogenic microorganisms. *Biomass Conversion and Biorefinery*, *13*(1), 417–430. doi:10.1007/s13399-022-03412-1

Abdel-Karim, R., & Waheed, A. F. (2013). Nanocoatings. In Modern Surface Engineering Treatments. IntechOpen.

Abdollahiyan, P., Oroojalian, F., & Mokhtarzadeh, A. (2021). The triad of nanotechnology, cell signalling, and scaffold implantation for the successful repair of damaged organs: An overview on soft-tissue engineering. *Journal of Controlled Release*, *332*, 460–492. doi:10.1016/j.jconrel.2021.02.036 PMID:33675876

Abdolrahimi, M., Vasilakaki, M., Slimani, S., Ntallis, N., Varvaro, G., Laureti, S., Meneghini, C., Trohidou, K. N., Fiorani, D., & Peddis, D. (2021). Magnetism of nanoparticles: Effect of the organic coating. *Nanomaterials (Basel, Switzerland)*, *11*(7), 1787. doi:10.3390/nano11071787 PMID:34361173

Abdul Khalil, H. P. S., Banerjee, A., Saurabh, C. K., Tye, Y. Y., Suriani, A. B., Mohamed, A., Karim, A. A., Rizal, S., & Paridah, M. T. (2018). Biodegradable films for fruits and vegetables packaging application: Preparation and properties. *Food Engineering Reviews*, *10*(3), 139–153. doi:10.1007/s12393-018-9180-3

Abdulraheem, M. I., Zhang, W., Li, S., Moshayedi, A. J., Farooque, A. A., & Hu, J. (2023). Advancement of Remote Sensing for Soil Measurements and Applications: A Comprehensive Review. Sustainability. doi:10.3390/su152115444

Abidi, N., Aminayi, P., Cabrales, L., & Hequet, E. (2012). Super-hydrophobic cotton fabric prepared using nanoparticles and molecular vapor deposition methods. *ACS Symposium Series, 1107*, 149–165. 10.1021/bk-2012-1107.ch008

Abimbola, O. F., Adetunji, C. O., Dare, B. J., Mathew, J. T., Inobeme, A., Ghazanfar, S., Olaniyan, O. T., Ajiboye, M. D., & Dauda, W. (2023). Packaging of Animal-Based Food Products: The Need for Intelligent Freshness Sensors. In Sensing and Artificial Intelligence Solutions for Food Manufacturing. CRC Press.

Abraham, A. M., & Venkatesan, S. (2023). A review on application of biomaterials for medical and dental implants. *Proceedings of the Institution of Mechanical Engineers, Part L: Journal of Materials: Design and Applications, 237*(2), 249-273. 10.1177/14644207221121981

Abubakar, I., Mann, A., & Mathew, J. T. (2015). Evaluation of Phytochemical, Anti-nutritional and Antioxidant Potentials of Flower and Seed Methanol Extracts of *Senna alata* L. Grown in Nigeria. *American Journal of Applied Chemistry, 3*(3), 93–100. doi:10.11648/j.ajac.20150303.12

Achachlouei, B. F., & Zahedi, Y. (2018). Fabrication and characterization of CMC-based nanocomposites reinforced with sodium montmorillonite and TiO2 nanomaterials. *Carbohydrate Polymers, 199*, 415–425. doi:10.1016/j.carbpol.2018.07.031 PMID:30143147

Adebayo, V. A., Adewale, O. B., Anadozie, S. O., Osukoya, O. A., Obafemi, T. O., Adewumi, D. F., Idowu, O. T., Onasanya, A., & Ojo, A. A. (2023). GC-MS analysis of aqueous extract of Nymphaea lotus and ameliorative potential of its biosynthesized gold nanoparticles against cadmium-induced kidney damage in rats. *Heliyon, 9*(6), e17124. doi:10.1016/j.heliyon.2023.e17124 PMID:37484424

Adetunji, C. O., Mathew, J. T., Singh, K. R. B., Inobeme, A., Olaniyan, O., Vanya, N., & Singh, J. (2021). Conducting Polymer-Based Microbial Fuel Cells. Conducting Polymers for Advanced Energy Applications, 18, 337-344.

Adetunji, C. O., Mathew, J. T., Singh, K. R. B., Inobeme, A., Vanya, N., Olaniyan, O., Singh, J., & Singh, R. P. (2022). Zero-dimensional carbon nanomaterials for targeted drug delivery and photothermal therapy applications: In Zero-dimensional Carbon Nanomaterials: Fundamentals and applications. IOP Publishing.

Adetunji, C. O., Ogundolie, F. A., Mathew, J. T., Inobeme, A., Olotu, T., Adetunji, J. B., Dauda, W. P., Aborode, A. T., Ajayi, O. O., Ghazanfar, S., Aliyu, A., Imoisi, S. E., Moses-Oke, R. O., Egbuna, C., Uzualu, E. J., Ajenifujah-Solebo, S. O., Akinbo, O., Adetuyi, B. O., & Edetalehn, O. I. (2023). Chapter Twenty-Six - Patenting protocols, toxicity, risk assessments, and policy issues of nanomaterials with diverse applications in food, biomedical, and other relevant sectors. In Evaporative Coolers for the Postharvest Management of Fruits and Vegetables. Academic Press.

Adetunji, C. O., Ogundolie, F. A., Mathew, J. T., Inobeme, A., Olotu, T., Olaniyan, O. T., Ghazanfar, S., Ijabadeniyi, O. A., Ajiboye, M. D., Ajayi, O. O., Dauda, W., & Adetunji, J. B. (2022). Nanotube platforms for effective drug delivery applications. In Novel Platforms for Drug Delivery Applications. Elsevier Ltd.

Adetunji, C. O., Ogundolie, F. A., Mathew, J. T., Inobeme, A., Olotu, T., Olaniyan, O. T., Ijabadeniyi, O. A., Ajiboye, M. D., Ajayi, O. O., Dauda, W., Ghazanfar, S., & Adetunji, J. B. (2022). Graphene-based nanomaterials for targeted drug delivery and tissue engineering. In Novel Platforms for Drug Delivery Applications. Elsevier Ltd.

Adetunji, C. O., Ogundolie, F. A., Olaniyan, O., Mathew, J. T., Inobeme, A., Olotu, T., Ghazanfar, S., Ijabadeniyi, O. A., Ajiboye, M. D., Ajayi, O. O., Dauda, W. P., & Adetunji, J. B. (2022). Nanobiomaterials for Food Packaging Sensor Applications. In Bio- and Nano-sensing Technologies for Food Processing and Packaging. The Royal Society of Chemistry.

Adetunji, C. O., Olaniyan, O. T., Shariati, M. A., Hefft, D. I., Olugbenga, S. M., Rebezov, M., Ajayi, O. O., Anichkina, O., Bodunrinde, R. E., Adetunji, R. B., Adeniyi, M. J., Inobeme, A., & Mathew, J. T. (2023). Medicinal Properties of Sennaalata (L.) Roxb. (Syn. Cassia alata L.) and Its Biological Activities. In Phytochemical Composition and Pharmacy of Medicinal Plants. Apple Academic Press, Inc.

Adetunji, C. O., Olaniyan, O. T., Shariati, M. A., Hefft, D. I., Olugbenga, S. M., Rebezov, M., Ajayi, O. O., Baibalinova, G., Bodunrinde, R. E., Adetunji, R. B., Adeniyi, M. J., Inobeme, A., Mathew, J. T., Elufisan, T. O., & Oyedara, O. O. (2023). Medicinal and Pharmacological Attributes of AcalyphahispidaBurm. F. In Phytochemical Composition and Pharmacy of Medicinal Plants. Apple Academic Press, Inc.

Adetunji, C. O., Olugbemi, T. O., Osikemekha, A. A., Inobeme, A., & Mathew, J. T. (2021). *Environmental Impact of Polyurethane Chemistry*. American Chemical Society.

Adeyemi, J. O., & Fawole, O. A. (2023). Metal-Based Nanoparticles in Food Packaging and Coating Technologies: A Review. *Biomolecules*, *13*(7), 1092. doi:10.3390/biom13071092 PMID:37509128

Adeyeye, S. A. O. (2019). Food packaging and nanotechnology: Safeguarding consumer health and safety. *Nutrition & Food Science*, *49*(6), 1164–1179. doi:10.1108/NFS-01-2019-0020

Adeyeye, S. A. O., & Ashaolu, T. J. (2021). Applications of nano-materials in food packaging: A review. *Journal of Food Process Engineering*, *44*(7), e13708. doi:10.1111/jfpe.13708

Afzal, A., Habib, A., Ulhasan, I., Shahib, M., & Rehman, A. (2021). Antireflective Self-Cleaning TiO_2 Coatings for Solar Energy Harvesting Applications. *Frontiers in Materials*, *8*, 687059. doi:10.3389/fmats.2021.687059

Afzal, O., Altamimi, A. S. A., Nadeem, M. S., Alzarea, S. I., Almalki, W. H., Tariq, A., Mubeen, B., Murtaza, B. N., Iftikhar, S., Riaz, N., & Kazmi, I. (2022). Nanoparticles in Drug Delivery: From History to Therapeutic Applications. *Nanomaterials (Basel, Switzerland)*, *12*(24), 4494. doi:10.3390/nano12244494 PMID:36558344

Agrawal, S., Kumar, V., Kumar, S., & Shahi, S. K. (2022). Plant development and crop protection using phytonanotechnology: A new window for sustainable agriculture. *Chemosphere, 299*, 134465. doi:10.1016/j.chemosphere.2022.134465 PMID:35367229

Aguda, O. N., & Lateef, A. (2022). Recent advances in functionalization of nanotextiles: A strategy to combat harmful microorganisms and emerging pathogens in the 21st century. *Heliyon, 8*(6), e09761. doi:10.1016/j.heliyon.2022.e09761 PMID:35789866

Agustin, S., Wahyuni, E. T., Suparmo, S., Supriyadi, S., & Cahyanto, M. N. (2021). Incorporation of pectin during biosynthesis of bacterial cellulose by *Gluconacetobacter xylinus* InaCC B404: Possibility for producing green food packaging. *Biodiversitas (Surakarta), 22*(5). Advance online publication. doi:10.13057/biodiv/d220512

Ahmad, A., Qurashi, A., & Sheehan, D. (2023). Nano packaging–Progress and future perspectives for food safety, and sustainability. *Food Packaging and Shelf Life, 35*, 100997. doi:10.1016/j.fpsl.2022.100997

Ahmad, A., Younas, M., Santulli, C., Kiyani, M. Z., Ali, R., Habbiba, O., & Zubair, M. (2021). Nanomedicine and tissue engineering. In *Nanomedicine Manufacturing and Applications* (pp. 261–277). Elsevier. doi:10.1016/B978-0-12-820773-4.00020-2

Ahmadi, M. H., Ghazvini, M., Nazari, M. A., Ahmadi, M. A., Pourfayaz, F., Lorenzini, G., & Ming, T. (2019). Renewable Energy Harvesting with the Application of Nanotechnology: A Review. *International Journal of Energy Research, 43*(4), 1387–1410. doi:10.1002/er.4282

Ahmed, W., AlHannan, M., Yusuf, S., & Jackson, M. J. (2016). Nanocoating for medical devices. *Surgical Tools and Medical Devices*, 519-543.

Ahmed, J., Arfat, Y. A., Al-Attar, H., Auras, R., & Ejaz, M. (2017). Rheological, structural, ultraviolet protection and oxygen barrier properties of linear low- density polyethylene films reinforced with zinc oxide (ZnO) nanoparticles. *Food Packaging and Shelf Life, 13*(March), 20–26. doi:10.1016/j.fpsl.2017.04.005

Ahmed, J., Hiremath, N., & Jacob, H. (2017). Antimicrobial efficacies of essential oils/nanoparticles incorporated polylactide films against *L. monocytogenes* and *S. typhimurium* on contaminated cheese. *International Journal of Food Properties, 20*(1), 53–67. doi:10.1080/10942912.2015.1131165

Aiacoboae, A., Gheorghe, T., Lungu, I. I., Curutiu, C., Chifiriuc, M. C., Grumezescu, A. M., & Holban, A. M. (2017). Applications of nanoscale drugs carriers in the treatment of chronic diseases. In *Nanostructures for Novel Therapy* (pp. 37–55). Elsevier. doi:10.1016/B978-0-323-46142-9.00002-5

Ainali, N. M., Kalaronis, D., Evgenidou, E., Bikiaris, D. N., & Lambropoulou, D. A. (2021). Insights into biodegradable polymer-supported titanium dioxide photocatalysts for environmental remediation. *Macromol, 1*(3), 201–233. doi:10.3390/macromol1030015

Akbari-Alavijeh, S., Shaddel, R., & Jafari, S. M. (2020). Encapsulation of food bioactives and nutraceuticals by various chitosan-based nanocarriers. *Food Hydrocolloids*, *105*, 105774. doi:10.1016/j.foodhyd.2020.105774

Akhila, V., & Badwaik, L. S. (2022). Recent advancement in improvement of properties of polysaccharides and proteins based packaging film with added nanoparticles: A review. *International Journal of Biological Macromolecules*, *203*, 515–525. doi:10.1016/j.ijbiomac.2022.01.181 PMID:35122798

Akinlabi, S., Ogbonna, O., Mashinini, P. M., Adediran, A., Fatoba, O., & Akinlabi, E. (2019). Titanium and Epoxy for Automobile Application. *RE:view*. Advance online publication. doi:10.15224/978-1-63248-170-2-09

Al-Amiery, A. A., Yousif, E., Isahak, W. N. R. W., & Al-Azzawi, W. K. (2023). A review of inorganic corrosion inhibitors: Types, mechanisms, and applications. *Tribology in Industry*, *44*(2), 313. doi:10.24874/ti.1456.03.23.06

Alamri, H. R., Rezk, H., Abd-Elbary, H., Ziedan, H. A., & Elnozahy, A. (2020). Experimental Investigation to Improve the Energy Efficiency of Solar PV Panels using Hydrophobic SiO_2 Nanomaterial. *Coatings*, *10*(5), 503. doi:10.3390/coatings10050503

Alarifi, I. M. (2023). Advanced Selection Materials in Solar Cell Efficiency and Their Properties – A Comprehensive Review. *Materials Today: Proceedings*, *81*(Part 2), 403–414. doi:10.1016/j.matpr.2021.03.427

Albertini, P., Newton, M., Wang, M., Lecina, O. S., Green, P., Stoian, D., & Buonsanti, R. (2023). Hybrid Oxide Coatings Generate Stable Cu. *Catalysts*, CO2.

Aldawoud, A., Aldawoud, A., Aryanfar, Y., El Haj Assad, M., Sharma, S., & Alayi, R. (2022). Reducing PV soiling and condensation using hydrophobic coating with brush and controllable curtains. *The International Journal of Low Carbon Technologies*, *17*, 919–930. doi:10.1093/ijlct/ctac056

Aldoasri, M. A., Darwish, S. S., Adam, M. A., Elmarzugi, N. A., & Ahmed, S. M. (2017). Protecting of marble stone facades of historic buildings using multifunctional TiO2 nanocoating. *Sustainability (Basel)*, *9*(11), 1–15. doi:10.3390/su9112002

Alghamdi, M. A., Fallica, A. N., Virzì, N., Kesharwani, P., Pittalà, V., & Greish, K. (2022). The Promise of Nanotechnology in Personalized Medicine. *Journal of Personalized Medicine*, *12*(5), 673. doi:10.3390/jpm12050673 PMID:35629095

Alharbi, K. K., & Al-Sheikh, Y. A. (2014). Role and implications of nanodiagnostics in the changing trends of clinical diagnosis. *Saudi Journal of Biological Sciences*, *21*(2), 109–117. doi:10.1016/j.sjbs.2013.11.001 PMID:24600302

Alinaqi, Z., Khezri, A., & Rezaeinia, H. (2021). Sustained release modeling of clove essential oil from the structure of starch-based bio-nanocomposite film reinforced by electrosprayed zein nanoparticles. *International Journal of Biological Macromolecules*, *173*, 193–202. doi:10.1016/j.ijbiomac.2021.01.118 PMID:33482206

Alizadeh Sani, M., Maleki, M., Eghbaljoo-Gharehgheshlaghi, H., Khezerlou, A., Mohammadian, E., Liu, Q., & Jafari, S. M. (2022). Titanium dioxide nanoparticles as multifunctional surface-active materials for smart/active nanocomposite packaging films. *Advances in Colloid and Interface Science*, *300*, 102593. doi:10.1016/j.cis.2021.102593 PMID:34971916

Alizadeh-Sani, M., Khezerlou, A., & Ehsani, A. (2018). Fabrication and characterization of the bionanocomposite film based on whey protein biopolymer loaded with TiO2 nanoparticles, cellulose nanofibers and rosemary essential oil. *Industrial Crops and Products*, *124*, 300–315. doi:10.1016/j.indcrop.2018.08.001

Aljdaeh, E., Kamwa, I., Hammad, W., Abuashour, M. I., Sweidan, T., Khalid, H. M., & Muyeen, S. M. (2021). Performance Enhancement of Self-Cleaning Hydrophobic Nanocoated Photovoltaic Panels in a Dusty Environment. *Energies*, *14*(20), 6800. doi:10.3390/en14206800

Allaka, G., & Yepuri, V. (2023). (in press). Inexpensive hydrophobic and infrared reflective coatings on the cotton and silk fabrics using sol-gel dip coating technique. *Materials Today: Proceedings*. Advance online publication. doi:10.1016/j.matpr.2023.06.042

Allehyani, E. S., Almulaiky, Y. Q., Al-Harbi, S. A., & El-Shishtawy, R. M. (2022). In Situ Coating of Polydopamine-AgNPs on Polyester Fabrics Producing Antibacterial and Antioxidant Properties. *Polymers*, *14*(18), 3794. Advance online publication. doi:10.3390/polym14183794 PMID:36145939

Al-Naamani, L., Dutta, J., & Dobretsov, S. (2018). Nanocomposite zinc oxide-chitosan coatings on polyethylene films for extending storage life of okra (*Abelmoschus esculentus*). *Nanomaterials (Basel, Switzerland)*, *8*(7), 479. doi:10.3390/nano8070479 PMID:29966220

Alotaibi, Y. K., & Federico, F. (2017). The impact of health information technology on patient safety. *Saudi Medical Journal*, *38*(12), 1173–1180. doi:10.15537/smj.2017.12.20631 PMID:29209664

Alshimaysawee, S., Fadhel Obaid, R., Al-Gazally, M. E., Alexis Ramírez-Coronel, A., & Bathaei, M. S. (2023). Recent Advancements in Metallic Drug-Eluting Implants. *Pharmaceutics*, *15*(1), 223. doi:10.3390/pharmaceutics15010223 PMID:36678852

Amjadi, S., Emaminia, S., Nazari, M., Davudian, S. H., Roufegarinejad, L., & Hamishehkar, H. (2019). Application of reinforced ZnO nanoparticle-incorporated gelatin bionanocomposite film with chitosan nanofiber for packaging of chicken fillet and cheese as food models. *Food and Bioprocess Technology*, *12*(7), 1205–1219. doi:10.1007/s11947-019-02286-y

Ammar, S., Ramesh, K., Ma, I. A. W., Farah, Z., Vengadaesvaran, B., Ramesh, S., & Arof, A. K. (2017). Studies on SiO2-hybrid polymeric nanocomposite coatings with superior corrosion protection and hydrophobicity. *Surface and Coatings Technology*, *324*, 536–545. doi:10.1016/j.surfcoat.2017.06.014

Amor, F., Baudys, M., Racova, Z., Scheinherrová, L., Ingrisova, L., & Hajek, P. (2022). Contribution of TiO2 and ZnO nanoparticles to the hydration of Portland cement and photocatalytic properties of High Performance Concrete. *Case Studies in Construction Materials*, *16*, e00965. Advance online publication. doi:10.1016/j.cscm.2022.e00965

Andreeva, D. V., Skorb, E. V., & Shchukin, D. G. (2010). Layer-by-Layer Polyelectrolyte/Inhibitor Nanostructures for Metal Corrosion Protection. *ACS Applied Materials & Interfaces*, *2*(7), 1954–1962. doi:10.1021/am1002712

Angellier, H., Molina-Boisseau, S., Dole, P., & Dufresne, A. (2006). Thermoplastic starch– waxy maize starch nanocrystals nanocomposites. *Biomacromolecules*, *7*(2), 531–539. doi:10.1021/bm050797s PMID:16471926

Angelopoulou, P., Giaouris, E., & Gardikis, K. (2022). Applications and Prospects of Nanotechnology in Food and Cosmetics Preservation. *Nanomaterials (Basel, Switzerland)*, *12*(7), 1196. doi:10.3390/nano12071196 PMID:35407315

Anjum, A. (2017). *Adsorption Technology for Removal of Toxic Pollutants*. doi:10.1007/978-3-319-61146-4_2

Anjum, A., Garg, R., Kashif, M., & Eddy, N. O. (2023). Nano-scale innovations in packaging: Properties, types, and applications of nanomaterials for the future. *Food Chemistry Advances*, *3*, 100560. doi:10.1016/j.focha.2023.100560

An, X., Li, W., Zhang, Y., Ding, J., Zhang, S., Hu, T., Li, S., Ren, Y., Liu, P., & Li, X. (2022). Antibacterial and superhydrophobic hemostatic properties of fluorinated random copolymer composite nanocoatings. *Progress in Organic Coatings*, *168*, 106869. doi:10.1016/j.porgcoat.2022.106869

Arif, M., Nawaz, A. F., Ullah Khan, S., Mueen, H., Rashid, F., Hemeg, H. A., & Rauf, A. (2023). Nanotechnology-based radiation therapy to cure cancer and the challenges in its clinical applications. *Heliyon*, *9*(6), e17252. doi:10.1016/j.heliyon.2023.e17252 PMID:37389057

Ariyarathna, I. R., Rajakaruna, R. M. P. I., & Karunaratne, D. N. (2017). The rise of inorganic nanomaterial implementation in food applications. In *Food Control* (Vol. 77, pp. 251–259). Elsevier Ltd. doi:10.1016/j.foodcont.2017.02.016

Arriagada, F., Günther, G., & Morales, J. (2020). Nanoantioxidant–based silica particles as flavonoid carrier for drug delivery applications. *Pharmaceutics*, *12*(4), 1–17. doi:10.3390/pharmaceutics12040302 PMID:32224905

Arsiwala, A. M., Raval, A. J., & Patravale, V. B. (2013). Nanocoating on implantable medical devices. *Pharmaceutical Patent Analyst*, *2*(4), 499–512. doi:10.4155/ppa.13.30 PMID:24237126

Arukalam, I. O., & Li, Y. (2019). Anticorrosion and barrier properties appraisal of poly(dimethylsiloxane)–ZnO nanocoating transition from superhydrophobic to hydrophobic state. *Journal of Coatings Technology and Research*, *16*(4), 1077–1088. doi:10.1007/s11998-018-00182-2

Arun, R., Shruthy, R., Preetha, R., & Sreejit, V. (2022). Biodegradable nano composite reinforced with cellulose nano fiber from coconut industry waste for replacing synthetic plastic food packaging. *Chemosphere*, *291*, 132786. doi:10.1016/j.chemosphere.2021.132786 PMID:34762882

Ashfaq, A., Khursheed, N., Fatima, S., Anjum, Z., & Younis, K. (2022). Application of nanotechnology in food packaging: Pros and Cons. *Journal of Agriculture and Food Research*, *7*, 100270. Advance online publication. doi:10.1016/j.jafr.2022.100270

Ashori, A., & Bahrami, R. (2014). Modification of physico-mechanical properties of chitosan-tapioca starch blend films using nano graphene. *Polymer-Plastics Technology and Engineering*, *53*(3), 312–318. doi:10.1080/03602559.2013.866246

Assis, R. Q., Lopes, S. M., Costa, T. M. H., Flôres, S. H., & de Oliveira Rios, A. (2017). Active biodegradable cassava starch films incorporated lycopene nanocapsules. *Industrial Crops and Products*, *109*, 818–827. doi:10.1016/j.indcrop.2017.09.043

Atta-ur-Rehman, Qudoos, A., Jakhrani, S. H., Kim, H. G., & Ryou, J.-S. (2019). Influence of Nano-silica on the Leaching Attack upon Photocatalytic Cement Mortars. *International Journal of Concrete Structures and Materials*, *13*(1), 35. Advance online publication. doi:10.1186/s40069-019-0348-x

Attia, N. F., Zakria, A. M., Nour, M. A., Abd El-Ghany, N. A., & Elashery, S. E. A. (2023). Rational strategy for construction of multifunctional coatings for achieving high fire safety, antibacterial, UV protection and electrical conductivity functions of textile fabrics. *Materials Today Sustainability*, *23*, 100450. doi:10.1016/j.mtsust.2023.100450

Atwah, A. A., & Khan, M. A. (2023). Influence of microscopic features on the self-cleaning ability of textile fabrics. *Textile Research Journal*, *93*(1-2), 450–467. doi:10.1177/00405175211069881

Awad Abouelata, A. M. (2019). Nanocomposite coatings on aluminum surfaces for several applications. In *Advances in Smart Coatings and Thin Films for Future Industrial and Biomedical Engineering Applications* (pp. 163–205). Elsevier. doi:10.1016/B978-0-12-849870-5.00002-1

Azaza, Y. B., Hamdi, M., Charmette, C., Jridi, M., Li, S., Nasri, M., & Nasri, R. (2022). Development and characterization of active packaging films based on chitosan and sardinella protein isolate: Effects on the quality and the shelf life of shrimps. *Food Packaging and Shelf Life*, *31*, 100796. doi:10.1016/j.fpsl.2021.100796

Azeez, S. O., Saheed, I. O., Adekola, F. A., & Salau, S. S. (2022). Preparation of tio2-activated kaolinite composite for photocatalytic degradation of rhodamine b dye. *Bulletin of the Chemical Society of Ethiopia*, *36*(1), 13–24. doi:10.4314/bcse.v36i1.2

Azevedo, V. M., Silva, E. K., Pereira, C. F. G., da Costa, J. M. G., & Borges, S. V. (2015). Whey protein isolate biodegradable films: Influence of the citric acid and montmorillonite clay nanoparticles on the physical properties. *Food Hydrocolloids*, *43*, 252–258. doi:10.1016/j.foodhyd.2014.05.027

Azizi-Lalabadi, M., Ehsani, A., Ghanbarzadeh, B., & Divband, B. (2020). Polyvinyl alcohol/gelatin nanocomposite containing ZnO, TiO_2 or ZnO/TiO_2 nanoparticles doped on 4A zeolite: Microbial and sensory qualities of packaged white shrimp during refrigeration. *International Journal of Food Microbiology*, *312*, 108375. doi:10.1016/j.ijfoodmicro.2019.108375 PMID:31669767

Babaee, M., Garavand, F., Rehman, A., Jafarazadeh, S., Amini, E., & Cacciotti, I. (2022). Biodegradability, physical, mechanical and antimicrobial attributes of starch nanocomposites containing chitosan nanoparticles. *International Journal of Biological Macromolecules*, *195*, 49–58. doi:10.1016/j.ijbiomac.2021.11.162 PMID:34856218

Bagale, U. D., Sonawane, S. H., Bhanvase, B. A., Kulkarni, R. D., & Gogate, P. R. (2018). Green synthesis of nanocapsules for self-healing anticorrosion coating using ultrasound-assisted approach. *Green Processing and Synthesis*, *7*(2), 147–159. doi:10.1515/gps-2016-0160

Bagherzadeh, M. R., & Mousavinejad, T. (2012). Preparation and investigation of anticorrosion properties of the water-based epoxy-clay nanocoating modified by Na +-MMT and Cloisite 30B. *Progress in Organic Coatings*, *74*(3), 589–595. doi:10.1016/j.porgcoat.2012.02.006

Baghi, F., Gharsallaoui, A., Dumas, E., & Ghnimi, S. (2022). Advancements in biodegradable active films for food packaging: Effects of nano/microcapsule incorporation. *Foods*, *11*(5), 760. doi:10.3390/foods11050760 PMID:35267394

Bai, B., Saranya, S., Dheepaasri, V., Muniasamy, S., Alharbi, N. S., Selvaraj, B., Undal, V. S., & Gnanamangai, B. M. (2022). Biosynthesized copper oxide nanoparticles (CuO NPs) enhances the anti-biofilm efficacy against K. pneumoniae and S. aureus. *Journal of King Saud University. Science*, *34*(6), 102120. doi:10.1016/j.jksus.2022.102120

Baino, F., Perero, S., Miola, M., & Ferraris, M. (2017). Antibacterial nanocoating for ocular applications. *Advances in Science and Technology (Owerri, Nigeria)*, *102*, 24–28. doi:10.4028/www.scientific.net/AST.102.24

Bains, D., Singh, G., Bhinder, J., Agnihotri, P. K., & Singh, N. (2020). Ionic Liquid-Functionalized Multiwalled Carbon Nanotube-Based Hydrophobic Coatings for Robust Antibacterial Applicatio. *ACS Applied Bio Materials*, *3*(4), 2092–2103. doi:10.1021/acsabm.9b01217 PMID:35025261

Bai, W., Yu, H., Liu, L., Pakdel, E., Tang, B., Su, H., Hurren, C., Liu, L., Wang, J., & Wang, X. (2023). The adsorption kinetics and mechanism of odorous gases onto textile fibers. *RSC Sustainability*, *1*(2), 357–367. doi:10.1039/D2SU00086E

Bajpai, V. K., Kamle, M., Shukla, S., Mahato, D. K., Chandra, P., Hwang, S. K., Kumar, P., Huh, Y. S., & Han, Y. K. (2018). Prospects of using nanotechnology for food preservation, safety, and security. *Yao Wu Shi Pin Fen Xi*, *26*(4), 1201–1214. doi:10.1016/j.jfda.2018.06.011 PMID:30249319

Bala, K., & Bhagat, M. (2023). Nanomaterial-based smart coatings as antimicrobials. *Antiviral and Antimicrobial Smart Coatings: Fundamentals and Applications*, 413–429. doi:10.1016/B978-0-323-99291-6.00010-4

Balázsi, K., Furkó, M., & Balázsi, C. (2021). Ceramic Matrix Graphene and Carbon Nanotube Composites. Encyclopedia of Materials: Technical Ceramics and Glasses: Volume 1-3, 2, 243–259. doi:10.1016/B978-0-12-818542-1.00070-9

Bansal, H., Singh, S., Sharma, A., Singh, J., & Mehta, S. K. (2023). Polymer nano-composite coatings and films: modern insights and emerging strategies to lengthen the lifespan of fruits and vegetables. *Food Materials Research, 3*(1). doi:10.48130/FMR-2023-0023

Bao, W., Deng, Z., Zhang, S., Ji, Z., & Zhang, H. (2019). Next-Generation Composite Coating System: Nanocoating. *Frontiers in Materials*, 6.

Barage, S., Lakkakula, J., Sharma, A., Roy, A., Alghamdi, S., Almehmadi, M., Hossain, M. J., Allahyani, M., & Abdulaziz, O. (2022). *Review Article Nanomaterial in Food Packaging: A Comprehensive Review*. doi:10.1155/2022/6053922

Barhoum, A., Pal, K., Rahier, H., Uludag, H., Kim, I. S., & Bechelany, M. (2019). Nanofibers as new-generation materials: From spinning and nano-spinning fabrication techniques to emerging applications. *Applied Materials Today, 17*, 1–35. doi:10.1016/j.apmt.2019.06.015

Barish, J. A., & Goddard, J. M. (2013). Anti-fouling surface modified stainless steel for food processing. *Food and Bioproducts Processing, 91*(4), 352–361. doi:10.1016/j.fbp.2013.01.003

Bashir, S., Lgaz, H., Chung, I. M., & Kumar, A. (2020). Effective green corrosion inhibition of aluminium using analgin in acidic medium: An experimental and theoretical study. *Chemical Engineering Communications, 0*(0), 1–10. doi:10.1080/00986445.2020.1752680

Bashir, S., Thakur, A., Lgaz, H., Chung, I. M., & Kumar, A. (2020). Corrosion inhibition efficiency of bronopol on aluminium in 0.5 M HCl solution: Insights from experimental and quantum chemical studies. *Surfaces and Interfaces, 20*(April), 100542. doi:10.1016/j.surfin.2020.100542

Bastarrachea, L. J., Wong, D. E., Roman, M. J., Lin, Z., & Goddard, J. M. (2015b). Active Packaging Coatings. Coatings, 771-791. doi:10.3390/coatings5040771

Bastarrachea, L. J., Denis-Rohr, A., & Goddard, J. M. (2015a). Antimicrobial food equipment coatings: Applications and challenges. *Annual Review of Food Science and Technology, 6*(1), 97–118. doi:10.1146/annurev-food-022814-015453 PMID:25422880

Bavithran, B., Nagarajan, V., & Vinod Kumar, K. (2018). Fire-resistant hydrophobic nanocoated composite partition sheets. *Journal of Thermoplastic Composite Materials*. Advance online publication. doi:10.1177/0892705718799814

Bayda, S., Adeel, M., Tuccinardi, T., Cordani, M., & Rizzolio, F. (2019). The History of Nanoscience and Nanotechnology: From Chemical-Physical Applications to Nanomedicine. *Molecules (Basel, Switzerland), 25*(1), 112. doi:10.3390/molecules25010112 PMID:31892180

Becerra, L. (2016). *CMF design: The fundamental principles of color, material, and finish design*. Frame Publishers.

Ben Abdelkader, M., Azizi, N., Salaün, F., Majdoub, M., & Chevalier, Y. (2023). Cosmetotextiles. In *Particle Technology and Textiles: Review of Applications* (pp. 235–274). De Gruyter. doi:10.1515/9783110670776-012

Ben John, P. R. (2022). Design and fabrication of nano textured superhydrophobic and anti-corrosive silane-grafted ZnO/bio-based polyurethane bilayer coating. *Surface and Coatings Technology*, *451*, 129036. doi:10.1016/j.surfcoat.2022.129036

Berger, L.-M. (2015). Application of hardmetals as thermal spray coatings. *International Journal of Refractory & Hard Metals*, *49*, 350–364. doi:10.1016/j.ijrmhm.2014.09.029

Berini, F., Orlandi, V., Gornati, R., Bernardini, G., & Marinelli, F. (2022). Nanoantibiotics to fight multidrug resistant infections by Gram-positive bacteria: Hope or reality? *Biotechnology Advances*, *57*, 107948. doi:10.1016/j.biotechadv.2022.107948 PMID:35337933

Bhalani, D. V., Nutan, B., Kumar, A., & Singh Chandel, A. K. (2022). Bioavailability Enhancement Techniques for Poorly Aqueous Soluble Drugs and Therapeutics. *Biomedicines*, *10*(9), 2055. doi:10.3390/biomedicines10092055 PMID:36140156

Bhardwaj, S., Lata, S., & Garg, R. (2022). Phyto-mediated green synthesis of silver nanoparticles using Acmella oleracea leaf extract: Antioxidant and catalytic activity. *Pharmacognosy Magazine*, *18*(77), 22. doi:10.4103/pm.pm_586_20

Bhatt, S., Pulpytel, J., & Arefi-Khonsari, F. (2015). *Low and atmospheric plasma polymerisation of nanocoatings for bio-applications*. Academic Press.

Bica, B. O., de Melo, J. V. S., & Gleize, P. J. P. (2022). Photocatalytic concrete blocks nanomodified with zinc oxide for paving: Mechanical performance and microstructural characteristics. *Journal of Testing and Evaluation*, *50*(1). Advance online publication. doi:10.1520/JTE20200326

Blum, A. P., Kammeyer, J. K., Rush, A. M., Callmann, C. E., Hahn, M. E., & Gianneschi, N. C. (2015). Stimuli-Responsive Nanomaterials for Biomedical Applications. *Journal of the American Chemical Society*, *137*(6), 2140–2154. doi:10.1021/ja510147n PMID:25474531

Blum, R., & Molian, P. (2009). Liquid-phase sintering of nanodiamond composite coatings on aluminum A319 using a focused laser beam. *Surface and Coatings Technology*, *204*(1), 1–14. doi:10.1016/j.surfcoat.2009.06.025

Boey, J. Y., Mohamad, L., Khok, Y. S., Tay, G. S., & Baidurah, S. (2021). A review of the applications and biodegradation of polyhydroxyalkanoates and poly (lactic acid) and its composites. *Polymers*, *13*(10), 1544. doi:10.3390/polym13101544 PMID:34065779

Bommana, M. M., Kirthivasan, B., Gupta, S. S., Shikhar, A., & Shah, A. (2019). Nanostructures in pharma: elixir to oral medicine. In *Nanoparticles in Pharmacotherapy* (pp. 23–44). William Andrew Publishing. doi:10.1016/B978-0-12-816504-1.00001-6

Bordin, G., Hristova, M., & Luque-Perez, E. (2020). *Horizon 2020-funded security research projects with dual-use potential: An overview (2014-2018)*. Publications Office of the European Union.

Boutamart, M., Jhabli, O., Nouneh, K., Rafqah, S., Eid, A. I., Briche, S., & Agzenai Ben Salem, Y. (2022). Environmental-friendly superhydrophobic bioactive nanocoatings. In A. Barhoum, J. Jeevanandam, & M. K. Danquah (Eds.), *Fundamentals of Bionanomaterials* (pp. 77–110). Elsevier. doi:10.1016/B978-0-12-824147-9.00004-2

Bugnicourt, E., Kehoe, T., Latorre, M., Serrano, C., Philippe, S., & Schmid, M. (2016). Recent Prospects in the Inline Monitoring of Nanocomposites and Nanocoatings by Optical Technologies. Nanomaterials. doi:10.3390/nano6080150

Bumbudsanpharoke, N., Choi, J., & Ko, S. (2015). Applications of nanomaterials in food packaging. *Journal of Nanoscience and Nanotechnology*, *15*(9), 6357–6372. doi:10.1166/jnn.2015.10847 PMID:26716190

Bumbudsanpharoke, N., & Ko, S. (2015). Nano-food packaging: An overview of market, migration research, and safety regulations. *Journal of Food Science*, *80*(5), R910–R923. doi:10.1111/1750-3841.12861 PMID:25881665

Burdusel, A. C., Neacsu, I. A., Birca, A. C., Chircov, C., Grumezescu, A. M., Holban, A. M., Curutiu, C., Ditu, L. M., Stan, M., & Andronescu, E. (2023). Microwave-Assisted Hydrothermal Treatment of Multifunctional Substituted Hydroxyapatite with Prospective Applications in Bone Regeneration. *Journal of Functional Biomaterials*, *14*(7), 378. doi:10.3390/jfb14070378 PMID:37504872

Butkinaree, S., Jinkarn, T., & Yoksan, R. (2008). Effects of biodegradable coating on barrier properties of paperboard food packaging. *J. Met. Mater. Miner*, *18*, 219–222.

Butler, J., Handy, R. D., Upton, M., & Besinis, A. (2023). Review of Antimicrobial Nanocoating in Medicine and Dentistry: Mechanisms of Action, Biocompatibility Performance, Safety, and Benefits Compared to Antibiotics. *ACS Nano*, *17*(8), 7064–7092. doi:10.1021/acsnano.2c12488 PMID:37027838

c, Z., Patel, F., Mastoi, N. R., &Saddiqa, A. (2016). Effect of sandblasting, annealing and hydrophobic treatment on the nano-mechanical and corrosion behaviour of n-TiO 2 -coated 316L stainless steel. *Current Science, 110*(3), 353–362. https://www.jstor.org/stable/24906780

Cabane, E., Zhang, X., Langowska, K., Palivan, C. G., & Meier, W. (2012). Stimuli-Responsive Polymers and Their Applications in Nanomedicine. *Biointerphases*, *7*(1), 9. doi:10.1007/s13758-011-0009-3 PMID:22589052

Cagri, A., Ustunol, Z., & Ryser, E. T. (2004). Antimicrobial edible films and coatings. *Journal of Food Protection*, *67*(4), 833–848. doi:10.4315/0362-028X-67.4.833 PMID:15083740

Cai, C., Sang, N., Teng, S., Shen, Z., Guo, J., Zhao, X., & Guo, Z. (2016). Superhydrophobic surface fabricated by spraying hydrophobic R974 nanoparticles and the drag reduction in water. *Surface and Coatings Technology*, *307*, 366–373. doi:10.1016/j.surfcoat.2016.09.009

Caixeta, M. P. (2023). Contaminants of Emerging Concern: A Review of Risk Assessment and Treatment Strategies. *U. Porto Journal of Engineering*, *9*(1), 191–228. doi:10.24840/2183-6493_009-001_001282

Campos, E. V. R., Pereira, A. E. S., de Oliveira, J. L., Carvalho, L. B., Guilger-Casagrande, M., de Lima, R., & Fraceto, L. F. (2020). How can nanotechnology help to combat COVID-19? Opportunities and urgent need. *Journal of Nanobiotechnology*, *18*(1), 125. doi:10.1186/s12951-020-00685-4 PMID:32891146

Cao, D., Ji, S., Zhao, J., Liu, Z., & Dai, H. (2022). Design and superhydrophobic performance analysis of the rectangular-groove microstructure coated by nanocoatings. *Materials Chemistry and Physics*, *291*, 126539. doi:10.1016/j.matchemphys.2022.126539

Cao, L., Huang, Y., Parakhonskiy, B., & Skirtach, A. G. (2022). Nanoarchitectonics beyond perfect order–not quite perfect but quite useful. *Nanoscale*, *14*(43), 15964–16002. doi:10.1039/D2NR02537J PMID:36278502

Castro-Mayorga, J. L., Fabra, M. J., & Lagaron, J. M. (2016). Stabilized nanosilver based antimicrobial poly(3-hydroxybutyrate-co-3-hydroxyvalerate) nanocomposites of interest in active food packaging. *Innovative Food Science & Emerging Technologies*, *33*, 524–533. doi:10.1016/j.ifset.2015.10.019

Cazón, P., Velazquez, G., Ramírez, J. A., & Vázquez, M. (2017). Polysaccharide-based films and coatings for food packaging: A review. *Food Hydrocolloids*, *68*, 136–148. doi:10.1016/j.foodhyd.2016.09.009

Celik, N., Sahin, F., Ozel, S. S., Sezer, G., Gunaltay, N., Ruzi, M., & Onses, M. (2023). Superhydrophobic coatings for food packaging applications: A review. *Food Packaging and Shelf Life*, *32*, 100823. doi:10.1016/j.fpsl.2022.100823

Cerqueira, M. Â., Pinheiro, A. C., Ramos, O. L., Silva, H., Bourbon, A. I., & Vicente, A. A. (2017). Chapter Two - Advances in Food Nanotechnology. In R. Busquets (Ed.), *Emerging Nanotechnologies in Food Science* (pp. 11–38). Elsevier. doi:10.1016/B978-0-323-42980-1.00002-9

Chadha, U., Bhardwaj, P., Selvaraj, S. K., Arasu, K., Praveena, S., Pavan, A., Khanna, M., Singh, P., Singh, S., Chakravorty, A., Badoni, B., Banavoth, M., Sonar, P., & Paramasivam, V. (2022). Current trends and future perspectives of nanomaterials in food packaging application. *Journal of Nanomaterials*, *2022*, 1–32. doi:10.1155/2022/2745416

Chagas, G. R., & Weibel, D. E. (2014). Development of Poly(propylene) Superhydrophobic Surfaces by Functionalised TiO2 Nanoparticles: Effect of Solvents and Dipping Times. *Macromolecular Symposia*, *344*(1), 55–62. doi:10.1002/masy.201300218

Chandio, A. D., Channa, I. A., Rizwan, M., Akram, S., Javed, M. S., Siyal, S. H., Saleem, M., Makhdoom, M., Ashfaq, T., Khan, S., Hussain, S., Albaqami, M., & Alotabi, R. G. (2021). Polyvinyl alcohol and nano-clay based solution processed packaging coatings. *Coatings*, *11*(8), 942. doi:10.3390/coatings11080942

Chandrasekar, M. S., Shanmugasigamani, & Malathy, P. (2010). Synergetic effects of pulse constraints and additives in electrodeposition of nanocrystalline zinc: Corrosion, structural and textural characterization. *Materials Chemistry and Physics*, *124*(1), 516–528. doi:10.1016/j.matchemphys.2010.07.004

Channa, I. A., Ashfaq, J., Gilani, S. J., Shah, A. A., Chandio, A. D., & Jumah, M. N. (2022). UV Blocking and Oxygen Barrier Coatings Based on Polyvinyl Alcohol and Zinc Oxide Nanoparticles for Packaging Applications. *Coatings*, *12*(7), 897. Advance online publication. doi:10.3390/coatings12070897

Chan, Y., Wu, X. H., Chieng, B. W., Ibrahim, N. A., & Then, Y. Y. (2021). Superhydrophobic nanocoating as intervention against biofilm-associated bacterial infections. *Nanomaterials (Basel, Switzerland)*, *11*(4), 1–33. doi:10.3390/nano11041046 PMID:33921904

Chaudhary, P., Ahamad, L., Chaudhary, A., Kumar, G., Chen, W.-J., & Chen, S. (2023). Nanoparticle-mediated bioremediation as a powerful weapon in the removal of environmental pollutants. *Journal of Environmental Chemical Engineering*, *11*(2), 109591. doi:10.1016/j.jece.2023.109591

Chaudhary, P., Fatima, F., & Kumar, A. (2020). Relevance of nanomaterials in food packaging and its advanced future prospects. *Journal of Inorganic and Organometallic Polymers and Materials*, *30*(12), 5180–5192. doi:10.1007/s10904-020-01674-8 PMID:32837459

Chauhan, D. S., Prasad, R., Srivastava, R., Jaggi, M., Chauhan, S. C., & Yallapu, M. M. (2020). Comprehensive Review on Current Interventions, Diagnostics, and Nanotechnology Perspectives against SARS-CoV-2. *Bioconjugate Chemistry*, *31*(9), 2021–2045. doi:10.1021/acs.bioconjchem.0c00323 PMID:32680422

Chausali, N., Saxena, J., & Prasad, R. (2022). Recent trends in nanotechnology applications of bio-based packaging. *Journal of Agriculture and Food Research*, *7*, 100257. doi:10.1016/j.jafr.2021.100257

Chavan Qi, L., Xu, Z., Jiang, X., Hu, C., & Zou, X. (2004). Preparation and antibacterial activity of chitosan nanoparticles. *Carbohydrate Research*, *339*(16), 2693–2700. doi:10.1016/j.carres.2004.09.007 PMID:15519328

Chavan, P., Sinhmar, A., Sharma, S., Dufresne, A., Thory, R., Kaur, M., Sandhu, K. S., Nehra, M., & Nain, V. (2022). Nanocomposite starch films: A new approach for biodegradable packaging materials. *Stärke*, *74*(5-6), 2100302. doi:10.1002/star.202100302

Chen, C., & Peng, X. (2023). Constructing superhydrophobic sands layer with PTFE nanocoating for desert water storage and oil/water separation. *Colloids and Surfaces A: Physicochemical and Engineering Aspects*, *657*(A), 130517.

Chen, C., & Peng, X. (2023). Constructing superhydrophobic sands layer with PTFE nanocoating for desert water storage and oil/water separation. *Colloids and Surfaces A: Physicochemical and Engineering Aspects*, *657*(Part A), 130517. ISSN 0927-7757. doi:10.1016/j.colsurfa.2022.130517

Chen, M. C., Koh, P. W., Ponnusamy, V. K., & Lee, S. L. (2022). Titanium dioxide and other nanomaterials based antimicrobial additives in functional paints and coatings [Review]. *Progress in Organic Coatings*, *163*, 106660. doi:10.1016/j.porgcoat.2021.106660

Chen, S., Wang, L., Lin, X., Ni, P., Liu, H., & Li, S. (2023). Catechol derivative urushiol's reactivity and applications beyond traditional coating. *Industrial Crops and Products*, *197*, 116598. doi:10.1016/j.indcrop.2023.116598

Chen, S., Zhang, Q., Hou, Y., Zhang, J., & Liang, X. J. (2013). Nanomaterials in medicine and pharmaceuticals: Nanoscale materials developed with less toxicity and more efficacy. *European Journal of Nanomedicine*, *5*(2), 61–79. doi:10.1515/ejnm-2013-0003

Chen, X., Chen, J., Zhu, J., Cai, T., & Guo, Z. (2021). Investigation on the Electrochemical Deposition of Nanocrystalline Zinc with Cationic Polyacrylamide (CPAM)-ZnSO(4) Electrolyte. *Micromachines*, 12. PMID:34577763

Chen, X., Wang, J., Xie, A., Wang, B., Wu, J., Chen, G., & Xing, T. (2023). Fabrication of transparent self-healing superhydrophobic coatings with high water contact angle stability. *Langmuir*, *39*(44), 15817–15827. doi:10.1021/acs.langmuir.3c02508 PMID:37877472

Chen, Y., Ding, Y., & Zheng, J. (2020). A polymer nanocomposite coating with enhanced hydrophilicity, antibacterial and antibiofouling properties: Role of polymerizable emulsifier/anionic ligand. *Chemical Engineering Journal*, *379*, 122268. doi:10.1016/j.cej.2019.122268

Chen, Z., Scharnagl, N., Zheludkevich, M. L., Ying, H., & Yang, W. (2023). Micro/nanocontainer-based intelligent coatings: Synthesis, performance and applications – A review. *Chemical Engineering Journal*, *451*, 138582. doi:10.1016/j.cej.2022.138582

Cherupurakal, N., Mozumder, M. S., Mourad, A.-H. I., & Lalwani, S. (2021). Recent Advances in Superhydrophobic Polymers for Antireflective Self-Cleaning Solar Panels. *Renewable & Sustainable Energy Reviews*, *151*, 111538. doi:10.1016/j.rser.2021.111538

Chiruvella, V., & Guddati, A. K. (2021). Ethical Issues in Patient Data Ownership. *Interactive Journal of Medical Research*, *10*(2), e22269. doi:10.2196/22269 PMID:34018968

Choe, E., & Min, D. B. (2009). Mechanisms of antioxidants in the oxidation of foods. *Comprehensive Reviews in Food Science and Food Safety*, *8*(4), 345–358. doi:10.1111/j.1541-4337.2009.00085.x

Chopra, H., Mohanta, Y. K., Rauta, P. R., Ahmed, R., Mahanta, S., Mishra, P. K., Panda, P., Rabaan, A. A., Alshehri, A. A., Othman, B., Alshahrani, M. A., Alqahtani, A. S., & Basha, A. L. (2023). An Insight into Advances in Developing Nanotechnology Based Therapeutics, Drug Delivery, Diagnostics and Vaccines: Multidimensional Applications in Tuberculosis Disease Management. *Pharmaceuticals (Basel, Switzerland)*, *16*(4), 581. doi:10.3390/ph16040581 PMID:37111338

Chougule, S. S., Gurme, S. T., Jadhav, J. P., Dongale, T. D., & Tiwari, A. P. (2021). Low density polyethylene films incorporated with Biosynthesised silver nanoparticles using Moringa oleifera plant extract for antimicrobial, food packaging, and photocatalytic degradation applications. *Journal of Plant Biochemistry and Biotechnology*, *30*(1), 208–214. doi:10.1007/s13562-020-00584-7

Christopher, G., Kulandainathan, M. A., & Harichandran, G. (2016). Biopolymers nanocomposite for material protection: Enhancement of corrosion protection using waterborne polyurethane nanocomposite coatings. *Progress in Organic Coatings*, *99*, 91–102. doi:10.1016/j.porgcoat.2016.05.012

Christy, P. N., Basha, S. K., Kumari, V. S., Bashir, A. K. H., Maaza, M., Kaviyarasu, K., Arasu, M. V., Al-Dhabi, N. A., & Ignacimuthu, S. (2020). Biopolymeric nanocomposite scaffolds for bone tissue engineering applications – A review. *Journal of Drug Delivery Science and Technology*, *55*, 101452. doi:10.1016/j.jddst.2019.101452

Chungsiriporn, J., Pongyeela, P., & Chairerk, N. (2020). Sol-gel self-cleaning superhydrophobic nanocoating for glass surface of solar cell. *Songklanakarin Journal of Science and Technology*, *42*(4), 923–927.

Cid, A., & Simal-Gandara, J. (2020). Synthesis, Characterization, and Potential Applications of Transition Metal Nanoparticles. In Journal of Inorganic and Organometallic Polymers and Materials (Vol. 30, Issue 4, pp. 1011–1032). Springer. doi:10.1007/s10904-019-01331-9

Contreras, C. B., Chagas, G., Strumia, M. C., & Weibel, D. E. (2014). Permanent superhydrophobic polypropylene nanocomposite coatings by a simple one-step dipping process. *Applied Surface Science*, *307*, 234–240. doi:10.1016/j.apsusc.2014.04.019

Corbett, J., McKeown, P. A., Peggs, G. N., & Whatmore, R. (2000). Nanotechnology: International Developments and Emerging Products. *CIRP Annals*, *49*(2), 523–545. doi:10.1016/S0007-8506(07)63454-4

Cordeiro, A. L. (2011). Stimuli-Responsive Polymer Nanocoatings. Nanotechnologies for the Life Sciences.

Coyle, S., Wu, Y., Lau, K.-T., De Rossi, D., Wallace, G., & Diamond, D. (2007). Smart Nanotextiles: A Review of Materials and Applications. *MRS Bulletin*, *32*(5), 434–442. doi:10.1557/mrs2007.67

Crespo, P., De la Presa, P., Marin, P., Multigner, M., Alonso, J. M., Rivero, G., ... Hernando, A. (2013). Magnetism in nanoparticles: Tuning properties with coatings. *Journal of Physics Condensed Matter*, *25*(48), 484006. doi:10.1088/0953-8984/25/48/484006 PMID:24201075

Cui, H., You, Y., Cheng, G.-W., Lan, Z., Zou, K.-L., Mai, Q.-Y., Han, Y.-H., Chen, H., Zhao, Y.-Y., & Yu, G.-T. (2023). Advanced materials and technologies for oral diseases. *Science and Technology of Advanced Materials*, *24*(1), 2156257. doi:10.1080/14686996.2022.2156257 PMID:36632346

Curulli, A. (2023). Functional Nanomaterials Enhancing Electrochemical Biosensors as Smart Tools for Detecting Infectious Viral Diseases. *Molecules (Basel, Switzerland)*, *28*(9), 3777. doi:10.3390/molecules28093777 PMID:37175186

Dahiya, S., Kumar, A. N., Shanthi Sravan, J., Chatterjee, S., Sarkar, O., & Mohan, S. V. (2018). Food waste biorefinery: Sustainable strategy for circular bioeconomy. In *Bioresource Technology* (Vol. 248, pp. 2–12). Elsevier Ltd., doi:10.1016/j.biortech.2017.07.176

Dai, D., Zhou, D., Xie, H., Wang, J., & Zhang, C. (2022). The design, construction and application of graphene family composite nanocoating on dental metal surface. *Biomaterials Advances*, *140*, 213087. doi:10.1016/j.bioadv.2022.213087 PMID:36029723

Damm, C., Münstedt, H., & Rösch, A. (2008). The antimicrobial efficacy of polyamide 6/silver-nano-and microcomposites. *Materials Chemistry and Physics*, *108*(1), 61–66. doi:10.1016/j.matchemphys.2007.09.002

Damodharan, J. (2021). Nanomaterials in medicine–An overview. *Materials Today: Proceedings*, *37*, 383–385. doi:10.1016/j.matpr.2020.05.380

Daoud, W. A., & Xin, J. H. (2004). Nucleation and growth of anatase crystallites on cotton fabrics at low temperatures. *Journal of the American Ceramic Society*, *87*(5), 953–955. doi:10.1111/j.1551-2916.2004.00953.x

Das, S., Kumar, S., Samal, S. K., Mohanty, S., & Nayak, S. K. (2018). A Review on Superhydrophobic Polymer Nanocoatings: Recent Development, Application. *Industrial & Engineering Chemistry Research*, *57*(8), 2727–2745. doi:10.1021/acs.iecr.7b04887

Dastkhoon, M., Ghaedi, M., Asfaram, A., & Alipanahpour Dil, E. (2020). Comparative study of ability of sonochemistry combined ZnS:Ni nanoparticles-loaded activated carbon in reductive of organic pollutants from environmental water samples. *Polyhedron*, *180*, 114341. doi:10.1016/j.poly.2019.114341

de Azeredo, H. M. (2009). Nanocomposites for food packaging applications. *Food Research International*, *42*(9), 1240–1253. doi:10.1016/j.foodres.2009.03.019

de Dio Ferreira, D., Galvão, T. D., & Appoloni, C. R. (2023). Washing and release of titanium nanoparticles from UV protective textiles. In *Advances in Healthcare and Protective Textiles* (pp. 447–466). Elsevier. doi:10.1016/B978-0-323-91188-7.00010-8

de Jesús Salas-Méndez, E., Vicente, A., Pinheiro, A. C., Ballesteros, L. F., Silva, P., Rodríguez-García, R., ... de Rodríguez, D. J. (2019). Application of edible nanolaminate coatings with antimicrobial extract of *Flourensia cernua* to extend the shelf-life of tomato (*Solanum lycopersicum L.*) fruit. *Postharvest Biology and Technology*, *150*, 19–27. doi:10.1016/j.postharvbio.2018.12.008

de Lima Barizão, C., Crepaldi, M. I., Oscar de Oliveira, S., de Oliveira, A. C., Martins, A. F., Garcia, P. S., & Bonafé, E. G. (2020). Biodegradable films based on commercial κ-carrageenan and cassava starch to achieve low production costs. *International Journal of Biological Macromolecules*, *165*, 582–590. doi:10.1016/j.ijbiomac.2020.09.150 PMID:32991902

de Santos, H., & Sanz-Bobi, M. Á. (2021). Research on the pollution performance and degradation of superhydrophobic nano-coatings for toughened glass insulators. *Electric Power Systems Research*, *191*, 106863. doi:10.1016/j.epsr.2020.106863

de Villiers, M. M., Otto, D. P., Strydom, S. J., & Lvov, Y. M. (2011). Introduction to nanocoatings produced by layer-by-layer (LbL) self-assembly. *Advanced Drug Delivery Reviews*, *63*(9), 701–715. doi:10.1016/j.addr.2011.05.011 PMID:21699936 del Mercato, L. L., Ferraro, M. M., Baldassarre, F., Mancarella, S., Greco, V., Rinaldi, R., & Leporatti, S. (2014). Biological applications of LbL multilayer capsules: From drug delivery to sensing. *Advances in Colloid and Interface Science*, *207*, 139–154. doi:10.1016/j.cis.2014.02.014 PMID:24625331

Dessale, M., Mengistu, G., & Mengist, H. M. (2022). Nanotechnology: A Promising Approach for Cancer Diagnosis, Therapeutics and Theragnosis. *International Journal of Nanomedicine*, *17*, 3735–3749. doi:10.2147/IJN.S378074 PMID:36051353

Dhiman, N. K., Sidhu, N., Agnihotri, S., Mukherjee, A., & Reddy, M. S. (2021). Role of nanomaterials in protecting building materials from degradation and deterioration. In Biodegradation and Biodeterioration at the Nanoscale (pp. 405–475). doi:10.1016/B978-0-12-823970-4.00024-5

Dhital, R., Joshi, P., Becerra-Mora, N., Umagiliyage, A., Chai, T., Kohli, P., & Choudhary, R. (2017). Integrity of edible nano-coatings and its effects on the quality of strawberries subjected to simulated in-transit vibrations. *Lebensmittel-Wissenschaft + Technologie*, *80*, 257–264. doi:10.1016/j.lwt.2017.02.033

Dickson, D., Bock, B. D., & Thome, J. R. (2024). Heat transfer of uncoated and nanostructure coated commercially micro-enhanced refrigeration tubes under pool boiling conditions. *Applied Thermal Engineering*, *236*, 121757. doi:10.1016/j.applthermaleng.2023.121757

Díez-Pascual, A. M. (2022). Surface Engineering of Nanomaterials with Polymers, Biomolecules, and Small Ligands for Nanomedicine. *Materials (Basel)*, *15*(9), 3251. doi:10.3390/ma15093251 PMID:35591584

Díez-Pascual, A. M., Sánchez, J. A. L., Capilla, R. P., & Díaz, P. G. (2018). Recent Developments in Graphene/Polymer Nanocomposites for Application in Polymer Solar Cells. *Polymers*, *10*(2), 217. doi:10.3390/polym10020217

Difonzo, M., Fliedel, L., Mignet, N., Andrieux, K., & Alhareth, K. (2022). How Could Nanomedicine Improve the Safety of Contrast Agents for Magnetic Resonance Imaging during Pregnancy? *Sci*, *4*(1), 11. doi:10.3390/sci4010011

Di, Y., Iannace, S., Di Maio, E., & Nicolais, L. (2003). Nanocomposites by melt intercalation based on polycaprolactone and organoclay. *Journal of Polymer Science. Part B, Polymer Physics*, *41*(7), 670–678. doi:10.1002/polb.10420

Dolatzadeh, F., Moradian, S., & Jalili, M. M. (2011). Influence of various surface treated silica nanoparticles on the electrochemical properties of SiO2/polyurethane nanocoatings. *Corrosion Science*, *53*(12), 4248–4257. doi:10.1016/j.corsci.2011.08.036

Du, C., Du, T., Chang, Z., Yin, C., & Cheng, Y. (2022). On the interface between biomaterials and two-dimensional materials for biomedical applications. *Advanced Drug Delivery Reviews*, *186*, 114314. doi:10.1016/j.addr.2022.114314 PMID:35568105

Dufrêne, Y. F., Ando, T., Garcia, R., Alsteens, D., Martinez-Martin, D., Engel, A., Gerber, C., & Müller, D. J. (2017). Imaging modes of atomic force microscopy for application in molecular and cell biology. *Nature Nanotechnology*, *12*(4), 295–307. doi:10.1038/nnano.2017.45 PMID:28383040

Du, J., Yuan, H., Xia, H., Kou, H., Zhang, Y., Xing, W., & Zhang, C. (2022). Improvement of superhydrophobicity and durability of EP+PDMS/SiO2 composite coatings by adjusting laser curing powers. *Materials Chemistry and Physics*, *289*, 126428. doi:10.1016/j.matchemphys.2022.126428

Duncan, T. V. (2011). Applications of nanotechnology in food packaging and food safety: Barrier materials, antimicrobials and sensors. *Journal of Colloid and Interface Science*, *363*(1), 1–24. doi:10.1016/j.jcis.2011.07.017 PMID:21824625

Dwivany, F. M., Aprilyandi, A. N., Suendo, V., & Sukriandi, N. (2020). Carrageenan edible coating application prolongs Cavendish banana shelf life. *International Journal of Food Sciences*, *2020*, 1–11. Advance online publication. doi:10.1155/2020/8861610 PMID:32908863

Eddy, N., Odiongenyi, A. O., Garg, R., & Garg, R. (2022). Plant Wastes as Alternative Sources of Sustainable and Green Corrosion Inhibitors in Different Environments. SSRN *Electronic Journal,* 1–13. doi:10.2139/ssrn.4051680

Eddy, N. O., Ibok, U. J., Garg, R., Garg, R., Iqbal, A., Amin, M., Mustafa, F., Egilmez, M., & Galal, A. M. (2022). A Brief Review on Fruit and Vegetable Extracts as Corrosion Inhibitors in Acidic Environments. *Molecules (Basel, Switzerland)*, *27*(9), 1–18. doi:10.3390/molecules27092991 PMID:35566341

Eddy, N. O., Odiongenyi, A. O., Garg, R., Ukpe, R. A., Garg, R., El Nemr, A., Ngwu, C. M., & Okop, I. J. (2023). Quantum and experimental investigation of the application of Crassostrea gasar (mangrove oyster) shell–based CaO nanoparticles as adsorbent and photocatalyst for the removal of procaine penicillin from aqueous solution. *Environmental Science and Pollution Research International*, *30*(23), 64036–64057. doi:10.1007/s11356-023-26868-8 PMID:37059957

Eessaa, A. K., & El-Shamy, A. M. (2023). Review on fabrication, characterization, and applications of porous anodic aluminum oxide films with tunable pore sizes for emerging technologies. *Microelectronic Engineering*, *279*, 112061. doi:10.1016/j.mee.2023.112061

Egodage, D. P., Jayalath, H. T. S., Samarasekara, A. M. P. B., Amarasinghe, D. A. S., Madushani, S. P. A., & Senerath, S. M. N. S. (2017, May). Novel antimicrobial nano coated polypropylene based materials for food packaging systems. In *2017 Moratuwa Engineering Research Conference (MERCon)* (pp. 88-92). IEEE. 10.1109/MERCon.2017.7980462

Ehsan, R. M., Simon, S. P., Kinattingal, S., Kumar, K. A., & Srihasrsha, I. (2022). Effects of Nanocoatings on the Temperature-dependent Cell Parameters and Power Generation of Photovoltaic Panels. *Applied Nanoscience*, *12*, 3945–3962. doi:10.1007/s13204-022-02633-0

Ehsan, R. M., Simon, S. P., Sundareswaran, K., Kumar, K. A., & Sriharsha, T. (2021). Effect of Soiling on Photovoltaic Modules and Its Mitigation using Hydrophobic Nanocoatings. *IEEE Journal of Photovoltaics*, *11*(3), 742–749. doi:10.1109/JPHOTOV.2021.3062023

El-Ashram, S., El-Samad, L. M., Basha, A. A., & El Wakil, A. (2021). Naturally-derived targeted therapy for wound healing: Beyond classical strategies. *Pharmacological Research*, *170*, 105749. doi:10.1016/j.phrs.2021.105749 PMID:34214630

El-Fattah, M. A., El Saeed, A. M., Dardir, M. M., & El-Sockary, M. A. (2015). Studying the effect of organo-modified nanoclay loading on the thermal stability, flame retardant, anti-corrosive and mechanical properties of polyurethane nanocomposite for surface coating. *Progress in Organic Coatings*, *89*, 212–219. doi:10.1016/j.porgcoat.2015.09.010

Elkalla, E., Khizar, S., Tarhini, M., Lebaz, N., Zine, N., Jaffrezic-Renault, N., Errachid, A., & Elaissari, A. (2023). Core-shell micro/nanocapsules: From encapsulation to applications. *Journal of Microencapsulation*, *40*(3), 125–156. doi:10.1080/02652048.2023.2178538 PMID:36749629

El-Naggar, M. E., Hasanin, M., & Hashem, A. H. (2022). Eco-Friendly Synthesis of Superhydrophobic Antimicrobial Film Based on Cellulose Acetate/Polycaprolactone Loaded with the Green Biosynthesized Copper Nanoparticles for Food Packaging Application. *Journal of Polymers and the Environment*, *30*(5), 1820–1832. doi:10.1007/s10924-021-02318-9

El-Nawasany, L. I., Sundookh, A., Kadoum, L. A., Yasin, M. A., AlSalem, H. S., Binkadem, M. S., Al-Goul, S. T., Zidan, N. S., & Shoueir, K. R. (2023). Ameliorating characteristics of magnetically sensitive TPU nanofibers-based food packaging film for long-life cheese preservation. *Food Bioscience*, *53*, 102633. doi:10.1016/j.fbio.2023.102633

Elsabee, M. Z., & Abdou, E. S. (2013). Chitosan based edible films and coatings: A review. *Materials Science and Engineering C*, *33*(4), 1819–1841. doi:10.1016/j.msec.2013.01.010 PMID:23498203

Elshami, A., & Ali, A. H. (2014). Using new coating materials for corrosion prevention of reinforced concrete. *MATEC Web of Conferences, 11*. 10.1051/matecconf/20141101026

Emmanuel Awe, F., Faruruwa, M. D., Abba, H., Dahiru Faruruwa, M., & Abba, H. (2021). Green Synthesis, Characterization and Antibacterial Activity of Zinc Oxide and Titanium Dioxide Nanoparticles Using Terminalia Catappa and Cymbopogon Citratus Leaf Extract. *Communication in Physical Sciences, 7*(4), 563–572. https://journalcps.com/index.php/volumes

Emmanuel, O. A., Fayomi, O. S. I., Agboola, O., Ayoola, A. A., Oloke, O. C., & Amusan, L. M. (2021). Short review on nanocomposite coating advances in the industry. *IOP Conference Series. Materials Science and Engineering*, *1107*(1), 012069. doi:10.1088/1757-899X/1107/1/012069

Ennaceri, H., Alami, H. E., Brik, H., Mokssit, O., & Khaldoun, A. (2014). Lotus effect and superhydrophobic coatings for concentrated solar power systems (CSP). *Proceedings of International Conference on Composite Materials & Renewable Energy Applications*, 14.

Erdonmez, D., Mosayyebi, S., Erkan, K., Salimi, K., Nagizade, N., Saglam, N., & Rzayev, Z. M. O. (2014). Nanofabrication and characterization of PVA–organofiller/Ag nanocoating on pMAD plasmids. *Applied Surface Science*, *318*, 127–131. doi:10.1016/j.apsusc.2014.02.007

Escobar, K., Garrido-Miranda, K. A., Pulido, R., Naveas, N., Manso-Silván, M., & Hernandez-Montelongo, J. (2023). Coatings of cyclodextrin/citric-acid biopolymer as drug delivery systems: A review. *Pharmaceutics*, *15*(1), 296. doi:10.3390/pharmaceutics15010296 PMID:36678924

Esposito, M. M., Glazer, J. R., & Turku, S. (2023). The Use of 3D Printing and Nanotechnologies to Prevent and Inhibit Biofilms on Medical Devices. *Hygiene*, *3*(3), 325–338. doi:10.3390/hygiene3030024

Esteban-arranz, A., de la Osa, A. R., García-lorefice, W. E., Sacristan, J., & Sánchez-silva, L. (2021). Long-term performance of nanomodified coated concrete structures under hostile marine climate conditions. *Nanomaterials (Basel, Switzerland)*, *11*(4), 869. Advance online publication. doi:10.3390/nano11040869 PMID:33805457

Fahmy, H., & Ibrahim, S. (2021). The effect Ecofriendly Super hydrophobic Nano composites Coating and Penetration of Printing Inks on Paperboard Packaging Materials. *International Design Journal*, *11*(1), 319–326. doi:10.21608/idj.2021.133025

Fakhri, M., & shahryari, E. (2021). The effects of nano zinc oxide (ZnO) and nano reduced graphene oxide (RGO) on moisture susceptibility property of stone mastic asphalt (SMA). *Case Studies in Construction Materials*, *15*, e00655. Advance online publication. doi:10.1016/j.cscm.2021.e00655

Falahaty marvast, M., Gholambargani, M., Ramezanzadeh, M., & Ramezanzadeh, B. (2023). Enhancing Self-Healing and Adhesion Bonding Properties: Investigating the Promising Potential of Ce-ZIF8 MOF Hybrid Thin Films as pH-Responsive Nanocoatings. *Langmuir*, *39*(27), 9579–9594. doi:10.1021/acs.langmuir.3c01363 PMID:37366573

Fallah, M., Ghashghaee, M., Rabiee, A., & Ershad-Langroudi, A. (2018). Spray-deposition of an organic/inorganic blend for fabrication of a superhydrophobic surface: Effect of admixing with silica aerogel and modified silica nanoparticles. *Protection of Metals and Physical Chemistry of Surfaces*, *54*(5), 909–916. doi:10.1134/S2070205118050064

Farag, A.A. (2020). *Applications of nanomaterials in corrosion protection coatings and inhibitors*. Academic Press.

Farag, A. A. (2020). Applications of nanomaterials in corrosion protection coatings and inhibitors. *Corrosion Reviews*, *38*(1), 67–86. doi:10.1515/corrrev-2019-0011

Farooq, S. A., Raina, A., Mohan, S., Arvind Singh, R., Jayalakshmi, S., & Irfan Ul Haq, M. (2022). Nanostructured Coatings: Review on Processing Techniques, Corrosion Behaviour and Tribological Performance. *Nanomaterials (Basel, Switzerland)*, *12*(8), 12. doi:10.3390/nano12081323 PMID:35458032

Farris, S., Silva De Sousa, M., Schlogl, A. E., Ramalho Estanislau, F., Gomes, V., Souza, L., Sélia, J., Coimbra, R., José, I., & Santos, B. (2023). Academic Editors: Swarna Jaiswal Nanotechnology in Packaging for Food Industry: Past, Present, and Future. doi:10.3390/coatings13081411

Farris, S., Unalan, I. U., Introzzi, L., Fuentes-Alventosa, J. M., & Cozzolino, C. A. (2014). Pullulan-based films and coatings for food packaging: Present applications, emerging opportunities, and future challenges. *Journal of Applied Polymer Science*, *131*(13), app.40539. Advance online publication. doi:10.1002/app.40539

Feng, F., Wang, R., Yuan, W., & Li, Y. (2023). Study on anti-icing performance of biogas-residue nano-carbon coating for wind-turbine blade. *Coatings*, *13*(5), 814. doi:10.3390/coatings13050814

Feng, X., Shi, Y., Wang, Y., Yue, G. R., & Yang, W. (2013). Preparation of superhydrophobic silver nano coatings with feather-like structures by electroless galvanic deposition. *Chinese Science Bulletin*, *58*(16), 1887–1891. doi:10.1007/s11434-013-5755-9

Feng, Y., Chen, S., & Frank Cheng, Y. (2018). Stearic acid modified zinc nano-coatings with superhydrophobicity and enhanced antifouling performance. *Surface and Coatings Technology*, *340*, 55–65. doi:10.1016/j.surfcoat.2018.02.053

Feng, Y., Zhong, L., Bilal, M., Tan, Z., Hou, Y., Jia, S., & Cui, J. (2019). Enzymes@ZIF-8 nanocomposites with protection nanocoating: Stability and acid-resistant evaluation. *Polymers*, *11*(1), 1–14. doi:10.3390/polym11010027 PMID:30960011

Ferrer, A., Pal, L., & Hubbe, M. (2017). Nanocellulose in packaging: Advances in barrier layer technologies. *Industrial Crops and Products*, *95*, 574–582. doi:10.1016/j.indcrop.2016.11.012

Ferri, M., Papchenko, K., Degli Esposti, M., Tondi, G., De Angelis, M. G., Morselli, D., & Fabbri, P. (2023). Fully Biobased Polyhydroxyalkanoate/Tannin Films as Multifunctional Materials for Smart Food Packaging Applications. *ACS Applied Materials & Interfaces*, *15*(23), 28594–28605. Advance online publication. doi:10.1021/acsami.3c04611 PMID:37265075

Flanagan, T. (2009). Alpine skiing technology: Faster, higher, stronger. *Sports Technology*, *2*(1-2), 5–6. doi:10.1080/19346182.2009.9648492

Floras, J. D., & Matsos, K. L. (2005). Introduction to modified atmosphere packaging. In *Innovations in Food Packaging*. Overiew. doi:10.1016/B978-012311632-1/50042-5

Fu, H., & Gray, K. A. (2021). The key to maximizing the benefits of antimicrobial and self-cleaning coatings is to fully determine their risks. *Current Opinion in Chemical Engineering*, *34*, 100761. doi:10.1016/j.coche.2021.100761 PMID:36569284

Future Market Insights. (n.d.). www.futuremarketinsights.com

Gago, Miguel, León, & Ibarra. (2016). *Graphene-based hydrophobic varnish for marine applications*. Academic Press.

Ganeshkumar, S., Kumar, A., Maniraj, J., Babu, Y. S., Ansu, A. K., Goyal, A., ... Hassan, A. M. (2023). Exploring the potential of nanotechnology: A assessment of nano-scale multi-layered-composite coatings for cutting tool performance. *Arabian Journal of Chemistry*, *16*(10), 105173. doi:10.1016/j.arabjc.2023.105173

Ganguly, S., & Margel, S. (2023). Superhydrophobic nanoscale materials for surface coatings. In S. Thomas & J. S. George (Eds.), *Polymer-Based Nanoscale Materials for Surface Coatings* (pp. 479–500). Elsevier. doi:10.1016/B978-0-32-390778-1.00029-3

Garcia, C. V., Shin, G. H., & Kim, J. T. (2018). Metal oxide-based nanocomposites in food packaging: Applications, migration, and regulations. *Trends in Food Science & Technology*, *82*, 21–31. doi:10.1016/j.tifs.2018.09.021

Garg, R., Garg, R., Sharma, R., Anjum, A., & Eddy, N. O. (2023). Introduction to Nanotechnology and Nanomaterials. In *Nanotechnology for Sustainable Agriculture, Food and Environment*. CRC Press. doi:10.1201/9781003397861-1

Garg, R., Rani, P., Garg, R., & Eddy, N. O. (2021). Study on potential applications and toxicity analysis of green synthesized nanoparticles. *Turkish Journal of Chemistry*, *45*(6), 1690–1706. doi:10.3906/kim-2106-59 PMID:38144602

Garg, R., Rani, P., Garg, R., Khan, M. A., Khan, N. A., Khan, A. H., & Américo-Pinheiro, J. H. P. (2022). Biomedical and catalytic applications of agri-based biosynthesized silver nanoparticles. *Environmental Pollution*, *310*(April), 119830. Advance online publication. doi:10.1016/j.envpol.2022.119830 PMID:35926739

Ge, L., Zhao, Y. S., Mo, T., Li, J. R., & Li, P. (2012). *Immobilization of glucose oxidase in electrospun nanofibrous membranes for food preservation*. doi:10.1016/j.foodcont.2012.01.022

Gebreslassie, Y. T., & Gebretnsae, H. G. (2021). Green and Cost-Effective Synthesis of Tin Oxide Nanoparticles: A Review on the Synthesis Methodologies, Mechanism of Formation, and Their Potential Applications. *Nanoscale Research Letters*, *16*(1), 97. Advance online publication. doi:10.1186/s11671-021-03555-6 PMID:34047873

Ghazal, H., Khaleed, N., & Abdelaziz, E. (2023). Significance Advantages, and disadvantages of nanotechnology in textile finishing. *Egyptian Journal of Chemistry*, *66*(12), 467–482. doi:10.21608/ejchem.2023.195121.7624

Gherardi, F., Colombo, A., Goidanich, S., &Toniolo, L. (2016). *Smart hydrophobic TiO2-nanocomposites for the protection of stone Cultural Heritage*. Academic Press.

Ghodrati, M., Mousavi-Kamazani, M., & Bahrami, Z. (2023). Synthesis of superhydrophobic coatings based on silica nanostructure modified with organosilane compounds by sol–gel method for glass surfaces. *Scientific Reports*, *13*(1), 548. doi:10.1038/s41598-023-27811-0 PMID:36631572

Ghosal, K. (2023). Tackling COVID-19 Using Antiviral Nanocoating's-Recent Progress and Future Challenges. *Particle & Particle Systems Characterization*, *40*(1), 2200154. doi:10.1002/ppsc.202200154 PMID:36711425

Ghosh, S., Sarkar, B., Thongmee, S., & Mostafavi, E. (2023). Hybrid antibacterial, antifungal, and antiviral smart coatings. In *Antiviral and antimicrobial smart coatings: Fundamentals and applications* (pp. 431–452). Elsevier. doi:10.1016/B978-0-323-99291-6.00016-5

Ghosh, T., & Katiyar, V. (2022). Nanochitosan functionalized hydrophobic starch/guar gum biocomposite for edible coating application with improved optical, thermal, mechanical, and surface property. *International Journal of Biological Macromolecules*, *211*, 116–127. doi:10.1016/j.ijbiomac.2022.05.079 PMID:35561853

Ghosh, T., Mondal, K., Giri, B. S., & Katiyar, V. (2021). Silk nanodisc based edible chitosan nanocomposite coating for fresh produces: A candidate with superior thermal, hydrophobic, optical, mechanical and food properties. *Food Chemistry*, *360*, 130048. doi:10.1016/j.foodchem.2021.130048 PMID:34034054

Ghosh, T., Nakano, K., Mulchandani, N., & Katiyar, V. (2021). Curcumin loaded iron functionalized biopolymeric nanofibre reinforced edible nanocoating for improved shelf life of cut pineapples. *Food Packaging and Shelf Life*, *28*, 100658. doi:10.1016/j.fpsl.2021.100658

Girigoswami, A., Thirumalai, A., Karthick, H., Pallavi, P., Gowtham, P., & Girigoswami, A. (2022). Nanotechnology driven improvement of smart food packaging. *Materials Research Innovations*.

Glenn, L. M., & Boyce, J. S. (2012). Regenerative nanomedicine: ethical, legal, and social issues. *Nanotechnology in Regenerative Medicine: Methods and Protocols*, 303-316.

Goffredo, G. B., Terlizzi, V., & Munafò, P. (2017). Multifunctional TiO2-based hybrid coatings on limestone: Initial performances and durability over time. *Journal of Building Engineering*, *14*, 134–149. doi:10.1016/j.jobe.2017.10.006

Goisauf, M., & Cano-Abadía, M. (2022). Ethics of AI in Radiology: A Review of Ethical and Societal Implications. *Frontiers in Big Data*, *5*, 850383. doi:10.3389/fdata.2022.850383 PMID:35910490

Gong, X., & He, S. (2020). Highly Durable Superhydrophobic Polydimethylsiloxane/Silica Nanocomposite Surfaces with Good Self-Cleaning Ability. *ACS Omega*, *5*(8), 4100–4108. doi:10.1021/acsomega.9b03775 PMID:32149238

Gopalakrishnan, K., & Mishra, P. (2023). Self-Healing Polymer a Dynamic Solution in Food Industry: A Comprehensive Review. *Food Biophysics*, 1–17. doi:10.1007/s11483-023-09780-z

Gopalakrishnan, V., Balasubramanian, A., Subramanian, L., Muhammed, R. I., Garg, R., & Eddy, N. O. (2023). Experimental and Theoretical Analysis on Mild Steel Corrosion Inhibition by Two Novel Compounds (FD and ACP) in Acidic Media. *Portugaliae Electrochimica Acta*, *41*(3), 223–246. doi:10.4152/pea.2023410304

Goudarzi, V., Shahabi-Ghahfarrokhi, I., & Babaei-Ghazvini, A. (2017). Preparation of ecofriendly UV-protective food packaging material by starch/TiO2 bio-nanocomposite: Characterization. *International Journal of Biological Macromolecules*, *95*, 306–313. doi:10.1016/j.ijbiomac.2016.11.065 PMID:27884670

Govardhana Reddy, P. V., Rajendra Prasad Reddy, B., Venkata Krishna Reddy, M., Raghava Reddy, K., Shetti, N. P., Saleh, T. A., & Aminabhavi, T. M. (2021). A review on multicomponent reactions catalysed by zero-dimensional/one-dimensional titanium dioxide (TiO2) nanomaterials: Promising green methodologies in organic chemistry. *Journal of Environmental Management*, *279*, 111603. doi:10.1016/j.jenvman.2020.111603 PMID:33172705

Goyal, N., & Jerold, F. (2023). Biocosmetics: Technological advances and future outlook. *Environmental Science and Pollution Research International*, *30*(10), 25148–25169. doi:10.1007/s11356-021-17567-3 PMID:34825334

Grewal, A. S., Lather, V., Sharma, N., Singh, S., Narang, R. S., Narang, J. K., & Pandita, D. (2018). Recent updates on nanomedicine based products: Current scenario and future opportunities. *Applied Clinical Research, Clinical Trials and Regulatory Affairs*, *5*(2), 132–144. doi:10.2174/2213476X05666180611115135

Gu, Y., Xia, K., Wu, D., Mou, J., & Zheng, S. (2020). Technical Characteristics and Wear-Resistant Mechanism of Nano Coatings: A Review. Coatings. doi:10.3390/coatings10030233

Guryanov, I., Naumenko, E., & Fakhrullin, R. (2022). Hair surface engineering: Combining nanoarchitectonics with hair topical and beauty formulations. *Applied Surface Science Advances*, *7*, 100188. https://doi.org/https://doi.org/10.1016/j.apsadv.2021.100188

Gutiérrez-Cano, V., Rodríguez, F., González, J. A., & Díaz, V. (2018). Thermal degradation of hydrophobic graphite-based thin film nano-coatings observed by Raman spectroscopy. *Thin Solid Films*, *648*, 8–11. doi:10.1016/j.tsf.2017.12.023

Haapanen, J., Aromaa, M., Teisala, H., Juuti, P., Tuominen, M., Sillanpää, M., Stepien, M., Saarinen, J. J., Toivakka, M., Kuusipalo, J., & Mäkelä, J. (2018). On the limit of superhydrophobicity: Defining the minimum amount of TiO2 nanoparticle coating. *Materials Research Express*, *6*(3), 035004. Advance online publication. doi:10.1088/2053-1591/aaf2ee

Hablot, E., Bordes, P., Pollet, E., & Avérous, L. (2008). Thermal and thermo-mechanical degradation of poly (3-hydroxybutyrate)-based multiphase systems. *Polymer Degradation & Stability*, *93*(2), 413–421. doi:10.1016/j.polymdegradstab.2007.11.018

Hall, K. D., Guo, J., Dore, M., & Chow, C. C. (2009). The progressive increase of food waste in America and its environmental impact. *PLoS One*, *4*(11), e7940. Advance online publication. doi:10.1371/journal.pone.0007940 PMID:19946359

Hamk, M., Akçay, F. A., & Avcı, A. (2023). Green synthesis of zinc oxide nanoparticles using *Bacillus subtilis* ZBP4 and their antibacterial potential against foodborne pathogens. *Preparative Biochemistry & Biotechnology, 53*(3), 255-264. https://doi.org/.2022.2076243 doi:10.1080/10826068

Han, J. H., Lee, D. S., Min, S. C., & Chung, M. S. (2012). Eco-design of food and beverage packaging. In *Emerging Food Packaging Technologies* (pp. 361–379). Woodhead Publishing. doi:10.1533/9780857095664.4.361

Han, J., Liu, L., Fan, Z., Zhang, Z., Yang, S., & Tang, Y. (2020). Grafting cosmetic active ingredients for the functionalization of cosmetotextiles. *IOP Conference Series. Materials Science and Engineering*, *782*(2), 022026. doi:10.1088/1757-899X/782/2/022026

Han, J., Ma, Q., An, Y., Wu, F., Zhao, Y., Wu, G., & Wang, J. (2023). The current status of stimuli-responsive nanotechnologies on orthopedic titanium implant surfaces. *Journal of Nanobiotechnology*, *21*(1), 277. doi:10.1186/s12951-023-02017-8 PMID:37596638

Hao, L., Wang, L., Ma, Y., Bottini, M., Li, L., Cheng, H.-B., Gao, S., Liang, X., & Zhang, J. (2024). Nanomaterials relieving hypoxia for enhanced tumor therapy. *Coordination Chemistry Reviews*, *499*, 215482. doi:10.1016/j.ccr.2023.215482

Harb, S. V., Trentin, A., Uvida, M. C., & Hammer, P. (2020). Advanced organic nanocomposite coatings for effective corrosion protection. In *Corrosion Protection at the Nanoscale* (pp. 315–343). Elsevier. doi:10.1016/B978-0-12-819359-4.00017-9

Harilal, M., Anandkumar, B., George, R. P., Albert, S. K., & Philip, J. (2023). High-performance eco-friendly ternary blended green concrete in seawater environment. *Hybrid Advances*, *3*(February), 100037. doi:10.1016/j.hybadv.2023.100037

Hasan, K. M. F., Champramary, S., Al Hasan, K. M. N., Indic, B., Ahmed, T., Pervez, M. N., Horváth, P. G., Bak, M., Sándor, B., Hofmann, T., Tolvaj, L., Horváth, A., Kóczán, Z., Sipos, G., Alpár, T., & Bejó, L. (2023). Eco-friendly production of cellulosic fibers from Scots pine wood and sustainable nanosilver modification: A path toward sustainability. *Results in Engineering*, *19*, 101244. doi:10.1016/j.rineng.2023.101244

Hasan, K., Yousuf, S. B., Tushar, M. S. H. K., Das, B. K., Das, P., & Islam, M. S. (2022). Effects of Different Environmental and Operational Factors on the PV Performance: A Comprehensive Review. *Energy Science & Engineering*, *10*(2), 656–675. doi:10.1002/ese3.1043

He, B., Hou, X., Liu, Y., Hu, J., Song, L., Tong, Z., . . . Zhang, Q. (2023). Design of fluorine-free waterborne fabric coating with robust hydrophobicity, water-resistant and breathability. *Separation and Purification Technology, 311*, 123308. https://doi.org/. 123308 doi:10.1016/j.seppur.2023

He, X., Deng, H., & Hwang, H. (2019). The current application of nanotechnology in food and agriculture. In Journal of Food and Drug Analysis (Vol. 27, Issue 1, pp. 1–21). Elsevier Taiwan LLC. doi:10.1016/j.jfda.2018.12.002

Hegyesi, N., Zhang, Y., Kohári, A., Polyák, P., Sui, X., & Pukánszky, B. (2019). Enzymatic degradation of PLA/cellulose nanocrystal composites. *Industrial Crops and Products*, *141*, 111799. doi:10.1016/j.indcrop.2019.111799

He, H., & Guo, Z. (2021). Superhydrophobic materials used for anti-icing Theory, application, and development. *iScience*, *24*(11), 103357. doi:10.1016/j.isci.2021.103357 PMID:34825136

He, Q., Paiva, J. M., Kohlscheen, J., Beake, B. D., & Veldhuis, S. C. (2020). An integrative approach to coating/carbide substrate design of CVD and PVD coated cutting tools during the machining of austenitic stainless steel. *Ceramics International, 46*(4), 5149–5158. doi:10.1016/j.ceramint.2019.10.259

HLPE. (2014). *Food Losses and Waste in the Context of Sustainable Food Systems*. HLPE.

Hong, R., Pan, T., Qian, J., & Li, H. (2006). Synthesis and surface modification of ZnO nanoparticles. *Chemical Engineering Journal, 119*(2–3), 71–81. doi:10.1016/j.cej.2006.03.003

Honorato, C., Kumar, V., Liu, J., Koivula, H., Xu, C., & Toivakka, M. (2015). Transparent nanocellulose-pigment composite films. *Journal of Materials Science, 50*(22), 7343–7352. doi:10.1007/s10853-015-9291-7

Hooda, A., Goyat, M. S., Gupta, R., Prateek, M., Agrawal, M., & Biswas, A. (2017). Synthesis of nano-textured polystyrene/ZnO coatings with excellent transparency and superhydrophobicity. *Materials Chemistry and Physics, 193*, 447–452. doi:10.1016/j.matchemphys.2017.03.011

Hossain, N., Islam, M. A., Chowdhury, M. A., & Alam, A. (2022). Advances of nanoparticles employment in dental implant applications. *Applied Surface Science Advances, 12*, 100341. https://doi.org/https://doi.org/10.1016/j.apsadv.2022.100341

Hou, Y., Wu, C., Yang, D., Wang, K., Ye, T., Brownlie, L., Wang, K., & Priya, S. (2020). Artemisinin (ART)-Induced "perovskite/perovskite" bilayer structured photovoltaics. *Nano Energy, 78*, 105133. doi:10.1016/j.nanoen.2020.105133

Hsieh, C. T., Lai, M. H., & Cheng, Y. S. (2009). Fabrication and superhydrophobicity of fluorinated titanium dioxide nanocoatings. *Journal of Colloid and Interface Science, 340*(2), 237–242. doi:10.1016/j.jcis.2009.08.029 PMID:19775698

Hsissou, R., Benzidia, B., Hajjaji, N., & Elharfi, A. (2018). Elaboration and electrochemical studies of the coating behavior of a new pentafunctional epoxy polymer (pentaglycidyl ether pentabisphenol phosphorus) on E24 carbon steel in 3.5% NaCl. *Journal of Chemical Technology and Metallurgy, 53*(5), 898–905.

Huang, B. S., Lai, G. H., Yang, T. I., Tsai, M. H., & Chou, Y. C. (2020). A novel electroactive imide oligomer and its application in anticorrosion coating. *Polymers, 12*(1), 91. Advance online publication. doi:10.3390/polym12010091 PMID:31947895

Huang, C., Pan, C.-H., & Liu, C.-H. (2010). Deposition of hydrophobic nano-coatings with low-pressure radio frequency CH2F2/Ar plasma processing. *Thin Solid Films, 518*(13), 3570–3574. doi:10.1016/j.tsf.2009.12.013

Huang, P. H., Laakso, M., Edinger, P., Hartwig, O., Duesberg, G. S., Lai, L. L., Mayer, J., Nyman, J., Errando-Herranz, C., Stemme, G., Gylfason, K. B., & Niklaus, F. (2023). Three-dimensional printing of silica glass with sub-micrometer resolution. *Nature Communications, 14*(1), 3305. doi:10.1038/s41467-023-38996-3 PMID:37280208

Huang, Z., Li, J., Chen, X., Yang, Q., Zeng, X., Bai, R., & Wang, L. (2023). Photothermal Sensitive 3D Printed Biodegradable Polyester Scaffolds with Polydopamine Coating for Bone Tissue Engineering. *Polymers*, *15*(2), 381. Advance online publication. doi:10.3390/polym15020381 PMID:36679260

Huang, Z.-S., Shen, C., Fan, L., Ye, X., Shi, H., Zhang, Y., Lai, Y., & Quan, Y.-Y. (2021). Experimental Investigation of the Anti-Soiling Performances of Different Wettability of Transparent Coatings: Superhydrophilic, hydrophilic, Hydrophobic and Superhydrophobic Coatings. *Solar Energy Materials and Solar Cells*, *225*, 111053. doi:10.1016/j.solmat.2021.111053

Hussein, H. S. (2023). The State of the Art of Nanomaterials and Its Applications in Energy Saving. *Bulletin of the National Research Center*, *47*, 7. doi:10.1186/s42269-023-00984-4

Hu, X., Lu, C., Tang, H., Pouri, H., Joulin, E., & Zhang, J. (2022). Active Food Packaging Made of Biopolymer-Based Composites. *Materials (Basel)*, *16*(1), 279. doi:10.3390/ma16010279 PMID:36614617

Ibrahim, N. I., Shahar, F. S., Sultan, M. T. H., Shah, A. U. M., Safri, S. N. A., & Mat Yazik, M. H. (2021). Overview of bioplastic introduction and its applications in product packaging. *Coatings*, *11*(11), 1423. doi:10.3390/coatings11111423

Ibrahim, S. H., & El-Tayeb, N. S. M. (2021). Effect of nano-silica/alumina hybrid coating on erosion resistance of glass fibre-reinforced polymer for the application of wind turbine blades. *Proceedings of the Institution of Mechanical Engineers. Part J, Journal of Engineering Tribology*, *236*(10), 2013–2031. doi:10.1177/13506501211052724

Idumah, C. I., Obele, C. M., Emmanuel, E. O., & Hassan, A. (2020a). Recently Emerging Nanotechnological Advancements in Polymer Nanocomposite Coatings for Anti-corrosion, Anti-fouling and Self-healing. *Surfaces and Interfaces*, *21*, 100734. doi:10.1016/j.surfin.2020.100734 PMID:34957345

İlhan, İ., Esen, M., Karaaslan, M., Akyürek, B. Y., & Çelikel, Ö. (2023). Investigation of EMI and UV-IR shielding properties of Kevlar and polyester/elastane nanocomposite fabrics coated by ECR and evaporation methods. *Journal of the Textile Institute*, 1–14. Advance online publication. doi:10.1080/00405000.2023.2200538

Inobeme, A., Nayak, V., Mathew, J. T., Okonkwo, S., Ekwoba, L., Ajai, A. I., Bernard, E., Inobeme, J., Agbugui, M. M., & Singh, K. R. B. (2022). Chemometric approach in environmental pollution analysis: A critical review. *Journal of Environmental Management*, *309*(114653), 1–18. doi:10.1016/j.jenvman.2022.114653 PMID:35176568

Islam, N. U., Amin, R., Shahid, M., & Amin, M. (2019). Gummy gold and silver nanoparticles of apricot (Prunus armeniaca) confer high stability and biological activity. *Arabian Journal of Chemistry*, *12*(8), 3977–3992. doi:10.1016/j.arabjc.2016.02.017

Jaafar, H. T., & Diaa, B. M. (2022). Hybrid Nanocoating Waterproof of Silicon Si and Nano-aluminum Oxide Al2O3 for Medical Lenses Applications. *Malaysian Journal of Medicine and Health Sciences*, *18*(SUPP11), 1–5.

Jafarzadeh, S., Nafchi, A. M., Salehabadi, A., Oladzad-Abbasabadi, N., & Jafari, S. M. (2021). Application of bio-nanocomposite films and edible coatings for extending the shelf life of fresh fruits and vegetables. *Advances in Colloid and Interface Science*, *291*, 102405. doi:10.1016/j.cis.2021.102405 PMID:33819726

Jafarzadeh, S., Nooshkam, M., Zargar, M., Garavand, F., Ghosh, S., Hadidi, M., & Forough, M. (2023). Green synthesis of nanomaterials for smart biopolymer packaging: challenges and outlooks. In *Journal of Nanostructure in Chemistry*. Springer Medizin. doi:10.1007/s40097-023-00527-3

Jagtiani, E. (2022). Advancements in nanotechnology for food science and industry. *Food Frontiers*, *3*(1), 56–82. doi:10.1002/fft2.104

Jahan, I. (2023). Advances and Applications of Nanomaterial-Assisted Phytoremediation. *Nanotechnology for Sustainable Agriculture, Food and Environment*, 159–170. doi:10.1201/9781003397861-12

Jain, K. K., & Jain, K. K. (2017). Ethical and Regulatory Aspects of Nanomedicine. The Handbook of Nanomedicine, 605-620. doi:10.1007/978-1-4939-6966-1_18

Jain, K. K. (2011). Nanobiotechnology and personalized medicine. *Progress in Molecular Biology and Translational Science*, *104*, 325–354. doi:10.1016/B978-0-12-416020-0.00008-5 PMID:22093223

Jambeck, J. R., Geyer, R., Wilcox, C., Siegler, T. R., Perryman, M., Andrady, A., Narayan, R., & Law, K. L. (2015). Plastic waste inputs from land into the ocean. *Science*, *347*(6223), 768–771. doi:10.1126/science.1260352 PMID:25678662

Janeni, J., & Adassooriya, N. M. (2023). Development of eco-friendly smart textiles from nanomaterials. In *Advances in Smart Nanomaterials and their Applications* (pp. 325–335). Elsevier. doi:10.1016/B978-0-323-99546-7.00002-1

Jang, E. J., Padhan, B., Patel, M., Pandey, J. K., Xu, B., & Patel, R. (2023). Antibacterial and biodegradable food packaging film from bacterial cellulose. *Food Control*, *109902*, 109902. Advance online publication. doi:10.1016/j.foodcont.2023.109902

Janjarasskul, T., and Suppakul, P. (2016). Active and intelligent packaging: the indication of quality and safety. *Crit. Rev. Food Sci. Nutr.* doi:10.1080/10408398.2016.1225278

Jaqt, R. A., Huq, T. N., Hill, S. A., Thway, M., Liu, T., Napari, M., Roose, B., Gałkowski, K., Li, W., Lin, S. F., Stranks, S. D., MacManus-Driscoll, J. L., & Joye, R. L. Z. (2020). Rapid Vapor-Phase Deposition of High-Mobility p-Type Buffer Layers on Perovskite Photovoltaics for Efficient Semi-Transparent Devices. *ACS Energy Letters*, *5*(8), 2456–2465. doi:10.1021/acsenergylett.0c00763

Jawad, A. H., Mubarak, N. S. A., Ishak, M. A. M., Ismail, K., & Nawawi, W. I. (2016). Kinetics of photocatalytic decolourization of cationic dye using porous TiO2 film. *Journal of Taibah University for Science : JTUSCI*, *10*(3), 352–362. doi:10.1016/j.jtusci.2015.03.007

Jayakumar, A., Heera, K. V., Sumi, T. S., Joseph, M., Mathew, S., Praveen, G., ... Radhakrishnan, E. K. (2019). Starch-PVA composite films with zinc-oxide nanoparticles and phytochemicals as intelligent pH sensing wraps for food packaging application. *International Journal of Biological Macromolecules, 136*, 395–403. doi:10.1016/j.ijbiomac.2019.06.018 PMID:31173829

Jayan, J. S., Sethulekshmi, A., Venu, G., Saritha, A., & Joseph, K. (2021). Sustainable nanotextiles: emerging antibacterial fabrics. In *Antimicrobial Textiles from Natural Resources* (pp. 619–651). Elsevier. doi:10.1016/B978-0-12-821485-5.00007-X

Jensin Joshua, J., Venkatanarayanan, P. S., & Singh, D. (2020). A literature review on composite materials filled with and without nanoparticles subjected to high/low velocity impact loads. *Materials Today: Proceedings, 33*, 4635–4641. doi:10.1016/j.matpr.2020.08.294

Jha, C. B., Santhosh, S., Singh, C., Bose, S., Manna, K., Varshney, R., & Mathur, R. (2023). 9 - Electrospun nanofiber a smart drug carriers: production methods, problems, solutions, and applications. In N. Sharma & B. S. Butola (Eds.), *Fiber and Textile Engineering in Drug Delivery Systems* (pp. 285–306). Woodhead Publishing. doi:10.1016/B978-0-323-96117-2.00002-9

Jiang. (2019). *Super hydrophobic nanostructured anti-stain surface coatings for smart green home applications.* ScholarBank@NUS Repository.

Jiang, J., Pi, J., & Cai, J. (2018). The Advancing of Zinc Oxide Nanoparticles for Biomedical Applications. *Bioinorganic Chemistry and Applications, 18*, 1–18. Advance online publication. doi:10.1155/2018/1062562 PMID:30073019

Jiang, W., Zhao, P., Song, W., Wang, M., & Yu, D. G. (2022). Electrospun zein/polyoxyethylene core-sheath ultrathin fibers and their antibacterial food packaging applications. *Biomolecules, 12*(8), 1110. doi:10.3390/biom12081110 PMID:36009003

Jose, A., Gizdavic-Nikolaidis, M., & Swift, S. (2023). Antimicrobial Coatings: Reviewing Options for Healthcare Applications. *Applied Microbiology, 3*(1), 145–174. doi:10.3390/applmicrobiol3010012

Joseph, T. M., Kar Mahapatra, D., Esmaeili, A., Piszczyk, Ł., Hasanin, M., Kattali, M., Haponiuk, J., & Thomas, S. (2023). Nanoparticles: Taking a Unique Position in Medicine. *Nanomaterials (Basel, Switzerland), 13*(3), 574. doi:10.3390/nano13030574 PMID:36770535

Joshi, M., Bhattacharyya, A., Agarwal, N., & Parmar, S. (2012). Nanostructured coatings for superhydrophobic textiles. *Bulletin of Materials Science, 35*(6), 933–938. Advance online publication. doi:10.1007/s12034-012-0391-6

Ju, J., Xie, Y., Guo, Y., Cheng, Y., Qian, H., & Yao, W. (2019). Application of edible coating with essential oil in food preservation. *Critical Reviews in Food Science and Nutrition, 59*(15), 2467–2480. doi:10.1080/10408398.2018.1456402 PMID:29580066

Julaeha, E., Pandiangan, N. S., Eddy, D. R., Permadi, N., Harja, A., Wahyudi, T., & Al-Anshori, J. (2023). Ethyl cellulose-based microcapsules of *Citrus aurantifolia* (Christm.) Swingle essential oil with an optimized emulsifier for antibacterial cosmetotextiles. *Polymer, 283*, 126265. Advance online publication. doi:10.1016/j.polymer.2023.126265

Kaew-on, N., Katemake, P., & Prasongsuk, S. (2020). Primer formulations with antibacterial properties for murals. *Progress in Organic Coatings, 138*, 105395. doi:10.1016/j.porgcoat.2019.105395

Kalisz, M., Grobelny, M., Świniarskib, M., & Firek, P. (2016). Comparison of the structural and corrosion properties of the graphene/SiN(200) coating system deposited on titanium alloy surfaces covered with SiN transition layers. *Surface and Coatings Technology, 299*, 65–70. doi:10.1016/j.surfcoat.2016.04.064

Kamal, A., Saba, M., Ullah, K., Almutairi, S. M., AlMunqedhi, B. M., & Ragab abdelGawwad, M. (2023). Mycosynthesis, Characterization of Zinc Oxide Nanoparticles, and Its Assessment in Various Biological Activities. *Crystals, 13*(2), 171. Advance online publication. doi:10.3390/cryst13020171

Kamalan Kirubaharan, A. M., & Kuppusami, P. (2020). Corrosion behavior of ceramic nanocomposite coatings at nanoscale. In *Corrosion Protection at the Nanoscale* (pp. 295–314). Elsevier. doi:10.1016/B978-0-12-819359-4.00016-7

Kamaleddin, M. A. (2017). Nano-ophthalmology: Applications and considerations. *Nanomedicine; Nanotechnology, Biology, and Medicine, 13*(4), 1459–1472. doi:10.1016/j.nano.2017.02.007 PMID:28232288

Kanmani, P., & Rhim, J. W. (2014). Development and characterization of carrageenan/grapefruit seed extract composite films for active packaging. *International Journal of Biological Macromolecules, 68*, 258–266. doi:10.1016/j.ijbiomac.2014.05.011 PMID:24832986

Kao, C. Y., Lin, T. L., Lin, Y. H., Lee, A. K. X., Ng, S. Y., Huang, T. H., & Hsu, T. T. (2022). Synergistic Effect of Static Magnetic Fields and 3D-Printed Iron-Oxide-Nanoparticle-Containing Calcium Silicate/Poly-ε-Caprolactone Scaffolds for Bone Tissue Engineering. *Cells, 11*(24), 3967. Advance online publication. doi:10.3390/cells11243967 PMID:36552731

Karim, M. A., Abdullah, M. Z., Deifalla, A. F., Azab, M., & Waqar, A. (2023). An assessment of the processing parameters and application of fiber-reinforced polymers (FRPs) in the petroleum and natural gas industries: A review. *Results in Engineering, 18*, 101091. doi:10.1016/j.rineng.2023.101091

Karimnezhad, H. (2017). *Fabrication of Hydrophobic Membrane for the Separation of n-Hexane/Water Mixture Using Novel Oleophilic Nanoparticle and Kevlar Fabric, as a Superior Support*. Academic Press.

Karimnezhad, H. (2017). Fabrication of Hydrophobic Membrane for the Separation of n-Hexane/Water Mixture Using Novel Oleophilic Nanoparticle and Kevlar Fabric, as a Superior Support. *Journal of Water and Environmental Nanotechnology*, *2*(3), 145–156. doi:10.22090/jwent.2017.03.002

Karmouch, R., Hor, H. E. (2017). Solar cells performance reduction under the effect of dust in Jazan Region. *Journal of Fundamental and Renewable Energy Applications, 7.*

Kashif, M., & Anjum, A. (2020). Chemical and electrochemical corrosion studies of ricinus communis oil based poly(urethane-ricinoleamide) coatings. *Journal of Physics: Conference Series*, *1706*(1), 012034. Advance online publication. doi:10.1088/1742-6596/1706/1/012034

Kathirvelu, S., D'Souza, L., & Dhurai, B. (2009). UV protection finishing of textiles using ZnO nanoparticles. *Indian Journal of Fibre and Textile Research*, *34*(3), 267–273.

Kauppinen, C., Pasanen, T. P., Isakov, K., Serué, M., Heinonen, J., Vähänissi, V., Lipsanen, H., & Savin, H. (2021). Grass-like alumina coated window harnesses the full omnidirectional potential of black silicon photodiodes. *Applied Optics*, *60*(33), 10415–10420. doi:10.1364/AO.442415 PMID:34807052

Kausar, A., Ahmad, I., Maaza, M., & Bocchetta, P. (2023). *Self-Healing Nanocomposites&mdash.* Advancements and Aerospace Applications. In Journal of Composites Science.

Kaya, S., Lgaz, H., Thakkur, A., Kumar, A., Özbakır, D., Karakuş, N., & Ben Ahmed, S. (2023). Molecular insights into the corrosion inhibition mechanism of omeprazole and tinidazole : A theoretical investigation. *Molecular Simulation*, *49*(17), 1–15. doi:10.1080/08927022.2023.2256888

Kaya, S., Thakur, A., & Kumar, A. (2023). The role of in Silico / DFT investigations in analyzing dye molecules for enhanced solar cell efficiency and reduced toxicity. *Journal of Molecular Graphics & Modelling*, *124*(June), 108536. doi:10.1016/j.jmgm.2023.108536 PMID:37300949

Keeney, M., Jiang, X. Y., Yamane, M., Lee, M., Goodman, S., & Yang, F. (2015). Nanocoating for biomolecule delivery using layer-by-layer self-assembly. *Journal of Materials Chemistry. B*, *3*(45), 8757–8770. doi:10.1039/C5TB00450K PMID:27099754

Keshtmand, Z., Khademian, E., Poorjafari Jafroodi, P., Abtahi, M. S., & Tavakkoli Yaraki, M. (2023). Green synthesis of selenium nanoparticles using Artemisia chamaemelifolia: toxicity effects through regulation of gene expression for cancer cells and bacteria. *Nano-Structures & Nano-Objects, 36*, 101049. https://doi.org/https://doi.org/10.1016/j.nanoso.2023.101049

Kesmez, Ö., Tamsü Selli, N., Tunalı, A., Akarsu, E., Akarsu, M., & Arpaç, E. (2017). Fingerprint resistant coatings for stainless steel substrates. Progress in Organic Coatings, 112, 51–56. https://doi.org/10.1016/J.PORGCOAT.2017.07.003

Khalid, H. M., Rafique, Z., Muyeen, S. M., Raqeeb, A., Said, Z., Saidur, R., & Sopian, K. (2023). Dust Accumulation and Aggregation on PV Panels: An Integrated Survey on Impacts, Mathematical Models, Cleaning Mechanisms, and Possible Sustainable Solution. *Solar Energy*, *251*, 261–285. doi:10.1016/j.solener.2023.01.010

Khalid, S., Yu, L., Feng, M., Meng, L., Bai, Y., Ali, A., Liu, H., & Chen, L. (2018). Development and characterization of biodegradable antimicrobial packaging films based on polycaprolactone, starch and pomegranate rind hybrids. *Food Packaging and Shelf Life*, *18*, 71–79. doi:10.1016/j.fpsl.2018.08.008

Khaneghah, A. M., Hashemi, S. M. B., & Limbo, S. (2018). Antimicrobial agents and packaging systems in antimicrobial active food packaging: An overview of approaches and interactions. *Food and Bioproducts Processing*, *111*, 1–19. doi:10.1016/j.fbp.2018.05.001

Khan, M. Z., Taghavian, H., Fijalkowski, M., Militky, J., Tomkova, B., Venkataraman, M., & Adach, K. (2023). Effect of microwave power on bactericidal and UV protection properties of the ZnO nanorods grown cotton fabrics. *Colloids and Surfaces. A, Physicochemical and Engineering Aspects*, *664*, 131135. doi:10.1016/j.colsurfa.2023.131135

Khan, S. A., Sezer, N., & Koç, M. (2020). Design, synthesis, and characterization of hybrid micro-nano surface coatings for enhanced heat transfer applications. *International Journal of Energy Research*, *44*(15), 12525–12534. doi:10.1002/er.5685

Khatoon, Z., McTiernan, C. D., Suuronen, E. J., Mah, T.-F., & Alarcon, E. I. (2018). Bacterial biofilm formation on implantable devices and approaches to its treatment and prevention. *Heliyon*, *4*(12), e01067. doi:10.1016/j.heliyon.2018.e01067 PMID:30619958

Khemthong, P., Phanthasri, J., Youngjan, S., Wanmolee, W., Samun, Y., Sosa, N., Rungnim, C., Kraithong, W., Sangkhun, W., Parnthong, J., Butburee, T., Thanee, K., Nakajima, H., Supruangnet, R., Towiwat, P., Chanvorachote, P., & Sukrong, S. (2023). Effect of the ethanol-to-water ratio on the properties of silica–carbon core–shell materials for prolonged antibacterial activity of thymol. *Applied Surface Science*, *635*, 157716. doi:10.1016/j.apsusc.2023.157716

Khodadadi, A., Liaghat, G., Taherzadeh-Fard, A., & Shahgholian-Ghahfarokhi, D. (2021). Impact characteristics of soft composites using shear thickening fluid and natural rubber–A review of current status. *Composite Structures*, *271*, 114092. doi:10.1016/j.compstruct.2021.114092

Kim, D. W., Eum, K., Kim, H., Kim, D., de Mello, M. D., Park, K., & Tsapatsis, M. (2019). Continuous ZIF-8/reduced graphene oxide nanocoating for ultrafast oil/water separation. *Chemical Engineering Journal*, *372*, 509–515. doi:10.1016/j.cej.2019.04.179

Kim, H. Y., Park, S. S., & Lim, S. T. (2015). Preparation, characterization and utilization of starch nanoparticles. *Colloids and Surfaces. B, Biointerfaces*, *126*, 607–620. doi:10.1016/j.colsurfb.2014.11.011 PMID:25435170

Kiran, S., Adeel, S., Abrar, S., Iqbal, S., Naz, S., & Amin, N. (2023). Textile applications of nanofibers and nanocomposites. In Handbook of: Nanofibers and Nanocomposites: Characteristics, Synthesis, and Applications in Textiles (pp. 149-167). Jenny Stanford Publishing. doi:10.1201/9781003432746-5

Kłosowska, A., Wawrzyńczak, A., & Feliczak-Guzik, A. (2023). Microencapsulation as a route for obtaining encapsulated flavors and fragrances. *Cosmetics*, *10*(1), 26. doi:10.3390/cosmetics10010026

Knausgård, K. (2012). *Superhydrophobic Anti-Ice Nanocoatings*. Institutt for konstruksjonsteknikk.

Koh, E., & Lee, Y. T. (2021). Development of hybrid hydrophobic molybdenum disulfide (MoS2) nanoparticles for super water repellent self-cleaning. *Progress in Organic Coatings*, *153*, 106161. doi:10.1016/j.porgcoat.2021.106161

Kokate, M., Garadkar, K., & Gole, A. (2013). One pot synthesis of magnetite–silica nanocomposites: Applications as tags, entrapment matrix and in water purification. *Journal of Materials Chemistry. A, Materials for Energy and Sustainability*, *1*(6), 2022–2029. doi:10.1039/C2TA00951J

Kolahalam, L. A., Kasi Viswanath, I. V., Diwakar, B. S., Govindh, B., Reddy, V., & Murthy, Y. L. N. (2019). Review on nanomaterials: Synthesis and applications. *Materials Today: Proceedings*, *18*, 2182–2190. doi:10.1016/j.matpr.2019.07.371

Koslowski, L. A. D., Nogueira, A. L., Licodiedoff, S., Comper, A. T., & Folgueras, M. V. (2018). Silver nanoparticles impregnated with polyamide-66 to disinfect drinking water. *Revista Ambiente & Água*, *13*(6), e1947. doi:10.4136/ambi-agua.1947

Kraśniewska, K., Pobiega, K., & Gniewosz, M. (2019). Pullulan–biopolymer with potential for use as food packaging. *International Journal of Food Engineering*, *15*(9), 20190030. doi:10.1515/ijfe-2019-0030

Kraśniewska, K., Ścibisz, I., Gniewosz, M., Mitek, M., Pobiega, K., & Cendrowski, A. (2017). Effect of pullulan coating on postharvest quality and shelf-life of highbush blueberry (*Vaccinium corymbosum L.*). *Materials (Basel)*, *10*(8), 965. doi:10.3390/ma10080965 PMID:28820473

Krella, A. K. (2023). Degradation and Protection of Materials from Cavitation Erosion: A Review. *Materials (Basel)*, *16*(5), 2058. doi:10.3390/ma16052058 PMID:36903173

Kryuchkov, M., Adamcik, J., & Katanaev, V. L. (2022). Bactericidal and Antiviral Bionic Metalized Nanocoating. *Nanomaterials (Basel, Switzerland)*, *12*(11), 1868. doi:10.3390/nano12111868 PMID:35683724

Kumar, A., & Thakur, A. (2023). Overview of the Properties, Applicability, and Recent Advancements of Some Natural Products Used as Potential Inhibitors in Various Corrosive Systems. In Handbook of Research on Corrosion Sciences and Engineering (pp. 275–277). doi:10.4018/978-1-6684-7689-5.ch010

Kumar, A., & Nanda, D. (2019). Methods and fabrication techniques of superhydrophobic surfaces. In S. K. Samal, S. Mohanty, & S. K. Nayak (Eds.), *Superhydrophobic Polymer Coatings* (pp. 43–75). Elsevier. doi:10.1016/B978-0-12-816671-0.00004-7

Kumar, A., Ryparovà, P., Petrič, M., Tywoniak, J., & Hajek, P. (2017). Coating of wood by means of electrospun nanofibers based on PVA/SiO2 and its hydrophobization with octadecyltrichlorosilane (OTS). *Holzforschung*, *71*(3), 225–231. doi:10.1515/hf-2016-0108

Kumar, A., & Thakur, A. (2020). Encapsulated nanoparticles in organic polymers for corrosion inhibition. In *Corrosion Protection at the Nanoscale*. Elsevier Inc. doi:10.1016/B978-0-12-819359-4.00018-0

Kumari, S. V. G., Pakshirajan, K., & Pugazhenthi, G. (2022). Recent advances and future prospects of cellulose, starch, chitosan, polylactic acid and polyhydroxyalkanoates for sustainable food packaging applications. *International Journal of Biological Macromolecules*, *221*, 163–182. Advance online publication. doi:10.1016/j.ijbiomac.2022.08.203 PMID:36067847

Kumar, K. R. T., Ramakrishna, M., & Sukumar, G. D. (2018). A Review on PV Cells and Nanocomposite-Coated PV Systems. *International Journal of Energy Research*, *42*(7), 2305–2319. doi:10.1002/er.4002

Kurniawan, F. H., Chinavinijkul, P., & Nasongkla, N. (2021). Hydrophobic and antibacterial bed sheet using ZnO nanoparticles: A large-scale technique. *Journal of Drug Delivery Science and Technology*, *62*, 102339. doi:10.1016/j.jddst.2021.102339

Kuswandi, B., & Moradi, M. (2019). Improvement of food packaging based on functional nanomaterial. *Nanotechnology: Applications in energy, drug and food*, 309-344. doi:10.1007/978-3-319-99602-8_16

Kuswandi, B. (2017). Environmental friendly food nano-packaging. *Environmental Chemistry Letters*, *15*(2), 205–221. doi:10.1007/s10311-017-0613-7

Kuznetsov, Y. I. (2011). Progress in inhibition of metal corrosion and modification of safety nanolayers on metals. *Protection of Metals and Physical Chemistry of Surfaces*, *47*(7), 821–829. doi:10.1134/S2070205111070100

La Russa, M. F., Rovella, N., Alvarez de Buergo, M., Belfiore, C. M., Pezzino, A., Crisci, G. M., & Ruffolo, S. A. (2016). Nano-TiO2 coatings for cultural heritage protection: The role of the binder on hydrophobic and self-cleaning efficacy. *Progress in Organic Coatings*, *91*, 1–8. doi:10.1016/j.porgcoat.2015.11.011

Lai, L., Wu, H., Mao, G., Li, Z., Zhang, L., & Liu, Q. (2022). Microstructure and Corrosion Resistance of Two-Dimensional TiO2/MoS2 Hydrophobic Coating on AZ31B Magnesium Alloy. *Coatings*, *12*(10), 1488. Advance online publication. doi:10.3390/coatings12101488

Law, M. K., Zhao, Y., Zhang, W., Wang, R., Shi, M., Zhang, Y., Chen, S., & Yang, J. (2022). Highly transparent and super-wettable nanocoatings hybridized with isocyanate-silane modified surfactant for multifunctional applications. *Nano Materials Science*, *4*(2), 151–168. doi:10.1016/j.nanoms.2021.09.001

Lee. (2020). *Synthesis and applications of hydrophobic nanostructured coatings for green buildings*. ScholarBank@NUS Repository.

Lee, J. J., Gandla, S., Lim, B., Kang, S., Kim, S., Lee, S., & Kim, S. (2020). Alcohol-based highly conductive polymer for conformal nanocoatings on hydrophobic surfaces toward a highly sensitive and stable pressure sensor. *NPG Asia Materials*, *12*(1), 65. doi:10.1038/s41427-020-00238-z

Lee, M., Hwang, S. Y., Koo, J. M., Jeon, H., Kim, H. J., Oh, D. X., & Park, J. (2023). Chitosan coating in the form of polymer and nanowhiskers on clothing fabrics for improved particulate matter removal efficiency in face mask filters. *Journal of Natural Fibers*, *20*(1), 2187507. doi:10.1080/15440478.2023.2187507

Levchuk, I., Homola, T., Singhal, G., Rueda-Márquez, J. J., Vida, J., Souček, P., Svoboda, T., Villar-Navarro, E., Levchuk, O., Dzik, P., Lähde, A., & Moreno-Andrés, J. (2022). UVA and solar driven photocatalysis with rGO/TiO2/polysiloxane for inactivation of pathogens in recirculation aquaculture systems (RAS) streams. *Chemical Engineering Journal Advances, 10*(December). doi:10.1016/j.ceja.2022.100243

Lhamo, P., & Mahanty, B. (2023). Tuning Material Properties of Microbially Synthesized Poly (3-hydroxybutyrate-co-hydroxyvalerate) and Their Composites for Various Applications. *Journal of Polymers and the Environment*, *31*(11), 1–21. doi:10.1007/s10924-023-02908-9

Li, T., Liu, Y., Liu, B., Guo, W., & Xu, L. (2017). Microstructure and Wear Behavior of FeCoCrNiMo0.2 High Entropy Coatings Prepared by Air Plasma Spray and the High Velocity Oxy-Fuel Spray Processes. Coatings.

Liang, S. S., Makamba, H., Huang, S. Y., & Chen, S. H. (2006). Nano-titanium dioxide composites for the enrichment of phosphopeptides. *Journal of Chromatography. A*, *1116*(1–2), 38–45. doi:10.1016/j.chroma.2006.03.012 PMID:16580007

Liang, Z., Zhou, Z., Zhao, L., Dong, B., & Wang, S. (2020). Fabrication of Transparent, Durable and Self-Cleaning Superhydrophobic Coatings for Solar Cells. *New Journal of Chemistry*, *44*(34), 14481–14489. doi:10.1039/D0NJ01402H

Lian, R., Ou, M., Guan, H., Cui, J., Piao, J., Feng, T., Ren, J., Wang, Y., Wang, Y., Liu, L., Chen, X., & Jiao, C. (2023). Facile fabrication of multifunctional energy-saving building materials with excellent thermal insulation, robust mechanical property, and ultrahigh flame retardancy. *Energy*, *277*, 127773. doi:10.1016/j.energy.2023.127773

Li, B., Bai, J., He, J., Ding, C., Dai, X., Ci, W., Zhu, T., Liao, R., & Yuan, Y. (2023). A Review on Superhydrophobic Surface with Anti-Icing Properties in Overhead Transmission Lines. *Coatings*, *13*(2), 301. doi:10.3390/coatings13020301

Li, C., Liu, W., Mao, J., Hu, L., Yun, Y., & Li, B. (2023). Superhydrophobic PVDF membrane modified by dopamine self-polymerized nanoparticles for vacuum membrane distillation. *Separation and Purification Technology*, *304*, 122182. doi:10.1016/j.seppur.2022.122182

Li, C., Xia, F., Ma, C., & Li, Q. (2021). Research on the Corrosion Behavior of Ni-SiC Nanocoating Prepared Using a Jet Electrodeposition Technique. *Journal of Materials Engineering and Performance*, *30*(8), 6336–6344. doi:10.1007/s11665-021-05891-1

Lichter, J. A., Van Vliet, K. J., & Rubner, M. F. (2009). Design of Antibacterial Surfaces and Interfaces: Polyelectrolyte Multilayers as a Multifunctional Platform. *Macromolecules*, *42*(22), 8573–8586. doi:10.1021/ma901356s

Li, D., Dai, D., Xiong, G., Lan, S., & Zhang, C. (2023). Composite Nanocoating of Biomedical Magnesium Alloy Implants: Advantages, Mechanisms, and Design Strategies. *Advanced Science (Weinheim, Baden-Wurttemberg, Germany)*, *10*(18), 2300658. doi:10.1002/advs.202300658

Li, J., Jiang, M., Zhou, H., Jin, P., Cheung, K. M. C., Chu, P. K., & Yeung, K. W. K. (2018). Vanadium Dioxide Nanocoating Induces Tumor Cell Death through Mitochondrial Electron Transport Chain Interruption. *Global Challenges (Hoboken, NJ)*, *3*(3), 1800058. doi:10.1002/gch2.201800058 PMID:31565366

Li, J., Tian, J., Gao, Y., Qin, R., Pi, H., Li, M., & Yang, P. (2021). All-natural superhydrophobic coating for packaging and blood-repelling materials. *Chemical Engineering Journal*, *410*, 128347. doi:10.1016/j.cej.2020.128347

Li, M., Su, B., Zhou, B., Wang, H., & Meng, J. (2020). One-pot synthesis and self-assembly of anti-wear octadecyltrichlorosilane/silica nanoparticles composite films on silicon. *Applied Surface Science*, *508*, 145187. doi:10.1016/j.apsusc.2019.145187

Lin, Y., Chen, H., Wang, G., & Liu, A. (2018). Recent Progress in Preparation and Anti-Icing Applications of Superhydrophobic Coatings. Coatings. doi:10.3390/coatings8060208

Lin, D., Yang, Y., Wang, J., Yan, W., Wu, Z., Chen, H., Zhang, Q., Wu, D., Qin, W., & Tu, Z. (2020). Preparation and characterization of TiO2-Ag loaded fish gelatin-chitosan antibacterial composite film for food packaging. *International Journal of Biological Macromolecules*, *154*, 123–133. doi:10.1016/j.ijbiomac.2020.03.070 PMID:32171840

Lin, J., Chen, H., Fei, T., & Zhang, J. (2013). Highly transparent superhydrophobic organic–inorganic nanocoating from the aggregation of silica nanoparticles. *Colloids and Surfaces. A, Physicochemical and Engineering Aspects*, *421*, 51–62. doi:10.1016/j.colsurfa.2012.12.049

Li, Q., Lu, H., Cui, J., An, M., & Li, D. (2016). Electrodeposition of nanocrystalline zinc on steel for enhanced resistance to corrosive wear. *Surface and Coatings Technology*, *304*, 567–573. doi:10.1016/j.surfcoat.2016.07.056

Li, Q., Wang, X., Wang, X., Zheng, L., Chen, P., & Zhang, B. (2023). Novel insights into versatile nanomaterials integrated bioreceptors toward zearalenone ultrasensitive discrimination. *Food Chemistry*, *410*, 135435. doi:10.1016/j.foodchem.2023.135435 PMID:36641913

Li, S., Wei, Y., & Huang, J. (2010). Facile fabrication of superhydrophobic cellulose materials by a nanocoating approach. *Chemistry Letters*, *39*(1), 20–21. doi:10.1246/cl.2010.20

Liu, B., Liu, Z., Li, Y., & Feng, F. (2023). A wind tunnel test of the anti-icing properties of MoS2/ZnO hydrophobic nano-coatings for wind turbine blades. *Coatings*, *13*(4), 686. doi:10.3390/coatings13040686

Liu, C. H., & Grodzinski, P. (2021). Nanotechnology for Cancer Imaging: Advances, Challenges, and Clinical Opportunities. *Radiology. Imaging Cancer*, *3*(3), e200052. doi:10.1148/rycan.2021200052 PMID:34047667

Liu, J., Janjua, Z. A., Roe, M., Xu, F., Turnbull, B., Choi, K. S., & Hou, X. (2016). Superhydrophobic/icephobic coatings based on silica nanoparticles modified by self-assembled monolayers. *Nanomaterials (Basel, Switzerland)*, *6*(12), 232. Advance online publication. doi:10.3390/nano6120232 PMID:28335360

Liu, N., Wu, Q., Liu, Y., Li, J., Ji, P., & Fu, G. (2021). Application of Nanomaterials in the Treatment and Diagnosis of Ophthalmology Diseases. *Current Stem Cell Research & Therapy*, *16*(1), 95–103. doi:10.2174/1574888X15666200210104449 PMID:32039688

Liu, S., Chen, X., Yu, M., Li, J., Liu, J., Xie, Z., Gao, F., & Liu, Y. (2022). Applications of Titanium Dioxide Nanostructure in Stomatology. *Molecules (Basel, Switzerland)*, *27*(12), 3881. doi:10.3390/molecules27123881 PMID:35745007

Liu, Y., Zhang, X., Li, C., Qin, Y., Xiao, L., & Liu, J. (2020). Comparison of the structural, physical and functional properties of κ-carrageenan films incorporated with pomegranate flesh and peel extracts. *International Journal of Biological Macromolecules*, *147*, 1076–1088. doi:10.1016/j.ijbiomac.2019.10.075 PMID:31739038

Li, W., Shi, J., Zhao, Y., Huo, Q., Sun, Y., Wu, Y., Tian, Y., & Jiang, Z. (2020). Superhydrophobic Metal–Organic Framework Nanocoating Induced by Metal-Phenolic Networks for Oily Water Treatment. *ACS Sustainable Chemistry & Engineering*, *8*(4), 1831–1839. doi:10.1021/acssuschemeng.9b05937

Li, W., Tan, X., Zhu, J., Xiang, P., Xiao, T., Tian, L., Yang, A., Wang, M., & Chen, X. (2019). Broadband Antireflective and Superhydrophobic Coatings for Solar Cells. *Materials Today. Energy*, *12*, 348–355. doi:10.1016/j.mtener.2019.03.006

Li, W., Yin, Z., Qi, L., Yu, B., & Xing, W. (2023). Scalable production of bioinspired MXene/black phosphorene nanocoatings for hydrophobic and fire-safe textiles with tunable electromagnetic interference and exceeding thermal management. *Chemical Engineering Journal*, *460*, 141870. doi:10.1016/j.cej.2023.141870

Li, Y., Zhang, Z., Wang, M., Men, X., & Xue, Q. (2017). Environmentally safe, substrate-independent and repairable nanoporous coatings: Large-scale preparation, high transparency and antifouling properties. *Journal of Materials Chemistry. A, Materials for Energy and Sustainability*, *5*(38), 20277–20288. doi:10.1039/C7TA05112C

Lotfi, M., Tajik, H., Moradi, M., Forough, M., Divsalar, E., & Kuswandi, B. (2018). Nanostructured chitosan/monolaurin film: Preparation, characterization and antimicrobial activity against *Listeria monocytogenes* on ultrafiltered white cheese. *Lebensmittel-Wissenschaft + Technologie, 92*, 576–583. doi:10.1016/j.lwt.2018.03.020

Loukanov, A., Nikolova, S., Filipov, C., & Nakabayashi, S. (2019). Nanomaterials for cancer medication: From individual nanoparticles toward nanomachines and nanorobots. *Pharmacia, 66*(3), 147–156. doi:10.3897/pharmacia.66.e37739

Lucia-Oana, S., Mirela, B., & Alexandra-Gabriela, E. (2023). Physical and mechanical characterization of cosmetic textiles with anti-acne effect. *Annals of the University of Oradea. Fascicle of Textiles. Leatherwork, 24*(1), 113–118.

Lu, H., & Zheng, C. (2022). Comparison of Dust Deposition Reduction Performance by Super-Hydrophobic and Super-Hydrophilic Coatings for Solar PV Cells. *Coatings, 12*(4), 502. doi:10.3390/coatings12040502

Luo, J., Yang, H. B., Zhuang, M., Liu, S., Wang, L., & Liu, B. (2020). Making Fully Printed Perovskite Solar Cells Stable Outdoor with Inorganic Superhydrophobic Coating. *Journal of Energy Chemistry, 50*, 332–338. doi:10.1016/j.jechem.2020.03.082

Lv, S., Liang, S., Zuo, J., Zhang, S., Wang, J., & Wei, D. (2023). Lignin-based anti-UV functional materials: Recent advances in preparation and application. *Iranian Polymer Journal, 32*(11), 1–21. doi:10.1007/s13726-023-01218-0

Maeztu, J. D., Rivero, P. J., Berlanga, C., Bastidas, D. M., Palacio, J. F., & Rodriguez, R. (2017). Effect of graphene oxide and fluorinated polymeric chains incorporated in a multilayered sol-gel nanocoating for the design of corrosion resistant and hydrophobic surfaces. *Applied Surface Science, 419*, 138–149. doi:10.1016/j.apsusc.2017.05.043

Magovac, E., Vončina, B., Budimir, A., Jordanov, I., Grunlan, J. C., & Bischof, S. (2021). Environmentally benign phytic acid-based nanocoating for multifunctional flame-retardant/antibacterial cotton. *Fibers (Basel, Switzerland), 9*(11), 69. Advance online publication. doi:10.3390/fib9110069

Maharjan, S., Liao, K.-S., Wang, A. J., Barton, K., Haldar, A., Alley, N. J., Byrne, H. J., & Curran, S. A. (2020). Self-Cleaning Hydrophobic Nanocoating on Glass: A Scalable Manufacturing Process. *Materials Chemistry and Physics, 239*, 122000. doi:10.1016/j.matchemphys.2019.122000

Mahesha, K. N., Guddaraddi, A., Reddy, H., & GV, S. K. (2023). Unleashing the Potential of Nanotechnology in Agriculture. *Advanced Innovative Technologies in Agricultural Engineering for Sustainable Agriculture*, 97.

Mahltig, B. (2023). 3 Finishing of textile materials. In *Particle Technology and Textiles: Review of Applications* (pp. 45–66). De Gruyter. doi:10.1515/9783110670776-003

Mahmoudi, A., Tavakoly Sany, S. B., Ahari Salmasi, M., Bakhshi, A., Bustan, A., heydari, S., Rezayi, M., & Gheybi, F. (2023). Application of nanotechnology in air purifiers as a viable approach to protect against Coronavirus. *IET Nanobiotechnology / IET*, *17*(4), 289–301. doi:10.1049/nbt2.12132 PMID:37096564

Maity, S., Singha, K., & Pandit, P. (2023). 1 - Introduction to functional and technical textiles. In S. Maity, K. Singha, & P. Pandit (Eds.), *Functional and Technical Textiles* (pp. 1–30). Woodhead Publishing. doi:10.1016/B978-0-323-91593-9.00021-3

Ma, J., Fan, J., Xia, Y., Kou, X., Ke, Q., & Zhao, Y. (2023). Preparation of aromatic β-cyclodextrin nano/microcapsules and corresponding aromatic textiles: A review. *Carbohydrate Polymers*, *308*, 120661. doi:10.1016/j.carbpol.2023.120661 PMID:36813345

Majid, I., Nayik, G. A., Dar, S. M., & Nanda, V. (2018). Novel food packaging technologies: Innovations and future prospective. *Journal of the Saudi Society of Agricultural Sciences*, *17*(4), 454–462. doi:10.1016/j.jssas.2016.11.003

Malekkhouyan, R., Nouri Khorasani, S., Esmaeely Neisiany, R., Torkaman, R., Koochaki, M. S., & Das, O. (2020). Preparation and Characterization of Electrosprayed Nanocapsules Containing Coconut-Oil-Based Alkyd Resin for the Fabrication of Self-Healing Epoxy Coatings. *Applied Sciences (Basel, Switzerland)*, *10*(9), 3171. doi:10.3390/app10093171

Malekmohammadi, S., Mohammed, R. U. R., Samadian, H., Zarebkohan, A., García-Fernández, A., Kokil, G. R., Sharifi, F., Esmaeili, J., Bhia, M., Razavi, M., Bodaghi, M., Kumeria, T., & Martínez-Máñez, R. (2022). Nonordered dendritic mesoporous silica nanoparticles as promising platforms for advanced methods of diagnosis and therapies. *Materials Today. Chemistry*, *26*, 101144. doi:10.1016/j.mtchem.2022.101144

Malhotra, S. P. K., & Mandal, T. (2019). Zinc oxide nanostructure and its application as agricultural and industrial material. *Contamin. Agric. Environ. Health Risks Remed*, *1*, 216–226. doi:10.26832/AESA-2019-CAE-0171-016

Malik, S., Muhammad, K., & Waheed, Y. (2023). a. Emerging Applications of Nanotechnology in Healthcare and Medicine. *Molecules (Basel, Switzerland)*, *28*(18), 6624. doi:10.3390/molecules28186624 PMID:37764400

Malik, S., Muhammad, K., & Waheed, Y. (2023). b. Nanotechnology: A Revolution in Modern Industry. *Molecules (Basel, Switzerland)*, *28*(2), 661. doi:10.3390/molecules28020661 PMID:36677717

Malik, S., Muhammad, K., & Waheed, Y. (2023). Nanotechnology: A Revolution in Modern Industry. *Molecules (Basel, Switzerland)*, 28. PMID:36677717

Malshe, A. P., Rajurkar, K. P., Virwani, K. R., Taylor, C. R., Bourell, D. L., Levy, G., Sundaram, M. M., McGeough, J. A., Kalyanasundaram, V., & Samant, A. N. (2010). Tip-based nanomanufacturing by electrical, chemical, mechanical and thermal processes. *CIRP Annals*, *59*(2), 628–651. doi:10.1016/j.cirp.2010.05.006

Manikandan, V., & Min, S. C. (2023). Roles of polysaccharides-based nanomaterials in food preservation and extension of shelf-life of food products: A review. *International Journal of Biological Macromolecules, 126381*, 126381. Advance online publication. doi:10.1016/j.ijbiomac.2023.126381 PMID:37595723

Mani, M., & Pillai, R. (2010). Impact of dust on solar photovoltaic (PV) performance: Research status, challenges, and recommendations. *Renewable & Sustainable Energy Reviews, 14*(9), 3124–3131. doi:10.1016/j.rser.2010.07.065

Manivannan, S., Praveen, R., Kim, K., & Ramaraj, R. (2024). Introduction: Nanomaterials for Sustainable Energy Applications. In Nanomaterials for Sustainable Energy Applications (pp. 1-24). CRC Press.

Mao, M., Xu, H., Guo, K.-Y., Zhang, J.-W., Xia, Q.-Q., Zhang, G.-D., ... Tang, L.-C. (2020). Mechanically flexible, super-hydrophobic and flame-retardant hybrid nano-silica/graphene oxide wide ribbon decorated sponges for efficient oil/water separation and fire warning response. *Composites. Part A, Applied Science and Manufacturing, 106191*. Advance online publication. doi:10.1016/j.compositesa.2020

Marilyn. (2019). *Water repellent hydrophobic surfaces and anti-fungal nanocoatings for building applications*. Academic Press.

Marshall, J. (2011). Draft guidelines for nanomedicine unveiled. *NATNews*.

Martinez, G., Shutthanandan, V., Thevuthasan, S., Chessa, J. F., & Ramana, C. V. (2014). Effect of thickness on the structure, composition and properties of titanium nitride nano-coatings. *Ceramics International, 40*(4), 5757–5764. doi:10.1016/j.ceramint.2013.11.014

Martínez-Morlanes, M. J., Pascual, F. J., Guerin, G., & Puértolas, J. A. (2021). Influence of processing conditions on microstructural, mechanical and tribological properties of graphene nanoplatelet reinforced UHMWPE. *Journal of the Mechanical Behavior of Biomedical Materials, 115*, 104248. doi:10.1016/j.jmbbm.2020.104248 PMID:33360486

Mashtalyar, D., Nadaraia, K., Sinebryukhov, S., & Gnedenkov, S. (2019). Polymer-Containing Layers Formed by PEO and Spray-Coating Method. *Materials Today: Proceedings, 11*, 150–154. doi:10.1016/j.matpr.2018.12.123

Maťátková, O., Michailidu, J., Miškovská, A., Kolouchová, I., Masák, J., & Čejková, A. (2022). Antimicrobial properties and applications of metal nanoparticles biosynthesized by green methods. *Biotechnology Advances, 58*, 107905. doi:10.1016/j.biotechadv.2022.107905 PMID:35031394

Mathew, J. T., Adetunji, C. O., Inobeme, A., Musah, M., Azeh, Y., Otori, A. A., Shaba, E. Y., Mamman, A., & Tanko, E. (2023). Removal of Heavy Metals Using Bio-remedial Techniques. Springer Nature Switzerland AG. doi:10.1007/978-3-031-24086-7_6

Mathew, J. T., Dauda, B. E. N., Mann, A., Ndamitso, M. M., Fadipe, A. L., & Shaba, E. Y. (2018). Assessment of Nutritional Properties of Fermented and Unfermented Seed of *Cissus populnae* From Niger State, Nigeria. *Assumption University-eJournal of Interdisciplinary Research, 3*(2), 70–77.

Maton, M., Gabut, S., Neut, C., Odou, P., Sacareau, C., Pinon, A., Vialette, M., Gerber, G., Martel, B., & Blanchemain, N. (2023). Antiviral functionalization of a polypropylene nonwoven textile structure as a self-decontaminating layer for respiratory masks. *Biomaterials Science*, *11*(10), 3502–3511. doi:10.1039/D2BM01988D PMID:36988417

Mazumder, M. K., Horenstein, M. N., Joglekar, N. R., Sayyah, A., Stark, J. W., Bernard, A. A. R., Garner, S. M., Yellowhair, J. E., Lin, H. Y., Eriksen, R. S., Griffin, A. C., Gao, Y., Centra, R. L., & Lloyd, A. H. (2017). Mitigation of dust impact on solar collectors by water-free cleaning with transparent electrodynamic films: Progress and challenges. *IEEE Journal of Photovoltaics*, *7*(5), 1342–1353. doi:10.1109/JPHOTOV.2017.2721462

McCann, J. (2023). 13 - Identification of design requirements for smart clothes and wearable technology. In J. McCann & D. Bryson (Eds.), *Smart Clothes and Wearable Technology* (2nd ed., pp. 327–369). Woodhead Publishing. doi:10.1016/B978-0-12-819526-0.00004-7

McDonald, T. O., Siccardi, M., Moss, D., Liptrott, N., Giardiello, M., Rannard, S., & Owen, A. (2015). The application of nanotechnology to drug delivery in medicine. In *Nanoengineering* (pp. 173–223). Elsevier. doi:10.1016/B978-0-444-62747-6.00007-5

Meda, U. S., Vora, K., Athreya, Y., & Mandi, U. A. (2022). Titanium dioxide based heterogeneous and heterojunction photocatalysts for pollution control applications in the construction industry. *Process Safety and Environmental Protection*, *161*, 771–787. doi:10.1016/j.psep.2022.03.066

Meereboer, K. W., Misra, M., & Mohanty, A. K. (2020). Review of recent advances in the biodegradability of polyhydroxyalkanoate (PHA) bioplastics and their composites. *Green Chemistry*, *22*(17), 5519–5558. doi:10.1039/D0GC01647K

Mehta, P., Al-Kinani, A. A., Arshad, M. S., Singh, N., van der Merwe, S. M., Chang, M.-W., Alany, R. G., & Ahmad, Z. (2019). Engineering and Development of Chitosan-Based Nanocoating for Ocular Contact Lenses. *Journal of Pharmaceutical Sciences*, *108*(4), 1540–1551. doi:10.1016/j.xphs.2018.11.036 PMID:30513319

Merugu, R., Sharma, S., & Rastogi, M. (2023). An overview of nanotechnological applications in biomedical engineering. *Materials Today: Proceedings*, *76*, 412–417. doi:10.1016/j.matpr.2022.12.028

Mesgari, M., Aalami, A. H., Sathyapalan, T., & Sahebkar, A. (2022). A comprehensive review of the development of carbohydrate macromolecules and copper oxide nanocomposite films in food nanopackaging. *Bioinorganic Chemistry and Applications*, *2022*, 1–28. Advance online publication. doi:10.1155/2022/7557825 PMID:35287316

Mihai, M. M., Dima, M. B., Dima, B., & Holban, A. M. (2019). Nanomaterials for wound healing and infection control. *Materials (Basel)*, *12*(13), 2176. doi:10.3390/ma12132176 PMID:31284587

Mihindukulasuriya, S. D. F., & Lim, L. T. (2014). Nanotechnology development in food packaging: A review. *Trends in Food Science & Technology*, *40*(2), 149–167. doi:10.1016/j.tifs.2014.09.009

Minh, D. Q., & Minh, H. N. (2020). Al2O3 Nano Coating and Hydrophobicity of Glass Surface. *Glass Physics and Chemistry*, *46*(6), 640–645. doi:10.1134/S1087659621010089

Mirabi, S. (2020). The use of self-cleaning materials and the study of their impact on the quality of the visual appearance of large industrial cities. *Journal of Architectural Research and Development*, *4*(6). Advance online publication. doi:10.26689/jard.v4i6.1435

Mirpoor, S. F., Corrado, I., Di Girolamo, R., Dal Poggetto, G., Panzella, L., Borselleca, E., Pezzella, C., & Giosafatto, C. V. L. (2023). Manufacture of active multilayer films made of functionalized pectin coated by polyhydroxyalkanoates: A fully renewable approach to active food packaging. *Polymer*, *126136*. Advance online publication. doi:10.1016/j.polymer.2023.126136

Mirza, A. Z., & Siddiqui, F. A. (2014). Nanomedicine and drug delivery: A mini review. *International Nano Letters*, *4*(1), 1–7. doi:10.1007/s40089-014-0094-7

Mishra, A., Bhatt, N., & Bajpai, A. K. (2019). Nanostructured superhydrophobic coatings for solar panel applications. In Nanomaterials-Based Coatings (pp. 397–424). doi:10.1016/B978-0-12-815884-5.00012-0

Mitelut, A. C., Tanase, E. E., Popa, V. I., & Popa, M. E. (2015). Sustainable alternative for food packaging: Chitosan biopolymer-a review. *AgroLife Scientific Journal*, *4*(2).

Miteu, G. D., Emmanuel, A. A., Addeh, I., Ojeokun, O., Olayinka, T., Godwin, J. S., ... Benneth, E. O. (2023). Nanoscience Technology as A Pivot For Sustainable Agriculture And Its One Health Approach Awareness. *Science in One Health*, 100020.

Moalla, S., Ammar, I., Fauconnier, M. L., Danthine, S., Blecker, C., Besbes, S., & Attia, H. (2021). Development and characterization of chitosan films carrying Artemisia campestris antioxidants for potential use as active food packaging materials. *International Journal of Biological Macromolecules*, *183*, 254–266. doi:10.1016/j.ijbiomac.2021.04.113 PMID:33892038

Mohammadi, M., Mousavi Shaegh, S. A., Alibolandi, M., Ebrahimzadeh, M. H., Tamayol, A., Jaafari, M. R., & Ramezani, M. (2018). Micro and nanotechnologies for bone regeneration: Recent advances and emerging designs. *Journal of Controlled Release*, *274*, 35–55. doi:10.1016/j.jconrel.2018.01.032 PMID:29410062

Mohammad, Z. H., & Ahmad, F. (2024). Nanocoating and its application as antimicrobials in the food industry: A review. *International Journal of Biological Macromolecules*, *254*, 127906. doi:10.1016/j.ijbiomac.2023.127906 PMID:37935295

Mohammed, M. T., & Hussein, S. M. (2019). Synthesis and characterization of nanocoatings derived by sol–gel onto a new surgical titanium surface. *Materials Research Express*, *6*(7), 076426. doi:10.1088/2053-1591/ab1722

MokhlespourEsfahani, M. I. (2021). Smart textiles in healthcare: a summary of history, types, applications, challenges, and future trends. In Micro and Nano Technologies, Nanosensors and Nanodevices for Smart Multifunctional Textiles (pp. 93-107). Elsevier. doi:10.1016/B978-0-12-820777-2.00006-6

Moncada, E., Quijada, R., & Retuert, J. (2007). Nanoparticles prepared by the sol–gel method and their use in the formation of nanocomposites with polypropylene. *Nanotechnology*, *18*(33), 335606. doi:10.1088/0957-4484/18/33/335606

Mondal, S. (2022). Nanomaterials for UV protective textiles. *Journal of Industrial Textiles*, *51*(4), 5592S–5621S. doi:10.1177/1528083721988949

Mondoc, B., & Pop, F. (2010). Factors influencing the performance of photovoltaic power plant. *Proceedings of Third International Conference on Modern Power Systems (MPS),* 271-276.

Montes, C., Villaseñor, M. J., & Ríos, Á. (2019). Analytical control of nanodelivery lipid-based systems for encapsulation of nutraceuticals: Achievements and challenges. *Trends in Food Science & Technology*, *90*, 47–62. doi:10.1016/j.tifs.2019.06.001

Moradi, M., & Rezaei, M. (2022). Construction of highly anti-corrosion and super-hydrophobic polypropylene/graphene oxide nanocomposite coatings on carbon steel: Experimental, electrochemical and molecular dynamics studies. *Construction & Building Materials*, *317*, 126136. doi:10.1016/j.conbuildmat.2021.126136

More, S. R., Bhatt, D. V., & Menghani, J. V. (2017). Resent Research Status on Laser Cladding as Erosion Resistance Technique - An Overview. *Materials Today: Proceedings*, *4*(9), 9902–9908. doi:10.1016/j.matpr.2017.06.291

Mosleh-Shirazi, S., Abbasi, M., Moaddeli, M. R., Vaez, A., Shafiee, M., Kasaee, S. R., Amani, A. M., & Hatam, S. (2022). Nanotechnology Advances in the Detection and Treatment of Cancer: An Overview. *Nanotheranostics*, *6*(4), 400–423. doi:10.7150/ntno.74613 PMID:36051855

Motelica, L., Ficai, D., Oprea, O. C., Ficai, A., & Andronescu, E. (2020). Smart Food Packaging Designed by Nanotechnological and Drug Delivery Approaches. Coatings. doi:10.3390/coatings10090806

Mott-Smith, H. D. (1971). History of 'plasmas'. *Nature*, *233*(5316), 219. doi:10.1038/233219a0 PMID:16063290

Mozafari, N., Mozafari, N., Dehshahri, A., & Azadi, A. (2023). Knowledge Gaps in Generating Cell-Based Drug Delivery Systems and a Possible Meeting with Artificial Intelligence. *Molecular Pharmaceutics*, *20*(8), 3757–3778. doi:10.1021/acs.molpharmaceut.3c00162 PMID:37428824

Mozumder, M. S., Mourad, A.-H. I., Pervez, H., & Surkatti, R. (2019). Recent Developments in Multifunctional Coatings for Solar Panel Applications: A Review. *Solar Energy Materials and Solar Cells*, *189*, 75–102. doi:10.1016/j.solmat.2018.09.015

Müller, K., Bugnicourt, E., Latorre, M., Jorda, M., Sanz, Y. E., Lagaron, J. M., Miesbauer, O., Bianchin, A., Hankin, S., Bölz, U., Pérez, G., Jesdinszki, M., Lindner, M., Scheuerer, Z., Castelló, S., & Schmid, M. (2017). Review on the Processing and Properties of Polymer Nanocomposites and Nanocoatings and Their Applications in the Packaging, Automotive and Solar Energy Fields. *Nanomaterials (Basel, Switzerland)*, *7*(4), 74. doi:10.3390/nano7040074

Mumtaz, Z., Rashid, Z., Ali, A., Arif, A., Ameen, F., AlTami, M. S., & Yousaf, M. Z. (2023). Prospects of Microfluidic Technology in Nucleic Acid Detection Approaches. *Biosensors (Basel), 13*(6), 584. doi:10.3390/bios13060584 PMID:37366949

Mustafa, F., & Andreescu, S. (2020). Nanotechnology-based approaches for food sensing and packaging applications. *RSC Advances, 10*(33), 19309–19336. doi:10.1039/D0RA01084G PMID:35515480

Muteeb, G. (2023). Nanotechnology—A Light of Hope for Combating Antibiotic Resistance. *Microorganisms, 11*(6), 1489. doi:10.3390/microorganisms11061489 PMID:37374990

MuthuKathija, M., Sheik Muhideen Badhusha, M., & Rama, V. (2023). Green synthesis of zinc oxide nanoparticles using Pisonia Alba leaf extract and its antibacterial activity. *Applied Surface Science Advances, 15*, 100400. doi:10.1016/j.apsadv.2023.100400

Naqvi, S. M. Z. A., Zhang, Y., Tahir, M. N., Ullah, Z., Ahmed, S., Wu, J., Raghavan, V., Abdulraheem, M. I., Ping, J., Hu, X., & Hu, J. (2023). Advanced strategies of the in-vivo plant hormone detection. *Trends in Analytical Chemistry, 166*, 117186. doi:10.1016/j.trac.2023.117186

Naresh, V., & Lee, N. (2021). A Review on Biosensors and Recent Development of Nanostructured Materials-Enabled Biosensors. *Sensors (Basel), 21*(4), 1109. doi:10.3390/s21041109 PMID:33562639

Nath, D., Santhosh, R., Pal, K., & Sarkar, P. (2022). Nanoclay-based active food packaging systems: A review. *Food Packaging and Shelf Life, 31*, 100803. doi:10.1016/j.fpsl.2021.100803

Nayashevsky, I., Xu, Q., Barahman, G., & Lyons, A.M. (2017). *Anti-reflective and anti-soiling properties of a KleanBoost™, a superhydrophobic nano-textured coating for solar glass*. Academic Press.

Nayshevsky, I. (2020). *Hydrophobic Fluoropolymer Nano-Coatings: Study of Anti-Reflective and Anti-Soiling Properties and Applications*. Academic Press.

Nazeer, A. A., & Madkour, M. (2018). Potential use of smart coatings for corrosion protection of metals and alloys: A review. *Journal of Molecular Liquids, 253*, 11–22. doi:10.1016/j.molliq.2018.01.027

Nazrin, A., Sapuan, S. M., Zuhri, M. Y. M., Ilyas, R. A., Syafiq, R. S. F. K. S., & Sherwani, S. F. K. (2020). Nanocellulose reinforced thermoplastic starch (TPS), polylactic acid (PLA), and polybutylene succinate (PBS) for food packaging applications. *Frontiers in Chemistry, 8*, 213. doi:10.3389/fchem.2020.00213 PMID:32351928

Negut, I., Grumezescu, V., Ficai, A., Grumezescu, A. M., Holban, A. M., Popescu, R. C., Savu, D., Vasile, B. S., & Socol, G. (2018). MAPLE deposition of Nigella sativa functionalized Fe3O4 nanoparticles for antimicrobial coatings. *Applied Surface Science, 455*, 513–521. doi:10.1016/j.apsusc.2018.05.202

Nejad, F. M., Tanzadeh, J., Soozangar, M., & Otadi, A. (2016). Performance Evaluation of the Impact of Modified Silica Nano-Materials on the Hydrophobicity of Hot-Mix Asphalt. *International Conference on Transportation and Development 2016*. 10.1061/9780784479926.101

Nešić, A., Cabrera-Barjas, G., Dimitrijević-Branković, S., Davidović, S., Radovanović, N., & Delattre, C. (2019). Prospect of polysaccharide-based materials as advanced food packaging. *Molecules (Basel, Switzerland)*, *25*(1), 135. doi:10.3390/molecules25010135 PMID:31905753

Nguyen-Tri, P., Tran, H. N., Plamondon, C. O., Tuduri, L., Vo, D.-V. N., Nanda, S., Mishra, A., Chao, H.-P., & Bajpai, A. K. (2019). Recent progress in the preparation, properties and applications of superhydrophobic nano-based coatings and surfaces: A review. *Progress in Organic Coatings*, *132*, 235–256. doi:10.1016/j.porgcoat.2019.03.042

Nile, S. H., Baskar, V., Selvaraj, D., Nile, A., Xiao, J., & Kai, G. (2020). Nanotechnologies in Food Science: Applications, Recent Trends, and Future Perspectives. *Nano-Micro Letters*, *12*(1), 45. doi:10.1007/s40820-020-0383-9 PMID:34138283

Nilsen-Nygaard, J., Fernández, E. N., Radusin, T., Rotabakk, B. T., Sarfraz, J., Sharmin, N., Sivertsvik, M., Sone, I., & Pettersen, M. K. (2021). Current status of biobased and biodegradable food packaging materials: Impact on food quality and effect of innovative processing technologies. *Comprehensive Reviews in Food Science and Food Safety*, *20*(2), 1333–1380. doi:10.1111/1541-4337.12715 PMID:33547765

Noah, N. M. (2020). Design and Synthesis of Nanostructured Materials for Sensor Applications. *Journal of Nanomaterials*, *2020*, 8855321. doi:10.1155/2020/8855321

Noeiaghaei, T., Dhami, N., & Mukherjee, A. (2017). Nanoparticles surface treatment on cemented materials for inhibition of bacterial growth. *Construction & Building Materials*, *150*, 880–891. doi:10.1016/j.conbuildmat.2017.06.046

Norazmi, F. S. (2019). *Synthesis and characterization of hydrophobic silica nanoparticles as anti-reflective coating for photovoltaic* [Masters thesis]]. Universiti Teknologi Malaysia, Faculty of Science.

Noshirvani, N., Ghanbarzadeh, B., Gardrat, C., Rezaei, M. R., Hashemi, M., Le Coz, C., & Coma, V. (2017). Cinnamon and ginger essential oils to improve antifungal, physical and mechanical properties of chitosan-carboxymethyl cellulose films. *Food Hydrocolloids*, *70*, 36–45. doi:10.1016/j.foodhyd.2017.03.015

Nouri, A., Yaraki, M. T., Lajevardi, A., Rahimi, T., Tanzifi, M., & Ghorbanpour, M. (2020). An investigation of the role of fabrication process in the physicochemical properties of κ-carrageenan-based films incorporated with *Zataria multiflora* extract and nanoclay. *Food Packaging and Shelf Life*, *23*, 100435. doi:10.1016/j.fpsl.2019.100435

Nunzi, F., & De Angelis, F. (2022). Modeling titanium dioxide nanostructures for photocatalysis and photovoltaics. *Chemical Science (Cambridge)*, *13*(33), 9485–9497. doi:10.1039/D2SC02872G PMID:36091912

Nyga, A., Czerwińska-Główka, D., Krzywiecki, M., Przystaś, W., Zabłocka-Godlewska, E., Student, S., Kwoka, M., Data, P., & Blacha-Grzechnik, A. (2021). Covalent immobilization of organic photosensitizers on the glass surface: Toward the formation of the light-activated antimicrobial nanocoating. *Materials (Basel), 14*(11), 3093. Advance online publication. doi:10.3390/ma14113093 PMID:34200077

Obšel, V., & Otřísal, P. (2022). Application of methods of preparation of plasmo-chemical nanocoatings designated for hydrophobic and oleophobic correction of surface working of textile materials for filtration protective suit. *Challenges to National Defence in Contemporary Geopolitical Situation, 2022*(1), 275–287. doi:10.47459/cndcgs.2022.35

Oliani, W. L., Pusceddu, F. H., & Parra, D. F. (2022). Silver-titanium polymeric nanocomposite non ecotoxic with bactericide activity. *Polymer Bulletin*, (12), 10949–10968. Advance online publication. doi:10.1007/s00289-021-04036-7 PMID:35035006

Ortega-Nieto, C., Losada-Garcia, N., Pessela, B. C., Domingo-Calap, P., & Palomo, J. M. (2023). Design and Synthesis of Copper Nanobiomaterials with Antimicrobial Properties. *ACS Bio & Med Chem Au, 3*(4), 349–358. doi:10.1021/acsbiomedchemau.2c00089 PMID:37599792

Otto, D. P., Otto, A., & de Villiers, M. M. (2022). In Vitro Skin Delivery of Griseofulvin by Layer-by-Layer Nanocoated Emulsions Stabilized by Whey Protein and Polysaccharides. *Pharmaceutics, 14*(3), 554. Advance online publication. doi:10.3390/pharmaceutics14030554 PMID:35335930

Oymaci, P., & Altinkaya, S. A. (2016). Improvement of barrier and mechanical properties of whey protein isolate based food packaging films by incorporation of zein nanoparticles as a novel bionanocomposite. *Food Hydrocolloids, 54*, 1–9. doi:10.1016/j.foodhyd.2015.08.030

Özen, İ., Demir, A., Bahtiyari, M. İ., Wang, X., Nilghaz, A., Wu, P., Shirvanimoghaddam, K., & Naebe, M. (2023). Multifaceted applications of thymol/carvacrol-containing polymeric fibrous structures. *Advanced Industrial and Engineering Polymer Research.* https://doi.org/https://doi.org/10.1016/j.aiepr.2023.09.001

Özpirin, M., & Ebil, Ö. (2018). Transparent block copolymer thin films for protection of optical elements via chemical vapor deposition. *Thin Solid Films, 674*, 1–5. doi:10.1016/j.tsf.2018.06.044

Özsevinç, A., & Alkan, C. (2023). Polyurethane shell medicinal lavender release microcapsules for textile materials: An environmentally friendly preparation. *Industrial Crops and Products, 192*, 116131. doi:10.1016/j.indcrop.2022.116131

Pakdel, E., & Wang, X. (2023). Thermoregulating textiles and fibrous materials for passive radiative cooling functionality. *Materials & Design, 231*, 112006. https://doi.org/.112006 doi:10.1016/j.matdes.2023

Pakdel, E., Fang, J., Fang, J., Sun, L., & Wang, X. (2018). Nanocoatings for Smart Textiles. In Smart Textiles (pp. 247–300). doi:10.1002/9781119460367.ch8

Pakdel, E., Wang, J., Allardyce, B. J., Rajkhowa, R., & Wang, X. (2016). Functionality of nano and 3D-microhierarchical TiO2 particles as coagulants for sericin extraction from the silk degumming wastewater. *Separation and Purification Technology*, *170*, 92–101. doi:10.1016/j.seppur.2016.06.025

Pal, A., Goswami, R., & Roy, D. N. (2021). A critical assessment on biochemical and molecular mechanisms of toxicity developed by emerging nanomaterials on important microbes. *Environmental Nanotechnology, Monitoring & Management*, *16*, 100485. doi:10.1016/j.enmm.2021.100485

Paladini, F., Di Franco, C., Panico, A., Scamarcio, G., Sannino, A., & Pollini, M. (2016). In vitro assessment of the antibacterial potential of silver nano-coatings on cotton gauzes for prevention of wound infections. *Materials (Basel)*, *9*(6), 1–14. doi:10.3390/ma9060411 PMID:28773531

Paladini, F., & Pollini, M. (2019). Antimicrobial Silver Nanoparticles for Wound Healing Application: Progress and Future Trends. *Materials (Basel)*, *12*(16), 2540. doi:10.3390/ma12162540 PMID:31404974

Pan, A., Lu, H., & Zhang, L.-Z. (2019). Experimental Investigation of Dust Deposition Reduction on Solar Cell Covering Glass by Different Self-Cleaning Coatings. *Energy*, *181*, 645–653. doi:10.1016/j.energy.2019.05.223

Pan, H., Jiang, B., Chen, J., & Jin, Z. (2014). Assessment of the physical, mechanical, and moisture-retention properties of pullulan-based ternary coblended films. *Carbohydrate Polymers*, *112*, 94–101. doi:10.1016/j.carbpol.2014.05.044 PMID:25129721

Parimalam, M., Islam, M. R., & Yunus, R. M. (2018). Effects of nanosilica, zinc oxide, titatinum oxide on the performance of epoxy hybrid nanocoating in presence of rubber latex. *Polymer Testing*, *70*, 197–207. doi:10.1016/j.polymertesting.2018.07.008

Parray, M. R., Kumar, H., Singh, A., Tomar, B., & Shiva, S. (2023). Synthesis of nanomaterial coatings for various applications. In *Nanomaterials for Sustainable Tribology* (pp. 167–190). CRC Press. doi:10.1201/9781003306276-11

Parveen, G., Bashir, S., Thakur, A., Saha, S. K., Banerjee, P., & Kumar, A. (2020). Experimental and computational studies of imidazolium based ionic liquid 1-methyl- 3-propylimidazolium iodide on mild steel corrosion in acidic solution Experimental and computational studies of imidazolium based ionic liquid 1-methyl- 3-propylimidazolium. *Materials Research Express*, *7*(1), 016510. doi:10.1088/2053-1591/ab5c6a

Pastorino, L., Dellacasa, E., Scaglione, S., Giulianelli, M., Sbrana, F., Vassalli, M., & Ruggiero, C. (2014). Oriented collagen nanocoating for tissue engineering. *Colloids and Surfaces. B, Biointerfaces*, *114*, 372–378. doi:10.1016/j.colsurfb.2013.10.026 PMID:24246194

Pathakoti, K., Manubolu, M., & Hwang, H. M. (2017). Nanostructures: Current uses and future applications in food science. *Yao Wu Shi Pin Fen Xi*, *25*(2), 245–253. doi:10.1016/j.jfda.2017.02.004 PMID:28911665

Pathan, S. I., Chundawat, N. S., Chauhan, N. P. S., & Singh, G. P. (2020). A review on synthetic approaches of heterocycles via insertion-cyclization reaction. *Synthetic Communications*, *50*(9), 1251–1285. doi:10.1080/00397911.2020.1712609

Paven, M., Mammen, L., & Vollmer, D. (2016). Smart Materials for Advanced Environmental Applications. In P. Wang (Ed.), *RSC Smart Materials No. 20.* The Royal Society of Chemistry. doi:10.1039/9781782622192-00209

Peng, C. W., Chang, K. C., Weng, C. J., Lai, M. C., Hsu, C. H., Hsu, S. C., Li, S. Y., Wei, Y., & Yeh, J. M. (2013). UV-curable nanocasting technique to prepare bio-mimetic super-hydrophobic non-fluorinated polymeric surfaces for advanced anticorrosive coatings. *Polymer Chemistry*, *4*(4), 926–932. doi:10.1039/C2PY20613G

Peng, X., Umer, M., Pervez, M. N., Hasan, K. M. F., Habib, M. A., Islam, M. S., Lin, L., Xiong, X., Naddeo, V., & Cai, Y. (2023). Biopolymers-based microencapsulation technology for sustainable textiles development: A short review. *Case Studies in Chemical and Environmental Engineering*, *7*, 100349. doi:10.1016/j.cscee.2023.100349

Perdani, C. G., & Gunawan, S. (2021, November). A short review: Nanocellulose for smart biodegradable packaging in the food industry. *IOP Conference Series. Earth and Environmental Science*, *924*(1), 012032. doi:10.1088/1755-1315/924/1/012032

Pereda, M., Amica, G., & Marcovich, N. E. (2012). Development and characterization of edible chitosan/olive oil emulsion films. *Carbohydrate Polymers*, *87*(2), 1318–1325. doi:10.1016/j.carbpol.2011.09.019

Perera, K. Y., Hopkins, M., Jaiswal, A. K., & Jaiswal, S. (2023). Nanoclays-containing bio-based packaging materials: Properties, applications, safety, and regulatory issues. *Journal of Nanostructure in Chemistry*, 1–23. doi:10.1007/s40097-023-00525-5 PMID:36747507

Perera, K. Y., Jaiswal, A. K., & Jaiswal, S. (2023). Biopolymer-Based Sustainable Food Packaging Materials: Challenges, Solutions, and Applications. *Foods*, *12*(12), 2422. doi:10.3390/foods12122422 PMID:37372632

Petkovska, J., Mladenovic, N., Marković, D., Radoičić, M., Chiang, H.-C., Palen, B., Mirakovski, D., Dimova, V., Radetić, M., Grunlan, J. C., & Jordanov, I. (2023). Environmentally benign few-bilayer intumescent nanocoating for flame retardant enzyme/plasma modified polyester fabric. *Polymer Degradation & Stability*, *214*, 110406. doi:10.1016/j.polymdegradstab.2023.110406

Petrik, P., Racz, A. S., & Menyhard, M. (2020). Complementary physicochemical analysis by ellipsometry and Auger spectroscopy of nano-sized protective coating layers. *Applied Surface Science*, *534*, 147593. doi:10.1016/j.apsusc.2020.147593

Pham, S. H., Choi, Y., & Choi, J. (2020). Stimuli-Responsive Nanomaterials for Application in Antitumor Therapy and Drug Delivery. *Pharmaceutics*, *12*(7), 12. doi:10.3390/pharmaceutics12070630 PMID:32635539

Phan, T., Jones, J. E., Chen, M., Bowles, D. K., Fay, W. P., & Yu, Q. (2022). A Biocompatibility Study of Plasma Nanocoating onto Cobalt Chromium L605 Alloy for Cardiovascular Stent Applications. *Materials (Basel), 15*(17), 5968. Advance online publication. doi:10.3390/ma15175968 PMID:36079346

Phisalaphong, M., & Jatupaiboon, N. (2008). Biosynthesis and characterization of bacteria cellulose–chitosan film. *Carbohydrate Polymers, 74*(3), 482–488. doi:10.1016/j.carbpol.2008.04.004

Pirozzi, A., Del Grosso, V., Ferrari, G., & Donsì, F. (2020). Edible coatings containing oregano essential oil nanoemulsion for improving postharvest quality and shelf life of tomatoes. *Foods, 9*(11), 1605. doi:10.3390/foods9111605 PMID:33158115

Pirozzi, A., Pataro, G., Donsì, F., & Ferrari, G. (2021). Edible coating and pulsed light to increase the shelf life of food products. *Food Engineering Reviews, 13*(3), 544–569. doi:10.1007/s12393-020-09245-w

Plazas-Tuttle, J., Rowles, L. S., Chen, H., Bisesi, J. H., Sabo-Attwood, T., & Saleh, N. B. (2015). *Dynamism of Stimuli-Responsive Nanohybrids*. Environmental Implications. In Nanomaterials.

Poggio, C., Colombo, M., Arciola, C. R., Greggi, T., Scribante, A., & Dagna, A. (2020). Copper-alloy surfaces and cleaning regimens against the spread of SARS-CoV-2 in dentistry and orthopedics. From fomites to anti-infective nanocoating. *Materials (Basel), 13*(15), 1–12. doi:10.3390/ma13153244 PMID:32707757

Polshettiwar, V. (2022). Dendritic Fibrous Nanosilica: Discovery, Synthesis, Formation Mechanism, Catalysis, and CO2 Capture-Conversion. *Accounts of Chemical Research, 55*(10), 1395–1410. doi:10.1021/acs.accounts.2c00031 PMID:35499964

Popović, S. Z., Lazić, V. L., Hromiš, N. M., Šuput, D. Z., & Bulut, S. N. (2018). Biopolymer packaging materials for food shelf-life prolongation. In *Biopolymers for food design* (pp. 223–277). Academic Press. doi:10.1016/B978-0-12-811449-0.00008-6

Popov, V., Kostyuk, G., Tymofyeyev, O., Kostyk, K., & Naboka, O. (2020). Design of new nanocoating based on hard alloy. In *Design, Simulation, Manufacturing: The Innovation Exchange* (pp. 522–531). Springer International Publishing. doi:10.1007/978-3-030-50794-7_51

Prabhakar, P., Sen, R. K., Patel, M., Shruti, Dwivedi, N., Singh, S., Kumar, P., Chouhan, M., Yadav, A. K., Mondal, D. P., Solanki, P. R., Srivastava, A. K., & Dhand, C. (2022). Development of copper impregnated bio-inspired hydrophobic antibacterial nanocoatings for textiles. *Colloids and Surfaces. B, Biointerfaces, 220*, 112913. doi:10.1016/j.colsurfb.2022.112913 PMID:36306694

Pradhan, L., Maiti, S., Mallick, A., Shahid, M., More, S. P., & Adivarekar, R. V. (2023). *3 Coating-and lamination-based smart textiles: techniques, features, and challenges. In Smart and Functional Textiles*. De Gruyter. doi:10.1515/9783110759747-003

Pradhan, R., & Grewal, H. S. (2023). Flexible and durable fluorine-free superhydrophobic films through a sustainable approach. *Chemical Engineering Journal, 474*, 145812. doi:10.1016/j.cej.2023.145812

Prakash, J., Sun, S., Swart, H. C., & Gupta, R. K. (2018). Noble metals-TiO2 nanocomposites: From fundamental mechanisms to photocatalysis, surface enhanced Raman scattering and antibacterial applications. *Applied Materials Today*, *11*, 82–135. doi:10.1016/j.apmt.2018.02.002

Prasad, S., Kumar, V., Kirubanandam, S., & Barhoum, A. (2018). Engineered nanomaterials: nanofabrication and surface functionalization. In A. Barhoum & A. S. H. Makhlouf (Eds.), *Emerging Applications of Nanoparticles and Architecture Nanostructures* (pp. 305–340). Elsevier. doi:10.1016/B978-0-323-51254-1.00011-7

Prathna, T. C. (2021). Nanotechnology in Imaging Applications: An Overview. *Nanotechnology in Medicine*, 75-86.

Pratim Das, P., Aditya Bachchan, A., Sahu, R., & Chaudhary, V. (2021). Whole body vibration: Effects on human body and role of biomaterials in repairing fracture joints and tissues. *Materials Today: Proceedings*, *43*, 141–147. doi:10.1016/j.matpr.2020.11.250

Primožič, M., Knez, Ž., & Leitgeb, M. (2021). (Bio) Nanotechnology in food science—Food packaging. *Nanomaterials (Basel, Switzerland)*, *11*(2), 292. doi:10.3390/nano11020292 PMID:33499415

Priyadarsini, G. B., & Sharma, A. K. (2016). Design of multi-layer anti-reflection coating for terrestrial solar panel glass. *Bulletin of Materials Science*, 39.

Puiu, R. A., Bîrcă, A. C., Grumezescu, V., Duta, L., Oprea, O. C., Holban, A. M., . . . Andronescu, E. (2023). Multifunctional polymeric biodegradable and biocompatible coatings based on silver nanoparticles: A comparative in vitro study on their cytotoxicity toward cancer and normal cell lines of cytostatic drugs versus essential-oil-loaded nanoparticles and on their antimicrobial and antibiofilm activities. *Pharmaceutics, 15*(7), 1882. https://doi.org/15071882 doi:10.3390/pharmaceutics

Qian, Y., Zhou, Y., Li, L., Liu, W., Yang, D., & Qiu, X. (2020). Facile preparation of active lignin capsules for developing self-healing and UV-blocking polyurea coatings. *Progress in Organic Coatings*, *138*, 105354. doi:10.1016/j.porgcoat.2019.105354

Qi, Z., Xue, X., Xu, X., Zhou, H., Li, W., Yang, G., & Xie, P. (2022). Detoxified and antimicrobial-enhanced olive mill wastewater phenols capping ZnO nanoparticles incorporated with carboxymethyl cellulose for fresh strawberry preservation. *Postharvest Biology and Technology*, *188*, 111891. doi:10.1016/j.postharvbio.2022.111891

Qu, L. L., Geng, Z. Q., Wang, W., Yang, K. C., Wang, W. P., Han, C. Q., Yang, G. H., Vajtai, R., Li, D. W., & Ajayan, P. M. (2019). Recyclable three-dimensional Ag nanorod arrays decorated with O-g-C3N4 for highly sensitive SERS sensing of organic pollutants. *Journal of Hazardous Materials*, *379*(January), 120823. doi:10.1016/j.jhazmat.2019.120823 PMID:31276918

Rad, F. H., Sharifan, A., Khodaiyan, F., & Shahabi-Ghahfarrokhi, I. (2016). Preparation and characterization of pullulan-soy protein concentrate blended film incorporated with *Zataria multiflora* and *Artemisia biennis* essential oils. *Jundishapur Journal of Natural Pharmaceutical Products*. Advance online publication. doi:10.17795/jjnpp37802

Radulescu, M., Andronescu, E., Dolete, G., Popescu, R. C., Fufă, O., Chifiriuc, M. C., Mogoantă, L., Bălşeanu, T. A., Mogoşanu, G. D., Grumezescu, A. M., & Holban, A. M. (2016). Silver nanocoating for reducing the exogenous microbial colonization of wound dressings. *Materials (Basel)*, *9*(5), 1–15. doi:10.3390/ma9050345 PMID:28773468

Raduly, F. M., Rădițoiu, V., Rădițoiu, A., Frone, A. N., Nicolae, C. A., Răut, I., Constantin, M., & Grapin, M. (2023). Multifunctional finishing of cotton fabric with curcumin derivatives coatings obtained by sol-gel method. *Gels (Basel, Switzerland)*, *9*(5), 369. doi:10.3390/gels9050369 PMID:37232961

Radu, R.-D., & Drăgănescu, D. (2023). Present and Future of ZrO2 Nanostructure as Reservoir for Drug Loading and Release. *Coatings*, *13*(7), 1273. doi:10.3390/coatings13071273

Raelyn, L. Y. L. (2020). *Synthesis and applications of hydrophobic nanostructured coatings for green buildings*. Academic Press.

Rai, M., Bonde, S., Yadav, A., Bhowmik, A., Rathod, S., Ingle, P., & Gade, A. (2021). Nanotechnology as a Shield against COVID-19: Current Advancement and Limitations. *Viruses*, *13*(7), 1224. doi:10.3390/v13071224 PMID:34202815

Rai, M., Ingle, A. P., Gupta, I., Pandit, R., Paralikar, P., Gade, A., Chaud, M. V., & dos Santos, C. A. (2019). Smart nanopackaging for the enhancement of food shelf life. *Environmental Chemistry Letters*, *17*(1), 277–290. doi:10.1007/s10311-018-0794-8

Raja Beryl, J., & Xavier, J. R. (2021). Influence of silane functionalized nanoclay on the barrier, mechanical and hydrophobic properties by clay nanocomposite films in an aggressive chloride medium. *Colloids and Surfaces. A, Physicochemical and Engineering Aspects*, *630*, 127625. doi:10.1016/j.colsurfa.2021.127625

Rajamohan, P. D. (2023). *A Review on Nanoparticles Based and Non-Nanoparticles Based Hydrophobic Nanocoatings and Its Applications*. Academic Press.

Rajendran, N. K., Kumar, S. S. D., Houreld, N. N., & Abrahamse, H. (2018). A review on nanoparticle based treatment for wound healing. *Journal of Drug Delivery Science and Technology*, *44*, 421–430. doi:10.1016/j.jddst.2018.01.009

Rajput, D. S., & Sudhakar, K. (2013). Effect of dust on the performance of solar PV panel. *International Journal of Chemtech Research*, *5*, 1083–1086.

Ramaratnam, K., Iyer, S. K., Kinnan, M. K., Chumanov, G., Brown, P. J., & Luzinov, I. (2008). Ultrahydrophobic Textiles Using Nanoparticles: Lotus Approach. *Journal of Engineered Fibers and Fabrics*, *3*(4). doi:10.1177/155892500800300402

Ramasubramanian, K., Arunachalam, N., & Ramachandra Rao, M. S. (2019). Wear performance of nano-engineered boron doped graded layer CVD diamond coated cutting tool for machining of Al-SiC MMC. *Wear*, *426-427*, 1536–1547. doi:10.1016/j.wear.2018.12.004

Ramezani, M., Mohd Ripin, Z., Pasang, T., & Jiang, C.-P. (2023). Surface Engineering of Metals: Techniques, Characterizations and Applications. *Metals*, *13*(7), 1299. doi:10.3390/met13071299

Rananavare, A. P., & Lee, J. (2022). Hydrophobic cotton fabric synthesized via dispersion polymerization from poly(glycidyl methacrylate) nanoparticles for self-cleaning applications. *Progress in Organic Coatings, 170.*

Rani, D., Singh, R., Kush, P., & Kumar, P. (2023). Nanomaterial-based smart coatings for antibacterial, antifungal, and antiviral activities. *Advances in Nanotechnology for Marine Antifouling*, 271–302. doi:10.1016/B978-0-323-91762-9.00008-3

Rastogi, A., Singh, A., Naik, K., Mishra, A., Chaudhary, S., Manohar, R., & Singh Parmar, A. (2022). A systemic review on liquid crystals, nanoformulations and its application for detection and treatment of SARS – CoV- 2 (COVID – 19). *Journal of Molecular Liquids*, *362*, 119795. doi:10.1016/j.molliq.2022.119795 PMID:35832289

Raveh, A., Zukerman, I., Shneck, R., Avni, R., & Fried, I. (2007). Thermal stability of nanostructured superhard coatings: A review. *Surface and Coatings Technology*, *201*(13), 6136–6142. doi:10.1016/j.surfcoat.2006.08.131

Ray, S., Raychaudhuri, U., & Chakraborty, R. (2016). An overview of encapsulation of active compounds used in food products by drying technology. *Food Bioscience*, *13*, 76–83. doi:10.1016/j.fbio.2015.12.009

Raza, Z. A., Abid, S., Azam, A., & Rehman, A. (2020). Synthesis of alpha-tocopherol encapsulated chitosan nanoassemblies and their impregnation on cellulosic fabric for potential antibacterial and antioxidant cosmetotextiles. *Cellulose (London, England)*, *27*(3), 1717–1731. doi:10.1007/s10570-019-02862-7

Reggente, M., Rossi, M., Angeloni, L., Tamburri, E., Lucci, M., Davoli, I., Terranova, M. L., & Passeri, D. (2015). Atomic Force Microscopy Techniques for Nanomechanical Characterization: A Polymeric Case Study. *Journal of the Minerals Metals & Materials Society*, *67*(4), 849–857. doi:10.1007/s11837-015-1340-9

Regulation (EC) No. 1935/2004 of The European Parliament and of The Council of 27 October 2004 on Materials and Articles Intended to Come into Contact with Food and Repealing Directives 80/590/EEC and 89/109/EEC. Available online: https://www.interpack.com/cgi-bin/md_interpack/lib/pub/tt.cgi/Food_ Safety.html?oid=55076&lang=2&ticket=g_u_e_s_t

Reuna, J., Kanter, R., Kandelin, N., Kiuru, K., Koivuluoto, H., & Guina, M. (2023). Ice resistance of hydrophobic fluoropolymerized nanostructured alumina films for antireflective coatings. *Nano Express*, *3*(4), 044002. doi:10.1088/2632-959X/acad19

Rhim, J. W., Park, H. M., & Ha, C. S. (2013). Bio-nanocomposites for food packaging applications. *Progress in Polymer Science*, *38*(10-11), 1629–1652. doi:10.1016/j.progpolymsci.2013.05.008

Riaz, A., Lagnika, C., Luo, H., Nie, M., Dai, Z., Liu, C., Abdin, M., Hashim, M. M., Li, D., & Song, J. (2020). Effect of Chinese chives (*Allium tuberosum*) addition to carboxymethyl cellulose based food packaging films. *Carbohydrate Polymers*, *235*, 115944. doi:10.1016/j.carbpol.2020.115944 PMID:32122483

Ristić, N., Marković-Nikolić, D., Zdravković, A., Mičić, I., & Ristić, I. (2022). Biofunctional textile materials: Cosmetic textiles. *Advanced Technologies*, *11*(1), 63–75. doi:10.5937/savteh2201063R

Ros-Lis, J. V., & Benitez Serra, M. (2023). Recent advancements in smart and intelligent packaging technologies: an overview. In B. P. Singh, S. Agnihotri, G. Singh, & V. K. Gupta (Eds.), *Postharvest Management of Fresh Produce* (pp. 241–263). Academic Press. doi:10.1016/B978-0-323-91132-0.00010-1

Rossi, G., Altavilla, C., Scarfato, P., Ciambelli, P., & Incarnato, L. (2015). Durability study of transparent and flexible nanolayer barrier for photovoltaic devices. *Polymer Degradation & Stability*, *112*, 160–166. doi:10.1016/j.polymdegradstab.2014.12.022

Rovera, C., Türe, H., Hedenqvist, M. S., & Farris, S. (2020). Water vapor barrier properties of wheat gluten/silica hybrid coatings on paperboard for food packaging applications. *Food Packaging and Shelf Life*, *26*, 100561. doi:10.1016/j.fpsl.2020.100561

Rühle, S. (2016). Tabulated Values of the Shockley-Queisser Limit for Single Junction Solar Cells. *Solar Energy*, *130*, 139–147. doi:10.1016/j.solener.2016.02.015

Ruzafa-Silvestre, C., Juan-Fernández, B., Carbonell-Blasco, M. P., Bañón-Gil, E., Orgilés-Calpena, E., & Arán-Ais, F. (2022). Organosilicon-Based Plasma Nanocoating on Crust Leather for Water-Repellent Footwear. *Materials (Basel)*, *15*(20), 7255. doi:10.3390/ma15207255 PMID:36295320

Sadabadi, H., Allahkaram, S. R., Kordijazi, A., & Rohatgi, P. K. (2022). Self-healing Coatings Loaded by Nano/microcapsules: A Review. *Protection of Metals and Physical Chemistry of Surfaces*, *58*(2), 287–307. doi:10.1134/S2070205122020162

Saemian, T., Gharagozlou, M., Hossaini Sadr, M., & Naghibi, S. (2019). Synthesis of CoFe2O4@Cu3(BTC)2 nanocomposite as a magnetic metal–organic framework. *Polyhedron*, *174*, 114163. doi:10.1016/j.poly.2019.114163

Safavi, M. S., Walsh, F. C., Visai, L., & Khalil-Allafi, J. (2022). Progress in Niobium Oxide-Containing Coatings for Biomedical Applications: A Critical Review. *ACS Omega*, *7*(11), 9088–9107. doi:10.1021/acsomega.2c00440 PMID:35356687

Saha, D. R., Kumar, K. R., Ansari, F., Ghosh, S. K., Datta, A., & Chakravorty, D. (2021). Stimuli-responsive coating by simple physical blending route. *Ceramics International*, *47*(18), 26357–26365. doi:10.1016/j.ceramint.2021.06.046

Saharudin, M. S., Ilyas, R. A., Awang, N., Hasbi, S., Shyha, I., & Inam, F. (2023). Advances in Sustainable Nanocomposites. *Sustainability (Basel)*, *15*(6), 5125. doi:10.3390/su15065125

Sahoo, J., Sarkhel, S., Mukherjee, N., & Jaiswal, A. (2022). Nanomaterial-Based Antimicrobial Coating for Biomedical Implants: New Age Solution for Biofilm-Associated Infections. *ACS Omega*, *7*(50), 45962–45980. doi:10.1021/acsomega.2c06211 PMID:36570317

Sahoo, M., Vishwakarma, S., Panigrahi, C., & Kumar, J. (2021). Nanotechnology: Current applications and future scope in food. *Food Frontiers*, *2*(1), 3–22. doi:10.1002/fft2.58

Saini, P., Kamalesu, L., & Manikanika. (2023). Review on nanotechnology "Impact on the food services industry." *Materials Today: Proceedings*. https://doi.org/https://doi.org/10.1016/j.matpr.2023.04.377

Saini, J., & Bhatt, R. (2020). Global warming-causes, impacts and mitigation strategies in agriculture. *Current Journal of Applied Science and Technology*, *39*, 93–107. doi:10.9734/cjast/2020/v39i730580

Saini, R. D. (2017). Biodegradable polymers. *Int. J. Appl. Chem*, *13*(2), 179–196. doi:10.5958/0974-4150.2018.00037.8

Salehi, H., Eshaghi, A., Rezazadeh, M., & Zabolian, H. (2022). Antireflective and anti-dust modified silica-based thin film on solar cell cover glass. *Journal of Alloys and Compounds*, *892*, 162228. doi:10.1016/j.jallcom.2021.162228

Saleh, T. A., Shetti, N. P., Shanbhag, M. M., Reddy, K. R., & Aminabhavi, T. M. (2020). Recent Trends in Functionalized Nanoparticles Loaded Polymeric Composites: An Energy Application. *Materials Science for Energy Technologies*, *3*, 515–525. doi:10.1016/j.mset.2020.05.005

Samrot, A. V., Sean, T. C., Kudaiyappan, T., Bisyarah, U., Mirarmandi, A., Faradjeva, E., ... Kumar, S. S. (2020). Production, characterization and application of nanocarriers made of polysaccharides, proteins, bio-polyesters and other biopolymers: A review. *International Journal of Biological Macromolecules*, *165*, 3088–3105. doi:10.1016/j.ijbiomac.2020.10.104 PMID:33098896

Samui, A. B., Hande, P. E., & Mondal, M. I. H. (2022). 25 - Care, maintenance and disposability of medical and protective textile products. In M. I. H. Mondal (Ed.), *Protective Textiles from Natural Resources* (pp. 793–837). Woodhead Publishing. doi:10.1016/B978-0-323-90477-3.00026-2

Sá, N. M., Mattos, A. L., Silva, L. M., Brito, E. S., Rosa, M. F., & Azeredo, H. M. (2020). From cashew byproducts to biodegradable active materials: Bacterial cellulose-lignin-cellulose nanocrystal nanocomposite films. *International Journal of Biological Macromolecules*, *161*, 1337–1345. doi:10.1016/j.ijbiomac.2020.07.269 PMID:32777430

Sani, I. K., Pirsa, S., & Tağı, Ş. (2019). Preparation of chitosan/zinc oxide/*Melissa officinalis* essential oil nano-composite film and evaluation of physical, mechanical and antimicrobial properties by response surface method. *Polymer Testing*, *79*, 106004. doi:10.1016/j.polymertesting.2019.106004

Sankar, S., Nair, B. N., Suzuki, T., Anilkumar, G. M., Padmanabhan, M., Hareesh, U. N., & Warrier, K. G. (2016). Hydrophobic and Metallophobic Surfaces: Highly Stable Non-wetting Inorganic Surfaces Based on Lanthanum Phosphate Nanorods. *Scientific Reports*, *6*(1), 22732. doi:10.1038/srep22732 PMID:26955962

Santhini, O. E., & Kirupha, S. D. (2020). Nanomaterials and ethical issues. *Applications of nanomaterials in human health*, 275-286.

Santos, I. M., Alexandre, M., Mihailetchi, V. D., Silva, J. A., Mateus, T., Mouquinho, A., Boane, J., Vicente, A. T., Nunes, D., Menda, U. D., Águas, H., Fortunato, E., Martins, R., & Mendes, M. J. (2023). Optically-Boosted Planar IBC Solar Cells with Electrically-Harmless Photonic Nanocoatings. *Advanced Optical Materials*, *11*(15), 2300276. doi:10.1002/adom.202300276

Sapper, M., Talens, P., & Chiralt, A. (2019). Improving functional properties of cassava starch-based films by incorporating xanthan, gellan, or pullulan gums. *International Journal of Polymer Science*. doi:10.1155/2019/5367164

Sarkin, A. S., Ekren, N., & Sağlam, S. (2020). A Review of Anti-Reflection and Self-Cleaning Coatings on Photovoltaic Panels. *Solar Energy*, *199*, 63–73. doi:10.1016/j.solener.2020.01.084

Sarode, A., Annapragada, A., Guo, J., & Mitragotri, S. (2020). Layered self-assemblies for controlled drug delivery: A translational overview. *Biomaterials*, *242*, 119929. doi:10.1016/j.biomaterials.2020.119929 PMID:32163750

Sartori, T., Feltre, G., Sobral, P. J. D. A., Cunha, R. L. D., & Menegalli, F. C. (2018). Properties of films produced from blends of pectin and gluten. *Food Packaging and Shelf Life*, *18*, 221–229. doi:10.1016/j.fpsl.2018.11.007

Sarwar, M. S., Niazi, M. B. K., Jahan, Z., Ahmad, T., & Hussain, A. (2018). Preparation and characterization of PVA/nanocellulose/Ag nanocomposite films for antimicrobial food packaging. *Carbohydrate Polymers*, *184*, 453–464. doi:10.1016/j.carbpol.2017.12.068 PMID:29352941

Sasidharan, A., & Monteiro-Riviere, N. A. (2015). Biomedical applications of gold nanomaterials: Opportunities and challenges. *Wiley Interdisciplinary Reviews. Nanomedicine and Nanobiotechnology*, *7*(6), 779–796. doi:10.1002/wnan.1341 PMID:25808787

Sathe, P., Laxman, K., Myint, M. T. Z., Dobretsov, S., Richter, J., & Dutta, J. (2017). Bioinspired nanocoatings for biofouling prevention by photocatalytic redox reactions. *Scientific Reports*, *7*(1), 3624. doi:10.1038/s41598-017-03636-6 PMID:28620218

Savaliya, R., Shah, D., Singh, R., Kumar, A., Shanker, R., Dhawan, A., & Singh, S. (2015). Nanotechnology in disease diagnostic techniques. *Current Drug Metabolism*, *16*(8), 645–661. doi:10.2174/138920021666150625121546 PMID:26560320

Saxomiddino'g'li, Z. S., & Marg'uba, Y. (2023). Anticorrosive Materials: Safeguarding the Future of Infrastructure. *Journal of Universal Science Research*, *1*(7), 71–76.

Schaefer, D., & Cheung, W. M. (2018). Smart packaging: Opportunities and challenges. *Procedia CIRP*, *72*, 1022–1027. doi:10.1016/j.procir.2018.03.240

Schiessl, S., Kucukpinar, E., Cros, S., Miesbauer, O., Langowski, H. C., & Eisner, P. (2022). Nanocomposite coatings based on polyvinyl alcohol and montmorillonite for high-barrier food packaging. *Frontiers in Nutrition*, *9*, 790157. doi:10.3389/fnut.2022.790157 PMID:35340548

Schrand, A. M., Rahman, M. F., Hussain, S. M., Schlager, J. J., Smith, D. A., & Syed, A. F. (2010). Metal-based nanoparticles and their toxicity assessment. *Wiley Interdisciplinary Reviews. Nanomedicine and Nanobiotechnology*, *2*(5), 544–568. doi:10.1002/wnan.103 PMID:20681021

Selim, M. S., Fatthallah, N. A., Shenashen, M. A., Higazy, S. A., Madian, H. R., Selim, M. M., & El-Safty, S. A. (2023). Bioinspired Graphene Oxide–Magnetite Nanocomposite Coatings as Protective Superhydrophobic Antifouling Surfaces. *Langmuir*, *39*(6), 2333–2346. doi:10.1021/acs.langmuir.2c03061 PMID:36719844

Selim, M. S., Shenashen, M. A., Elmarakbi, A., Fatthallah, N. A., Hasegawa, S., & El-Safty, S. A. (2017). Synthesis of ultrahydrophobic and thermally stable inorganic–organic nanocomposites for self-cleaning foul release coatings. *Chemical Engineering Journal*, *320*, 653–666. doi:10.1016/j.cej.2017.03.067

Selim, M. S., Shenashen, M. A., El-Sockary, M. A., Fatthallah, N. A., Higazy, S. A., El-Saeed, A. M., El-Azabawy, O. E., & El-Safty, S. A. (2023). Polymer/graphene-derived nanocomposites as advanced marine antifouling coatings. In *Micro and Nano Technologies: Advances in Nanotechnology for Marine Antifouling*. Elsevier. doi:10.1016/B978-0-323-91762-9.00007-1

Shaban, M., & Hasanzadeh, M. (2020). Biomedical applications of dendritic fibrous nanosilica (DFNS): Recent progress and challenges. *RSC Advances*, *10*(61), 37116–37133. doi:10.1039/D0RA04388E PMID:35521236

Shahini, M. H., Ramezanzadeh, B., & Mohammadloo, H. E. (2021). Recent advances in biopolymers/carbohydrate polymers as effective corrosion inhibitive macro-molecules: A review study from experimental and theoretical views. *Journal of Molecular Liquids*, *325*, 115110. doi:10.1016/j.molliq.2020.115110

Shah, M. A., Pirzada, B. M., Price, G., Shibiru, A. L., & Qurashi, A. (2022). Applications of nanotechnology in smart textile industry: A critical review. *Journal of Advanced Research*, *38*, 55–75. doi:10.1016/j.jare.2022.01.008 PMID:35572402

Shaik, J., & Kumar, A. C. (2021). Nanostructures for biomedical devices. In *Nanoscale Processing* (pp. 299–326). Elsevier. doi:10.1016/B978-0-12-820569-3.00011-6

Shaju, A., & Chacko, R. (2018). Soiling of Photovoltaic Modules – Review. *IOP Conference Series. Materials Science and Engineering*, *396*, 012050. doi:10.1088/1757-899X/396/1/012050

Shakil, U. A., Abu Hassan, S., Yahya, M. Y., Mujiyono, & Nurhadiyanto, D. (2021). A review of properties and fabrication techniques of fiber reinforced polymer nanocomposites subjected to simulated accidental ballistic impact. *Thin-walled Structures*, *158*, 107150. doi:10.1016/j.tws.2020.107150

Shanmugam, R., Mayakrishnan, V., Kesavan, R., Shanmugam, K., Veeramani, S., & Ilangovan, R. (2022). Mechanical, barrier, adhesion and antibacterial properties of pullulan/graphene bio nanocomposite coating on spray coated nanocellulose film for food packaging applications. *Journal of Polymers and the Environment*, *30*(5), 1–9. doi:10.1007/s10924-021-02311-2

Shao, J., Sheng, W., Wang, C., & Ye, Y. (2021). Solvent-free fabrication of tough self-crosslinkable short-fluorinated copolymer nanocoatings for ultradurable superhydrophobic fabrics. *Chemical Engineering Journal*, *416*, 128043. doi:10.1016/j.cej.2020.128043

Sharifi, S., Behzadi, S., Laurent, S., Forrest, M. L., Stroeve, P., & Mahmoudi, M. (2012). Toxicity of nanomaterials. *Chemical Society Reviews*, *41*(6), 2323–2343. doi:10.1039/C1CS15188F PMID:22170510

Sharma & Thakur. (2023). Synthesis, Electrochemical, Morphological, Computational and Corrosion Inhibition Studies of 3-(5-Naphthalen-2-yl-[1,3,4]oxadiazol-2-yl)-pyridine against Mild Steel in 1 M HCl. *Asian Journal of Chemistry, 35*(5), 1079–1088. https://doi.org/https://doi.org/10.14233/ajchem.2023.27711

Sharma, C., Dhiman, R., Rokana, N., & Panwar, H. (2017). Nanotechnology: An Untapped Resource for Food Packaging. *Frontiers in Microbiology*, *8*, 8. doi:10.3389/fmicb.2017.01735 PMID:28955314

Sharma, D., Thakur, A., Kumar, M., Kumar, A., & Jakhar, K. (2023). A convenient synthesis, electrochemical profiling, and morphological studies of a pyridine-based 1, 3, 4-oxadiazole hybrid : A promising study for corrosion mitigation of mild steel in strongly acidic environment. *Inorganic Chemistry Communications*, *158*(P1), 111554. doi:10.1016/j.inoche.2023.111554

Sharma, P., Pant, S., Dave, V., Tak, K., Sadhu, V., & Reddy, K. R. (2019). Green synthesis and characterization of copper nanoparticles by Tinospora cardifolia to produce nature-friendly copper nano-coated fabric and their antimicrobial evaluation. *Journal of Microbiological Methods*, *160*, 107–116. doi:10.1016/j.mimet.2019.03.007 PMID:30871999

Sharma, R., Garg, R., & Kumari, A. (2020). A review on biogenic synthesis, applications and toxicity aspects of zinc oxide nanoparticles. *EXCLI Journal*, *19*, 1325–1340. doi:10.17179/excli2020-2842 PMID:33192216

Sharma, S., Kumar, A., Bashir, S., Ramesh, K., & Ramesh, S. (2022). A comprehensive review: Super hydrophobic graphene nanocomposite coatings for underwater and wet applications to enhance corrosion resistance. *FlatChem*, *31*, 100326. doi:10.1016/j.flatc.2021.100326

Shen, G. X., Du, R. G., Chen, Y. C., Lin, C. J., & Scantlebury, D. (2005). Study on hydrophobic nano-titanium dioxide coatings for improvement in corrosion resistance of type 316L stainless steel. *Corrosion*, *61*(10), 943–950. doi:10.5006/1.3280894

Sheng, Z., Liu, Z., Hou, Y., Jiang, H., Li, Y., Li, G., & Zhang, X. (2023). The rising aerogel fibers: Status, challenges, and opportunities. *Advancement of Science*, *10*(9), 2205762. doi:10.1002/advs.202205762 PMID:36658735

Sheta, S. M., & El-Sheikh, S. M. (2022). Nanomaterials and metal-organic frameworks for biosensing applications of mutations of the emerging viruses. *Analytical Biochemistry*, *648*, 114680. doi:10.1016/j.ab.2022.114680 PMID:35429447

Shi, A. M., Wang, L. J., Li, D., & Adhikari, B. (2013). Characterization of starch films containing starch nanoparticles: Part 1: Physical and mechanical properties. *Carbohydrate Polymers*, *96*(2), 593–601. doi:10.1016/j.carbpol.2012.12.042 PMID:23768605

Shie, M.-Y., Shen, Y.-F., Astuti, S. D., Lee, A. K.-X., Lin, S.-H., Dwijaksara, N. L. B., & Chen, Y.-W. (2019). Review of polymeric materials in 4D printing biomedical applications. *Polymers*, *11*(11), 1864. doi:10.3390/polym11111864 PMID:31726652

Shirvanimoghaddam, K., Akbari, M. K., Yadav, R., Al-Tamimi, A. K., & Naebe, M. (2021). Fight against COVID-19: The case of antiviral surfaces. *APL Materials*, *9*(3), 031112. doi:10.1063/5.0043009 PMID:33842101

Shi, X.-H., Liu, Q.-Y., Li, X.-L., Du, A.-K., Niu, J.-W., Li, Y.-M., Li, Z., Wang, M., & Wang, D.-Y. (2022). Construction phosphorus/nitrogen-containing flame-retardant and hydrophobic coating toward cotton fabric via layer-by-layer assembly. *Polymer Degradation & Stability*, *197*, 109839. doi:10.1016/j.polymdegradstab.2022.109839

Shiya, L., Xiaoke, J., Wei, T., Yanqing, L., Chengyan, Z., & Hongxia, Z. (2023). Research status and development trend of vitamin E loaded cosmeto-textiles. *Advanced Textile Technology*, *31*(1), 293. doi:10.19398/j.att.202206002

Shockley, W., & Queisser, H. J. (1961). Detailed Balance Limit of Efficiency of p-n Junction Solar Cells. *Journal of Applied Physics*, *32*(3), 510–519. doi:10.1063/1.1736034

Shojaee-Aliabadi, S., Hosseini, H., Mohammadifar, M. A., Mohammadi, A., Ghasemlou, M., Ojagh, S. M., Hosseini, S. M., & Khaksar, R. (2013). Characterization of antioxidant-antimicrobial κ-carrageenan films containing *Satureja hortensis* essential oil. *International Journal of Biological Macromolecules*, *52*, 116–124. doi:10.1016/j.ijbiomac.2012.08.026 PMID:22959956

Silva, M. R. F., Alves, M. F. R. P., Cunha, J. P. G. Q., Costa, J. L., Silva, C. A., Fernandes, M. H. V., Vilarinho, P. M., & Ferreira, P. (2023). Nanostructured transparent solutions for UV-shielding: Recent developments and future challenges. *Materials Today Physics*, *35*(February), 101131. Advance online publication. doi:10.1016/j.mtphys.2023.101131

Simos, Y. V., Spyrou, K., Patila, M., Karouta, N., Stamatis, H., Gournis, D., Dounousi, E., & Peschos, D. (2021). Trends of nanotechnology in type 2 diabetes mellitus treatment. *Asian Journal of Pharmaceutical Sciences*, *16*(1), 62–76. doi:10.1016/j.ajps.2020.05.001 PMID:33613730

Simpson, J. T., Hunter, S. R., & Aytug, T. (2015). Superhydrophobic materials and coatings: A review. *Reports on Progress in Physics*, *78*(8), 1–14. doi:10.1088/0034-4885/78/8/086501 PMID:26181655

Singh, A., Mittal, S., Mudgal, D., & Gupta, P. (2018). Design, development and application of nanocoating. *Nanomaterials and Their Applications*, 191-207.

Singh, H., & Kaur, K. (2023). Role of nanotechnology in research fields: Medical sciences, military & tribology- A review on recent advancements, grand challenges and perspectives. *Materials Today: Proceedings*. https://doi.org/https://doi.org/10.1016/j.matpr.2023.02.061

Singh, N., Gaur, S., Chawla, S., Singh, S., & Husen, A. (2023). Use of smart nanomaterials in food packaging. *Advances in Smart Nanomaterials and Their Applications*, 233–245. doi:10.1016/B978-0-323-99546-7.00008-2

Singh, A., Dogra, T. S., Mandal, U. K., & Narang, R. K. (2019). Novel approaches for ocular drug delivery: A review. Int. J. Bio-. *Pharmaceutical Research*, *8*, 2722–2732.

Singh, A., Mittal, S., Mudgal, D., & Gupta, P. (2018). Design, Development and Application of Nanocoatings. In Z. H. Khan (Ed.), *Nanomaterials and Their Applications* (pp. 191–207). Springer Singapore. doi:10.1007/978-981-10-6214-8_7

Singh, P., Ali, S. W., & Kale, R. D. (2023). Antimicrobial nanomaterials as advanced coatings for self-sanitizing of textile clothing and personal protective equipment. *ACS Omega*, *8*(9), 8159–8171. doi:10.1021/acsomega.2c06343 PMID:36910928

Singh, R., Dutt, S., Sharma, P., Sundramoorthy, A. K., Dubey, A., Singh, A., & Arya, S. (2023). Future of Nanotechnology in Food Industry: Challenges in Processing, Packaging, and Food Safety. *Global Challenges (Hoboken, NJ)*, *7*(4), 2200209. doi:10.1002/gch2.202200209 PMID:37020624

Singh, Y., Kumar, U., Panigrahi, S., Balyan, P., Mehla, S., Sihag, P., Sagwal, V., Singh, K. P., White, J. C., & Dhankher, O. P. (2023). Nanoparticles as novel elicitors in plant tissue culture applications: Current status and future outlook. *Plant Physiology and Biochemistry*, *203*, 108004. doi:10.1016/j.plaphy.2023.108004 PMID:37714027

Singla, R., Abidi, S. M., Dar, A. I., & Acharya, A. (2019). Nanomaterials as potential and versatile platform for next generation tissue engineering applications. *Journal of Biomedical Materials Research. Part B, Applied Biomaterials*, *107*(7), 2433–2449. doi:10.1002/jbm.b.34327 PMID:30690870

Siró, I., & Plackett, D. (2010). Microfibrillated cellulose and new nanocomposite materials: A review. *Cellulose (London, England)*, *17*(3), 459–494. doi:10.1007/s10570-010-9405-y

Si, Y., & Guo, Z. (2015). Superhydrophobic Nanocoatings: From Materials to Fabrications and to Applications. *Nanoscale*, *7*(14), 5922–5946. doi:10.1039/C4NR07554D

Si, Y., Zhu, H., Chen, L., Jiang, T., & Guo, Z. (2015). A multifunctional transparent superhydrophobic gel nanocoating with self-healing properties. *Chemical Communications*, *51*(94), 16794–16797. doi:10.1039/C5CC06977G PMID:26434728

Smirnova, V. V., Krasnoiarova, O. V., Pridvorova, S. M., Zherdev, A. V., Gmoshinskiĭ, I. V., Kazydub, G. V., Popov, K. I., & Khotimchenko, S. A. (2012). Characterization of silver nanoparticles migration from package materials destined for contact with foods. *Voprosy Pitaniia*, *81*, 34–39. PMID:22774476

Smitha, S., Shajesh, P., Mukundan, P., & Warrier, K. G. K. (2007). Antiwetting silica–gelatin nanohybrid and transparent nano coatings synthesised through an aqueous sol–gel process. *Journal of Sol-Gel Science and Technology*, *42*(2), 157–163. doi:10.1007/s10971-007-1535-9

Solanki, R., Shankar, A., Modi, U., & Patel, S. (2023). New insights from nanotechnology in SARS-CoV-2 detection, treatment strategy and prevention. *Materials Today. Chemistry*, *29*, 101478. doi:10.1016/j.mtchem.2023.101478 PMID:36950312

Sothornvit, R. (2019). Nanostructured materials for food packaging systems: New functional properties. *Current Opinion in Food Science*, *25*, 82–87. doi:10.1016/j.cofs.2019.03.001

Souza, J. C. M., Sordi, M. B., Kanazawa, M., Ravindran, S., Henriques, B., Silva, F. S., Aparicio, C., & Cooper, L. F. (2019). Nano-scale modification of titanium implant surfaces to enhance osseointegration. *Acta Biomaterialia*, *94*, 112–131. doi:10.1016/j.actbio.2019.05.045 PMID:31128320

Srinivasan, S., Praveen, V. K., Philip, R., & Ajayaghosh, A. (2008). Bioinspired Superhydrophobic Coatings of Carbon Nanotubes and Linear π Systems Based on the "Bottom-up" Self-Assembly Approach. *Angewandte Chemie International Edition*, *47*(31), 5750–5754. doi:10.1002/anie.200802097 PMID:18604864

Srinivasa, P. C., Ramesh, M. N., & Tharanathan, R. N. (2007). Effect of plasticizers and fatty acids on mechanical and permeability characteristics of chitosan films. *Food Hydrocolloids*, *21*(7), 1113–1122. doi:10.1016/j.foodhyd.2006.08.005

Sriramulu, D., Reed, E. L., Annamalai, M., Venkatesan, T. V., & Valiyaveettil, S. (2016). Synthesis and Characterization of Superhydrophobic, Self-cleaning NIR-reflective Silica Nanoparticles. *Scientific Reports*, *6*(1), 35993. doi:10.1038/srep35993 PMID:27824064

Strudwick, A. J., Weber, N. E., Schwab, M. G., Kettner, M., Weitz, R. T., Wünsch, J. R., Müllen, K., & Sachdev, H. (2015). Chemical vapor deposition of high quality graphene films from carbon dioxide atmospheres. *ACS Nano*, *9*(1), 31–42. doi:10.1021/nn504822m PMID:25398132

Subbaiyan, R., Ganesan, A., & Varadharajan, V. (2023). Bioprospecting and Exploration of the Natural Antifouling Approaches against Marine Foulers. *Journal of Pure & Applied Microbiology*, *17*(3), 1374–1390. doi:10.22207/JPAM.17.3.02

Su, C., Huang, X., Zhang, L., Zhang, Y., Yu, Z., Chen, C., Ye, Y., & Guo, S. (2023). Robust superhydrophobic wearable piezoelectric nanogenerators for self-powered body motion sensors. *Nano Energy*, *107*, 108095. doi:10.1016/j.nanoen.2022.108095

Suellen Ferro de Oliveira, C., & Kekhasharú Tavaria, F. (2023). The impact of bioactive textiles on human skin microbiota. *European Journal of Pharmaceutics and Biopharmaceutics*, *188*, 66–77. doi:10.1016/j.ejpb.2023.05.004 PMID:37182552

Suhag, D., Chauhan, M., Shakeel, A., & Das, S. (2020). Emerging trends in nanotheranostics. *Nanobiomedicine*, 335–368.

Sumit, Jeon, & Lim. (2022). Superhydrophobic sponge decorated with hydrophobic MOF-5 nanocoating for efficient oil-water separation and antibacterial applications. *Sustainable Materials and Technologies, 33*, 468–477. doi:10.1016/j.cej.2017.03.052

Sundeep, D., Varadharaj, E. K., Umadevi, K., & Jhansi, R. (2023). Role of Nanomaterials in Screenprinted Electrochemical Biosensors for Detection of COVID-19 and Post-Covid Syndromes. *ECS Advances*, *2*(1), 016502. doi:10.1149/2754-2734/acb832

Sun, J., Gao, L., Wang, L., & Sun, X. (2021). Recent advances in single-cell analysis: Encapsulation materials, analysis methods and integrative platform for microfluidic technology. *Talanta, 234*, 122671. doi:10.1016/j.talanta.2021.122671 PMID:34364472

Sun, K., Lu, L., Jiang, Y., Wang, Y., Zhou, K., & He, Z. (2018). Integrated Effects of PM2.5 Deposition, Module Surface Conditions and Nanocoatings on Solar PV Surface Glass Transmittance. *Renewable & Sustainable Energy Reviews, 82*(Part 3), 4107–4120. doi:10.1016/j.rser.2017.10.062

Sun, X., Gu, S., Wang, L., Wang, H., Xiong, S., Yin, X., & Yang, S. (2024). Multifunctional liquid-like magnetic nanofluids mediated coating with anticorrosion and self-healing performance. *Journal of Colloid and Interface Science, 654*, 25–35. doi:10.1016/j.jcis.2023.09.182 PMID:37832232

Sun, Z., Yang, L., Zhang, D., & Song, W. (2019). High performance, flexible and renewable nano-biocomposite artificial muscle based on mesoporous cellulose/ ionic liquid electrolyte membrane. *Sensors and Actuators. B, Chemical, 283*, 579–589. doi:10.1016/j.snb.2018.12.073

Su, S., & Kang, M, P. (. (2020). Recent advances in nanocarrier-assisted therapeutics delivery systems. *Pharmaceutics, 12*(9), 837. doi:10.3390/pharmaceutics12090837 PMID:32882875

Swain, S. S., Paidesetty, S. K., Padhy, R. N., & Hussain, T. (2023). Nano-technology platforms to increase the antibacterial drug suitability of essential oils: A drug prospective assessment. *OpenNano, 9*, 100115. https://doi.org/https://doi.org/10.1016/j.onano.2022.100115

Swaroop, C., & Shukla, M. (2018). Nano-magnesium oxide reinforced polylactic acid biofilms for food packaging applications. *International Journal of Biological Macromolecules, 113*, 729–736. doi:10.1016/j.ijbiomac.2018.02.156 PMID:29499267

Syafiq, A., Pandey, A. K., Balakrishnan, V., Shahabuddin, S., & Abd Rahim, N. (2018). Organic-inorganic composite nanocoatings with superhydrophobicity and thermal stability. *Pigment & Resin Technology.*

Syafiq, A., Balakrishnan, V., Ali, M. S., Dhoble, S. J., Rahim, N. A., Omar, A., & Bakar, A. H. A. (2022). Application of Transparent Self-Cleaning Coating for Photovoltaic Panel: A Review. *Current Opinion in Chemical Engineering, 36*, 100801. doi:10.1016/j.coche.2022.100801

Syafiq, A., Pandey, A. K., Balakrishnan, V., Shahabuddin, S., & Abd Rahim, N. (2018). Organic-inorganic composite nanocoating with superhydrophobicity and thermal stability. *Pigment & Resin Technology.*

Szőke, Á. F., Szabó, G. S., Hórvölgyi, Z., Albert, E., Gaina, L., & Muresan, L. M. (2019). Eco-friendly indigo carmine-loaded chitosan coatings for improved anti-corrosion protection of zinc substrates. *Carbohydrate Polymers, 215*, 63–72. doi:10.1016/j.carbpol.2019.03.077 PMID:30981371

Taghvaei, E., Moosavi, A., Nouri-Borujerdi, A., Daeian, M. A., & Vafaeinejad, S. (2017). Superhydrophobic surfaces with a dual-layer micro- and nanoparticle coating for drag reduction. *Energy, 125*, 1–10. doi:10.1016/j.energy.2017.02.117

Tahir, R., Albargi, H. B., Ahmad, A., Qadir, M. B., Khaliq, Z., Nazir, A., ... Jalalah, M. (2023). Development of sustainable hydrophilic *Azadirachta indica* loaded PVA nanomembranes for cosmetic facemask applications. *Membranes*, *13*(2), 156. doi:10.3390/membranes13020156 PMID:36837659

Taieb, A. H., & Mnejja, H. (2023). Design of microencapsulated nigelle sativa oil impregnated in textile structure. *Journal of Engineering Research and Reports*, *25*(5), 81–87. doi:10.9734/jerr/2023/v25i5913

Talebi, F., Misaghi, A., Khanjari, A., Kamkar, A., Gandomi, H., & Rezaeigolestani, M. (2018). Incorporation of spice essential oils into poly-lactic acid film matrix with the aim of extending microbiological and sensorial shelf life of ground beef. *Lebensmittel-Wissenschaft + Technologie*, *96*, 482–490. doi:10.1016/j.lwt.2018.05.067

Taleuzzaman, M., Gupta, D. K., Kala, C., Gilani, S. J., & Beg, S. (2021). Nanotechnology-assisted medical devices in cancer treatment. In *Nanoformulation Strategies for Cancer Treatment* (pp. 195–205). Elsevier. doi:10.1016/B978-0-12-821095-6.00010-0

Tanaka, Y. (2023). Photoprotective Ability of Sunscreens against Ultraviolet, Visible Light and Near-Infrared Radiation. *Optics and Photonics Journal*, *13*(6), 140–146. doi:10.4236/opj.2023.136012

Tanasa, F., Teaca, C.-A., Nechifor, M., Ignat, M., Duceac, I. A., & Ignat, L. (2023). Highly specialized textiles with antimicrobial functionality—Advances and challenges. *Textiles*, *3*(2), 219–245. doi:10.3390/textiles3020015

Tang, K., Xue, J., Zhu, Y., & Wu, C. (2023). Design and synthesis of bioinspired nanomaterials for biomedical application. *WIREs Nanomedicine and Nanobiotechnology*.

Tang, Y., Liu, L., Han, J., Zhang, Z., Yang, S., Li, S., Fan, Z., & Zhao, H. (2021). Fabrication and characterization of multiple herbal extracts-loaded nanofibrous patches for topical treatment of acne vulgaris. *Fibers and Polymers*, *22*(2), 323–333. doi:10.1007/s12221-021-0156-1

Tansaoui, H., Bouazizi, N., Behary, N., Campagne, C., El-Achari, A., & Vieillard, J. (2023). Assessing alternative pretreatment methods to promote essential oil fixation into cotton and polyethylene terephthalate Fiber: A comparative study. *Polymers*, *15*(6), 1362. doi:10.3390/polym15061362 PMID:36987143

Taş, H., & Soykok, I. F. (2021). Puncture resistance of STF impregnated MWCNTs-grafted carbon fabrics. *Materials Today. Communications*, *28*, 102692. doi:10.1016/j.mtcomm.2021.102692

Tawfik, S. M., Azizov, S., Elmasry, M. R., Sharipov, M., & Lee, Y. I. (2020). Recent advances in nanomicelles delivery systems. *Nanomaterials (Basel, Switzerland)*, *11*(1), 70. doi:10.3390/nano11010070 PMID:33396938

Tayel, S. A., Abu El-Maaty, A. E., Mostafa, E. M., & Elsaadawi, Y. F. (2022). Enhance the performance of photovoltaic solar panels by a self-cleaning and hydrophobic nanocoating. *Scientific Reports*, *12*(1), 1–13. doi:10.1038/s41598-022-25667-4 PMID:36481954

Teisala, H., Tuominen, M., Aromaa, M., Mäkelä, J. M., Stepien, M., Saarinen, J. J., Toivakka, M., & Kuusipalo, J. (2010). Development of superhydrophobic coating on paperboard surface using the Liquid Flame Spray. *Surface and Coatings Technology*, *205*(2), 436–445. doi:10.1016/j.surfcoat.2010.07.003

Teow, Y., Asharani, P. V., Hande, M. P., & Valiyaveettil, S. (2011). Health impact and safety of engineered nanomaterials. *Chemical Communications*, *47*(25), 7025–7038. doi:10.1039/c0cc05271j PMID:21479319

Thakur, A., Kumar, A., Kaya, S., Vo, D. V. N., & Sharma, A. (2022). Suppressing inhibitory compounds by nanomaterials for highly efficient biofuel production: A review. *Fuel*, *312*(September), 122934. doi:10.1016/j.fuel.2021.122934

Thakur, A., Savaş, K., Kumar, A., Kaya, S., & Kumar, A. (2023). Recent Trends in the Characterization and Application Progress of Nano - Modified Coatings in Corrosion Mitigation of Metals and Alloys. *Applied Sciences*, *13*, 730. https://doi.org/https://doi.org/10.3390/app13020730

Thakur, N., Singh, S. B., & Anshuman. (2022). a Nanotechnology Approach on Self-Cleaning Material: a Review. *Indian Concrete Journal*, *96*(4), 40–48.

Thakur, A., Kaya, S., Abousalem, A. S., & Kumar, A. (2022). Experimental, DFT and MC simulation analysis of Vicia Sativa weed aerial extract as sustainable and eco-benign corrosion inhibitor for mild steel in acidic environment. *Sustainable Chemistry and Pharmacy*, *29*(July), 100785. doi:10.1016/j.scp.2022.100785

Thakur, A., Kaya, S., Abousalem, A. S., Sharma, S., Ganjoo, R., Assad, H., & Kumar, A. (2022). Computational and experimental studies on the corrosion inhibition performance of an aerial extract of Cnicus Benedictus weed on the acidic corrosion of mild steel. *Process Safety and Environmental Protection*, *161*, 801–818. doi:10.1016/j.psep.2022.03.082

Thakur, A., Kaya, S., & Kumar, A. (2021). Recent Innovations in Nano Container-Based Self-Healing Coatings in the Construction Industry. *Current Nanoscience*, *18*(2), 203–216. doi:10.2174/1573413717666210216120741

Thakur, A., Kaya, S., & Kumar, A. (2023). Recent Trends in the Characterization and Application Progress of Nano-Modified Coatings in Corrosion Mitigation of Metals and Alloys. *Applied Sciences (Basel, Switzerland)*, *13*(2), 730. doi:10.3390/app13020730

Thakur, A., & Kumar, A. (2021). Sustainable Inhibitors for Corrosion Mitigation in Aggressive Corrosive Media: A Comprehensive Study. *Journal of Bio- and Tribo-Corrosion*, *7*(2), 1–48. doi:10.1007/s40735-021-00501-y

Thakur, A., & Kumar, A. (2022). Recent advances on rapid detection and remediation of environmental pollutants utilizing nanomaterials-based (bio)sensors. *The Science of the Total Environment*, *834*(January), 155219. doi:10.1016/j.scitotenv.2022.155219 PMID:35421493

Thakur, A., & Kumar, A. (2022). Self-healing nanocoatings for automotive application. In H. Song, T. A. Nguyen, G. Yasin, N. B. Singh, & R. K. Gupta (Eds.), *Nanotechnology in the Automotive Industry* (pp. 403–427). Elsevier. doi:10.1016/B978-0-323-90524-4.00019-0

Thakur, A., & Kumar, A. (2023a). Ecotoxicity Analysis and Risk Assessment of Nanomaterials for the Environmental Remediation. *Macromolecular Symposia*, *410*(1), 1–23. doi:10.1002/masy.202100438

Thakur, A., & Kumar, A. (2023b). Recent trends in nanostructured carbon-based electrochemical sensors for the detection and remediation of persistent toxic substances in real- time analysis. *Materials Research Express*, *10*(3), 034001. https://doi.org/https://doi.org/10.1088/2053-1591/acbd1a. doi:10.1088/2053-1591/acbd1a

Thakur, A., Kumar, A., Kaya, S., Benhiba, F., Sharma, S., Ganjoo, R., & Assad, H. (2023). Electrochemical and computational investigations of the Thysanolaena latifolia leaves extract : An eco-benign solution for the corrosion mitigation of mild steel. *Results in Chemistry*, *6*(September), 101147. doi:10.1016/j.rechem.2023.101147

Thakur, A., Kumar, A., Kaya, S., Marzouki, R., Zhang, F., & Guo, L. (2022). Recent Advancements in Surface Modification, Characterization and Functionalization for Enhancing the Biocompatibility and Corrosion Resistance of Biomedical Implants. *Coatings*, *12*(10), 1459. doi:10.3390/coatings12101459

Thakur, A., Kumar, A., Sharma, S., Ganjoo, R., & Assad, H. (2022). Computational and experimental studies on the efficiency of Sonchus arvensis as green corrosion inhibitor for mild steel in 0. 5 M HCl solution. *Materials Today: Proceedings*, *66*, 609–621. doi:10.1016/j.matpr.2022.06.479

Thakur, A., Sharma, S., Ganjoo, R., Assad, H., & Kumar, A. (2022). Anti-Corrosive Potential of the Sustainable Corrosion Inhibitors Based on Biomass Waste: A Review on Preceding and Perspective Research. *Journal of Physics: Conference Series*, *2267*(1), 012079. doi:10.1088/1742-6596/2267/1/012079

Thomas, T. M., & Mahapatra, P. S. (2021). Condensation of Humid Air on Superhydrophobic Surfaces: Effect of Nanocoatings on a Hierarchical Interface. *Langmuir*, *37*(44), 12767–12780. doi:10.1021/acs.langmuir.1c01012 PMID:34714651

Thongsuwan, W., Sroila, W., Kumpika, T., Kantarak, E., & Singjai, P. (2022). Antireflective, Photocatalytic, and Superhydrophilic Coating prepared by Facile Sparking Process for Photovoltaic Panels. *Scientific Reports*, *12*, 1675. doi:10.1038/s41598-022-05733-7

Thwala, L. N., Ndlovu, S. C., Mpofu, K. T., Lugongolo, M. Y., & Mthunzi-Kufa, P. (2023). Nanotechnology-Based Diagnostics for Diseases Prevalent in Developing Countries: Current Advances in Point-of-Care Tests. *Nanomaterials (Basel, Switzerland)*, *13*(7), 1247. doi:10.3390/nano13071247 PMID:37049340

Tonyali, B., McDaniel, A., Amamcharla, J., Trinetta, V., & Yucel, U. (2020). Release kinetics of cinnamaldehyde, eugenol, and thymol from sustainable and biodegradable active packaging films. *Food Packaging and Shelf Life*, *24*, 100484. doi:10.1016/j.fpsl.2020.100484

Tracey, S. R., Smyth, P., Barelle, C. J., & Scott, C. J. (2021). Development of next generation nanomedicine-based approaches for the treatment of cancer: We've barely scratched the surface. *Biochemical Society Transactions*, *49*(5), 2253–2269. doi:10.1042/BST20210343 PMID:34709394

Trentin, A., Uvida, M. C., de Araújo Almeida, A., de Souza, T. A. C., & Hammer, P. (2022). Self-healing nanocoatings. In H. Song, T. A. Nguyen, G. Yasin, N. B. Singh, & R. K. Gupta (Eds.), *Nanotechnology in the Automotive Industry* (pp. 371–401). Elsevier. doi:10.1016/B978-0-323-90524-4.00018-9

Trivedi, M. U., Greczynski, G., Kanth, P. C., Pandey, M. K., Ivanov, I. G., Syväjärvi, M., & Yazdi, G. R. (2021). Study of Cucurbit[7]uril nanocoating on epitaxial graphene to design a versatile sensing platform. *Applied Surface Science*, *563*, 150096. doi:10.1016/j.apsusc.2021.150096

Tsai, Y.-T., Maggay, I. V., Venault, A., & Lin, Y.-F. (2021). Fluorine-free and hydrophobic/oleophilic PMMA/PDMS electrospun nanofibrous membranes for gravity-driven removal of water from oil-rich emulsions. *Separation and Purification Technology*, *279*, 119720. doi:10.1016/j.seppur.2021.119720

Turalija, M., Bischof, S., Budimir, A., & Gaan, S. (2016). Antimicrobial PLA films from environment friendly additives. *Composites. Part B, Engineering*, *102*, 94–99. doi:10.1016/j.compositesb.2016.07.017

Ubaid, F., Radwan, A. B., Naeem, N., Shakoor, R. A., Ahmad, Z., Montemor, M. F., Kahraman, R., Abdullah, A. M., & Soliman, A. (2019). Multifunctional self-healing polymeric nanocomposite coatings for corrosion inhibition of steel. *Surface and Coatings Technology*, *372*, 121–133. doi:10.1016/j.surfcoat.2019.05.017

Ulaeto, S. B., Pancrecious, J. K., Rajan, T. P. D., & Pai, B. C. (2019). Smart Coatings. In S. Mohapatra, T. A. Nguyen, & P. Nguyen-Tri (Eds.), *Noble Metal-Metal Oxide Hybrid Nanoparticles* (pp. 341–372). Woodhead Publishing. doi:10.1016/B978-0-12-814134-2.00017-6

Unni, R., Reshmy, R., Latha, M. S., Philip, E., Sindhu, R., Binod, P., Pandey, A., & Awasthi, M. K. (2022). Enhancement of mechanical and thermal properties of Ixora coccinea L. plant root derived nanocellulose using polyethylene glycol-glutaraldehyde system. *Chemosphere*, *298*, 134324. doi:10.1016/j.chemosphere.2022.134324 PMID:35307393

Vahdati, S. N., Behboudi, H., Navasatli, S. A., Tavakoli, S., & Safavi, M. (2022). New insights into the inhibitory roles and mechanisms of D-amino acids in bacterial biofilms in medicine, industry, and agriculture. *Microbiological Research*, *263*, 127107. doi:10.1016/j.micres.2022.127107 PMID:35843196

Vasile, C. (2018). *Polymeric Nanocomposites and Nanocoatings for Food Packaging: A Review*. Materials. doi:10.3390/ma11101834

Vasilopoulou, M., da Silva, W. J., Soultati, A., Kim, H. P., Kim, B. S., Reo, Y., Gavim, A. E. X., Conforto, J., Schneider, F. S., Felippi, M., & Palilis, L. C. (2022). Photonic Nanostructures Mimicking Floral Epidermis for Perovskite Solar Cells. *Cell Reports. Physical Science*, *3*(9), 101019. doi:10.1016/j.xcrp.2022.101019

Vats, M. K., Swami, M., Sheokand, B., Rawat, V., Negi, D. S., & Kumar, A. (2023). Chapter 12 - Antimicrobial textiles based on nanoparticles and composite, antiviral and antimicrobial coatings based on functionalized nanomaterials. In S. ul Islam, C. M. Hussain, & S. K. Shukla (Eds.), Antiviral and Antimicrobial Coatings Based on Functionalized Nanomaterials (pp. 389-420). Elsevier. doi:10.1016/B978-0-323-91783-4.00013-9

Velickova, E., Winkelhausen, E., Kuzmanova, S., Alves, V. D., & Moldão-Martins, M. (2013). Impact of chitosan-beeswax edible coatings on the quality of fresh strawberries (Fragaria ananassa cv Camarosa) under commercial storage conditions. *Lebensmittel-Wissenschaft + Technologie*, *52*(2), 80–92. doi:10.1016/j.lwt.2013.02.004

Venkatraman, P. D., Butler, J. A., & Britten, N. S. (2023). 2 - Advances in medical textiles. In S. Maity, K. Singha, & P. Pandit (Eds.), *Functional and Technical Textiles* (pp. 31–70). Woodhead Publishing. doi:10.1016/B978-0-323-91593-9.00003-1

Verma, C. (2021). *Recent advancements in corrosion inhibitor systems through carbon allotropes : Past, present, and future*. doi:10.1002/nano.202100039

Verma, C., Thakur, A., Ganjoo, R., Sharma, S., & Assad, H. (2023). Coordination bonding and corrosion inhibition potential of nitrogen-rich heterocycles : Azoles and triazines as specific examples. *Coordination Chemistry Reviews, 488*(November), 215177. doi:10.1016/j.ccr.2023.215177

Verma, C., Quraishi, M. A., Alfantazi, A., & Rhee, K. Y. (2023). *Biodegradable Synthetic Polymers in Sustainable Corrosion Protection: Present and Future Scenarios*. Advanced Industrial and Engineering Polymer Research., doi:10.1016/j.aiepr.2023.04.005

Versino, F., Lopez, O. V., Garcia, M. A., & Zaritzky, N. E. (2016). Starch-based films and food coatings: An overview. *Stärke*, *68*(11-12), 1026–1037. doi:10.1002/star.201600095

Víg, P., Tóth, J., & Seres, I. (2023). Optical effects of water droplets on the performance of solar cell covered with a hydrophobic thin layer. *Solar Energy*, *251*, 366–381. doi:10.1016/j.solener.2023.01.028

Vijayan, P. P., & Puglia, D. (2019). Biomimetic multifunctional materials: A review. *Emergent Materials*, *2*(4), 391–415. doi:10.1007/s42247-019-00051-7

Vilarinho, F., Sanches Silva, A., Vaz, M. F., & Farinha, J. P. (2018). Nanocellulose in green food packaging. *Critical Reviews in Food Science and Nutrition*, *58*(9), 1526–1537. doi:10.1080/10408398.2016.1270254 PMID:28125279

Visan, A. I., & Cristescu, R. (2023). Polysaccharide-Based Coatings as Drug Delivery Systems. *Pharmaceutics*, *15*(9), 2227. doi:10.3390/pharmaceutics15092227 PMID:37765196

Vishvakarma, V., Kaur, M., Nagpal, M., & Arora, S. (2023). Role of Nanotechnology in Taste Masking: Recent Updates. *Current Drug Research Reviews*, *15*(1), 1–14. doi:10.2174/2589977514666220526091259 PMID:35619251

Vishwakarma, V. (2020). Synthesis, design, and characterization of controlled nanoparticles for the construction industry. In C. M. Hussain (Ed.), *Handbook of Nanomaterials for Manufacturing Applications* (pp. 323–341). Elsevier. doi:10.1016/B978-0-12-821381-0.00015-6

Voigtländer, B. (2019). Technical Aspects of Atomic Force Microscopy. In B. Voigtländer (Ed.), *Atomic Force Microscopy* (pp. 35–67). Springer International Publishing. doi:10.1007/978-3-030-13654-3_3

Wadullah, H. M., Ajeel, S. A., & Abbass, M. K. (2019). Synthesis and Characterization of Nanocoatings Thin films by Atomic Layer Deposition for Medical Applications. *IOP Conference Series. Materials Science and Engineering*, *518*(3), 032057. doi:10.1088/1757-899X/518/3/032057

Wadullah, H. M., Talib Mohammed, M., & Khalid Abdulrazzaq, T. (2022). Structure and characteristics of Nb2O5 nanocoating thin film for biomedical applications. *Materials Today: Proceedings*, *62*, 3076–3080. doi:10.1016/j.matpr.2022.03.229

Walz, K. A., Hoege, T. D., Duensing, J. W., Zeltner, W. A., & Anderson, M. A. (2023). Field Tests of a Self-Sintering, Anti-Soiling, Self-Cleaning, Nanoporous Metal Oxide, Transparent Thin Film Coating for Solar Photovoltaic Modules. *Solar Energy Materials and Solar Cells*, *262*, 112560. doi:10.1016/j.solmat.2023.112560

Wang, D., Li, Y., Wen, Y., Li, X., & Du, X. (2021). Simple and Low Cost Fabrication of Large Area Nanocoatings with Mechanical Robustness, Enhanced Broadband Transmittance and Antifogging. *Colloids and Surfaces. A, Physicochemical and Engineering Aspects*, *629*, 127522. doi:10.1016/j.colsurfa.2021.127522

Wang, H., Kumar, R., & Memon, H. (2020). Strongly Hydrophobic and Superoleophilic PMMA Based Nanocoated Cotton Fabrics. *Coatings*, *10*(10), 943. doi:10.3390/coatings10100943

Wang, H., Zhou, Y., Sun, Q., Zhou, C., Hu, S., Lenahan, C., Xu, W., Deng, Y., Li, G., & Tao, S. (2021). Update on nanoparticle-based drug delivery system for anti-inflammatory treatment. *Frontiers in Bioengineering and Biotechnology*, *9*, 630352. doi:10.3389/fbioe.2021.630352 PMID:33681167

Wang, J., Tang, B., Bai, W., Lu, X., & Wang, X. (2020). Deodorizing for fiber and fabric: Adsorption, catalysis, source control and masking. *Advances in Colloid and Interface Science*, *283*, 102243. doi:10.1016/j.cis.2020.102243 PMID:32854016

Wang, J., Xue, Z., Yu, T., Liu, Z., & Mu, T. (2017). Activation of cellulose assisted by CO2 for the preparation of a superhydrophobic nanocoating. *Chemistry, an Asian Journal*, *12*(14), 1773–1779. doi:10.1002/asia.201700335 PMID:28471506

Wang, K., Bu, N., Zhen, Q., Liu, J., & Bashir, S. (2023). Modified nano-SiO2/TiO2 hybrid fluorinated B-72 as antimicrobial and hydrophobic coatings for the conservation of ancient bricks. *Construction & Building Materials*, *365*, 130090. doi:10.1016/j.conbuildmat.2022.130090

Wang, M., Huang, X., Zheng, H., Tang, Y., Zeng, K., Shao, L., & Li, L. (2021). Nanomaterials applied in wound healing: Mechanisms, limitations and perspectives. *Journal of Controlled Release*, *337*, 236–247. doi:10.1016/j.jconrel.2021.07.017 PMID:34273419

Wang, P., Yan, X., Zeng, J., Luo, C., & Wang, C. (2022). Anti-reflective Superhydrophobic Coatings with Excellent Durable and Self-Cleaning Properties for Solar Cells. *Applied Surface Science*, *602*, 154408. doi:10.1016/j.apsusc.2022.154408

Wang, S., Du, X., Deng, S., Fu, X., Du, Z., Cheng, X., & Wang, H. (2019). A polydopamine-bridged hierarchical design for fabricating flame-retarded, superhydrophobic, and durable cotton fabric. *Cellulose (London, England)*, *26*(11), 7009–7023. doi:10.1007/s10570-019-02586-8

Wang, X., Chai, Y., & Liu, J. (2013). Formation of highly hydrophobic wood surfaces using silica nanoparticles modified with long-chain alkylsilane. *Holzforschung*, *67*(6), 667–672. Advance online publication. doi:10.1515/hf-2012-0153

Wang, X., Han, Z., Liu, Y., & Wang, Q. (2020). Micro-nano surface structure construction and hydrophobic modification to prepare efficient oil-water separation melamine formaldehyde foam. *Applied Surface Science*, *505*, 144577. doi:10.1016/j.apsusc.2019.144577

Wang, X., Liu, S., Chang, H., & Liu, J. (2014). Sol-Gel Deposition of TiO2 Nanocoatings on Wood Surfaces with Enhanced Hydrophobicity and Photostability. Wood and Fiber Science. *Journal of the Society of Wood Science and Technology*, *46*, 109–117.

Wang, X., Pan, J., Li, Q., Dong, X., Shi, L., & Chang, S. (2022). Preparation, Corrosion Resistance, and Electrochemical Properties of MnO2/TiO2 Coating on Porous Titanium. *Coatings*, *12*(10), 1381. Advance online publication. doi:10.3390/coatings12101381

Wang, X., Zhang, J., Bao, L., Yang, W., Zhou, F., & Liu, W. (2020). Enhancement of the ballistic performance of aramid fabric with polyurethane and shear thickening fluid. *Materials & Design*, *196*, 109015. doi:10.1016/j.matdes.2020.109015

Wang, Y., Li, B., Xu, C., & Nie, R. (2013). Superhydrophobic ZnONanocoatings for Different Substrates. *Asian Journal of Chemistry*, *25*(10), 5571–5572. doi:10.14233/ajchem.2013.OH21

Wang, Y., Liu, K., Zhang, M., Xu, T., Du, H., Pang, B., & Si, C. (2023). Sustainable polysaccharide-based materials for intelligent packaging. *Carbohydrate Polymers*, *313*, 120851. doi:10.1016/j.carbpol.2023.120851 PMID:37182951

Wang, Y., Wang, Y., Li, X., Li, J., Su, L., Zhang, X., & Du, X. (2018). Dendritic Silica Particles with Well-Dispersed Ag Nanoparticles for Robust Antireflective and Antibacterial Nanocoatings on Polymeric Glass. *ACS Sustainable Chemistry & Engineering*, *6*(11), 14071–14081. doi:10.1021/acssuschemeng.8b02617

Wang, Y., Yu, L., Kong, X., & Sun, L. (2017). Application of nanodiagnostics in point-of-care tests for infectious diseases. *International Journal of Nanomedicine*, *12*, 4789–4803. doi:10.2147/IJN.S137338 PMID:28740385

Wang, Y., Zhang, R., Ahmed, S., Qin, W., & Liu, Y. (2019). Preparation and characterization of corn starch bio-active edible packaging films based on zein incorporated with orange-peel oil. *Antioxidants*, *8*(9), 391. doi:10.3390/antiox8090391 PMID:31514341

Wang, Z., Chen, X., Gong, Y., Zhang, B., & Li, H. (2017). Superhydrophobic nanocoatings prepared by a novel vacuum cold spray process. *Surface and Coatings Technology*, *325*, 52–57. doi:10.1016/j.surfcoat.2017.06.044

Wani, M. Y., Ganie, N. A., Dar, K. A., Dar, S. Q., Khan, A. H., Khan, N. A., Zahmatkesh, S., Manzar, M. S., & Banerjee, R. (2023). Nanotechnology future in food using carbohydrate macromolecules: A state-of-the-art review. *International Journal of Biological Macromolecules*, *239*, 124350. doi:10.1016/j.ijbiomac.2023.124350 PMID:37028631

Wani, N. R., Dar, A. H., Dash, K. K., Pandey, V. K., Srivastava, S., Jan, S. Y., Deka, P., & Sabahi, N. (2023). Recent advances in the production of bionanomaterials for development of sustainable food packaging: A comprehensive review. *Environmental Research*, *116948*, 116948. Advance online publication. doi:10.1016/j.envres.2023.116948 PMID:37611789

Wardhono, E. Y., Pinem, M. P., Kustiningsih, I., Agustina, S., Oudet, F., Lefebvre, C., Clausse, D., Saleh, K., & Guénin, E. (2019). Cellulose nanocrystals to improve stability and functional properties of emulsified film based on chitosan nanoparticles and beeswax. *Nanomaterials (Basel, Switzerland)*, *9*(12), 1707. doi:10.3390/nano9121707 PMID:31795284

Wasti, S., Lee, I. H., Kim, S., Lee, J. H., & Kim, H. (2023). Ethical and legal challenges in nanomedical innovations: A scoping review. *Frontiers in Genetics*, *14*, 1163392. doi:10.3389/fgene.2023.1163392 PMID:37252668

Wei, H., Wang, Y., Guo, J., Shen, N. Z., Jiang, D., Zhang, X., Yan, X., Zhu, J., Wang, Q., Shao, L., Lin, H., Wei, S., & Guo, Z. (2015). Advanced micro/nanocapsules for self-healing smart anticorrosion coatings. *Journal of Materials Chemistry. A, Materials for Energy and Sustainability*, *3*(2), 469–480. doi:10.1039/C4TA04791E

Wen, P., Zhu, D. H., Wu, H., Zong, M. H., Jing, Y. R., & Han, S. Y. (2016). Encapsulation of cinnamon essential oil in electrospun nanofibrous film for active food packaging. *Food Control*, *59*, 366–376. doi:10.1016/j.foodcont.2015.06.005

Williams, A. G., Moore, E., Thomas, A., & Johnson, J. A. (2023). Graphene-Based Materials in Dental Applications: Antibacterial, Biocompatible, and Bone Regenerative Properties. *International Journal of Biomaterials*. doi:10.1155/2023/8803283

Wimalasiri, V. K., Weerathunga, H. U., Kottegoda, N., & Karunaratne, V. (2017). Silica Based Superhydrophobic Nanocoatings for Natural Rubber Surfaces. *Journal of Nanomaterials*, *2102467*, 1–14. Advance online publication. doi:10.1155/2017/2102467

Wiwanitkit, V. (2017). Ethics in Nanomedicine: A Concern on New Nanotechnology. *Med One*, *2*(5).

Wouters, M., Rentrop, C., & Willemsen, P. (2010). Surface structuring and coating performance: Novel biocidefree nanocomposite coatings with anti-fouling and fouling-release properties. *Progress in Organic Coatings*, *68*(1-2), 4–11. doi:10.1016/j.porgcoat.2009.10.005

Wu, J., Tu, J., Yu, S., Wu, H., Xie, Y., Yang, Y., Xv, Z., & Zhang, Q. (2023). Hollow core-shell nanocoatings with gradient refractive index structure for enhanced photovoltaic performance. *Colloids and Surfaces. A, Physicochemical and Engineering Aspects*, *667*, 131424. doi:10.1016/j.colsurfa.2023.131424

Wu, Y. S., Zeng, D. C., Liu, Z. W., Qiu, W. Q., Zhong, X. C., Yu, H. Y., & Li, S. Z. (2011). Microstructure and sliding wear behavior of nanostructured Ni60–TiB2 composite coating sprayed by HVOF technique. *Surface and Coatings Technology*, *206*(6), 1102–1108. doi:10.1016/j.surfcoat.2011.07.096

Wypij, M., Trzcińska-Wencel, J., Golińska, P., Avila-Quezada, G. D., Ingle, A. P., & Rai, M. (2023). The strategic applications of natural polymer nanocomposites in food packaging and agriculture: Chances, challenges, and consumers' perception. *Frontiers in Chemistry*, *10*, 1106230. doi:10.3389/fchem.2022.1106230 PMID:36704616

Xia, F., Li, Q., Ma, C., & Guo, X. (2020). Preparation and characterization of Ni–AlN nanocoatings deposited by magnetic field assisted electrodeposition technique. *Ceramics International*, *46*(2), 2500–2509. doi:10.1016/j.ceramint.2019.09.244

Xiang, Y., Si, Y., Xin, Y., & Guo, Z. (2017). One-step strategy to prepare utility ZnO–stearic acid (STA) superhydrophobic nanocoating. *Chemistry Letters*, *46*(9), 1393–1395. doi:10.1246/cl.170545

Xiao, J., & Huang, Y. (2011). Design of sustainable multifunctional nanocoating: A goal-driven multiscale systems approach. *Chinese Journal of Chemical Engineering*, *19*(4), 666–673. doi:10.1016/S1004-9541(11)60039-6

Xiao, Y., Liu, Y., Kang, S., Wang, K., & Xu, H. (2020). Development and evaluation of soy protein isolate-based antibacterial nanocomposite films containing cellulose nanocrystals and zinc oxide nanoparticles. *Food Hydrocolloids*, *106*, 105898. doi:10.1016/j.foodhyd.2020.105898

Xie, Q., Hao, T., Zhang, J., Wang, C., Zhang, R., & Qi, H. (2019). Anti-Icing Performance of a Coating Based on Nano/Microsilica Particle-Filled Amino-Terminated PDMS-Modified Epoxy. Coatings.

Xing, Y., Wu, Z., Yang, J., Wang, X., & Liu, L. (2020). LIPSS combined with ALD MoS2 nano-coatings for enhancing surface friction and hydrophobic performances. *Surface and Coatings Technology*, *385*, 125396. doi:10.1016/j.surfcoat.2020.125396

Xin, J. H., Daoud, W. A., & Kong, Y. Y. (2004). A new approach to UV-blocking treatment for cotton fabrics. *Textile Research Journal*, *74*(2), 97–100. doi:10.1177/004051750407400202

Xin, M. (2022). Crystal Structure and Optical Properties of ZnO:Ce Nano Film. *Molecules (Basel, Switzerland)*, *27*(16), 1–9. doi:10.3390/molecules27165308 PMID:36014546

Xiong, J., Luo, H., Gao, D., Zhou, X., Cui, P., Thangavel, G., Parida, K., & Lee, P. S. (2019). Self-restoring, waterproof, tunable microstructural shape memory triboelectric nanogenerator for self-powered water temperature sensor. *Nano Energy*, *61*, 584–593. doi:10.1016/j.nanoen.2019.04.089

Xi, R., Wang, Y., Li, X., Zhang, X., & Du, X. (2020). A facile strategy to form a three-dimensional network structure for mechanically robust superhydrophobic nanocoatings with enhanced transmittance. *Journal of Colloid and Interface Science*, *563*, 42–53. doi:10.1016/j.jcis.2019.12.049 PMID:31865047

Xu, C., Chen, C., & Wu, D. (2018). The starch nanocrystal filled biodegradable poly (ε-caprolactone) composite membrane with highly improved properties. *Carbohydrate Polymers*, *182*, 115–122. doi:10.1016/j.carbpol.2017.11.001 PMID:29279105

Xu, C., Ma, J., Liu, Z., Wang, W., Liu, X., Qian, S., Chen, L., Gu, L., Sun, C., Hou, J., & Jiang, Z. (2023). Preparation of shell-core fiber-encapsulated Lactobacillus rhamnosus 1.0320 using coaxial electrospinning. *Food Chemistry*, *402*, 134253. doi:10.1016/j.foodchem.2022.134253 PMID:36162172

Xu, C., Zhou, S., Song, H., Hu, H., Yang, Y., Zhang, X., Ma, S., Feng, X., Pan, Y., Gong, S., Fan, F., Chen, P., & Chu, Q. (2023). Green tea polyphenols-derived hybrid materials in manufacturing, environment, food and healthcare. *Nano Today*, *52*, 101990. doi:10.1016/j.nantod.2023.101990

Xue, C.-H., Chen, J., Yin, W., Jia, S.-T., & Ma, J.-Z. (2012). Superhydrophobic conductive textiles with antibacterial property by coating fibers with silver nanoparticles. *Applied Surface Science*, *258*(7), 2468–2472. doi:10.1016/j.apsusc.2011.10.074

Xue, C.-H., Li, Y.-R., Hou, J.-L., Zhang, L., Ma, J.-Z., & Jia, S.-T. (2015). Self-roughened superhydrophobic coatings for continuous oil–water separation. *Journal of Materials Chemistry. A, Materials for Energy and Sustainability*, *3*(19), 10248–10253. doi:10.1039/C5TA01014D

Xue, L., Wang, W., Guo, Y., Liu, G., & Wan, P. (2017). Flexible polyaniline/carbon nanotube nanocomposite film-based electronic gas sensors. *Sensors and Actuators. B, Chemical*, *244*, 47–53. doi:10.1016/j.snb.2016.12.064

Xu, H., Chai, Y., & Zhang, G. (2012). Synergistic effect of oleic acid and glycerol on zein film plasticization. *Journal of Agricultural and Food Chemistry*, *60*(40), 10075–10081. doi:10.1021/jf302940j PMID:22989052

Xu, Q. F., Wang, J. N., & Sanderson, K. D. (2010). Organic–Inorganic Composite Nanocoatings with Superhydrophobicity, Good Transparency, and Thermal Stability. *ACS Nano*, *4*(4), 2201–2209. doi:10.1021/nn901581j PMID:20302323

Xu, Q., Sun, Z., Feng, F., Li, Y., Tong, G., & Noor, R. S. (2020). Preparation and anti-icing of hydrophobic polypyrrole coatings on wind turbine blade. *International Journal of Rotating Machinery*, *8626457*, 1–9. Advance online publication. doi:10.1155/2020/8626457

Xu, T., Zhang, K., Cai, Q., Wang, N., Wu, L., He, Q., Wang, H., Zhang, Y., Xie, Y., Yao, Y., & Chen, Y. (2022). Advances in synthesis and applications of boron nitride nanotubes: A review. Chemical Engineering Journal, 431, 134118. https://doi.org/10.1016/J.CEJ.2021.134118

Xu, Y.-D., Zhu, Z.-Y., Xu, T.-Z., Abadikhah, H., Wang, J.-W., Xu, X., & Agathopoulos, S. (2018). Fabrication and characterization of robust hydrophobic lotus leaf-like surface on Si3N4 porous membrane via polymer-derived SiNCO inorganic nanoparticle modification. *Ceramics International, 44*(14), 16443–16449. doi:10.1016/j.ceramint.2018.06.058

Yaashikaa, P. R., Kamalesh, R., Senthil Kumar, P., Saravanan, A., Vijayasri, K., & Rangasamy, G. (2023). Recent advances in edible coatings and their application in food packaging. *Food Research International, 173*, 113366. doi:10.1016/j.foodres.2023.113366 PMID:37803705

Yadav, A., Mangaraj, S., Singh, R., Kumar, N., & Arora, S. (2018). *Biopolymers as packaging material in food and allied industry*. Academic Press.

Yang, G., Zhang, Z., Li, C., & Hu, J. (2023). A facile approach to fabricate omniphobic and robust polyurethane coatings for anti-smudge, anti-ink. *Progress in Organic Coatings, 179*, 107488. doi:10.1016/j.porgcoat.2023.107488

Yang, H. M. (2021). Role of organic and eco-friendly inhibitors on the corrosion mitigation of steel in acidic environments—A state-of-art review. *Molecules (Basel, Switzerland), 26*(11), 3473. Advance online publication. doi:10.3390/molecules26113473 PMID:34200445

Yang, J., Qian, H., Wang, J., Ju, P., Lou, Y., Li, G., & Zhang, D. (2021). Mechanically durable antibacterial nanocoatings based on zwitterionic copolymers containing dopamine segments. *Journal of Materials Science and Technology, 89*, 233–241. doi:10.1016/j.jmst.2020.11.031

Yang, Z., Zhang, D., Wang, K., He, J., Li, J., & Huang, C. (2022). Investigating graphdiyne based materials for rechargeable batteries. *Nano Today, 46*, 101588. doi:10.1016/j.nantod.2022.101588

Yao, L., & He, J. (2014). Recent progress in antireflection and self-cleaning technology–From surface engineering to functional surfaces. *Progress in Materials Science, 61*, 94–143. doi:10.1016/j.pmatsci.2013.12.003

Yao, Y., Zhou, Y., Liu, L., Xu, Y., Chen, Q., Wang, Y., Wu, S., Deng, Y., Zhang, J., & Shao, A. (2020). Nanoparticle-Based Drug Delivery in Cancer Therapy and Its Role in Overcoming Drug Resistance. *Frontiers in Molecular Biosciences, 7*, 193. doi:10.3389/fmolb.2020.00193 PMID:32974385

Yasin, G., Arif, M., Mehtab, T., Shakeel, M., Khan, M. A., & Khan, W. Q. (2020). Metallic nanocomposite coatings. In *Corrosion Protection at the Nanoscale* (pp. 245–274). Elsevier. doi:10.1016/B978-0-12-819359-4.00014-3

Yazdani-Ahmadabadi, H., Rastegar, S., & Ranjbar, Z. (2016). Acidic hydrophobic nanosilica particle-induced mechanical dissipation and adhesion promotion of acrylate-based photopolymericnanocoatings. *Progress in Organic Coatings, 100*, 173–177. doi:10.1016/j.porgcoat.2016.05.004

Yazdani, J., Ahmadian, E., Sharifi, S., Shahi, S., & Maleki Dizaj, S. (2018). A short view on nanohydroxyapatite as coating of dental implants. *Biomedicine and Pharmacotherapy*, *105*, 553–557. doi:10.1016/j.biopha.2018.06.013 PMID:29886376

Yedra, Á., Vilar, N., González, M., & Manteca, C. (2014). *Advanced Materials and Nanomaterials Unit*. Fundación Centro Tecnológico de Componentes (CTC).

Ye, H., Zhu, L., Li, W., Liu, H., & Chen, H. (2017). Constructing Fluorine-Free and Cost-Effective Superhydrophobic Surface with Normal-Alcohol-Modified Hydrophobic SiO2 Nanoparticles. *ACS Applied Materials & Interfaces*, *9*(1), 858–867. doi:10.1021/acsami.6b12820 PMID:27957841

Yeong, Y. H., Davis, A., Steele, A., Loth, E., & Bayer, I. S. (2014). Spray deposition effects on superhydrophobicity and durability of nanocoatings. *Surface Innovations*, *2*(2), 70–78. doi:10.1680/si.13.00044

Yetisen, A. K., Qu, H., Manbachi, A., Butt, H., Dokmeci, M. R., Hinestroza, J. P., Skorobogatiy, M., Khademhosseini, A., & Yun, S. H. (2016). Nanotechnology in Textiles. *ACS Nano*, *10*(3), 3042–3068. doi:10.1021/acsnano.5b08176 PMID:26918485

Ye, Z., Sang, T., Li, K., Fischer, N. G., Mutreja, I., Echeverría, C., & Aparicio, C. (2022). Hybrid nanocoating of self-assembled organic-inorganic amphiphiles for prevention of implant infections. *Acta Biomaterialia*, *140*, 338–349. doi:10.1016/j.actbio.2021.12.008 PMID:34896631

Yilbas, B. S., Keles, O., & Toprakli, A. Y. (2017). Surface engineering towards self-cleaning applications: Laser textured silicon surface. *Procedia Engineering*, *184*, 716–724. doi:10.1016/j.proeng.2017.04.147

Yilmaz, O., & Yorgancioglu, A. (2019). Nanocoating: preparation, properties, and biomedical applications. In Polymeric Nanomaterials In Nanotherapeutics (pp. 299-331). Elsevier.

Ying. (2018). Hydrophobic polymer-based nano-coatings for building applications. Academic Press.

Yingngam, B., Chiangsom, A., Pharikarn, P., Vonganakasame, K., Kanoknitthiran, V., Rungseevijitprapa, W., & Prasitpuriprecha, C. (2019). Optimization of menthol-loaded nanocapsules for skin application using the response surface methodology. *Journal of Drug Delivery Science and Technology*, *53*, 101138. doi:10.1016/j.jddst.2019.101138

Yingngam, B., Kacha, W., Rungseevijitprapa, W., Sudta, P., Prasitpuriprecha, C., & Brantner, A. (2019). Response surface optimization of spray-dried citronella oil microcapsules with reduced volatility and irritation for cosmetic textile uses. *Powder Technology*, *355*, 372–385. doi:10.1016/j.powtec.2019.07.065

Yong, L. X., & Calautit, J. K. (2023). A Comprehensive Review on the Integration of Antimicrobial Technologies onto Various Surfaces of the Built Environment. *Sustainability (Basel)*, *15*(4), 3394. doi:10.3390/su15043394

Young, E., Mirosa, M., & Bremer, P. (2020). A Systematic Review of Consumer Perceptions of Smart Packaging Technologies for Food. *Frontiers in Sustainable Food Systems*, *4*, 63. doi:10.3389/fsufs.2020.00063

Yousefi, F., Mousavi, S. B., Heris, S. Z., & Naghash-Hamed, S. (2023). UV-shielding properties of a cost-effective hybrid PMMA-based thin film coatings using TiO2 and ZnO nanoparticles: A comprehensive evaluation. *Scientific Reports*, *13*(1), 1–19. doi:10.1038/s41598-023-34120-z PMID:37130903

Yu, F., Gao, J., Liu, C., Chen, Y., Zhong, G., Hodges, C., Chen, M., & Zhang, H. (2020). Preparation and UV aging of nano-SiO2/fluorinated polyacrylate polyurethane hydrophobic composite coating. *Progress in Organic Coatings*, *141*, 105556. doi:10.1016/j.porgcoat.2020.105556

Yurtsever, M. (2019). Glitters as a source of primary microplastics: An approach to environmental responsibility and ethics. *Journal of Agricultural & Environmental Ethics*, *32*(3), 459–478. doi:10.1007/s10806-019-09785-0

Yusuf, A., Almotairy, A. R. Z., Henidi, H., Alshehri, O. Y., & Aldughaim, M. S. (2023). Nanoparticles as Drug Delivery Systems: A Review of the Implication of Nanoparticles' Physicochemical Properties on Responses in Biological Systems. *Polymers*, *15*(7), 1596. doi:10.3390/polym15071596 PMID:37050210

Yu, Y., Zhang, J., Wang, J., & Sun, B. (2021). MicroRNAs: The novel mediators for nutrient-modulating biological functions. *Trends in Food Science & Technology*, *114*, 167–175. doi:10.1016/j.tifs.2021.05.028

Yu, Z., Dhital, R., Wang, W., Sun, L., Zeng, W., Mustapha, A., & Lin, M. (2019). Development of multifunctional nanocomposites containing cellulose nanofibrils and soy proteins as food packaging materials. *Food Packaging and Shelf Life*, *21*, 100366. doi:10.1016/j.fpsl.2019.100366

Zaitseva, O., Khudyakov, A., Sergushkina, M., Solomina, O., & Polezhaeva, T. (2020). Pectins as a universal medicine. *Fitoterapia*, *146*, 104676. doi:10.1016/j.fitote.2020.104676 PMID:32561422

Zambonino, M. C., Quizhpe, E. M., Mouheb, L., Rahman, A., Agathos, S. N., & Dahoumane, S. A. (2023). Biogenic Selenium Nanoparticles in Biomedical Sciences: Properties, Current Trends, Novel Opportunities and Emerging Challenges in TheranosticNanomedicine. *Nanomaterials (Basel, Switzerland)*, *13*(3), 424. doi:10.3390/nano13030424 PMID:36770385

Zamora-Ledezma, C., Clavijo, D. F. C., Medina, E., Sinche, F., Vispo, N. S., Dahoumane, S. A., & Alexis, F. (2020). Biomedical science to tackle the COVID-19 pandemic: Current status and future perspectives. *Molecules (Basel, Switzerland)*, *25*(20), 4620. Advance online publication. doi:10.3390/molecules25204620 PMID:33050601

Zemajtis, F., Hasan, A. B. M. R., Yetik, O., Trtik, P., Pillai, K. M., & Sobolev, K. (2023). From superhydrophilicity to superhydrophobicity: High-resolution neutron imaging and modeling of water imbibition through porous surfaces treated with engineered nano-coatings. *Scientific Reports*, *13*(1), 11176. doi:10.1038/s41598-023-38324-1 PMID:37429919

Zhang, D., Chen, L., Cai, J., Dong, Q., Din, Z. U., Hu, Z. Z., Wang, G.-Z., Ding, W.-P., He, J.-R., & Cheng, S. Y. (2021). Starch/tea polyphenols nanofibrous films for food packaging application: From facile construction to enhance mechanical, antioxidant and hydrophobic properties. *Food Chemistry*, *360*, 129922. doi:10.1016/j.foodchem.2021.129922 PMID:33965711

Zhang, D., Williams, B. L., Liu, J., Hou, Z., Smith, A. T., Nam, S., Nasir, Z., Patel, H., Partyka, A., Becher, E. M., Lofink, B. J., Santos, V. H., Peng, X., & Sun, L. (2021). An environmentally-friendly sandwich-like structured nanocoating system for wash durable, flame retardant, and hydrophobic cotton fabrics. *Cellulose (London, England)*, *28*(16), 10277–10289. doi:10.1007/s10570-021-04177-y

Zhang, D., Yu, G., Long, Z., Xiao, H., & Qian, L. (2016). Bio-wax latex-modified paper as antimicrobial and water-vapor-resistant packaging material. *Journal of Wood Chemistry and Technology*, *36*(3), 182–191. doi:10.1080/02773813.2015.1104545

Zhang, L., Bai, R., Li, B., Ge, C., Du, J., Liu, Y., Le Guyader, L., Zhao, Y., Wu, Y., He, S., Ma, Y., & Chen, C. (2011). Rutile TiO2 particles exert size and surface coating dependent retention and lesions on the murine brain. *Toxicology Letters*, *207*(1), 73–81. doi:10.1016/j.toxlet.2011.08.001 PMID:21855616

Zhang, L.-Z., Pan, A.-J., Cai, R.-R., & Lu, H. (2019). Indoor Experiments of Dust Deposition Reduction on Solar Cell Covering Glass by Transparent Super-Hydrophobic Coating with Different Tilt Angles. *Solar Energy*, *188*, 1146–1155. doi:10.1016/j.solener.2019.07.026

Zhang, N., Wang, W., Zhou, M., Yu, Y., Wang, P., & Wang, Q. (2023). A controllable, universal, and natural supramolecular assembly coating strategy for multifunctionality textiles of antibacterial properties, UV resistance, antioxidant, and secondary reactivity. *Industrial Crops and Products*, *198*, 116637. doi:10.1016/j.indcrop.2023.116637

Zhang, X., Wang, H., Zhang, X., Zhao, Z., & Zhu, Y. (2019). A multifunctional super-hydrophobic coating based on PDA modified MoS2 with anti-corrosion and wear resistance. *Colloids and Surfaces. A, Physicochemical and Engineering Aspects*, *568*, 239–247. doi:10.1016/j.colsurfa.2019.02.016

Zhang, Y., Cui, L., Xu, H., Feng, X., Wang, B., Pukánszky, B., Mao, Z., & Sui, X. (2019). Poly (lactic acid)/cellulose nanocrystal composites via the Pickering emulsion approach: Rheological, thermal and mechanical properties. *International Journal of Biological Macromolecules*, *137*, 197–204. doi:10.1016/j.ijbiomac.2019.06.204 PMID:31255621

Zhang, Y., Ma, X., Yang, M., Lan, D., Li, X., Gao, Z., & Ji, T. (2023). Ablation resistance of ethylene propylene diene monomer insulation materials reinforced with liquid hyperbranched polycarbosilane coated aramid fibers. *Polymer Degradation & Stability*, *217*, 110510. doi:10.1016/j.polymdegradstab.2023.110510

Zhang, Y., Xia, X., Ma, K., Xia, G., Wu, M., Cheung, Y. H., ... Farha, O. K. (2023). Functional textiles with smart properties: Their fabrications and sustainable applications. *Advanced Functional Materials*, *33*(33), 2301607. doi:10.1002/adfm.202301607

Zhao, W., & Lu, H. (2021). Self-Cleaning Performance of Super-Hydrophilic Coatings for Dust Deposition Reduction on Solar Photovoltaic Cells. *Coatings*, *11*(9), 1059. doi:10.3390/coatings11091059

Zhao, Y., Hu, J., Hu, X., Zhu, F., Su, J., & Han, J. (2023). A novel strategy for fabrication of antistatic and antibacterial fabric via layer-by-layer self-assembly. *Surface and Coatings Technology*, *453*, 129143. doi:10.1016/j.surfcoat.2022.129143

Zhao, Y., Li, B., Li, C., Xu, Y., Luo, Y., Liang, D., & Huang, C. (2021). Comprehensive review of polysaccharide-based materials in edible packaging: A sustainable approach. *Foods*, *10*(8), 1845. doi:10.3390/foods10081845 PMID:34441621

Zhao, Y., Zhang, Y., Dong, H., Wu, W., Yang, X., & He, Q. (2023). Functional biopolymers for food packaging: Formation mechanism and performance improvement of chitosan-based composites. *Food Bioscience*, *54*, 102927. doi:10.1016/j.fbio.2023.102927

Zheng, G., Cui, Y., Jiang, Z., Zhou, M., Wang, P., Yu, Y., & Wang, Q. (2023). Multifunctional composite coatings with hydrophobic, UV-resistant, antioxidative, and photothermal performance for healthcare. *Colloids and Surfaces. A, Physicochemical and Engineering Aspects*, *667*, 131367. doi:10.1016/j.colsurfa.2023.131367 PMID:37025928

Zhou, C., Chen, Q., Chen, Q., Yin, H., Wang, S., & Hu, C. (2022). Preparation of TiO2 Superhydrophobic Composite Coating and Studies on Corrosion Resistance. *Frontiers in Chemistry*, *10*(July), 1–9. doi:10.3389/fchem.2022.943055 PMID:35873046

Zhou, J., Wen, Y., Marshall, M. R., Zhao, J., Gui, H., Yang, Y., Zeng, Z., Jones, D. L., & Zang, H. (2021). Microplastics as an emerging threat to plant and soil health in agroecosystems. *The Science of the Total Environment*, *787*, 147444. Advance online publication. doi:10.1016/j.scitotenv.2021.147444

Zhou, L., & Wang, Y. (2021). Physical and antimicrobial properties of zein and methyl cellulose composite films with plasticizers of oleic acid and polyethylene glycol. *Lebensmittel-Wissenschaft + Technologie*, *140*, 110811. doi:10.1016/j.lwt.2020.110811

Zhou, Z., Pourhashem, S., Wang, Z., Duan, J., Zhang, R., & Hou, B. (2022). Distinctive roles of graphene oxide, ZnO quantum dots, and their nanohybrids in anti-corrosion and anti-fouling performance of waterborne epoxy coatings. *Chemical Engineering Journal*, *439*, 135765. doi:10.1016/j.cej.2022.135765

Zhu, G., Wang, J., Yuan, X., & Yuan, B. (2022). Hydrophobic and fire-safe polyurethane foam coated with chitosan and nano-montmorillonite via layer-by-layer assembly for emergency absorption of oil spill. *Materials Letters*, *316*, 132009. doi:10.1016/j.matlet.2022.132009

Zhu, L., Yao, K. L., & Liu, Z. L. (2006a). First-principles study of the polar (111) surface of Fe3 O4. *Physical Review B: Condensed Matter and Materials Physics*, *74*(3), 035409. Advance online publication. doi:10.1103/PhysRevB.74.035409

Zolriasatein, A., & Abdollahzadeh, S., & MotakefKazemi, N. (2023). An Examination of the Application of Polyurethane-Based Hydrophobic Nanocoatings Inspired by Natural Structures in Extracting Water from Air Humidity. *Journal of Water and Environmental Nanotechnology*, *8*(2), 151–163. doi:10.22090/jwent.2023.02.005

Zulfiqar, U., Awais, M., Hussain, S. Z., Hussain, I., Husain, S., & Subhani, T. (2017). Durable and self-healing superhydrophobic surfaces for building materials. *Materials Letters*, *192*, 56–59. doi:10.1016/j.matlet.2017.01.070

About the Contributors

Rajni Garg is working as Associate Professor (Chemistry) in the Department of Applied Science and Humanities at Galgotias College of Engineering and Technology, Greater Noida (Uttar Pradesh), India. She received her B.Sc. (Medical) at the D.A.V. College, Abohar (Punjab), India; M.Sc. (H.S.) in Chemistry, Panjab University, India; M.Phil. in Chemistry at CDLU, India and Ph.D. (Chemistry) at the GKU, India. Her areas of interest include nanochemistry, green chemistry and supramolecular chemistry. She has authored and co-authored many scientific papers published in international and national journals, and proceedings of international and national scientific conferences. She has also authored three textbooks and edited two reference books. She has published five patents on the potential applications of nanomaterials. Currently she is working on green synthesis and potential applications of nanoparticles in environmental remediation.

Ansar Anjum is the Dean Student Welfare and Professor in the Department of Applied Science and Humanities at Galgotias College of Engineering and Technology, Greater Noida, India. She is a doctorate from University of Delhi, India. With a robust tenure of 14 years dedicated to the realms of teaching and research, she has cultivated a profound expertise. Her scholarly pursuits center around the intricate domains of water remediation, nanocomposites, green chemistry, materials science, and analytical techniques. She has authored and co-authored numerous scientific papers published in renowned international and national journals and proceedings, demonstrating her expertise and commitment to environmental sustainability. Beyond her research endeavors, she has also served as a reviewer for many scientific journals. Currently, Dr. Anjum is actively engaged in research on the green synthesis and potential applications of nanoparticles, striving to develop eco-friendly and sustainable nanotechnologies.

* * *

Mukhtar Iderawumi Abdulraheem is a member of the Centre for Advanced Agri-Optoelectronic Research (CAAOR), Henan Agricultural University, Zhengzhou, China, where he conducted various research works and experiments. Abdulraheem is among the few Nigerians who can attend all levels of education in Nigeria (i.e. Primary, Secondary, College of Education, Polytechnic and Universities). He attended the Federal College of Education (Special), Oyo for his NCE; the Federal College of Agriculture, Ibadan for HND. Federal University of Technology, Akure, for PGD as well as M. agric. Tech; Ladoke Akintola University of Technology, Ogbomoso, for Certificate in MIS/Data Processing and the National Open University of Nigeria NOUN for B.Sc. He is a renowned researcher and academician with over sixty (60) published articles in reputed journals. Abdulraheem made presentations at seminars and Conferences and successfully completed six (8) conference paper proceedings, Four (4) chapters in edited books; Three (3) seminar papers and attended over forty-five (45) workshops/conference/webinars. He is a member of several professional bodies, among which are; Scholars Academic and

Charles Adetunji is Professor of Applied Microbiology, Biotechnology and Nanotechnology Laboratory, Department of Microbiology, Edo University Uzairue, PMB 04, Auchi, Edo State, Nigeria.

Betül Bay Yılmaz is currently working as a researcher at Yıldız Technical University, Türkiye. She received her B.Sc., M.Sc., and Ph.D. in Food Science and Technology. Her expertise encompasses various areas, including functional foods, anthocyanins, reaction kinetics, cereal science, bioactivity, and food packaging.

Hongjun Chen is a member of the Centre for Advanced Agri-Optoelectronic Research (CAAOR) and is currently a PhD Scholar (Agricultural Engineering) at the College of Electrical and Mechanical Engineering, Henan Agricultural University, Zhengzhou, China

Yongqi Chen is currently a postgraduate (Master's) student at the College of Electrical and Mechanical Engineering, Henan Agricultural University, Zhengzhou, China.

Sana Dahmani is an Environmental Engineer specialized in Hydraulics Engineering and Rural Management with an MBA from the University of Applied Sciences for Engineering and Economics of Berlin. Her career is steered towards harnessing research and cutting-edge technologies for sustainability topics to support companies in their digital transformation journey through Green Tech expertise, life cycle assessment, and ESG integration. She is active in organizations of civil society: She is in the position of Rapporteur at the UNESCO Groundwater Youth Network and an active member at the Institute of Electrical and Electronics Engineers (IEEE).

Boopathy G. completed his Ph.D. from Department of Aeronautical Engineering, St.Peter's Institute of Higher Education and Research, Chennai, Master of Engineering in Aeronautical from Madras Institute of Technology, Chennai and UG degree in Mechanical Engineering from University of Madras. He has published 15 International journals, guided 8 PG Scholars and 90 UG scholars for their major projects. He is a Life Member of Aeronautical Society of India.

Rishav Garg is working as Head, Department of Civil Engg. at Galgotias College of Engg and Tech, Greater Noida (UP) India. He received his B.tech. in Civil Engg. at the RIC, Kurukshetra India; M.Tech in Structural Engg. at PTU, Kaurthala, India; Ph.D. in Civil Engg. at PTU, Kapurthala, India. He has authored and co-authored more than 50 scientific papers printed in international and national journals, and proceedings of international and national scientific conferences.

Deepak Gupta is a faculty member at the Department of Applied Chemistry, Delhi Technological University (India). With a distinguished career path, he has conducted significant research at renowned international institutions, including the University of Delhi, Freie Universitat (Berlin), Indian Institute of Technology (Delhi), and Catholique Universite de Louvain (Belgium). Specializing in electrochemical conversions and the development of sustainable materials for energy storage, his impactful research has been disseminated through numerous publications with a cumulative impact factor exceeding 100. His pioneering work has garnered recognition and support from prestigious funding bodies such as the Council of Scientific and Industrial Research (India), DAAD (Germany), European Research Council (Belgium), and the Science and Engineering Research Board (India). As a testament to his expertise, he holds membership in the Royal Society of Chemistry and actively contributes as a reviewer for various high-impact journals. His commitment to advancing scientific knowledge and sustainable energy solutions marks him as a leading figure in the field.

Jiandong Hu is a Principal Investigator and founder of the Center for Advanced Agri-Optoelectronic Research; and Henan International Joint Laboratory of Laser Technology in Agriculture Science. He is a Professor at the College of Electrical and Mechanical Engineering, Henan Agricultural University, Zhengzhou, China. He has many research publications in reputed journals and attended several conferences/workshops.

Abel Inobeme obtained his B.Ed. from ABU Zaria. His M.Tech as well as Ph.D. degrees from the Federal University of Technology Minna, Niger State in Nigeria, (), which is rated as one of the best Universities in Nigeria., I graduated from the university in 2008 as a first degree holder and in 2019 as a PhD holder affording him a three-year PhD degree experience in Analytical Chemistry. Through his years of study, he became fascinated by the research field of chemistry using analytical techniques.

Gurusami K. serves as an Associate Professor in the Department of Mechanical Engineering at St. Peter's Institute of Higher Education and Research in Chennai, India. In the year 2016, he successfully completed his Ph.D. in the area of Laser shock peening of dental restorative materials and characterization at Anna University. Furthermore, he obtained his M.E. in Computer Integrated Manufacturing from Anna University in 2008, and his B.E. in Mechanical Engineering from the same institution in 2003. He has contributed significantly to the field, with over 20 research articles published in reputable journals. Notably, he has successfully supervised one Ph.D. candidate, while three more scholars are currently under his guidance. Additionally, in 2017, he obtained his MBA from Madras University. As part of his academic responsibilities, he has supervised the final year projects of approximately 150 MBA students. Furthermore, he has obtained NDT Level-2 certification in Radiographic, Ultrasonic, Magnetic particle, and Dye penetrant testing.

Vijayakumar K. R. is working as Professor in Department of Mechanical Engineering at Dr. MGR Educational and Research Institute, Tamil Nadu, India. He did his PhD in Composite Materials at Anna University in the year 2011. He did his M.E(Engineering Design) at Anna University in the year 2001 and B.E(Mechanical Engineering) at Bharathi University in the year 1993. He has published more than 30 articles in various reputed journals. He has guided 6 Ph.D 's. He successfully completed three defense funded projects as co-investigator. His areas of specialization is composites materials, nano technology and engineering design.

Ashish Kumar is Professor and head of department - Applied Science and Humanities, Nalanda College of Engineering, Bihar Engineering University. His research work is related to Corrosion Inhibition, Material chemistry, thermodynamic and interaction studies in liquid medium and homogeneous catalysis. He is listed in the "World Ranking of Top 2% Scientists-2023" database released by Stanford University, USA and Elsevier.

Mohit Hemanth Kumar completed his Ph.D. from Department of Mechanical Engineering, National Institute of Technology Tiruchirappalli, Master of Engineering in Thermal Engineering with specialization in Refrigeration and Air Conditioning Engineering and Bachelor of Engineering in Mechanical Engineering. He has done Post-Doctoral Research from King Mongkut's University of Technology North Bangkok, Thailand. He is a life member of Institution of Engineers (India), Powder Metallurgy Association of India (India), Society for Failure Analysis (India) and Science and Technology Research Association (Singapore).

John Tsado Mathew obtained his BTech., M.Tech as well as Ph.D. degrees from the Federal University of Technology Minna, Niger State in Nigeria, which is rated as one of the best Universities in Nigeria., I graduated from the university in 2010 as a first degree holder, 2014 M.Tech and in 2018 as a PhD holder affording him a three-year PhD degree experience in Analytical Chemistry. Through his years of study, he became fascinated by the research field of chemistry using analytical techniques.

Abiodun Yusuff Moshood is a member of the Centre for Advanced Agri-Optoelectronic Research (CAAOR) and is currently a PhD Scholar (Agricultural Engineering) at Henan Agricultural University, Zhengzhou, China.

Rahul Singhal possesses over 14 years of experience in the area of clay polymer nanocomposite and its application in Oral drug delivery. He also has expertise in fabrication, characterization and optimization of clay polymer nanocomposite and its application in invitro drug delivery using XRD, HPLC using analytical techniques. Prof. Rahum possesses proven project management and organizational skills with experience in assuming ownership of assigned tasks, successfully completing the tasks, and mentoring/supervising UG and PG students. He has many research papers in International and National journals as well as and conferences.

Abhinay Thakur did his Ph.D. under Dr. Ashish Kumar .His research work is related to Corrosion Inhibition, Material chemistry, thermodynamic and interaction studies in liquid medium.

Rafael Vargas-Bernal received a Bachelor's degree in Communications and Electronics Engineering from the University of Guanajuato in 1995, and the degrees of Master of Science and Doctorate in Science with Specialty in Electronics from the National Institute of Astrophysics, Optics and Electronics (INAOE) in 1997 and 2000, respectively. He works at the University of Guanajuato as a professor-researcher. He has edited 2 books and authored 19 articles in journals, 45 chapters in books, and about 100 conference articles. He is a member of the National System of Researchers (SNI-Mexico). He regularly serves as a reviewer of scientific articles in RSC, Hindawi, Elsevier, IEEE, Wiley, World Scientific Publishing, and MDPI as well as Standards in Semiconductor Equipment and Materials International (SEMI). His research interests include two-dimensional materials, nanomaterials, space materials, composite materials, gas sensors, and biosensors.

Bancha Yingngam, an Associate Professor at Ubon Ratchathani University, Thailand, specializes in Pharmaceutical Sciences and Technology. He earned his pharmacy degree in 2005 and a Ph.D. in 2011 from the same institution. During a postdoctoral fellowship at Karl-Franzens-Universität Graz, Austria, from 2013 to 2021, he expanded his research to include optimization and machine learning algorithms, earning a Scopus H-index of 10. He has published over 20 scholarly works and currently focuses on artificial intelligence applications in green chemistry and pharmacy. Dr. Yingngam has received the "Outstanding Research Award in Health Science" twice and has mentored 11 graduate students in Austria. He has also reviewed over 100 Scopus-listed articles, underscoring his commitment to advancing his field. https://orcid.org/0000-0001-7215-9123.028abe51-615e-466f-a319-307c6044b3a5

Haiyan Zhang is currently at the College of Agronomy, Henan Agricultural University, Zhengzhou, China.

Index

A

Active Packaging 19, 79, 117, 141, 144, 146, 152, 157, 164, 168
Antibacterial Nanocoatings 55, 137, 177, 189, 199, 297, 337, 342
Antifogging Nanocoating 112
Anti-Fouling 2-3, 24, 60-61, 65-66, 69, 78, 81, 87, 101, 212, 288, 308, 313, 316
Antiicing Nanocoating 112
Antimicrobial 17, 22, 27-28, 31, 48, 54, 58-59, 65, 67, 79, 83, 102, 113-114, 118, 124-128, 130, 132-133, 135-137, 140-147, 151-152, 154-155, 157-158, 163, 166-168, 176, 178, 186, 190, 193, 195-197, 199-201, 203-204, 210-212, 215, 220-222, 224-226, 228, 230-231, 237, 239, 241-245, 251, 254, 257-258, 260, 264, 266, 270, 274, 277, 279, 288-289, 293-296, 298-302, 341
Antimicrobial Activtiy 168
Anti-Reflective Nanocoatings 96, 104
Antoxidant Activity 168
Applications 1-10, 13-15, 17-18, 22-23, 25-29, 33-34, 36, 38-48, 50-54, 56-59, 62-68, 70-86, 95, 98, 101, 104, 106, 108-112, 117-118, 123-124, 126-130, 132, 135-138, 141, 144-146, 148-150, 153-165, 167, 170, 177, 179, 181-182, 184-186, 192, 194-198, 200, 202-205, 207, 210, 212, 214-215, 217-223, 225, 228-230, 232, 237, 241-242, 244-246, 248-251, 253-259, 261, 265-272, 274, 276-279, 282-288, 290-302, 305-310, 312-340, 342, 344

B

Bactericidal 142, 168, 197, 269, 285, 297
Bio-Based 20, 22, 113-115, 117-119, 121-122, 124, 132, 134, 138, 141, 144-145, 153, 161, 168, 327
Biodegradable Packaging 113, 119, 121, 123-127, 130, 154, 161, 168
biodegradation 49, 113, 121, 139, 153, 190
Biomedical Materials 25, 266, 287
Bionanomaterials 117, 165, 168, 327
biopackaging 113
Bio-Packaging 20-21, 168
Bio-Plastic 168

C

Ceramics 5, 11, 22, 40, 44-45, 47, 55, 60, 65, 81-82, 84, 86, 99, 106, 184, 297, 343
Chemical Vapor Deposition 54, 60, 63, 96, 267, 313-314, 322, 336
Cosmetic Textiles 170-172, 197, 199
Cosmetotextiles 170-180, 182-183, 185-186, 189-197, 199
Cytotoxicity 136, 148-149, 168, 199, 210, 240, 266, 298

D

Deacetylation 128, 168
Delivery 3, 12, 18, 59, 67-68, 80, 83-85, 89, 114, 126, 175, 183, 185, 196-197, 201, 203-205, 208-209, 211, 213-215, 217-219, 221, 224, 226, 228-229, 232-238, 250-251, 253-254, 258, 265-268, 270,

272-274, 277-279, 282-283, 285-287, 294, 299, 332
DPPH 126, 133, 147, 168, 236
Drug 3, 24, 59, 68, 80, 83-85, 114, 158, 183, 196-197, 201, 203-205, 208-209, 211, 213-215, 217-219, 221, 224-226, 228-230, 232-236, 238-239, 250-251, 253-254, 261, 266-268, 270, 272-274, 277, 279, 281-283, 285-287, 294, 299, 332
Drug Delivery Systems 59, 196-197, 205, 209, 211, 213-214, 224, 226, 228, 233-234, 266, 270, 272-273, 279, 283

E

Electronics 1-3, 18, 59-60, 63, 66-69, 74, 93, 171, 206-207, 302, 306-307
Electrospinning 133, 147, 168, 183, 185, 263, 275, 318
Encapsulation 67, 89, 104, 139, 165, 186, 196, 236, 252, 258, 260-261, 273
Energies 92, 96, 99, 107, 111, 290, 292-293, 302, 306, 326
Enzymatic Hydrolysis 123, 169

F

food packaging 3, 12, 14, 19-23, 26-27, 47-48, 59, 78, 81, 83, 85-86, 114-115, 117-119, 121, 123-133, 135-138, 141-167, 211, 219, 251, 254-255, 263-264, 293, 308, 328, 338
Functional Properties 11, 96, 98, 106, 112, 125, 158, 163-165, 175

H

Healing 65, 186, 224, 228, 230, 237-239, 241, 251, 254, 265, 277-278, 283, 286, 288, 297
Health Perspectives 228
Healthcare 1-2, 17, 59-60, 63, 66-67, 69, 74-75, 77, 177, 184, 189, 195, 202, 204-205, 207, 211, 216-217, 222-223, 228-232, 237-238, 241-243, 245, 251-252, 263, 265-268, 281-283, 290, 293-294, 298, 300, 302, 335
Hydrophilic Nanocoatings 175
Hydrophobic 28, 30, 36-37, 52-54, 61, 65, 68, 96-98, 100-102, 104-110, 112, 124, 128-129, 133-134, 142, 166, 175, 195, 199, 202, 240, 268, 290-295, 297-302, 304-327, 329-341, 343-345
Hydrophobic Coating 28, 52, 98, 291, 299, 322, 324-326, 339
Hydrophobic Nanocoating 54, 97, 102, 109-110, 112, 291-294, 299-300, 302, 305-306, 309-310, 313-314, 325, 334, 340, 345

I

Implants 8-9, 65-66, 205, 209-211, 214, 218, 220, 223, 225, 245, 250, 262, 264-270, 272, 274-275, 278-280, 294-296, 298

L

Lifecycle Analysis 172, 190-191

M

Material 9, 45, 64-65, 125, 239
Materials Science 9, 15, 23, 51-52, 55, 76, 86, 109-110, 116, 155, 170, 196, 226, 293, 331, 337
Medicinal 198, 203-205, 219, 228, 230, 264, 266, 272-273, 279, 282
Medicinal Applications 203-205, 228, 230, 266
Medicine 3, 171, 204-206, 209, 217, 219-220, 222-223, 227, 229-230, 232-233, 235-236, 246, 249-252, 263-264, 266-267, 272-273, 275, 281-282, 284-286, 294, 331
Migration 11, 13, 85, 149, 153, 155, 169, 240, 308

N

Nanocoating 1-2, 4, 11-12, 15-16, 28, 30-31, 41, 48-49, 52-54, 57, 59, 62, 71,

74-76, 78, 82, 86, 96-98, 102-103, 105, 109-110, 112, 116, 122, 126, 128, 138-139, 159, 170, 175, 181, 186-187, 189-190, 204-217, 220, 222-223, 229-239, 241-246, 249-255, 257-260, 262, 266-278, 280-288, 290-294, 296, 298-299, 301-302, 305-306, 308-310, 313-317, 321-323, 325, 327-328, 331-334, 336-337, 339-342, 344-345

Nanocoatings 1-3, 13, 16-17, 28-35, 38-40, 42-43, 45-46, 52, 55, 57-74, 76-80, 82, 85-86, 88-91, 95-98, 100-107, 109-113, 116-118, 123, 126, 128, 130-138, 141-143, 148, 159, 164, 167, 170-173, 175-181, 184-186, 189-194, 199, 204, 228, 265, 272, 290-300, 302, 304-307, 309-313, 315-329, 331-340, 342-345

Nanocomposites 1-2, 6, 8-10, 18, 26, 33, 48, 59, 62, 79, 81-83, 86, 88, 101, 104, 106-107, 109, 115-117, 124, 133, 135-136, 139, 143, 145-146, 150-152, 154-155, 159, 162, 164-166, 170, 194, 197, 254, 259, 264, 288, 290, 311-312, 319, 323, 338

Nanomaterials 2-3, 12-14, 21-24, 26, 29-31, 33, 37, 39-40, 47, 49-50, 52-53, 55-56, 69-70, 76, 79-81, 83-86, 88, 90, 96, 101, 104, 108-109, 114, 116, 120-121, 126, 129, 141-142, 144, 148-151, 153, 158-159, 161, 164-165, 170-171, 173-175, 178, 192-194, 197, 200-201, 204, 206, 210, 212, 216-223, 225-227, 253-258, 262, 265-269, 276, 278-280, 282-288, 294-295, 300, 305, 334, 344

Nanoparticles 2-3, 5, 10-13, 18, 21, 23-24, 30-31, 33-34, 37-45, 47-55, 59, 62-63, 65, 67, 69, 78, 82-86, 101-102, 104-105, 109, 113, 116, 125, 151-152, 154-157, 160, 163-166, 169, 175, 177-178, 180-186, 189-190, 193, 195-196, 198-199, 201, 209-211, 213-215, 219, 222, 224, 226-227, 229-230, 232, 234-237, 239-240, 244-245, 247, 251-253, 255-261, 264, 268-270, 273, 275-276, 278-280, 282-285, 294-300, 302, 304-305, 308-309, 311-316, 318-319, 321-323, 326-330, 332-334, 336-337, 341-343

Nanotechnology 1-3, 11-15, 18, 21-24, 26-27, 29, 31, 50-51, 54, 56, 58, 69, 74, 78-83, 85-86, 106, 111, 113-114, 116, 141, 146-150, 152-154, 158-159, 161-162, 164, 167, 170, 172-173, 178, 191-192, 196, 203-204, 208, 215-217, 219-226, 229-230, 233, 250-251, 253-254, 258, 260, 263, 265-267, 275-276, 278-279, 281, 283, 285-289, 291, 293, 302, 310, 327, 332, 337-338, 344-345

P

Packaging 1-3, 12-14, 18-24, 26-27, 37, 41, 47-48, 59, 69, 78-79, 81, 83-87, 104, 109, 114-119, 121, 123-138, 140-147, 149-169, 211, 218-219, 242, 251, 254-255, 263-264, 292-293, 299, 307-309, 328-329, 333, 338

Pickering Emulsion 134-135, 166, 169

Polymerization 134, 168-169, 300, 317, 320, 337

S

Self-Cleaning Nanocoating 112

Self-Healing Nanocoating 287

Smart and Intelligent 11, 57, 60, 65, 68, 70, 74-77, 84

Smart Coating 28

Smart Packaging 19, 87, 148, 163, 168-169

Solar Cells 88-91, 94-96, 98, 100-101, 103-104, 106-112, 304-305, 332, 341, 345

Sol-gel Methods 57, 60, 62-63

Solvent Casting Method 125, 135, 169

Substrate 3, 8, 28, 31, 42, 44, 61-63, 65, 71-72, 76, 81, 96-97, 100, 102-103, 105, 112, 139, 181-184, 230, 268, 314, 318

Sustainability 4, 13, 16, 18, 20, 25, 58-60, 74, 78, 82, 86, 113, 118, 121, 151, 162, 170-173, 178-180, 190-195, 226, 243, 252, 255, 290-291, 293, 297, 307, 315, 324, 343

Sustainable 2, 8, 12-14, 18-21, 23-24, 28, 49-51, 56, 77, 87, 107-108, 110-111, 114-115, 117, 119, 121, 146, 149, 157,

159, 161-162, 164-166, 170-171, 173, 185, 190-191, 193-194, 197, 199-200, 202, 223-225, 236, 242, 252, 255, 261-263, 282, 286, 288, 290, 295, 299-301, 304, 307, 315, 317, 319-320, 323, 328, 333, 335, 340, 342

T

Tear Strength 124, 127, 169
Tensile Strength 71, 123, 125-127, 129-135, 137-138, 140, 143, 169, 244, 308
Textile 2, 38, 52, 85, 170-172, 174-175, 177-187, 189-193, 195-201, 244, 290, 292-293, 297, 299-302, 336, 342
Textile Industry 2, 85, 170-171, 178, 293
Tissue Engineering 183, 218, 228, 230-232, 245-246, 248-249, 251, 254-256, 265-266, 271-272, 277-278, 282-283, 286-287
Transparent Nanocoating 112

U

UV Protection 2, 28, 38, 52, 65, 172-173, 176-177, 180, 185-186, 189, 195, 197

W

Water Vapor Permeability 169

www.ingramcontent.com/pod-product-compliance
Lightning Source LLC
LaVergne TN
LVHW080120011026
849160LV00013B/1143